T0182203

Graduate Texts in Mathematics 279

Graduate Texts in Mathematics

Graduate Texts in Mathematics bridge the gap between passive study and creative understanding, offering graduate-level introductions to advanced topics in mathematics. The volumes are carefully written as teaching aids and highlight characteristic features of the theory. Although these books are frequently used as textbooks in graduate courses, they are also suitable for individual study.

More information about this series at http://www.springer.com/series/136

Jürgen Herzog • Takayuki Hibi • Hidefumi Ohsugi

Binomial Ideals

 Springer

Jürgen Herzog
Fakultät für Mathematik
Universität Duisburg-Essen
Essen, Germany

Hidefumi Ohsugi
Department of Mathematical Sciences
School of Science and Technology
Kwansei Gakuin University
Sanda, Hyogo, Japan

Takayuki Hibi
Department of Pure & Applied Mathematics
Graduate School of Information Science
and Technology
Osaka University
Suita, Osaka, Japan

ISSN 0072-5285   ISSN 2197-5612   (electronic)
Graduate Texts in Mathematics
ISBN 978-3-030-07019-9   ISBN 978-3-319-95349-6   (eBook)
https://doi.org/10.1007/978-3-319-95349-6

Mathematics Subject Classification (2010): 13-01, 05B50, 05C25, 13F20, 13P10, 13P25, 52B20

This Springer imprint is published by the registered company Springer Nature Switzerland AG
The registered company address is: Gewerbestrasse 11, 6330 Cham, Switzerland

To our wives Maja, Kumiko, and Naoko, our children Susanne, Ulrike, Masaki, Ayako, Takuya, and Tomoya, and our grandchildren Paul, Jonathan, Vincent, Nelson, Sofia, and Jesse

Preface

Historically, commutative algebra, whose foundations were laid by Dedekind, Hilbert, Noether, and Krull, has developed in step with algebraic geometry, number theory, representation theory, combinatorics, and, recently, with statistics. The development of modern commutative algebra has been very much influenced by the work of Kaplansky [126], Zariski–Samuel [220], Nagata [151], and Matsumura [145].

The trend of combining commutative algebra with combinatorics originated in the pioneering work by Richard Stanley [194] in 1975, where squarefree monomial ideals played an important role. Since then, the study of squarefree monomial ideals from viewpoints of both commutative algebra and combinatorics has become a very active area of research. The standard references regarding this area include the monographs of Stanley [199], Hibi [105], Bruns–Herzog [27], Miller–Sturmfels [146], Bruns–Gubeladze [25], and Herzog–Hibi [94].

Since the early 1990s binomial ideals became gradually fashionable. They now appear in various areas of commutative algebra and combinatorics as well as of statistics. A comprehensive analysis of the algebraic properties of binomial ideals, including their primary decompositions, was given by Eisenbud–Sturmfels [58]. Among the binomial ideals, toric ideals form a distinguished class which has first been considered and studied by Conti–Traverso [41] in the algebraic study of integer programming by using the theory of Gröbner bases. Sturmfels, in his influential monograph [202], presented a first systematic treatment of toric ideals. Exciting applications of the theory of toric ideals and their Gröbner bases to statistics were first explored by Diaconis and Sturmfels [53] with creating a new area of research, called computational algebraic statistics.

The present text invites the reader to become acquainted with current trends in the combinatorial and statistical aspects of commutative algebra with the main emphasis on binomial ideals. Apart from a few exceptions, where we refer to the books [27, 135, 145], only basic knowledge of commutative algebra is required to follow most of the text. Part I consists of a self-contained quick introduction to the modern theory of Gröbner bases (Chapter 1) and of reviews on several concepts of commutative algebra (Chapter 2) which are frequently used in later chapters. Part II supplies the reader with the ABC of binomial ideals (Chapter 3) and with that of convex polytopes (Chapter 4). Part III provides several aspects of the theory of binomial ideals. Topics include edge rings and edge polytopes (Chapter 5), join-meet ideals of finite lattices (Chapter 6), binomial edge ideals (Chapter 7), ideals generated by 2-minors (Chapter 8), and binomial ideals arising from statistics (Chapter 9). Each chapter of Part III may be read independently.

We are now in the position to discuss the contents of each chapter of the present text in detail.

Chapter 1 summarizes the fundamental material on Gröbner bases. Starting with Dickson's Lemma which is a classical result in combinatorics, the definition of Gröbner bases is introduced and the division algorithm is discussed. We then come to the highlights of the foundation of Gröbner bases, viz., Buchberger's criterion and Buchberger's algorithm. Furthermore, the elimination theorem and its application are presented and the notion of universal Gröbner bases is considered.

Chapter 2 introduces the nonspecialist to the algebraic and homological concepts from commutative algebra, which are relevant for the material presented in this text. Topics include graded rings and modules, Hilbert functions, finite free resolutions, Betti numbers, linear resolutions, linear quotients, dimension and depth, Cohen–Macaulay rings, and Gorenstein rings. Gröbner basis techniques in the study of ideals and algebras are discussed with a focus on Koszul algebras.

Chapter 3 provides a short introduction to the main topics of the present text, namely, binomials and binomial ideals. Some of the elementary properties of binomial ideals are discussed, including the important fact that the reduced Gröbner basis of a binomial ideal consists of binomials. Toric ideals are identified as those binomial ideals which are prime ideals. Special attention is paid to lattice ideals. Finally Graver bases, Lawrence ideals, and squarefree divisor complexes are introduced and studied.

Chapter 4 is a quick introduction to the fundamental theory of convex polytopes. Integral convex polytopes are mainly studied. After recalling some basic definitions and facts on convex polytopes, the integer decomposition property and normality of convex polytopes are discussed. Then unimodular coverings together with unimodular triangulations of convex polytopes are introduced. Especially, the role of Gröbner bases in the modern analysis of convex polytopes is emphasized.

Chapter 5 deals with edge polytopes and edge rings of finite graphs. The problem when the edge polytope of a finite graph is normal as well as the problem when the toric ideal of an edge ring is generated by quadratic binomials is mainly studied. These problems, whose solutions are provided in the language of finite graphs, were the starting point on the research of edge polytopes and edge rings. Furthermore, a characterization for the edge ring of a bipartite graph to be Koszul is supplied.

Chapter 6 offers the study on a special class of binomial ideals, the so-called join-meet ideals, which arise from finite lattices. In the algebraic study of join-meet ideals, Birkhoff's fundamental structure theorem for finite distributive lattices together with a characterization of distributive lattices due to Dedekind is indispensable. One of the basic facts is that the join-meet ideal of a finite lattice is a prime ideal if and only if the lattice is distributive. An example of a modular lattice

Notation

$A(\mathscr{P})$	Configuration associated with an integral polytope $\mathscr{P}$
$A_{r_1 \cdots r_m}(\Delta)$	Model matrix of a hierarchical model given by Δ
$\beta_{ij}(M)$	Graded Betti numbers of M
$B(\mathscr{P})$	Border of the polyomino $\mathscr{P}$
$b(T)$	Number of bipartite components of $G_{([n]\setminus T)}$
$\text{char}(K)$	Characteristic of a field
$\chi^2(T_0)$	χ^2 statistics of T_0
$\mathscr{C}(G)$	The set of all $T \subset V(G)$ such that T has the cut point property for G
$c(W)$	Number of connected components of $G_{[n]\setminus W}$
C_n	Cycle of length n
$\widetilde{\mathscr{C}}(\Delta; K)$	Augmented oriented chain complex of Δ (with coefficients in K)
$\text{conv}(X)$	Convex hull of X
$\deg x$	Degree of x
Δ	Simplicial complex
$\Delta_{\mathbf{a}}$	Squarefree divisor complex of H (or of A)
$\Delta(G)$	Clique complex of the graph G
$\Delta^{\vee}$	Alexander dual of Δ
$\text{depth } M$	Depth of M
$\dim$	Dimension of a face/simplicial complex
$\dim_K$	Vector space dimension
$\dim M$	Krull dimension of M
$E(G)$	Set of edges of the graph G
$\mathscr{F}$	Koszul filtration
f_α	Binomial defined by an admissible labeling α
$\mathscr{F}_{T_0}$	Set of tables with the same marginals as T_0
$f(\mathscr{P})$	f-vector of $\mathscr{P}$
$g(A)$	Graver complexity of A
$G \setminus \{e\}$	Graph G with edge e omitted
$G(\mathscr{P})$	Bipartite graph associated with the polyomino $\mathscr{P}$

$G\|_W$	Induced subgraph of G on W
H_0	Null hypothesis
$\widetilde{H}_i(\Delta; K)$	i-th reduced simplicial homology of Δ
$\widetilde{H}^i(\Delta; K)$	i-th reduced simplicial cohomology of Δ
$\widetilde{H}(\Delta, \Gamma; K)$	Relative simplicial homology
$H(M, i)$	Dimension of the i-th graded component of M
$H(\mathbf{f}, M)$	Koszul homology of M with respect to the sequence $\mathbf{f}$
height I	Height of the ideal I
$\mathrm{Hilb}_M(t)$	Hilbert series of M
I_A	Toric ideal of an integer matrix A
$I_{\mathscr{B}}$	Lattice basis ideal of a basis $\mathscr{B}$
$I_{\mathscr{C}}$	Ideal generated by the set $\mathscr{C}$ of 2-minors
I_G	Toric ideal of $K[G]$
I_Σ	Stanley–Reisner ideal of Σ
$I_2(X)$	Ideal generated by the 2-minors of the matrix X
I_L	Lattice ideal of a lattice L
$I : J$	Colon ideal of I with respect to J
$I : J^\infty$	Saturation of I with respect to J
$I : f^\infty$	Saturation of I with respect to f
$I : \mathfrak{m}^\infty$	Saturation of I
$\mathrm{in}_<(f)$	Initial monomial of a polynomial f
$\mathrm{in}_<(I)$	Initial ideal of an ideal I
$\mathrm{indmatch}(\Gamma)$	Induced matching number of the graph Γ
$i(\mathscr{P}, N)$	Normalized Ehrhart function of $\mathscr{P}$
$I_{r_1\cdots r_m}(\Delta)$	Toric ideal of $A_{r_1\cdots r_m}(\Delta)$
J_G	Binomial edge ideal of a graph
$K[A_{\tau,\mathbf{a},\mathbf{b},\mathbf{r},\mathbf{s}}]$	Algebra of Segre–Veronese type
$K[A]$	Toric ring of A
$K[G]$	Edge ring of G
$K[\Sigma]$	Stanley–Reisner ring of Σ over K
$\mathrm{Ker}\,\varphi$	Kernel of the map φ
$K[t_1^{\pm 1}, \ldots, t_d^{\pm 1}]$	Laurent polynomial ring over K in the variables $t_1, \ldots, t_d$
$K[x_1, \ldots, x_n]$	Polynomial ring over K
$K[\mathscr{P}]$	Coordinate ring of the polyomino $\mathscr{P}$
$K(\mathbf{f}, M)$	Koszul complex of RM with respect to the sequence $\mathbf{f}$
K_n	Complete graph on $[n]$
$K_{n,m}$	Complete bipartite graph with bipartition $[n]\cup[m]$
$\Lambda(A)$	Lawrence lifting of A
$\Lambda^{(r)}(A)$	rth Lawrence lifting of A
$\mathrm{lcm}(u, v)$	Least common multiple of u and v
L_G	Lovász–Saks–Schrijver ideal of a graph
$\mathfrak{m}$	Graded maximal ideal $(x_1, \ldots, x_n)$ of $K[x_1, \ldots, x_n]$
$m(A)$	Markov complexity of A
$\mathrm{mult}(w)$	Number of k for which $x_k y_k$ divides the monomial w
$[n]$	Set of integers $\{1, \ldots, n\}$

$N^<(k)$	Set of edges j, k with $j < k$
$N^>(k)$	Set of edges j, k with $j > k$
$N(G; T)$	Set of those i for which there is $j \in T$ with $\{i, j\} \in E(G)$
$N\mathscr{P}$	Dilated polytope
$N_G(v)$	Neighborhood of the vertex v in the graph G
proj dim M	Projective dimension of the module M
$\mathscr{P}(G)$	Set of all w-admissible paths of G
Π_G	Permanental edge ideal of a graph
$P_R^M(s, t)$	Graded Poincaré series
P_n	Path graph on $[n]$
$P_W(G)$	Prime ideal attached to the graph G and the subset $W \subset [n]$
p_X	Permanent of X
reg M	Castelnuovo–Mumford regularity of the module M
$\sqrt{I}$	Radical of an ideal I
$\sqrt{u}$	Radical of a monomial u
$r(R)$	Cohen–Macaulay type of R
$S(f, g)$	S-polynomial of f and g
supp($\mathbf{a}$)	Support of the vector $\mathbf{a}$
supp(f)	Support of a polynomial f
type($\mathbf{b}$)	Type (rth Lawrence lifting)
$V(G)$	Set of vertices of a graph G
$V(\mathscr{P})$	Set of vertices of a polytope $\mathscr{P}$
$\bigwedge^j F$	jth exterior power of F
$\mathbb{Z}_{>0}$	Set of nonnegative integers
$\mathbb{Z}^{d \times n}$	Set of $d \times n$-matrices $A = (a_{ij})_{\substack{1 \leq i \leq d \\ 1 \leq j \leq n}}$ with each $a_{ij} \in \mathbb{Z}$

Part I
Basic Concepts

Chapter 1
Polynomial Rings and Gröbner Bases

Abstract The purpose of Chapter 1 is to provide the reader with sufficient knowledge of the basic theory of Gröbner bases which is required for reading the later chapters. In Section 1.1, we study Dickson's Lemma, which is a classical result in combinatorics. Gröbner bases are then introduced and Hilbert's Basis Theorem and Macaulay's Theorem follow. In Section 1.2, the division algorithm, which is the framework of Gröbner bases, is discussed with a focus on the importance of the remainder when performing division. The highlights of the fundamental theory of Gröbner bases are Buchberger's criterion and Buchberger's algorithm presented in Section 1.3. Furthermore, in Section 1.4, elimination theory will be introduced. This theory is very useful for solving a system of polynomial equations. Finally, in Section 1.5, we discuss the universal Gröbner basis of an ideal. This is a finite set of polynomials which is a Gröbner basis for the ideal with respect to any monomial order.

1.1 Dickson's Lemma and Gröbner Bases

A *monomial* u in the variables $x_1, x_2, \ldots, x_n$ is a product of the form

$$u = \prod_{i=1}^{n} x_i^{a_i} = x_1^{a_1} x_2^{a_2} \cdots x_n^{a_n},$$

where each a_i is a nonnegative integer. We often use the notation $u = \mathbf{x}^{\mathbf{a}}$, where $\mathbf{a} = (a_1, a_2, \ldots, a_n) \in \mathbb{Z}_{\geq 0}^n$. The *degree* of u is $\sum_{i=1}^{n} a_i$. For example, the degree of $x_2^5 x_4^3 x_6$ is 9. In particular $1 \ (= x_1^0 x_2^0 \cdots x_n^0)$ is a monomial of degree 0. A *term* is a monomial together with a nonzero *coefficient*. For example, $-7x_2^5 x_4^3 x_6$ is a term of degree 9 with -7 its coefficient. A *constant term* is the monomial 1 together with a nonzero coefficient.

A *polynomial* is a finite sum of terms. For example,

$$f = -5x_1^2 x_2 x_3^2 + \frac{2}{3} x_2 x_4^3 x_5^2 - x_3^3 - 7$$

is a polynomial with 4 terms. The monomials appearing in f are

$$x_1^2 x_2 x_3^2, \; x_2 x_4^3 x_5^2, \; x_3^3, \; 1$$

and the coefficients of f are

$$-5, \; \frac{2}{3}, \; -1, \; -7.$$

The degree of a polynomial is defined to be the maximal degree of monomials which appears in the polynomial. For example, the degree of the above polynomial f is 6. With an exception 0 is regarded as a polynomial, but the degree of 0 is undefined. If the degree of all monomials appearing in a polynomial is equal to q, then the polynomial is called a *homogeneous* polynomial of degree q. For example,

$$-7x_1^2 x_3 + \frac{3}{5} x_2 x_4 x_5 - x_4^3 + x_1 x_3 x_5$$

is a homogeneous polynomial of degree 3.

Let K be a field and $S = K[x_1, x_2, \ldots, x_n]$ the set of all polynomials in the variables $x_1, x_2, \ldots, x_n$ with coefficients in K. If f and g are polynomials belonging to $K[x_1, x_2, \ldots, x_n]$, then the sum $f + g$ and the product fg can be defined in the obvious way. It then turns out that S is a commutative algebra, which is called the *polynomial ring* in n variables over K.

Let $\mathcal{M}_n$ denote the set of monomials in the variables $x_1, x_2, \ldots, x_n$. When we deal with monomials, we often use u, v, and w instead of $\prod_{i=1}^n x_i^{a_i}$ unless confusion arises.

We say that a monomial $u = \prod_{i=1}^n x_i^{a_i}$ divides $v = \prod_{i=1}^n x_i^{b_i}$ if one has $a_i \leq b_i$ for all $1 \leq i \leq n$. We write $u \mid v$ if u divides v.

Let M be a nonempty subset of $\mathcal{M}_n$. A monomial $u \in M$ is called a *minimal element* of M if the following condition is satisfied: If $v \in M$ and $v \mid u$, then $v = u$.

Example 1.1

(a) Let $n = 1$. Then a minimal element of a nonempty subset M of $\mathcal{M}_1$ is unique. In fact, if q is the minimal degree of monomials belonging to M, then the monomial x_1^q is the unique minimal element of M.

(b) Let $n = 2$ and M a nonempty subset of $\mathcal{M}_2$. Then the number of minimal elements of M is at most finite. To see why this is true, suppose that $u_1 = x_1^{a_1} x_2^{b_1}, u_2 = x_1^{a_2} x_2^{b_2}, \ldots$ are the minimal elements of M with $a_1 \leq a_2 \leq \ldots$. If $a_i = a_{i+1}$, then either u_i or u_{i+1} cannot be minimal. Hence $a_1 < a_2 < \ldots$. Since u_i cannot divide u_{i+1}, one has $b_i > b_{i+1}$. Thus $b_1 > b_2 > \ldots$. Hence the number of minimal elements of M is finite, as desired.

Example 1.1 (b) will turn out to be true for every $n \geq 1$. This fact is stated in *Dickson's Lemma*, which is a classical result in combinatorics and which can be proved easily by using induction. On the other hand, however, Dickson's Lemma plays an essential role in the foundation of the theory of Gröbner bases. It guarantees that several important algorithms terminate after a finite number of steps.

Theorem 1.2 (Dickson's Lemma) *The set of minimal elements of a nonempty subset M of $\mathcal{M}_n$ is finite.*

Proof We work with induction on the number of variables. First of all, it follows from Example 1.1 that Dickson's Lemma is true for $n = 1$ and $n = 2$. Let $n > 2$ and suppose that Dickson's Lemma is true for $n - 1$. Let $y = x_n$. Let N denote the set of monomials u in the variables $x_1, x_2, \ldots, x_{n-1}$ satisfying the condition that there exists $b \geq 0$ with $u y^b \in M$. Clearly $N \neq \emptyset$. The induction hypothesis says that the number of minimal elements of N is finite. Let $u_1, u_2, \ldots, u_s$ denote the minimal elements of N. Then by the definition of N, it follows that, for each u_i, there is $b_i \geq 0$ with $u_i y^{b_i} \in M$. Let b be the largest integer among $b_1, b_2, \ldots, b_s$. Moreover, given $0 \leq c < b$, we define a subset N_c of N by setting

$$N_c = \{u \in N : u y^c \in M\}.$$

Again, the induction hypothesis says that the number of minimal elements of N_c is finite. Let $u_1^{(c)}, u_2^{(c)}, \ldots, u_{s_c}^{(c)}$ denote the minimal elements of N_c. Then we claim that a monomial belonging to M can be divided by one of the monomials listed below:

$$u_1 y^{b_1}, \ldots, u_s y^{b_s}$$
$$u_1^{(0)}, \ldots, u_{s_0}^{(0)}$$
$$u_1^{(1)} y, \ldots, u_{s_1}^{(1)} y$$
$$\cdots$$
$$u_1^{(b-1)} y^{b-1}, \ldots, u_{s_{b-1}}^{(b-1)} y^{b-1}$$

In fact, for a monomial $w = u y^e \in M$, where u is a monomial in $x_1, x_2, \ldots, x_{n-1}$, one has $u \in N$. Hence if $e \geq b$, then w is divided by one of $u_1 y^{b_1}, \ldots, u_s y^{b_s}$. On the other hand, if $0 \leq e < b$, then, since $u \in N_e$, it follows that w can be divided by one of $u_1^{(e)} y^e, \ldots, u_{s_e}^{(e)} y^e$. Hence each minimal element of M must appear in the above list of monomials. In particular, the number of minimal elements of M is finite, as required. $\qquad\square$

A nonempty subset I of S is called an *ideal* of S if the following conditions are satisfied:

- If $f \in I$, $g \in I$, then $f + g \in I$;
- If $f \in I$, $g \in S$, then $gf \in I$.

Example 1.3 The ideals of the polynomial ring $K[x]$ ($= K[x_1]$) in one variable can be easily determined. Let $I \subset K[x]$ be an ideal with at least one nonzero polynomial and d the smallest degree of nonzero polynomials belonging to I. Let $g \in I$ be a polynomial of degree d. Given an arbitrary polynomial $f \in I$, the division algorithm of $K[x]$, which is learned in the elementary algebra, guarantees the existence of unique polynomials q and r such that $f = qg + r$, where either $r = 0$ or the degree of r is less than d. Since f and g belong to the ideal I, it follows that $r = f - qg$ also belongs to I. If $r \neq 0$, then r is a nonzero polynomial belonging to I whose degree is less than d. This contradicts the choice of d. Hence $r = 0$. Thus

$$I = \{ qg \ : \ q \in K[x] \}.$$

Let $\{ f_\lambda \ : \ \lambda \in \Lambda \}$ be a nonempty subset of S. It then follows that the set of polynomials of the form

$$\sum_{\lambda \in \Lambda} g_\lambda f_\lambda,$$

where $g_\lambda \in S$ is 0 except for a finite number of λ's, is an ideal of S, which is called the ideal *generated by* $\{ f_\lambda \ : \ \lambda \in \Lambda \}$ and is written as

$$(\{ f_\lambda \ : \ \lambda \in \Lambda \}).$$

Conversely, given an arbitrary ideal $I \subset S$, there exists a subset $\{ f_\lambda \ : \ \lambda \in \Lambda \}$ of S with $I = (\{ f_\lambda \ : \ \lambda \in \Lambda \})$. The subset $\{ f_\lambda \ : \ \lambda \in \Lambda \}$ is called a *system of generators* of the ideal I. In particular, if $\{ f_\lambda \ : \ \lambda \in \Lambda \}$ is a finite set $\{ f_1, f_2, \ldots, f_s \}$, then $(\{ f_1, f_2, \ldots, f_s \})$ is abbreviated as

$$(f_1, f_2, \ldots, f_s).$$

A *finitely generated* ideal is an ideal with a system of generators consisting of a finite number of polynomials. In particular, an ideal with a system of generators consisting of only one polynomial is called a *principal ideal*. Example 1.3 says that every ideal of the polynomial ring in one variable is principal. However, the ideal $(x_1, x_2, \ldots, x_n)$ of $S = K[x_1, x_2, \ldots, x_n]$ with $n \geq 2$ cannot be a principal ideal.

Now, a *monomial ideal* is an ideal with a system of generators consisting of monomials.

Lemma 1.4 *Every monomial ideal is finitely generated. More precisely if I is a monomial ideal and if $\{ u_\lambda \ : \ \lambda \in \Lambda \}$ is its system of generators consisting of monomials, then there exists a finite subset $\{ u_{\lambda_1}, u_{\lambda_2}, \ldots, u_{\lambda_s} \}$ of $\{ u_\lambda \ : \ \lambda \in \Lambda \}$ such that $I = (u_{\lambda_1}, u_{\lambda_2}, \ldots, u_{\lambda_s})$.*

Proof It follows from Theorem 1.2 that the number of minimal elements of the set of monomials $\{ u_\lambda \ : \ \lambda \in \Lambda \}$ is finite. Let $\{ u_{\lambda_1}, u_{\lambda_2}, \ldots, u_{\lambda_s} \}$ be the set of its minimal elements. We claim $I = (u_{\lambda_1}, u_{\lambda_2}, \ldots, u_{\lambda_s})$. In fact, each $f \in I$ can be expressed as $f = \sum_{\lambda \in \Lambda} g_\lambda u_\lambda$, where $g_\lambda \in S$ is 0 except for a finite number of λ's. Then, for each λ with $g_\lambda \neq 0$, we choose u_{λ_i} which divides u_λ and set $h_\lambda = g_\lambda(u_\lambda/u_{\lambda_i})$. Thus $g_\lambda u_\lambda = h_\lambda u_{\lambda_i}$. Hence f can be expressed as $f = \sum_{i=1}^{s} f_i u_{\lambda_i}$ with each $f_i \in S$. □

Let I be a monomial ideal. A system of generators of I consisting of a finite number of monomials is called a *system of monomial generators* of I.

Lemma 1.5 *Let* $I = (u_1, u_2, \ldots, u_s)$ *be a monomial ideal, where* $u_1, u_2, \ldots, u_s$ *are monomials. Then a monomial* u *belongs to* I *if and only if one of* u_i's *divides* u.

Proof The sufficiency is clear. We prove the necessity. A monomial u belonging to I can be expressed as $u = \sum_{i=1}^{s} f_i u_i$ with each $f_i \in S$. Let $f_i = \sum_{j=1}^{s_i} a_j^{(i)} v_j^{(i)}$, where $0 \neq a_j^{(i)} \in K$ and where each $v_j^{(i)}$ is a monomial. Since $u = \sum_{i=1}^{s} f_i u_i = \sum_{i=1}^{s} (\sum_{j=1}^{s_i} a_j^{(i)} v_j^{(i)}) u_i$, there exist i and j with $u = v_j^{(i)} u_i$. In other words, there is u_i which divides u, as desired. □

A system of generators of a monomial ideal does not necessarily consist of monomials. For example, $\{x_1^2 + x_2^3, x_2^2\}$ is a system of generators of the monomial ideal (x_1^2, x_2^2).

Corollary 1.6 *Among all systems of monomial generators of a monomial ideal, there exists a unique system of monomial generators which is minimal with respect to inclusion.*

Proof Lemma 1.4 guarantees the existence of a system of monomial generators of a monomial ideal I. If it is not minimal, then removing redundant monomials yields a minimal system of monomials generators.

Now, suppose that $\{u_1, u_2, \ldots, u_s\}$ and $\{v_1, v_2, \ldots, v_t\}$ are minimal systems of monomial generators of I. It follows from Lemma 1.5 that, for each $1 \leq i \leq s$, there is v_j which divides u_i. Similarly, there is u_k which divides v_j. Consequently, u_k divides u_i. Since $\{u_1, u_2, \ldots, u_s\}$ is minimal, one has $i = k$. Thus $u_i = v_j$. Hence $\{u_1, u_2, \ldots, u_s\} \subset \{v_1, v_2, \ldots, v_t\}$. Since $\{v_1, v_2, \ldots, v_t\}$ is minimal, it follows that $\{u_1, u_2, \ldots, u_s\}$ coincides with $\{v_1, v_2, \ldots, v_t\}$, as required. □

Let, in general, I and J be ideals of the polynomial ring $S = K[x_1, \ldots, x_n]$. Then the sum $I + J$, the intersection $I \cap J$, the colon ideal $I : J$ of I with respect to J, and the radical $\sqrt{I}$ of I are defined as follows:

$$I + J = \{f + h : f \in I, h \in J\},$$

$$I \cap J = \{f \in S : f \in I, f \in J\},$$

$$I : J = \{f \in S : fg \in I \text{ for all } g \in J\},$$

$$\sqrt{I} = \{f \in S : f^k \in I \text{ for some } k\}.$$

Then all of them are ideals of S. An ideal I of S is called *radical* if we have $I = \sqrt{I}$. Let $\{f_1, f_2, \ldots\}$ be a system of generators of I and $\{h_1, h_2, \ldots\}$ that of J. Then

$$\{f_1, f_2, \ldots, h_1, h_2, \ldots\}$$

is a system of generators of $I + J$. However, to find a system of generators of $I \cap J$ is rather difficult, see Section 1.4.

Recall that a *partial order* on a set Σ is a binary relation $\leq$ on Σ such that, for all $a, b, c \in \Sigma$, one has

(i) $a \leq a$ (reflexivity);
(ii) $a \leq b$ and $b \leq a \Rightarrow a = b$ (antisymmetry);
(iii) $a \leq b$ and $b \leq c \Rightarrow a \leq c$ (transitivity).

A *partially ordered set* is a set Σ with a partial order $\leq$ on Σ. It is custom to write $a < b$ if $a \leq b$ and $a \neq b$. A *total order* on Σ is a partial order $\leq$ on Σ such that, for any two elements a and b belonging to Σ, one has either $a \leq b$ or $b \leq a$. A *totally ordered set* is a set Σ with a total order $\leq$ on Σ.

Example 1.7

(a) Let T be a nonempty set and $\mathscr{B}_T$ the set of subsets of T. If A and B belong to $\mathscr{B}_T$, then we define $A \leq B$ if $A \subset B$. It turns out that $\leq$ is a partial order on $\mathscr{B}_T$, which is called a partial order by inclusion.
(b) Let $N > 0$ be an integer and $\mathscr{D}_N$ the set of divisors of N. If a and b are divisors of N, then we define $a \leq b$ if a divides b. Then $\leq$ is a partial order on $\mathscr{D}_N$, which is called a partial order by divisibility. If $p_1, p_2, \ldots, p_d$ are prime numbers with $p_1 < p_2 < \cdots < p_d$ and if $N = p_1 p_2 \cdots p_d$, then $\mathscr{D}_N$ coincides with $\mathscr{B}_{[d]}$.

Let, as before $S = K[x_1, x_2, \ldots, x_n]$ be the polynomial ring in n variables over K and $\mathscr{M}_n$ the set of monomials in the variables $x_1, x_2, \ldots, x_n$. A *monomial order* on S is a total order $<$ on $\mathscr{M}_n$ such that

(i) $1 < u$ for all $1 \neq u \in \mathscr{M}_n$;
(ii) if $u, v \in \mathscr{M}_n$ and $u < v$, then $uw < vw$ for all $w \in \mathscr{M}_n$.

Example 1.8

(a) Let $u = x_1^{a_1} x_2^{a_2} \cdots x_n^{a_n}$ and $v = x_1^{b_1} x_2^{b_2} \cdots x_n^{b_n}$ be monomials. We define the total order $<_{\text{lex}}$ on $\mathscr{M}_n$ by setting $u <_{\text{lex}} v$ if either (i) $\sum_{i=1}^n a_i < \sum_{i=1}^n b_i$, or (ii) $\sum_{i=1}^n a_i = \sum_{i=1}^n b_i$ and the leftmost nonzero component of the vector $(b_1 - a_1, b_2 - a_2, \ldots, b_n - a_n)$ is positive. It follows that $<_{\text{lex}}$ is a monomial order on S, which is called the *lexicographic order* on S induced by the ordering $x_1 > x_2 > \cdots > x_n$.
(b) Let $u = x_1^{a_1} x_2^{a_2} \cdots x_n^{a_n}$ and $v = x_1^{b_1} x_2^{b_2} \cdots x_n^{b_n}$ be monomials. We define the total order $<_{\text{rev}}$ on $\mathscr{M}_n$ by setting $u <_{\text{rev}} v$ if either (i) $\sum_{i=1}^n a_i < \sum_{i=1}^n b_i$, or (ii) $\sum_{i=1}^n a_i = \sum_{i=1}^n b_i$ and the rightmost nonzero component of the vector $(b_1 - a_1, b_2 - a_2, \ldots, b_n - a_n)$ is negative. It follows that $<_{\text{rev}}$ is a monomial order on S, which is called the *reverse lexicographic order* on S induced by the ordering $x_1 > x_2 > \cdots > x_n$.

(c) Let $u = x_1^{a_1} x_2^{a_2} \cdots x_n^{a_n}$ and $v = x_1^{b_1} x_2^{b_2} \cdots x_n^{b_n}$ be monomials. We define the total order $<_{\text{purelex}}$ on $\mathcal{M}_n$ by setting $u <_{\text{purelex}} v$ if the leftmost nonzero component of the vector $(b_1 - a_1, b_2 - a_2, \ldots, b_n - a_n)$ is positive. It follows that $<_{\text{purelex}}$ is a monomial order on S, which is called the *pure lexicographic order* on S induced by the ordering $x_1 > x_2 > \cdots > x_n$.

Let $\pi = i_1 i_2 \cdots i_n$ be a permutation of $[n] = \{1, 2, \ldots, n\}$. How can we define the lexicographic order (or the reverse lexicographic order) induced by the ordering $x_{i_1} > x_{i_2} > \cdots > x_{i_n}$? First, given a monomial $u = x_1^{a_1} x_2^{a_2} \cdots x_n^{a_n} \in \mathcal{M}_n$, we set

$$u^\pi = x_1^{b_1} x_2^{b_2} \cdots x_n^{b_n}, \quad \text{where} \quad b_j = a_{i_j}.$$

Second, we introduce the total order $<_{\text{lex}}^\pi$ (resp. $<_{\text{rev}}^\pi$) on $\mathcal{M}_n$ by setting $u <_{\text{lex}}^\pi v$ (resp. $u <_{\text{rev}}^\pi v$) if $u^\pi <_{\text{lex}} v^\pi$ (resp. $u^\pi <_{\text{rev}} v^\pi$), where $u, v \in \mathcal{M}_n$. It then follows that $<_{\text{lex}}^\pi$ (reps. $<_{\text{rev}}^\pi$) is a monomial order on S. The monomial order $<_{\text{lex}}^\pi$ (reps. $<_{\text{rev}}^\pi$) is called the lexicographic order (resp. reverse lexicographic order) on S induced by the ordering $x_{i_1} > x_{i_2} > \cdots > x_{i_n}$.

Unless otherwise stated, we usually consider monomial orders satisfying

$$x_1 > x_2 > \cdots > x_n.$$

Example 1.9 Fix a nonzero vector $\mathbf{w} = (w_1, w_2, \ldots, w_n)$ with each $w_i \geq 0$. Let $<$ be a monomial order on S. We then define the total order $<_{\mathbf{w}}$ on $\mathcal{M}_n$ as follows: If $u = x_1^{a_1} x_2^{a_2} \cdots x_n^{a_n}$ and $v = x_1^{b_1} x_2^{b_2} \cdots x_n^{b_n}$ are monomials, then we define $u <_{\mathbf{w}} v$ if either (i) $\sum_{i=1}^n a_i w_i < \sum_{i=1}^n b_i w_i$, or (ii) $\sum_{i=1}^n a_i w_i = \sum_{i=1}^n b_i w_i$ and $u < v$. It follows that $<_{\mathbf{w}}$ is a monomial order on S.

Lemma 1.10 *Let $<$ be a monomial order on S. Let u and v be monomials with $u \neq v$ and suppose that u divides v. Then $u < v$.*

Proof Let w be a monomial with $v = wu$. Since $u \neq v$, one has $w \neq 1$. The definition of monomial orders says that $1 < w$. Hence, again, the definition of monomial orders says that $1 \cdot u < w \cdot u$. Thus $u < v$, as desired.   □

Lemma 1.11 *Let $<$ be a monomial order on S. Then there exists no infinite descending sequence of the form*

$$u_0 > u_1 > u_2 > \cdots,$$

where $u_0, u_1, u_2, \ldots$ are monomials.

Proof Suppose on the contrary that such an infinite descending sequence exists. Let $M = \{u_0, u_1, u_2, \ldots\}$. Theorem 1.2 then guarantees that the number of minimal elements of M is finite. Let $u_{i_1}, u_{i_2}, \ldots, u_{i_s}$ be the minimal elements of M, where $i_1 < i_2 < \cdots < i_s$. Now, if $j > i_s$, then u_j must be divided by one of the minimal elements. Let, say, u_{i_k} divide u_j. Then Lemma 1.10 says that $u_{i_k} < u_j$. However, since $j > i_s \geq i_k$, it follows $u_{i_k} > u_j$ and a contradiction arises.   □

We fix a monomial order $<$ on the polynomial ring $S = K[x_1, x_2, \ldots, x_n]$. Given a nonzero polynomial

$$f = a_1 u_1 + a_2 u_2 + \cdots + a_t u_t$$

of S, where $0 \neq a_i \in K$ and where $u_1, u_2, \ldots, u_t$ are monomials with

$$u_1 > u_2 > \cdots > u_t,$$

the *support* of f is the set of monomials appearing in f. It is written as supp(f). The *initial monomial* of f with respect to $<$ is the largest monomial belonging to supp(f) with respect to $<$. It is written as in$_<(f)$. Thus

$$\text{supp}(f) = \{u_1, u_2, \ldots, u_t\}$$

and

$$\text{in}_<(f) = \{u_1\}.$$

Example 1.12 Let $n = 4$ and $f = x_1 x_4 - x_2 x_3$. Then supp(f) $= \{x_1 x_4, x_2 x_3\}$. One has in$_{<_{\text{lex}}}(f) = \{x_1 x_4\}$ and in$_{<_{\text{rev}}}(f) = \{x_2 x_3\}$.

Let f and g be nonzero polynomials of S. Then in$_<(fg) = $ in$_<(f) \cdot$ in$_<(g)$. In particular if w is a monomial, then in$_<(wg) = w \cdot$ in$_<(g)$, see Problem 1.5. Using this fact, we have a result on the radical of a monomial ideal. If $u = x_1^{b_1} \cdots x_n^{b_n}$ is a monomial of S, then its *radical* $\sqrt{u}$ is

$$\sqrt{u} = \prod_{b_i > 0} x_i.$$

For example, $\sqrt{x_1^3 x_2 x_4^5} = x_1 x_2 x_4$. Thus in particular one has $\sqrt{u} = u$ if and only if each $b_i \leq 1$. A monomial u is called *squarefree* if $\sqrt{u} = u$. We say that a monomial ideal I is *squarefree* if $I = \sqrt{I}$.

Lemma 1.13 *Let $\{u_1, u_2, \ldots, u_s\}$ be the minimal system of monomials generators of the monomial ideal $I = (u_1, u_2, \ldots, u_s)$ of S. Then $\sqrt{I} = (\sqrt{u_1}, \ldots, \sqrt{u_s})$. Furthermore, I is squarefree if and only if u_i is squarefree for all $1 \leq i \leq s$.*

Proof Let $u = x_1^{b_1} x_2^{b_2} \cdots x_n^{b_n} \in I$ be a monomial and $N = \max\{b_1, b_2, \ldots, b_n\}$. Then $\sqrt{u}^N \in I$ and $u \in \sqrt{I}$. Thus each of $\sqrt{u_1}, \ldots, \sqrt{u_s}$ belongs to $\sqrt{I}$.

We now show that $\sqrt{I} \subset (\sqrt{u_1}, \ldots, \sqrt{u_s})$. Let $<$ be a monomial order on S. Let $0 \neq f \in \sqrt{I}$ and write $f = \sum_{k=1}^{\ell} c_k w_k$, where $0 \neq c_k \in K$ and w_k is a monomial with $w_1 = $ in$_<(f)$. If $f^N \in I$, then one can write $f^N = \sum_{i=1}^{s} h_i u_i$ with each $h_i \in S$. Thus in$_<(f^N) = $ in$_<(f)^N = w_1^N$ is divided by one of $u_1, \ldots, u_s$. Thus

$w_1 \in (\sqrt{u_1}, \ldots, \sqrt{u_s})$. Hence $f - c_1 w_1 \in \sqrt{I}$. Now, by using induction on ℓ, it follows that $f - c_1 w_1 \in (\sqrt{u_1}, \ldots, \sqrt{u_s})$. Thus $f \in (\sqrt{u_1}, \ldots, \sqrt{u_s})$, as desired.

For the second part, we only need to show that if I is squarefree, then $u_i = \sqrt{u_i}$ for $1 \leq i \leq s$. Since $\sqrt{u_i}$ divides u_i, each u_j with $i \neq j$ cannot divide $\sqrt{u_i}$. Hence, if $u_i \neq \sqrt{u_i}$, one has $\sqrt{u_i} \notin I$. Thus $I \neq \sqrt{I}$.   □

Let I be an ideal of the polynomial ring S with $I \neq (0)$. The monomial ideal generated by $\{\mathrm{in}_<(f) \,:\, 0 \neq f \in I\}$ is called the *initial ideal* of I with respect to $<$ and is written as $\mathrm{in}_<(I)$. In other words,

$$\mathrm{in}_<(I) = (\{\mathrm{in}_<(f) \,:\, 0 \neq f \in I\}).$$

In general, however, even if $I = (\{f_\lambda\}_{\lambda \in \Lambda})$, it is not necessarily true that $\mathrm{in}_<(I)$ coincides with $(\{\mathrm{in}_<(f_\lambda)\}_{\lambda \in \Lambda})$.

Example 1.14 Let $n = 7$. Let $f = x_1 x_4 - x_2 x_3$, $g = x_4 x_7 - x_5 x_6$ and $I = (f, g)$. Then $\mathrm{in}_{<_{\mathrm{lex}}}(f) = x_1 x_4$, $\mathrm{in}_{<_{\mathrm{lex}}}(g) = x_4 x_7$. Let $h = x_7 f - x_1 g = x_1 x_5 x_6 - x_2 x_3 x_7$. Since $h \in I$, it follows that $\mathrm{in}_{<_{\mathrm{lex}}}(h) = x_1 x_5 x_6 \in \mathrm{in}_{<_{\mathrm{lex}}}(I)$. However, $x_1 x_5 x_6 \notin (x_1 x_4, x_4 x_7)$. Hence $(x_1 x_4, x_4 x_7) \neq \mathrm{in}_{<_{\mathrm{lex}}}(I)$.

Now, Lemma 1.4 says that the monomial ideal $\mathrm{in}_<(I)$ is finitely generated. Thus there exists a finite subset

$$\{\mathrm{in}_<(f_1), \mathrm{in}_<(f_2), \ldots, \mathrm{in}_<(f_s)\}$$

of $\{\mathrm{in}_<(f) \,:\, 0 \neq f \in I\}$ which is a system of monomial generators of $\mathrm{in}_<(I)$.

Definition 1.15 Let $S = K[x_1, x_2, \ldots, x_n]$ be the polynomial ring and fix a monomial order $<$ on the polynomial ring. Let I be an ideal of S with $I \neq (0)$. Then a *Gröbner basis* of I with respect to $<$ is a finite set $\{g_1, g_2, \ldots, g_s\}$ of nonzero polynomials belonging to I such that $\{\mathrm{in}_<(g_1), \mathrm{in}_<(g_2), \ldots, \mathrm{in}_<(g_s)\}$ is a system of monomial generators of the initial ideal $\mathrm{in}_<(I)$.

A Gröbner basis exists. However, a Gröbner basis cannot be unique. In fact, if $\{g_1, g_2, \ldots, g_s\}$ is a Gröbner basis of I, then any finite subset of $I \setminus \{0\}$ which contains $\{g_1, g_2, \ldots, g_s\}$ is again a Gröbner basis of I.

Corollary 1.6 says that the monomial ideal $\mathrm{in}_<(I)$ possesses a unique minimal system of monomial generators. We say that a Gröbner basis $\{g_1, g_2, \ldots, g_s\}$ of I is a *minimal Gröbner basis* of I if $\{\mathrm{in}_<(g_1), \mathrm{in}_<(g_2), \ldots, \mathrm{in}_<(g_s)\}$ is a minimal system of monomial generators of $\mathrm{in}_<(I)$ and if the coefficient of $\mathrm{in}_<(g_i)$ coincides with 1 for all $1 \leq i \leq s$. A minimal Gröbner basis exists. However, a minimal Gröbner basis may not be unique. For example, if $\{g_1, g_2, g_3, \ldots, g_s\}$, where $s > 1$, is a minimal Gröbner basis of I with $\mathrm{in}_<(g_1) < \mathrm{in}_<(g_2)$, then $\{g_1, g_2 + g_1, g_3, \ldots, g_s\}$ is again a minimal Gröbner basis of I.

Theorem 1.16 *Every Gröbner basis of an ideal $I \subset S$ is a system of generators of I.*

Proof Let I be an ideal of S and $\{g_1, g_2, \ldots, g_s\}$ a Gröbner basis of I with respect to a monomial order $<$. Then

$$\mathrm{in}_<(I) = (\mathrm{in}_<(g_1), \mathrm{in}_<(g_2), \ldots, \mathrm{in}_<(g_s)).$$

We claim $I = (g_1, g_2, \ldots, g_s)$.

Let $0 \neq f \in I$. Since $\mathrm{in}_<(f) \in \mathrm{in}_<(I)$, there exists a monomial w together with $1 \leq i \leq s$ such that $\mathrm{in}_<(f) = w \cdot \mathrm{in}_<(g_i)$. Thus $\mathrm{in}_<(f) = \mathrm{in}_<(wg_i)$. Let c_i be the coefficient of $\mathrm{in}_<(g_i)$ in g_i and c the coefficient of $\mathrm{in}_<(f)$ in f. Let $f^{(1)} = c_i f - cwg_i \in I$. If $f^{(1)} = 0$, then $f = (c/c_i)wg_i \in (g_1, g_2, \ldots, g_s)$.

Let $f^{(1)} \neq 0$. Then $\mathrm{in}_<(f^{(1)}) < \mathrm{in}_<(f)$. In the case that $f^{(1)} \neq 0$, the same technique as we used for f can be applied to $f^{(1)}$ and we obtain $f^{(2)} \in I$. If $f^{(2)} = 0$, then $f^{(1)}$ belongs to $(g_1, g_2, \ldots, g_s)$ and $f \in (g_1, g_2, \ldots, g_s)$. If $f^{(2)} \neq 0$, then $\mathrm{in}_<(f^{(2)}) < \mathrm{in}_<(f^{(1)})$. In general, if $f^{(k-1)} \neq 0$, then the same technique as we used for f can be applied to $f^{(k-1)}$ and we obtain $f^{(k)} \in I$. If $f^{(k)} = 0$, then $f^{(k-1)}, f^{(k-2)}, \ldots, f^{(1)}$ belong to $(g_1, g_2, \ldots, g_s)$ and $f \in (g_1, g_2, \ldots, g_s)$. If $f^{(k)} \neq 0$, then $\mathrm{in}_<(f^{(k)}) < \mathrm{in}_<(f^{(k-1)})$.

Now, suppose that $f^{(k)} \neq 0$ for all $k \geq 1$. Then the infinite sequence

$$\mathrm{in}_<(f) > \mathrm{in}_<(f^{(1)}) > \cdots > \mathrm{in}_<(f^{(k-1)}) > \mathrm{in}_<(f^{(k)}) > \cdots$$

arises. However, Lemma 1.11 rejects the existence of such a sequence. In other words, there is $q > 0$ with $f^{(q)} = 0$, as desired.   □

Since a Gröbner basis is a finite set, Theorem 1.16 yields the so-called *Hilbert Basis Theorem*.

Corollary 1.17 (Hilbert Basis Theorem) *Every ideal of the polynomial ring is finitely generated. More precisely, given a system of generators $\{f_\lambda : \lambda \in \Lambda\}$ of an ideal I of S, there exists a finite subset of $\{f_\lambda : \lambda \in \Lambda\}$ which is a system of generators of I.*

Proof Theorem 1.16 guarantees that every ideal of the polynomial ring is finitely generated. Let $I = (\{f_\lambda : \lambda \in \Lambda\})$ be an ideal of S and $\{f_1, f_2, \ldots, f_s\}$ a system of generators of I consisting of a finite number of polynomials. Then, for each $1 \leq i \leq s$, there exists an expression of the form $f_i = \sum_{\lambda \in \Lambda} h_\lambda^{(i)} f_\lambda$, where $h_\lambda^{(i)} \in S$ is 0 except for a finite number of λ's. Let

$$\Lambda_i = \{\lambda \in \Lambda : h_\lambda^{(i)} \neq 0\}.$$

Then the finite set

$$\{f_\lambda : \lambda \in \cup_{i=1}^s \Lambda_i\}$$

is a system of generators of I.   □

Example 1.18 Let $n = 10$ and I the ideal of $K[x_1, x_2, \ldots, x_{10}]$ generated by

$$f_1 = x_1 x_8 - x_2 x_6, \quad f_2 = x_2 x_9 - x_3 x_7, \quad f_3 = x_3 x_{10} - x_4 x_8,$$

$$f_4 = x_4 x_6 - x_5 x_9, \quad f_5 = x_5 x_7 - x_1 x_{10}.$$

We claim that there exists *no* monomial order $<$ on $K[x_1, x_2, \ldots, x_{10}]$ such that $\{f_1, f_2, \ldots, f_5\}$ is a Gröbner basis of I with respect to $<$.

Suppose on the contrary that there exists a monomial order $<$ on $K[x_1, x_2, \ldots, x_{10}]$ such that $\mathscr{G} = \{f_1, f_2, \ldots, f_5\}$ is a Gröbner basis of I with respect to $<$. First, routine computation says that each of the five polynomials

$$x_1 x_8 x_9 - x_3 x_6 x_7, \quad x_2 x_9 x_{10} - x_4 x_7 x_8, \quad x_2 x_6 x_{10} - x_5 x_7 x_8,$$

$$x_3 x_6 x_{10} - x_5 x_8 x_9, \quad x_1 x_9 x_{10} - x_4 x_6 x_7$$

belongs to I. Let, say, $x_1 x_8 x_9 > x_3 x_6 x_7$. Since $x_1 x_8 x_9 \in \mathrm{in}_<(I)$, there is $g \in \mathscr{G}$ such that $\mathrm{in}_<(g)$ divides $x_1 x_8 x_9$. Such $g \in \mathscr{G}$ must be f_1. Hence $x_1 x_8 > x_2 x_6$. Thus $x_2 x_6 \notin \mathrm{in}_<(I)$. Hence there exists no $g \in \mathscr{G}$ such that $\mathrm{in}_<(g)$ divides $x_2 x_6 x_{10}$. Hence $x_2 x_6 x_{10} < x_5 x_7 x_8$. Thus $x_5 x_7 > x_1 x_{10}$. Continuing these arguments yields

$$x_1 x_8 x_9 > x_3 x_6 x_7, \quad x_2 x_9 x_{10} > x_4 x_7 x_8, \quad x_2 x_6 x_{10} < x_5 x_7 x_8,$$

$$x_3 x_6 x_{10} > x_5 x_8 x_9, \quad x_1 x_9 x_{10} < x_4 x_6 x_7$$

and

$$x_1 x_8 > x_2 x_6, \quad x_2 x_9 > x_3 x_7, \quad x_3 x_{10} > x_4 x_8,$$

$$x_4 x_6 > x_5 x_9, \quad x_5 x_7 > x_1 x_{10}.$$

Hence

$$(x_1 x_8)(x_2 x_9)(x_3 x_{10})(x_4 x_6)(x_5 x_7) > (x_2 x_6)(x_3 x_7)(x_4 x_8)(x_5 x_9)(x_1 x_{10}).$$

However, both sides of the above inequality coincide with $x_1 x_2 \cdots x_{10}$. This is a contradiction.

Theorem 1.19 (Macaulay) *Let I be an ideal of the polynomial ring $S = K[x_1, \ldots, x_n]$ and fix a monomial order $<$ on S. Let $\mathscr{B}$ denote the set of monomials w of S with $w \notin \mathrm{in}_<(I)$. Then $\mathscr{B}$ is a K-basis of the residue ring S/I as a vector space over K.*

Proof First we show that $\mathscr{B}$ is linearly independent in S/I. Let

$$f = c_1 u_1 + \cdots + c_r u_r \in I,$$

where each $0 \neq c_i \in K$ and where each u_i is a monomial of S with $u_i \notin \mathrm{in}_<(I)$. Let $u_1 < \cdots < u_r$. Since $0 \neq f \in I$, it follows that $u_i = \mathrm{in}_<(f)$ must belong to $\mathrm{in}_<(I)$. This contradicts $u_i \in \mathscr{B}$.

Second, in order to prove that the vector space S/I is spanned by $\mathscr{B}$, we write $(\mathscr{B})$ for the subspace of S/I spanned by $\mathscr{B}$. Let $0 \neq f \in S$. We then show that, by using induction on $\mathrm{in}_<(f)$, f belongs to $(\mathscr{B})$. Let $u = \mathrm{in}_<(f)$ and $c \in K$ the coefficient of u in f. If $u \in \mathscr{B}$ then, by assumption of induction, one has $f - c \cdot u \in (\mathscr{B})$. Hence $f \in (\mathscr{B})$. Let $u \notin \mathscr{B}$. Since $u \in \mathrm{in}_<(I)$, there is a polynomial $g \in I$ with $u = \mathrm{in}_<(g)$. Let $c' \in K$ be the coefficient of u in g. Then, again by assumption of induction, it follows that $c'f - cg \in (\mathscr{B})$. However, in S/I, the two polynomials $c'f - cg$ and $c'f$ coincide. Since $c' \neq 0$, one has $f \in (\mathscr{B})$. $\qquad\square$

In Theorem 1.19 each monomial belonging to $\mathscr{B}$ is called *standard* with respect to $<$.

Problems

1.1 Let I and J be ideals of the polynomial ring $S = K[x_1, \ldots, x_n]$. Show that the sum $I + J$, the intersection $I \cap J$, the colon ideal $I : J$ of I with respect to J, and the radical $\sqrt{I}$ of I are ideals of S.

1.2 Show that there is a unique monomial order on the polynomial ring $K[x]$ in one variable.

1.3 Show that orders given in Examples 1.8 and 1.9 are monomial orders.

1.4 There are 20 monomials of degree ≤ 3 belonging to $S = K[x_1, x_2, x_3]$. Order them with respect to the following monomial orders:

(a) the lexicographic order on S induced by the ordering $x_1 > x_2 > x_3$;
(b) the reverse lexicographic order on S induced by the ordering $x_1 > x_2 > x_3$;
(c) the pure lexicographic order on S induced by the ordering $x_1 > x_2 > x_3$.

1.5 Let f and g be nonzero polynomials of $S = K[x_1, \ldots, x_n]$.

(a) Show that $\mathrm{in}_<(fg) = \mathrm{in}_<(f) \cdot \mathrm{in}_<(g)$.
(b) Show that, if w is a monomial, then $\mathrm{in}_<(wg) = w \cdot \mathrm{in}_<(g)$.

1.6 Let $S = K[x_1, \ldots, x_n]$ be the polynomial ring. For a polynomial $f = \sum_i f_i \in S$ with f_i homogeneous of degree i, each f_i is called a *homogeneous component* of f. Suppose that an ideal $I \subset S$ is *graded*, that is, for each $f \in I$, all homogeneous components of f belong to I. Let $<$ denote the reverse lexicographic order induced by $x_1 > x_2 > \cdots > x_n$. Show that for $i = 1, \ldots, n$,

$$\mathrm{in}_<(I, x_{i+1}, \ldots, x_n) = (\mathrm{in}_<(I), x_{i+1}, \ldots, x_n).$$

1.2   The Division Algorithm

The *division algorithm* plays a fundamental role in the theory of Gröbner bases. In order to aid understanding of the proof of Theorem 1.20, the reader may wish to read Example 1.22.

Theorem 1.20 (The division algorithm) *We work with a fixed monomial order $<$ on the polynomial ring $S = K[x_1, x_2, \ldots, x_n]$ and with nonzero polynomials $g_1, g_2, \ldots, g_s$ belonging to S. Then, given a polynomial $0 \neq f \in S$, there exist $f_1, f_2, \ldots, f_s$ and f' belonging to S with*

$$f = f_1 g_1 + f_2 g_2 + \cdots + f_s g_s + f' \tag{1.1}$$

such that the following conditions are satisfied:

- *If $f' \neq 0$ and $u \in \mathrm{supp}(f')$, then none of the initial monomials $\mathrm{in}_<(g_i)$, $1 \leq i \leq s$, divides u. In other words, if $f' \neq 0$, then no monomial $u \in \mathrm{supp}(f')$ belongs to the monomial ideal $(\mathrm{in}_<(g_1), \mathrm{in}_<(g_2), \ldots, \mathrm{in}_<(g_s))$.*
- *If $f_i \neq 0$, then*

$$\mathrm{in}_<(f) \geq \mathrm{in}_<(f_i g_i).$$

Definition 1.21 The right-hand side of the Equation (1.1) is said to be a *standard expression* of f with respect to $g_1, g_2, \ldots, g_s$ and f' a *remainder* of f with respect to $g_1, g_2, \ldots, g_s$.

Proof (of Theorem 1.20) Let $I = (\mathrm{in}_<(g_1), \mathrm{in}_<(g_2), \ldots, \mathrm{in}_<(g_s))$. If no monomial $u \in \mathrm{supp}(f)$ belongs to I, then the desired expression can be obtained by setting $f' = f$ and $f_1 = f_2 = \cdots = f_s = 0$.

Now, suppose that a monomial $u \in \mathrm{supp}(f)$ belongs to I and write u_0 for the monomial which is biggest with respect to $<$ among the monomials belonging to $\mathrm{supp}(f) \cap I$. Let, say, $\mathrm{in}_<(g_{i_0})$ divide u_0 and $w_0 = u_0/\mathrm{in}_<(g_{i_0})$. We rewrite

$$f = c'_0 c_{i_0}^{-1} w_0 g_{i_0} + h_1,$$

where c'_0 is the coefficient of u_0 in f and c_{i_0} is that of $\mathrm{in}_<(g_{i_0})$ in g_{i_0}. Then

$$\mathrm{in}_<(w_0 g_{i_0}) = w_0 \cdot \mathrm{in}_<(g_{i_0}) = u_0 \leq \mathrm{in}_<(f).$$

If either $h_1 = 0$ or if $h_1 \neq 0$ and no monomial $u \in \mathrm{supp}(h_1)$ belongs to I, then $f = c'_0 c_{i_0}^{-1} w_0 g_{i_0} + h_1$ is a standard expression of f with respect to $g_1, g_2, \ldots, g_s$ and h_1 is a remainder of f.

If a monomial $u \in \mathrm{supp}(h_1)$ belongs to I and if u_1 is the monomial which is biggest with respect to $<$ among the monomials belonging to $\mathrm{supp}(h_1) \cap I$, then

$$u_1 < u_0$$

In fact, if a monomial u with $u > u_0$ $(= \text{in}_<(w_0 g_{i_0}))$ belongs to $\text{supp}(h_1)$, then u must belong to $\text{supp}(f)$. This is impossible. Moreover, since the coefficient of u_0 in f coincides with that in $c'_0 c_{i_0}^{-1} w_0 g_{i_0}$, it follows that u_0 cannot belong to $\text{supp}(h_1)$.

Let, say, $\text{in}_<(g_{i_1})$ divide u_1 and $w_1 = u_1 / \text{in}_<(g_{i_1})$. Again, we rewrite

$$f = c'_0 c_{i_0}^{-1} w_0 g_{i_0} + c'_1 c_{i_1}^{-1} w_1 g_{i_1} + h_2,$$

where c'_1 is the coefficient of u_1 in h_1 and c_{i_1} is that of $\text{in}_<(g_{i_1})$ in g_{i_1}. Then

$$\text{in}_<(w_1 g_{i_1}) < \text{in}_<(w_0 g_{i_0}) \leq \text{in}_<(f).$$

Continuing these procedures yields the descending sequence

$$u_0 > u_1 > u_2 > \cdots$$

Lemma 1.11 thus guarantees that these procedures will stop after a finite number of steps, say N steps, and we obtain an expression

$$f = \sum_{q=0}^{N-1} c'_q c_{i_q}^{-1} w_q g_{i_q} + h_N,$$

where either $h_N = 0$ or, in case of $h_N \neq 0$, no monomial $u \in \text{supp}(h_N)$ belongs to I. Moreover, for each $1 \leq q \leq N - 1$, one has

$$\text{in}_<(w_q g_{i_q}) < \cdots < \text{in}_<(w_0 g_{i_0}) \leq \text{in}_<(f).$$

Thus, by letting $\sum_{i=1}^{s} f_i g_i = \sum_{q=0}^{N-1} c'_q c_{i_q}^{-1} w_q g_{i_q}$ and $f' = h_N$, we obtain a standard expression $f = \sum_{i=1}^{s} f_i g_i + f'$ of f, as desired.   □

Example 1.22 Let $<_{\text{lex}}$ denote the lexicographic order on $K[x, y, z]$ induced by $x > y > z$. Let $g_1 = x^2 - z$, $g_2 = xy - 1$ and $f = x^3 - x^2 y - x^2 - 1$. Each of

$$f = x^3 - x^2 y - x^2 - 1 = x(g_1 + z) - x^2 y - x^2 - 1$$
$$= x g_1 - x^2 y - x^2 + xz - 1 = x g_1 - (g_1 + z)y - x^2 + xz - 1$$
$$= x g_1 - y g_1 - x^2 + xz - yz - 1 = x g_1 - y g_1 - (g_1 + z) + xz - yz - 1$$
$$= (x - y - 1)g_1 + (xz - yz - z - 1)$$

and

$$f = x^3 - x^2 y - x^2 - 1 = x(g_1 + z) - x^2 y - x^2 - 1$$
$$= x g_1 - x^2 y - x^2 + xz - 1 = x g_1 - x(g_2 + 1) - x^2 + xz - 1$$

$$= xg_1 - xg_2 - x^2 + xz - x - 1 = xg_1 - xg_2 - (g_1 + z) + xz - x - 1$$
$$= (x - 1)g_1 - xg_2 + (xz - x - z - 1)$$

is a standard expression of f with respect to g_1 and g_2, and each of $xz - yz - z - 1$ and $xz - x - z - 1$ is a remainder of f.

Example 1.22 shows that in the division algorithm a remainder of f is, in general, not unique. However,

Lemma 1.23 *If a finite set $\{g_1, g_2, \ldots, g_s\}$ consisting of polynomials belonging to S is a Gröbner basis of the ideal $I = (g_1, g_2, \ldots, g_s)$, then any nonzero polynomial $f \in S$ has a unique remainder with respect to $g_1, g_2, \ldots, g_s$.*

Proof Suppose that each of the polynomials f' and f'' is a remainder of f with respect to $g_1, \ldots, g_s$. Let $f' \neq f''$. Since $0 \neq f' - f'' \in I$, the initial monomial $w = \mathrm{in}_<(f' - f'')$ belongs to $\mathrm{in}_<(I)$. On the other hand, since w belongs to either $\mathrm{supp}(f')$ or $\mathrm{supp}(f'')$, it follows that w cannot belong to $(\mathrm{in}_<(g_1), \mathrm{in}_<(g_2), \ldots, \mathrm{in}_<(g_s))$. However, since $\{g_1, \ldots, g_s\}$ is a Gröbner basis, the initial ideal $\mathrm{in}_<(I)$ coincides with $(\mathrm{in}_<(g_1), \mathrm{in}_<(g_2), \ldots, \mathrm{in}_<(g_s))$. This is a contradiction. □

Corollary 1.24 *Suppose that a finite set $\{g_1, g_2, \ldots, g_s\}$ consisting of polynomials belonging to S is a Gröbner basis of the ideal $I = (g_1, g_2, \ldots, g_s)$ of S. Then a polynomial $0 \neq f \in S$ belongs to I if and only if the unique remainder of f with respect to $g_1, g_2, \ldots, g_s$ is 0.*

Proof In general, if a remainder of a polynomial $0 \neq f \in S$ with respect to $g_1, g_2, \ldots, g_s$ is 0, then f belongs to the ideal $I = (g_1, g_2, \ldots, g_s)$.

Now, suppose that $0 \neq f \in S$ belongs to I and that a standard expression of f with respect to $g_1, g_2, \ldots, g_s$ is $f = f_1 g_1 + f_2 g_2 + \cdots + f_s g_s + f'$. Since $f \in I$, one has $f' \in I$. If $f' \neq 0$, then $\mathrm{in}_<(f') \in \mathrm{in}_<(I)$. Since $\{g_1, g_2, \ldots, g_s\}$ is a Gröbner basis of I, one has $\mathrm{in}_<(I) = (\mathrm{in}_<(g_1), \mathrm{in}_<(g_2), \ldots, \mathrm{in}_<(g_s))$. However, since f' is a remainder, $\mathrm{in}_<(f') \in \mathrm{supp}(f')$ cannot belong to $(\mathrm{in}_<(g_1), \mathrm{in}_<(g_2), \ldots, \mathrm{in}_<(g_s))$. This is a contradiction. □

We work with a fixed monomial order $<$ on the polynomial ring $S = K[x_1, \ldots, x_n]$. A Gröbner basis $\{g_1, g_2, \ldots, g_s\}$ of an ideal of S is called *reduced* if the following conditions are satisfied:

- The coefficient of $\mathrm{in}_<(g_i)$ in g_i is 1 for all $1 \leq i \leq s$;
- If $i \neq j$, then none of the monomials belonging to $\mathrm{supp}(g_j)$ is divided by $\mathrm{in}_<(g_i)$.

A reduced Gröbner basis is a minimal Gröbner basis. However, the converse is false. See Problem 1.9.

Theorem 1.25 *A reduced Gröbner basis exists and is uniquely determined.*

Proof (Existence) Let $\{g_1, g_2, \ldots, g_s\}$ be a minimal Gröbner basis of an ideal I of S. Then $\{\text{in}_<(g_1), \text{in}_<(g_2), \ldots, \text{in}_<(g_s)\}$ is the unique minimal system of monomial generators of the initial ideal $\text{in}_<(I)$. Thus, if $i \neq j$, then $\text{in}_<(g_i)$ cannot be divided by $\text{in}_<(g_j)$.

First, let h_1 be a remainder of g_1 with respect to $g_2, g_3, \ldots, g_s$. Since $\text{in}_<(g_1)$ can be divided by none of $\text{in}_<(g_j)$, $2 \leq j \leq s$, it follows that $\text{in}_<(h_1)$ coincides with $\text{in}_<(g_1)$. Thus $\{h_1, g_2, \ldots, g_s\}$ is a minimal Gröbner basis of I and each monomial belonging to $\text{supp}(h_1)$ can be divided by none of $\text{in}_<(g_j)$, $2 \leq j \leq s$.

Second, let h_2 be a remainder of g_2 with respect to $h_1, g_3, \ldots, g_s$. Since $\text{in}_<(g_2)$ can be divided by none of $\text{in}_<(h_1)(= \text{in}_<(g_1))$, $\text{in}_<(g_3), \ldots, \text{in}_<(g_s)$, it follows that $\text{in}_<(h_2)$ coincides with $\text{in}_<(g_2)$ and $\{h_1, h_2, g_3, \ldots, g_s\}$ is a minimal Gröbner basis of I with the property that each monomial belonging to $\text{supp}(h_1)$ can be divided by none of $\text{in}_<(h_2)$, $\text{in}_<(g_3), \ldots, \text{in}_<(g_s)$ and each monomial belonging to $\text{supp}(h_2)$ can be divided by none of $\text{in}_<(h_1)$, $\text{in}_<(g_3), \ldots, \text{in}_<(g_s)$.

Continuing these procedures yields polynomials $h_3, h_4, \ldots, h_s$ and we obtain a reduced Gröbner basis $\{h_1, h_2, \ldots, h_s\}$ of I.

(Uniqueness) If $\{g_1, g_2, \ldots, g_s\}$ and $\{g'_1, g'_2, \ldots, g'_t\}$ are reduced Gröbner bases of I, then $\{\text{in}_<(g_1), \text{in}_<(g_2), \ldots, \text{in}_<(g_s)\}$ and $\{\text{in}_<(g'_1), \text{in}_<(g'_2), \ldots, \text{in}_<(g'_t)\}$ are minimal system of monomial generators of $\text{in}_<(I)$. Lemma 1.6 then says that $s = t$ and, after rearranging the indices, we may assume that $\text{in}_<(g_i) = \text{in}_<(g'_i)$ for all $1 \leq i \leq s (= t)$. Let, say $g_i - g'_i \neq 0$. Then $\text{in}_<(g_i - g'_i) < \text{in}_<(g_i)$. Since $\text{in}_<(g_i - g'_i)$ belongs to either $\text{supp}(g_i)$ or $\text{supp}(g'_i)$, it follows that none of $\text{in}_<(g_j)$, $j \neq i$, can divide $\text{in}_<(g_i - g'_i)$. Hence $\text{in}_<(g_i - g'_i) \notin \text{in}_<(I)$. This contradicts the fact that $g_i - g'_i$ belongs to I. Hence $g_i = g'_i$ for all $1 \leq i \leq s$. $\square$

We write $\mathcal{G}_{\text{red}}(I; <)$ for *the* reduced Gröbner basis of an ideal I of S with respect to a monomial order $<$.

Corollary 1.26 *Let I and J be ideals of S. Then $I = J$ if and only if $\mathcal{G}_{\text{red}}(I; <) = \mathcal{G}_{\text{red}}(J; <)$.*

Problems

1.7 Consider the polynomials $f = x^2y^2z + xyz^2 + xy^4$, $g_1 = x^2 - xyz + y^3$, and $g_2 = xz^2 - y^2z$ in $S = K[x, y, z]$. Give a standard expression of f with respect to g_1 and g_2 for the following monomial orders:

(a) the lexicographic order on S induced by the ordering $x > y > z$;
(b) the reverse lexicographic order on S induced by the ordering $x > y > z$;
(c) the pure lexicographic order on S induced by the ordering $x > y > z$.

1.8 Let $\mathcal{G}$ be a Gröbner basis of an ideal I of $S = K[x_1, \ldots, x_n]$.

(a) Let r be a remainder of $f \in S$ with respect to $\mathcal{G}$. Show that f belongs to $\sqrt{I}$ if and only if r belongs to $\sqrt{I}$.

(b) Show that I is radical (that is, $I = \sqrt{I}$), if the initial ideal $\mathrm{in}_<(I)$ is generated by squarefree monomials,

1.9 Show that any reduced Gröbner basis is a minimal Gröbner basis. Give a counterexample of the converse of the above statement.

1.3   Buchberger's Criterion

The highlights of the theory of Gröbner bases must be Buchberger's criterion and Buchberger's algorithm. A Gröbner basis of an ideal is its system of generators. It is then natural to ask: Given a system of generators of an ideal, how can we decide whether they form its Gröbner basis or not? The answer is Buchberger's criterion, which also yields an algorithm called Buchberger's algorithm. Starting from a system of generators of an ideal, the algorithm supplies the effective procedure to compute a Gröbner basis of the ideal. The discovery of the algorithm is one of the most important achievements of Buchberger.

Let, as before, $S = K[x_1, \ldots, x_n]$ denote the polynomial ring over K. We work with a fixed monomial order $<$ on S and, for simplicity, omit the phrase "with respect to $<$", if there is no danger of confusion.

The least common multiple $\mathrm{lcm}(u, v)$ of two monomials $u = x_1^{a_1} x_2^{a_2} \cdots x_n^{a_n}$ and $v = x_1^{b_1} x_2^{b_2} \cdots x_n^{b_n}$ is the monomial $x_1^{c_1} x_2^{c_2} \cdots x_n^{c_n}$ with each $c_i = \max\{a_i, b_i\}$.

Let f and g be nonzero polynomials of S. Let c_f be the coefficient of $\mathrm{in}_<(f)$ in f and c_g that of $\mathrm{in}_<(g)$ in g. Then the polynomial

$$S(f, g) = \frac{\mathrm{lcm}(\mathrm{in}_<(f), \mathrm{in}_<(g))}{c_f \cdot \mathrm{in}_<(f)} f - \frac{\mathrm{lcm}(\mathrm{in}_<(f), \mathrm{in}_<(g))}{c_g \cdot \mathrm{in}_<(g)} g$$

is called the *S-polynomial* of f and g.

In other words, the S-polynomial of f and g can be obtained by canceling the initial monomials of f and g. For example, if $f = x_1x_4 - x_2x_3$ and $g = x_4x_7 - x_5x_6$, then with respect to $<_{\mathrm{lex}}$ one has

$$S(f, g) = x_7 f - x_1 g = x_1 x_5 x_6 - x_2 x_3 x_7,$$

and with respect to $<_{\mathrm{rev}}$ one has

$$S(f, g) = -x_5x_6 f + x_2x_3 g = x_2x_3x_4x_7 - x_1x_4x_5x_6.$$

We say that f *reduces to* 0 with respect to $g_1, g_2, \ldots, g_s$ if there is a standard expression (1.1) of f with respect to $g_1, g_2, \ldots, g_s$ with $f' = 0$.

Lemma 1.27 *Let f and g be nonzero polynomials of S and suppose that $\mathrm{in}_<(f)$ and $\mathrm{in}_<(g)$ are relatively prime, i.e., $\mathrm{lcm}(\mathrm{in}_<(f), \mathrm{in}_<(g)) = \mathrm{in}_<(f)\mathrm{in}_<(g)$. Then $S(f, g)$ reduces to 0 with respect to f, g.*

Proof To simplify the notation, we assume that each of the coefficients of $\text{in}_<(f)$ in f and $\text{in}_<(g)$ in g is 1. Let $f = \text{in}_<(f) + f_1$ and $g = \text{in}_<(g) + g_1$. Since $\text{in}_<(f)$ and $\text{in}_<(g)$ are relatively prime, it follows that

$$
\begin{aligned}
S(f, g) &= \text{in}_<(g)f - \text{in}_<(f)g \\
&= (g - g_1)f - (f - f_1)g \\
&= f_1 g - g_1 f.
\end{aligned}
$$

We claim that $\text{in}_<(f_1)\text{in}_<(g)$ cannot coincide with $\text{in}_<(g_1)\text{in}_<(f)$. In fact, if we have $\text{in}_<(f_1)\text{in}_<(g) = \text{in}_<(g_1)\text{in}_<(f)$, then, since $\text{in}_<(f)$ and $\text{in}_<(g)$ are relatively prime, it follows that $\text{in}_<(f)$ divides $\text{in}_<(f_1)$. However, since $\text{in}_<(f_1) < \text{in}_<(f)$, this is impossible. Let, say, $\text{in}_<(f_1 g) < \text{in}_<(g_1 f)$. Then $\text{in}_<(S(f, g)) = \text{in}_<(g_1 f)$. Hence $S(f, g) = f_1 g - g_1 f$ is a standard expression of $S(f, g)$ with respect to f, g with a remainder 0. Thus $S(f, g)$ reduces to 0 with respect to f, g. □

We now come to the most important theorem in the theory of Gröbner bases.

Lemma 1.28 *Let w be a monomial and $f_1, f_2, \ldots, f_s$ polynomials with $\text{in}_<(f_i) = w$ for all $1 \leq i \leq s$. Let $g = \sum_{i=1}^{s} b_i f_i$ with each $b_i \in K$ and suppose that $\text{in}_<(g) < w$. Then there exist $c_{jk} \in K$ with*

$$
g = \sum_{1 \leq j, k \leq s} c_{jk} S(f_j, f_k).
$$

Proof Let c_i be the coefficient of $w = \text{in}_<(f_i)$ in f_i. Then $\sum_{i=1}^{s} b_i c_i = 0$. Let $g_i = (1/c_i)f_i$. Then

$$
S(f_j, f_k) = g_j - g_k, \quad 1 \leq j, k \leq s.
$$

Hence

$$
\sum_{i=1}^{s} b_i f_i = \sum_{i=1}^{s} b_i c_i g_i
$$

$$
= b_1 c_1 (g_1 - g_2) + (b_1 c_1 + b_2 c_2)(g_2 - g_3) + (b_1 c_1 + b_2 c_2 + b_3 c_3)(g_3 - g_4)
$$
$$
+ \cdots + (b_1 c_1 + \cdots + b_{s-1} c_{s-1})(g_{s-1} - g_s) + (b_1 c_1 + \cdots + b_s c_s)g_s.
$$

Since $\sum_{i=1}^{s} b_i c_i = 0$, it follows that

$$
\sum_{i=1}^{s} b_i f_i = \sum_{i=2}^{s} (b_1 c_1 + \cdots + b_{i-1} c_{i-1}) S(f_{i-1}, f_i),
$$

as desired.   □

Theorem 1.29 (Buchberger's criterion) *Let I be an ideal of the polynomial ring S and $\mathscr{G} = \{g_1, g_2, \ldots, g_s\}$ a system of generators of I. Then $\mathscr{G}$ is a Gröbner basis of I if and only if the following condition is satisfied:*

($\star$) *For all $i \neq j$, $S(g_i, g_j)$ reduces to 0 with respect to $g_1, g_2, \ldots, g_s$.*

Proof ("Only If") Suppose that a system of generators $\mathscr{G} = \{g_1, g_2, \ldots, g_s\}$ is a Gröbner basis of I. Since the S-polynomial $S(g_i, g_j)$ of g_i and g_j belongs to the ideal (g_i, g_j), we have, in particular, $S(g_i, g_j) \in I$. Since $\mathscr{G}$ is a Gröbner basis of I, Corollary 1.24 guarantees that $S(g_i, g_j)$ reduces to 0 with respect to $g_1, g_2, \ldots, g_s$, as required.

("If") Let $\mathscr{G} = \{g_1, g_2, \ldots, g_s\}$ be a system of generators of I which satisfies the condition ($\star$).

(First Step) If a nonzero polynomial f belongs to I, then we write $\mathscr{H}_f$ for the set of sequences $(h_1, h_2, \ldots, h_s)$ with each $h_i \in S$ such that

$$f = \sum_{i=1}^{s} h_i g_i. \tag{1.2}$$

Since $\mathscr{G} = \{g_1, g_2, \ldots, g_s\}$ is a system of generators of I, it follows that $\mathscr{H}_f$ is nonempty. We associate each sequence $(h_1, h_2, \ldots, h_s) \in \mathscr{H}_f$ with the monomial

$$\delta_{(h_1, h_2, \ldots, h_s)} = \max\{\mathrm{in}_<(h_i g_i) : h_i g_i \neq 0\}.$$

Then

$$\mathrm{in}_<(f) \leq \delta_{(h_1, h_2, \ldots, h_s)}. \tag{1.3}$$

Now, among all of the monomials $\delta_{(h_1, h_2, \ldots, h_s)}$ with $(h_1, h_2, \ldots, h_s) \in \mathscr{H}_f$, we are especially interested in the monomial

$$\delta_f = \min_{(h_1, h_2, \ldots, h_s) \in \mathscr{H}_f} \delta_{(h_1, h_2, \ldots, h_s)}.$$

Then the inequality (1.3) says that

$$\mathrm{in}_<(f) \leq \delta_f.$$

In the following discussion, we will assume that the monomial $\delta_{(h_1, h_2, \ldots, h_s)}$ arising from the equality (1.2) coincides with δ_f.

(Second Step) Suppose for a while that $\mathrm{in}_<(f) = \delta_f$. Then, in the right-hand side of the equality (1.2), there is $h_i g_i \neq 0$ with $\mathrm{in}_<(f) = \mathrm{in}_<(h_i g_i)$. In particular $\mathrm{in}_<(f)$ belongs to the monomial ideal generated by $\mathrm{in}_<(g_1), \mathrm{in}_<(g_2), \ldots, \mathrm{in}_<(g_s)$.

Hence, if we can prove that $\mathrm{in}_<(f) = \delta_f$ for any nonzero polynomial $f \in I$, then

$$\mathrm{in}_<(I) = (\mathrm{in}_<(g_1), \mathrm{in}_<(g_2), \ldots, \mathrm{in}_<(g_s))$$

and $\mathcal{G}$ turns out to be a Gröbner basis of I.

(Third Step) Now, suppose that there is a nonzero polynomial $f \in I$ with $\mathrm{in}_<(f) < \delta_f$. If we can get a contradiction, then our proof finishes.

We rewrite the right-hand side of the equality (1.2) as

$$(\sharp) \quad f = \sum_{\mathrm{in}_<(h_i g_i) = \delta_f} h_i g_i + \sum_{\mathrm{in}_<(h_i g_i) < \delta_f} h_i g_i$$

$$= \sum_{\mathrm{in}_<(h_i g_i) = \delta_f} c_i \cdot \mathrm{in}_<(h_i) g_i$$

$$+ \sum_{\mathrm{in}_<(h_i g_i) = \delta_f} (h_i - c_i \cdot \mathrm{in}_<(h_i)) g_i + \sum_{\mathrm{in}_<(h_i g_i) < \delta_f} h_i g_i,$$

where $c_i \in K$ is the coefficient of $\mathrm{in}_<(h_i)$ in h_i. The first equality is clear. The second equality is the consequence of the simple rewriting

$$h_i = c_i \cdot \mathrm{in}_<(h_i) + (h_i - c_i \cdot \mathrm{in}_<(h_i)).$$

A crucial fact is that every monomial u belonging to the support of

$$\sum_{\mathrm{in}_<(h_i g_i) = \delta_f} (h_i - c_i \cdot \mathrm{in}_<(h_i)) g_i + \sum_{\mathrm{in}_<(h_i g_i) < \delta_f} h_i g_i$$

satisfies $u < \delta_f$. Hence, the hypothesis that $\mathrm{in}_<(f) < \delta_f$ guarantees that

$$\mathrm{in}_<\left(\sum_{\mathrm{in}_<(h_i g_i) = \delta_f} c_i \cdot \mathrm{in}_<(h_i) g_i \right) < \delta_f.$$

However, since $\mathrm{in}_<(h_i g_i) = \delta_f$, one has

$$\mathrm{in}_<(\mathrm{in}_<(h_i) g_i) = \delta_f.$$

It then follows from Lemma 1.28 that, by using those S-polynomials

$$S(\mathrm{in}_<(h_j) g_j, \mathrm{in}_<(h_k) g_k)$$

with $\mathrm{in}_<(h_j g_j) = \mathrm{in}_<(h_k g_k) = \delta_f$ and $c_{jk} \in K$, we can rewrite the first sum in the right-hand side of the second equality of $(\sharp)$ as follows:

$$\sum_{\mathrm{in}_<(h_i g_i)=\delta_f} c_i \cdot \mathrm{in}_<(h_i)g_i = \sum_{j,k} c_{jk} S(\mathrm{in}_<(h_j)g_j, \mathrm{in}_<(h_k)g_k). \tag{1.4}$$

Since $\mathrm{in}_<(h_j g_j) = \mathrm{in}_<(h_k g_k) = \delta_f$, it follows that

$$S(\mathrm{in}_<(h_j)g_j, \mathrm{in}_<(h_k)g_k) = (1/b_j)\mathrm{in}_<(h_j)g_j - (1/b_k)\mathrm{in}_<(h_k)g_k,$$

where b_j is the coefficient of $\mathrm{in}_<(g_j)$ in g_j. Here each monomial u belonging to the support of $S(\mathrm{in}_<(h_j)g_j, \mathrm{in}_<(h_k)g_k)$ satisfies $u < \delta_f$.

Let

$$u_{jk} = \delta_f/\mathrm{lcm}(\mathrm{in}_<(g_j), \mathrm{in}_<(g_k)).$$

Then

$$
\begin{aligned}
u_{jk}S(g_j, g_k) &= u_{jk}\left[\frac{\mathrm{lcm}(\mathrm{in}_<(g_j), \mathrm{in}_<(g_k))}{b_j \cdot \mathrm{in}_<(g_j)}g_j - \frac{\mathrm{lcm}(\mathrm{in}_<(g_j), \mathrm{in}_<(g_k))}{b_k \cdot \mathrm{in}_<(g_k)}g_k\right] \\
&= \delta_f\left[\frac{1}{b_j \cdot \mathrm{in}_<(g_j)}g_j - \frac{1}{b_k \cdot \mathrm{in}_<(g_k)}g_k\right] \\
&= \frac{\mathrm{in}_<(h_j)}{b_j}g_j - \frac{\mathrm{in}_<(h_k)}{b_k}g_k \\
&= S(\mathrm{in}_<(h_j)g_j, \mathrm{in}_<(h_k)g_k).
\end{aligned}
$$

By using the equality (1.4), there exists an expression of the form

$$\sum_{\mathrm{in}_<(h_i g_i)=\delta_f} c_i \cdot \mathrm{in}_<(h_i)g_i = \sum_{j,k} c_{jk} u_{jk} S(g_j, g_k), \quad c_{jk} \in K \tag{1.5}$$

with

$$\mathrm{in}_<(u_{jk}S(g_j, g_k)) < \delta_f.$$

The condition $(\star)$ guarantees the existence of an expression of $S(g_j, g_k)$ of the form

$$S(g_j, g_k) = \sum_{i=1}^{s} p_i^{jk}g_i, \quad \mathrm{in}_<(p_i^{jk}g_i) \le \mathrm{in}_<(S(g_j, g_k)), \tag{1.6}$$

where $p_i^{jk} \in S$. Combining (1.6) with (1.5) yields

$$\sum_{\mathrm{in}_<(h_i g_i)=\delta_f} c_i \cdot \mathrm{in}_<(h_i)g_i = \sum_{j,k} c_{jk} u_{jk} \left(\sum_{i=1}^{s} p_i^{jk}g_i\right). \tag{1.7}$$

We rewrite the right-hand side of the equality (1.7) as $\sum_{i=1}^{s} h_i' g_i$. Then

$$\mathrm{in}_<(h_i' g_i) < \delta_f.$$

Finally, by virtue of (1.7) together with the second equality of ($\sharp$), it turns out that there exists an expression of f of the form

$$f = \sum_{i=1}^{s} h_i'' g_i, \quad \mathrm{in}_<(h_i'' g_i) < \delta_f.$$

The existence of such an expression contradicts the definition of δ_f, as desired. $\square$

In applying Buchberger's criterion it is not always necessary to check whether *all* S-polynomials $S(g_i, g_j)$ with $i \neq j$ reduce to 0 with respect to $g_1, \ldots, g_s$. In fact, Lemma 1.27 says that if $\mathrm{in}_<(g_i)$ and $\mathrm{in}_<(g_j)$ are relatively prime, then $S(g_i, g_j)$ reduces to 0 with respect to g_i, g_j. Thus in particular $S(g_i, g_j)$ reduces to 0 with respect to $g_1, g_2, \ldots, g_s$. Hence we only check those S-polynomials $S(g_i, g_j)$ with $i \neq j$ such that $\mathrm{in}_<(g_i)$ and $\mathrm{in}_<(g_j)$ possess at least one common variable.

Corollary 1.30 *If $g_1, \ldots, g_s$ are nonzero polynomials belonging to S such that $\mathrm{in}_<(g_i)$ and $\mathrm{in}_<(g_j)$ are relatively prime for all $i \neq j$, then $\{g_1, \ldots, g_s\}$ is a Gröbner basis of $I = (g_1, \ldots, g_s)$.*

Example 1.31 Let $n = 7$ and consider the reverse lexicographic order $<_{\mathrm{rev}}$. Let $f = x_1 x_4 - x_2 x_3$, $g = x_4 x_7 - x_5 x_6$ and $I = (f, g)$. Then, since $\mathrm{in}_{<_{\mathrm{rev}}}(f) = x_2 x_3$ and $\mathrm{in}_{<_{\mathrm{rev}}}(g) = x_5 x_6$ are relatively prime, it follows that $\{f, g\}$ is a Gröbner basis of I with respect to $<_{\mathrm{rev}}$.

Example 1.32 Let $f = x_1 x_4 - x_2 x_3$, $g = x_4 x_7 - x_5 x_6$ and $I = (f, g)$. Example 1.14 shows that $\{f, g\}$ cannot be a Gröbner basis of I with respect to the lexicographic order $<_{\mathrm{lex}}$. On the other hand, if $h = S(f, g) = x_1 x_5 x_6 - x_2 x_3 x_7$, then $\{f, g, h\}$ is a Gröbner basis of I with respect to $<_{\mathrm{lex}}$. To see why this is true, we must check the criterion ($\star$) for $S(f, g)$, $S(g, h)$, and $S(f, h)$. First, $S(f, g) = h$ reduces to 0 with respect to h. Since $\mathrm{in}_{<_{\mathrm{lex}}}(g)$ and $\mathrm{in}_{<_{\mathrm{lex}}}(h)$ are relatively prime, $S(g, h)$ reduces to 0 with respect to g, h. Moreover, since

$$S(f, h) = x_5 x_6 f - x_4 h = x_2 x_3 x_4 x_7 - x_2 x_3 x_5 x_6 = x_2 x_3 g,$$

it follows that $S(f, h)$ reduces to 0 with respect to g.

One of the advantages of Buchberger's criterion is that it yields an algorithm, called Buchberger's algorithm, which supplies a procedure for computing a Gröbner basis of an ideal I of S from a system of generators of I.

- Let I be an ideal of the polynomial ring S and $\mathcal{G} = \{g_1, g_2, \ldots, g_s\}$ its system of generators. If each S-polynomial $S(g_i, g_j)$, $1 \leq i < j \leq s$, reduces to 0 with respect to $g_1, g_2, \ldots, g_s$, then Buchberger's criterion guarantees that $\mathcal{G}$ is a Gröbner basis of I.

- Otherwise there is $S(g_i, g_j)$ with nonzero remainder g_{s+1}. It follows from the definition of a remainder that none of $\text{in}_<(g_i) \in \mathcal{G}$ divides $\text{in}_<(g_{s+1})$. Hence the monomial ideal

$$(\text{in}_<(g_1), \text{in}_<(g_2), \ldots, \text{in}_<(g_s))$$

is strictly contained in the monomial ideal

$$(\text{in}_<(g_1), \text{in}_<(g_2), \ldots, \text{in}_<(g_s), \text{in}_<(g_{s+1})).$$

- Since $S(g_i, g_j) \in I$, it follows that $g_{s+1} \in I$. Now, replace $\mathcal{G}$ with

$$\mathcal{G}' = \mathcal{G} \cup \{g_{s+1}\},$$

which is a system of generators of I with a redundant polynomial g_{s+1}. We then apply Buchberger's criterion to $\mathcal{G}'$. If each $S(g_i, g_j)$, $1 \leq i < j \leq s+1$, reduces to 0 with respect to $g_1, g_2, \ldots, g_s, g_{s+1}$, then Buchberger's criterion guarantees that $\mathcal{G}'$ is a Gröbner basis of I.

- Otherwise there is $S(g_k, g_\ell)$ with nonzero remainder g_{s+2} and

$$(\text{in}_<(g_1), \text{in}_<(g_2), \ldots, \text{in}_<(g_s), \text{in}_<(g_{s+1}))$$

is strictly contained in

$$(\text{in}_<(g_1), \text{in}_<(g_2), \ldots, \text{in}_<(g_s), \text{in}_<(g_{s+1}), \text{in}_<(g_{s+2})).$$

- Again, the remainder g_{s+2} belongs to I. We thus apply Buchberger's criterion to $\mathcal{G}'' = \mathcal{G}' \cup \{g_{s+2}\}$, which is a system of generators of I with redundant polynomials g_{s+1} and g_{s+2}.
- By virtue of Theorem 1.2, it follows that the above procedure will terminate after a finite number of steps, and a Gröbner basis of I can be obtained.
- In fact, if the above procedure will eternally persist, then there exists a strictly increasing infinite sequence of monomial ideals

$$(\text{in}_<(g_1), \ldots, \text{in}_<(g_s)) \subset (\text{in}_<(g_1), \ldots, \text{in}_<(g_s), \text{in}_<(g_{s+1}))$$

$$\subset \cdots \subset (\text{in}_<(g_1), \ldots, \text{in}_<(g_s), \text{in}_<(g_{s+1}), \ldots, \text{in}_<(g_{s+k})) \subset \cdots$$

Theorem 1.2 says that the set of minimal elements of the set of monomials

$$\mathcal{M} = \{\text{in}_<(g_1), \ldots, \text{in}_<(g_s), \text{in}_<(g_{s+1}), \ldots\}$$

is finite. If

$$\text{in}_<(g_{i_1}), \text{in}_<(g_{i_2}), \ldots, \text{in}_<(g_{i_q}), \qquad i_1 < i_2 < \cdots < i_q,$$

are the minimal elements of $\mathcal{M}$, then for all $j > i_q$ one has

$$(\text{in}_<(g_{i_1}), \text{in}_<(g_{i_2}), \dots, \text{in}_<(g_{i_q}))$$
$$= (\text{in}_<(g_1), \text{in}_<(g_2), \dots, \text{in}_<(g_{i_q}), \text{in}_<(g_{i_q+1}), \dots, \text{in}_<(g_j)),$$

which is a contradiction.

The reader may have observed that the basic fact which guarantees that the above procedure terminates after a finite number of steps is again Theorem 1.2. The above algorithm which, starting from a system of generators of I, enables us to find a Gröbner basis of I is said to be *Buchberger's algorithm*

Example 1.33 We follow Example 1.18. Let $n = 10$ and $I = (f_1, f_2, f_3, f_4, f_5)$ the ideal of $K[x_1, x_2, \dots, x_{10}]$, where

$$f_1 = x_1 x_8 - x_2 x_6, \quad f_2 = x_2 x_9 - x_3 x_7, \quad f_3 = x_3 x_{10} - x_4 x_8,$$
$$f_4 = x_4 x_6 - x_5 x_9, \quad f_5 = x_5 x_7 - x_1 x_{10}.$$

In Example 1.18 it is shown that there exists no monomial order $<$ such that $\mathcal{F} = \{f_1, f_2, f_3, f_4, f_5\}$ is a Gröbner basis of I. In what follows, by using Buchberger's algorithm, we compute a Gröbner basis of I with respect to the lexicographic order as well as that with respect to the reverse lexicographic order.

(Lexicographic order) The initial monomials of f_1, f_2, f_3, f_4, f_5 are

$$x_1 x_8, \quad x_2 x_9, \quad x_3 x_{10}, \quad x_4 x_6, \quad x_1 x_{10},$$

respectively. Recall that if $\text{in}_{<_{\text{lex}}}(f_i)$ and $\text{in}_{<_{\text{lex}}}(f_j)$ with $i \neq j$ are relatively prime, then $S(f_i, f_j)$ reduces to 0. Thus the S-polynomials which we must check are

$$S(f_1, f_5) = x_{10} f_1 + x_8 f_5 = x_5 x_7 x_8 - x_2 x_6 x_{10},$$
$$S(f_3, f_5) = x_1 f_3 + x_3 f_5 = x_3 x_5 x_7 - x_1 x_4 x_8.$$

One has

$$S(f_3, f_5) = -x_4 f_1 - x_2 x_4 x_6 + x_3 x_5 x_7$$
$$= -x_4 f_1 - x_2 f_4 - x_2 x_5 x_9 + x_3 x_5 x_7$$
$$= -x_4 f_1 - x_2 f_4 - x_5 f_2,$$

which reduces to 0. On the other hand, $S(f_1, f_5)$ itself is a remainder with respect to f_1, f_2, f_3, f_4, f_5. Thus, letting

$$f_6 = x_5 x_7 x_8 - x_2 x_6 x_{10},$$

we consider $\mathscr{F}' = \{f_1, f_2, f_3, f_4, f_5, f_6\}$ to be a system of generators of I (with a redundant polynomial f_6) and apply Buchberger's criterion to $\mathscr{F}'$. Since $\mathrm{in}_{<_{\mathrm{lex}}}(f_6) = x_2x_6x_{10}$, the S-polynomials which we must check are

$$S(f_2, f_6) = x_6x_{10}f_2 + x_9 f_6 = x_5x_7x_8x_9 - x_3x_6x_7x_{10}$$
$$= x_7(x_5x_8x_9 - x_3x_6x_{10}) = x_7(-x_6f_3 - x_4x_6x_8 + x_5x_8x_9)$$
$$= -x_7(x_6f_3 + x_8f_4),$$

$$S(f_3, f_6) = x_2x_6f_3 + x_3 f_6 = x_3x_5x_7x_8 - x_2x_4x_6x_8$$
$$= x_8(x_3x_5x_7 - x_2x_4x_6) = x_8(-x_2f_4 - x_2x_5x_9 + x_3x_5x_7)$$
$$= -x_8(x_5f_2 + x_2f_4),$$

$$S(f_4, f_6) = x_2x_{10}f_4 + x_4 f_6 = x_4x_5x_7x_8 - x_2x_5x_9x_{10}$$
$$= x_5(x_4x_7x_8 - x_2x_9x_{10}) = x_5(-x_{10}f_2 - x_3x_7x_{10} + x_4x_7x_8)$$
$$= -x_5(x_{10}f_2 + x_7f_3),$$

$$S(f_5, f_6) = -x_2x_6f_5 + x_1 f_6 = x_1x_5x_7x_8 - x_2x_5x_6x_7$$
$$= x_5x_7f_1.$$

Each of them reduces to 0. Thus $\mathscr{F}'$ is a Gröbner basis of I with respect to the lexicographic order.

(Reverse lexicographic order) The initial monomials of f_1, f_2, f_3, f_4, f_5 are

$$x_2x_6, \quad x_3x_7, \quad x_4x_8, \quad x_4x_6, \quad x_5x_7,$$

respectively. Thus the S-polynomials which we must check are

$$S(f_1, f_4) = -x_4f_1 - x_2 f_4 = x_2x_5x_9 - x_1x_4x_8,$$
$$S(f_2, f_5) = -x_5f_2 - x_3 f_5 = x_1x_3x_{10} - x_2x_5x_9,$$
$$S(f_3, f_4) = -x_6f_3 - x_8 f_4 = x_5x_8x_9 - x_3x_6x_{10}.$$

Since

$$S(f_1, f_4) = x_1 f_3 + x_2x_5x_9 - x_1x_3x_{10},$$

its remainder is $-S(f_2, f_5)$. Thus, letting

$$f_6 = x_2x_5x_9 - x_1x_3x_{10},$$
$$f_7 = x_5x_8x_9 - x_3x_6x_{10},$$

we consider $\mathscr{F}'' = \{f_1, f_2, f_3, f_4, f_5, f_6, f_7\}$ to be a system of generators of I and apply Buchberger's criterion to $\mathscr{F}''$. The initial monomials of f_6 and f_7 are $x_2x_5x_9$ and $x_5x_8x_9$, respectively. Thus the S-polynomials which we must check are

$$S(f_1, f_6) = -x_5x_9 f_1 - x_6 f_6 = x_1x_3x_6x_{10} - x_1x_5x_8x_9$$
$$= x_1(x_3x_6x_{10} - x_5x_8x_9) = -x_1 f_7,$$
$$S(f_3, f_7) = -x_5x_9 f_3 - x_4 f_7 = x_3x_4x_6x_{10} - x_3x_5x_9x_{10}$$
$$= x_3x_{10}(x_4x_6 - x_5x_9) = x_3x_{10} f_4,$$
$$S(f_5, f_6) = x_2x_9 f_5 - x_7 f_6 = x_1x_3x_7x_{10} - x_1x_2x_9x_{10}$$
$$= x_1x_{10}(x_3x_7 - x_2x_9) = -x_1x_{10} f_2,$$
$$S(f_5, f_7) = x_8x_9 f_5 - x_7 f_7 = x_3x_6x_7x_{10} - x_1x_8x_9x_{10}$$
$$= x_{10}(x_3x_6x_7 - x_1x_8x_9) = x_{10}(-x_6 f_2 + x_2x_6x_9 - x_1x_8x_9)$$
$$= -x_{10}(x_6 f_2 + x_9 f_1).$$

Each of them reduces to 0. Thus $\mathscr{F}''$ is a Gröbner basis of I with respect to the reverse lexicographic order.

Problems

1.10 Let $I = (x^2 - xyz + y^3, \ xz^2 - y^2z)$ be an ideal of $S = K[x, y, z]$. Using Buchberger's algorithm, compute a Gröbner basis of I with respect to the following monomial orders:

(a) the lexicographic order on S induced by the ordering $x > y > z$;
(b) the reverse lexicographic order on S induced by the ordering $x > y > z$;
(c) the pure lexicographic order on S induced by the ordering $x > y > z$.

1.4   Elimination

Let $S = K[x_1, x_2, \ldots, x_n]$ be the polynomial ring and write $B_{i_1 i_2 \cdots i_m}$ for the subset of S consisting of those $f \in S$ such that each monomial belonging to $\mathrm{supp}(f)$ is a monomial in the variables $x_{i_1}, x_{i_2}, \ldots, x_{i_m}$, where $1 \leq i_1 < i_2 < \cdots < i_m \leq n$. Thus

$$B_{i_1 i_2 \cdots i_m} = K[x_{i_1}, x_{i_2}, \ldots, x_{i_m}].$$

If f and g belong to $B_{i_1 i_2 \cdots i_m}$, then the sum and the product of f and g again belong to $B_{i_1 i_2 \cdots i_m}$. Thus $B_{i_1 i_2 \cdots i_m}$ is a polynomial ring.

A monomial order $<$ on S naturally induces a monomial order $<'$ on $B_{i_1 i_2 \cdots i_m}$. More precisely, for monomials u and v belonging to $B_{i_1 i_2 \cdots i_m}$, one has $u <' v$ if and only if $u < v$ in S. Unless confusion arises, the monomial order $<'$ on $B_{i_1 i_2 \cdots i_m}$ induced by a monomial order $<$ on S will be also written as $<$.

In general, if I is an ideal of S, then $I \cap B_{i_1 i_2 \cdots i_m}$ is an ideal of $B_{i_1 i_2 \cdots i_m}$, see Problem 1.11. It is then natural to ask, for a given Gröbner basis $\mathcal{G}$ of I, whether $\mathcal{G} \cap B_{i_1 i_2 \cdots i_m}$ is a Gröbner basis of $I \cap B_{i_1 i_2 \cdots i_m}$ or not.

Theorem 1.34 (The elimination theorem) *Let $<$ be a monomial order on S and $\mathcal{G}$ a Gröbner basis of an ideal I of S with respect to $<$. Suppose that*

$$\text{For each } g \in \mathcal{G}, \text{ one has } g \in B_{i_1 i_2 \cdots i_m} \text{ if } \mathrm{in}_<(g) \in B_{i_1 i_2 \cdots i_m}. \tag{1.8}$$

Then $\mathcal{G} \cap B_{i_1 i_2 \cdots i_m}$ is a Gröbner basis of $I \cap B_{i_1 i_2 \cdots i_m}$ with respect to $<$ on $B_{i_1 i_2 \cdots i_m}$.

Proof What we must prove is that the initial ideal $\mathrm{in}_<(I \cap B_{i_1 i_2 \cdots i_m})$ of the ideal $I \cap B_{i_1 i_2 \cdots i_m}$ is generated by

$$\{\mathrm{in}_<(g) \ : \ g \in \mathcal{G} \cap B_{i_1 i_2 \cdots i_m}\}.$$

Let u be a monomial belonging to $\mathrm{in}_<(I \cap B_{i_1 i_2 \cdots i_m})$. Then there is $0 \neq f \in I \cap B_{i_1 i_2 \cdots i_m}$ with $\mathrm{in}_<(f) = u$. Since $f \in I$, one has $u \in \mathrm{in}_<(I)$. Now, since $\mathcal{G}$ is a Gröbner basis of I, there is $g \in \mathcal{G}$ such that $\mathrm{in}_<(g)$ divides u. Since $u \in B_{i_1 i_2 \cdots i_m}$ and since $\mathrm{in}_<(g)$ divides u, it follows that $\mathrm{in}_<(g) \in B_{i_1 i_2 \cdots i_m}$. Hence the condition (1.8) guarantees that g belongs to $B_{i_1 i_2 \cdots i_m}$. Consequently, for any monomial u belonging to the initial ideal $\mathrm{in}_<(I \cap B_{i_1 i_2 \cdots i_m})$, there is $g \in \mathcal{G} \cap B_{i_1 i_2 \cdots i_m}$ such that $\mathrm{in}_<(g)$ divides u. Hence $\mathrm{in}_<(I \cap B_{i_1 i_2 \cdots i_m})$ is generated by $\{\mathrm{in}_<(g) \ : \ g \in \mathcal{G} \cap B_{i_1 i_2 \cdots i_m}\}$, as desired. $\square$

Corollary 1.35 *Let $<_{\text{purelex}}$ denote the pure lexicographic order on S and*

$$B_{\geq p} = K[x_p, x_{p+1}, \ldots x_n].$$

Let $\mathcal{G}$ be a Gröbner basis of an ideal I of S with respect to $<_{\text{purelex}}$. Then $\mathcal{G} \cap B_{\geq p}$ is a Gröbner basis of $I \cap B_{\geq p}$ with respect to $<_{\text{purelex}}$.

Proof We must prove the condition (1.8) is satisfied. If $g \in \mathcal{G}$ and if its initial monomial $\mathrm{in}_{<_{\text{purelex}}}(g)$ belongs to $B_{\geq p}$, then $\mathrm{in}_{<_{\text{purelex}}}(g)$ is a monomial in the variables $x_p, x_{p+1}, \ldots x_n$. Hence by the definition of the pure lexicographic order $<_{\text{purelex}}$ it follows that each monomial belonging to the support of g is a monomial in $x_p, x_{p+1}, \ldots x_n$. Thus $g \in B_{\geq p}$, as desired. $\square$

As one of the typical applications of Corollary 1.35, we discuss the problem of computing the intersection of ideals. With adding a new variable t to S, we consider the polynomial ring

$$S[t] = K[t, x_1, x_2, \ldots, x_n]$$

in $n+1$ variables. If I and J are ideals of S, then we introduce ideals tI and $(1-t)J$ of $S[t]$ as follows:

$$tI = (\{tf : f \in I\}),$$
$$(1-t)J = (\{(1-t)f : f \in J\}).$$

Lemma 1.36 *As ideals of S one has*

$$I \cap J = (tI + (1-t)J) \cap S.$$

Proof Let $f \in S$ belong to $I \cap J$. Since $f \in I$ one has $tf \in tI$, and since $f \in J$, one has $(1-t)f \in (1-t)J$. Hence $f = tf + (1-t)f \in tI + (1-t)J$.

On the other hand, if a polynomial $f(\mathbf{x}) = f(x_1, \ldots, x_n) \in S$ belongs to $tI + (1-t)J$, then there exist $f_i \in I$, $f'_j \in J$ and $h_i, h'_j \in S[t]$ such that

$$f = t \sum_i f_i(\mathbf{x})h_i(t, \mathbf{x}) + (1-t) \sum_j f'_j(\mathbf{x})h'_j(t, \mathbf{x}).$$

Letting $t = 0$ one has $f = \sum_j f'_j(\mathbf{x})h'_j(0, \mathbf{x}) \in J$, and letting $t = 1$ one has $f = \sum_i f_i(\mathbf{x})h_i(1, \mathbf{x}) \in I$. Hence $f \in I \cap J$, as required. $\square$

Let $<_{\text{purelex}}$ be the pure lexicographic order on the polynomial ring $S[t] = K[t, x_1, x_2, \ldots, x_n]$ induced by the ordering $t > x_1 > x_2 > \cdots > x_n$. Let I and J be ideal of S. If $\{f_1, f_2, \ldots\}$ is a system of generators of I and $\{h_1, h_2, \ldots\}$ that of J, then a system of generators of the ideal $tI + (1-t)J$ of $K[t, \mathbf{x}]$ is

$$\{tf_1, tf_2, \ldots, (1-t)h_1, (1-t)h_2, \ldots\}.$$

Now Buchberger's algorithm gives a Gröbner basis $\mathscr{G}$ of $tI + (1-t)J$ with respect to $<_{\text{purelex}}$. Corollary 1.35 then guarantees that

$$\mathscr{G}' = \{g \in \mathscr{G} : t \text{ does not appear in } g\}$$

is a Gröbner basis of $(tI + (1-t)J) \cap S$. Hence Lemma 1.36 says that $\mathscr{G}'$ is a Gröbner basis of $I \cap J$ with respect to the pure lexicographic order on S induced by $x_1 > x_2 > \cdots > x_n$. Thus in particular $\mathscr{G}'$ is a system of generators of $I \cap J$.

Example 1.37 Let $n = 2$. Let $I = (x^2)$ and $J = (xy)$ be ideals of $K[x, y]$. We compute $I \cap J$. We apply Buchberger's algorithm to the system of generators

$\{tx^2, (1-t)xy\}$ of the ideal $tI + (1-t)J$ of $K[t, x, y]$. The S-polynomial of tx^2 and $(1-t)xy$ is x^2y. We then apply Buchberger's criterion to the system of generators $\{tx^2, (1-t)xy, x^2y\}$ of $tI + (1-t)J$. The S-polynomial of tx^2 and x^2y is 0. The S-polynomial of $(1-t)xy$ and x^2y is x^2y. Thus $\{tx^2, (1-t)xy, x^2y\}$ is a Gröbner basis of $tI + (1-t)J$. Hence $I \cap J = (x^2y)$.

Example 1.38 Let $n = 1$. Let $I = (x(x-1))$ and $J = (x^3)$ be ideals of $K[x]$. In order to compute $I \cap J$, Buchberger's algorithm can be applied to the system of generators $\{tx(1-x), (1-t)x^3\}$ of the ideal $tI + (1-t)J$ of $K[t, x]$. A routine computation shows that

$$\{tx(1-x),\ (1-t)x^3,\ (t-x^2)x,\ x^5 - x^3,\ x^4 - x^3\}$$

is a Gröbner basis of $tI + (1-t)J$. In particular the initial ideal of $tI + (1-t)J$ is (x^4, tx). Hence the reduced Gröbner basis of $tI + (1-t)J$ is $\{(t-x^2)x, x^4 - x^3\}$. Thus $I \cap J = (x^4 - x^3)$.

Let I be a graded ideal of S and $\mathfrak{m} = (x_1, \ldots, x_n)$ the graded maximal ideal of S. The ideal I is called *saturated* if $I : \mathfrak{m} = I$. The *saturation* of I is the ideal

$$I : \mathfrak{m}^\infty = \bigcup_{k=1}^\infty (I : \mathfrak{m}^k).$$

For a graded ideal I of S and a polynomial $f \in S$, the *saturation* of I with respect to f is the ideal

$$I : f^\infty = \{g \in S : \text{ there exists } i > 0 \text{ such that } f^i g \in I\}.$$

Then

$$I : \mathfrak{m}^\infty = \bigcap_{i=1}^n (I : x_i^\infty),$$

see Problem 1.13. Hence the following proposition is important.

Proposition 1.39 *Let I be an ideal of S and f a polynomial of S. Then*

$$I : f^\infty = \tilde{I} \cap S,$$

where $\tilde{I}$ is the ideal generated in $S[t]$ by I and the polynomial $1 - ft$.

Proof Let g be a nonzero polynomial in $I : f^\infty$. Then $f^i g \in I$ for some $i > 0$. Since

$$g = f^i g t^i + (1 - f^i t^i)g = f^i g t^i + (1 - ft)(1 + ft + \cdots + f^{i-1}t^{i-1})g$$

belongs to $(I, 1 - ft)$, we have $I : f^\infty \subset \tilde{I} \cap S$.

Let g be a polynomial in $\widetilde{I} \cap S$. Suppose that I is generated by $f_1, \ldots, f_s$. Then

$$g = a_1 f_1 + \cdots + a_s f_s + v(1 - tf) \tag{1.9}$$

for some $a_1, \ldots, a_s, v \in S[t]$. Substituting t by $1/f$ in the Equation (1.9), we have

$$g = a_1(1/f, x_1, \ldots, x_n) f_1 + \cdots + a_s(1/f, x_1, \ldots, x_n) f_s.$$

For a large enough i, $f^i a_j(1/f, x_1, \ldots, x_n)$ belongs to S for all $1 \leq j \leq s$. Then $f^i g = f^i a_1(1/f, x_1, \ldots, x_n) f_1 + \cdots + f^i a_s(1/f, x_1, \ldots, x_n) f_s$ belongs to I. Hence g belongs to $I : f^\infty$. Thus $I : f^\infty \supset \widetilde{I} \cap S$, as desired.   □

On the other hand, there is another method to compute $(I : x_i^\infty)$.

Proposition 1.40 *Let I be a graded ideal of S and $\mathscr{G}$ be the reduced Gröbner basis of I with respect to the reverse lexicographic order induced by $x_1 > x_2 > \cdots > x_n$. Then*

$$\mathscr{G}' = \{g/x_n^k : g \in \mathscr{G}, \ k \in \mathbb{Z}_{\geq 0}, \ x_n^k \text{ divides } g, \ x_n^{k+1} \text{ does not divide } g\}$$

is a Gröbner basis of $(I : x_n^\infty)$.

Proof Let f be a nonzero polynomial in $(I : x_n^\infty)$. Then $x_n^i f \in I$ for some $i > 0$. Since $\mathscr{G}$ is a Gröbner basis of I, there exists $g \in \mathscr{G}$ such that $\mathrm{in}(g)$ divides $\mathrm{in}(x_n^i f) = x_n^i \mathrm{in}(f)$. Let $k \geq 0$ be an integer such that x_n^k divides g, and x_n^{k+1} does not divide g. Then $h = g/x_n^k$ belongs to $\mathscr{G}'$. Since x_n is the smallest variable, it follows that x_n^k divides $\mathrm{in}(g)$ and x_n^{k+1} does not divide $\mathrm{in}(g)$. Hence x_n does not divide $\mathrm{in}(h) = \mathrm{in}(g)/x_n^k$. Thus $\mathrm{in}(h)$ divides $\mathrm{in}(f)$ as desired.   □

Problems

1.11 Let I be an ideal of $S = K[x_1, \ldots, x_n]$ and let $B_{i_1 i_2 \cdots i_m} = K[x_{i_1}, x_{i_2}, \ldots, x_{i_m}]$, where $1 \leq i_1 < i_2 < \cdots < i_m \leq n$. Show that $I \cap B_{i_1 i_2 \cdots i_m}$ is an ideal of $B_{i_1 i_2 \cdots i_m}$.

1.12 Let $I = (x^2 + y^2 + z^2, \ xy + xz + yz, \ xyz)$ be an ideal of $S = K[x, y, z]$. By using the elimination theorem, compute a set of generators of $I \cap K[y, z]$.

1.13 Let I be a graded ideal of $S = K[x_1, \ldots, x_n]$ and $\mathfrak{m}$ the graded maximal ideal of S. Show

$$I : \mathfrak{m}^\infty = \bigcap_{i=1}^{n} (I : x_i^\infty).$$

1.14 Let $I = (x_1x_5 - x_2x_4, x_2x_6 - x_3x_5)$ be an ideal of $S = K[x_1, \ldots, x_6]$. Compute a set of generators of $I : (x_1 \cdots x_6)^\infty = (\cdots((I : x_1^\infty) : x_2^\infty) \cdots) : x_6^\infty$ by using

(a) Proposition 1.39;
(b) Proposition 1.40.

1.5 Universal Gröbner Bases

For an ideal $I \subset K[\mathbf{x}]$, a finite set of polynomials of I is called a *universal Gröbner basis* of I if it is a Gröbner basis of I with respect to any monomial order. By the following theorem, a universal Gröbner basis always exists.

Theorem 1.41 *Let* $(0) \neq I \subset K[\mathbf{x}]$ *be an ideal. Then, there exist only finitely many initial ideals for* I.

Proof Let $\Sigma_0 = \{\text{in}_<(I) \ : \ < \text{ is a monomial order on } K[\mathbf{x}]\}$. Suppose that Σ_0 is an infinite set. We choose a nonzero polynomial $f_1 \in I$. Then, since f_1 has only finitely many monomials, there exists a monomial m_1 appearing in f_1 such that $\Sigma_1 = \{M \in \Sigma_0 \ : \ m_1 \in M\}$ is an infinite set. Then there exists a monomial order $<$ such that $m_1 \in \text{in}_<(I) \in \Sigma_1$. Suppose that $\text{in}_<(I) = (m_1)$. Then we have $\text{in}_<(I) = (m_1) \subset \text{in}_{<'}(I)$ for any $\text{in}_{<'}(I)$ belonging to Σ_1. By Macaulay's Theorem 1.19, $\text{in}_<(I) = (m_1) = \text{in}_{<'}(I)$ for any $\text{in}_{<'}(I)$ belonging to Σ_1. Thus, $\Sigma_1 = \{\text{in}_<(I)\}$, which is a contradiction. Hence, $(m_1) \subsetneq \text{in}_<(I)$. By Macaulay's Theorem 1.19 again, this means that the set of monomials $w \notin (m_1)$ is linearly dependent in $K[\mathbf{x}]/I$. Thus, there exists a nonzero polynomial $f_2 \in I$ such that no monomials in f_2 belong to (m_1). Since f_2 has only finitely many monomials, there exists a monomial m_2 in f_2 such that $\Sigma_2 = \{M \in \Sigma_1 \ : \ m_2 \in M\}$ is an infinite set. Then, by Macaulay's Theorem 1.19 and by using a similar argument as before, it follows that there exists a monomial order $<$ such that $(m_1, m_2) \subsetneq \text{in}_<(I) \in \Sigma_2$. Thus, there exists a nonzero polynomial $f_3 \in I$ such that no monomial in f_3 belongs to (m_1, m_2). By repeating such arguments, we have an infinite ascending chain of monomial ideals

$$(m_1) \subsetneq (m_1, m_2) \subsetneq (m_1, m_2, m_3) \subsetneq \cdots.$$

Let J be a monomial ideal of $K[\mathbf{x}]$ generated by $\{m_k \ : \ 0 < k \in \mathbb{Z}\}$. By Lemma 1.4, J is generated by a finite set $\{m_{\lambda_1}, \ldots, m_{\lambda_s}\}$. Let $\lambda = \max(\lambda_1, \ldots, \lambda_s)$. Since $J = (m_1, m_2, \ldots, m_k)$ for all $k \geq \lambda$, this contradicts the above infinite ascending chain. $\square$

Corollary 1.42 *For any ideal* $(0) \neq I \subset K[\mathbf{x}]$, *there exists a universal Gröbner basis of* I.

Proof Let, as before, $\mathcal{G}_{\text{red}}(I; <)$ be the reduced Gröbner basis of I with respect to a monomial order $<$. Then, by Theorem 1.41, the union

$$\bigcup_{<:\ \text{monomial order}} \mathcal{G}_{\text{red}}(I; <)$$

is a finite set. Moreover, since this set contains the reduced Gröbner basis with respect to an arbitrary monomial order, it is a Gröbner basis of I with respect to an arbitrary monomial order.   □

We call a universal Gröbner basis given in the proof of Corollary 1.42 *the universal Gröbner basis* of I.

Problems

1.15 Let $I = (x - y, x - z)$ be an ideal of $S = K[x, y, z]$. Compute a universal Gröbner basis of I.

Notes

In the 1960s, Buchberger invented the notion of Gröbner bases in his PhD thesis [30]. Hironaka [114] independently introduced a similar notion "standard bases" for formal power series rings. Standard textbooks for Gröbner bases are, e.g., Adams–Loustaunau [1], Becker–Weispfenning [13], and Cox–Little–O'Shea [44]. Buchberger's algorithm is an important method to compute Gröbner bases. However, it is very difficult to compute Gröbner bases by hand in practice. One can use various mathematical software to compute Gröbner bases. For example,

- Macaulay2: available at http://www.math.uiuc.edu/Macaulay2
- SINGULAR: available at https://www.singular.uni-kl.de
- CoCoA: available at http://cocoa.dima.unige.it
- Risa/Asir: available at http://www.math.kobe-u.ac.jp/Asir/asir.html

Universal Gröbner bases were introduced by Weispfenning [217]. In Chapters 3 and 4, the notion of Graver bases for binomial ideals is introduced. Graver bases are universal Gröbner bases, as shown by Sturmfels–Thomas [204] for toric ideals and by Sturmfels–Weismantel–Ziegler [205] for lattice ideals. For universal Gröbner bases of general ideals, a state polytope was introduced by Bayer–Morrison [11] and its normal fan is called a Gröbner fan which was introduced by Mora–Robbiano [149]. See [106, Chapter 5] for details on state polytopes and Gröbner fans.

Chapter 2
Review of Commutative Algebra

Abstract In this chapter we recall basis concepts from commutative algebra which are relevant for the subjects treated in the later chapters. We begin with a review on graded rings, Hilbert functions, and Hilbert series, and introduce the multiplicity and the a-invariant of a graded module. The Krull dimension of a graded module will be defined in terms of its Hilbert series. We will give various characterizations of the depth of a module and its relation to the Krull dimension. These considerations lead to Cohen–Macaulay modules and Gorenstein rings. We then describe the relationship, known as Auslander–Buchsbaum formula, between the depth of a graded S-module M and its projective dimension, where S is a polynomial ring, and study in more detail the finite minimal graded free S-resolution of M. The regularity of M will be defined via this resolution. Koszul algebras are standard graded K-algebras whose graded maximal ideal has a linear resolution. Unless this graded ring is a polynomial ring, this resolution is infinite. We discuss various necessary and sufficient conditions for Koszulness. The methods involved include Gröbner bases and Koszul filtrations.

2.1 Graded Rings and Hilbert Functions

Algebras and modules which are introduced in combinatorial contexts, in particular toric rings, usually admit a natural graded structure. In this section we recall the basis concepts and facts related to graded rings and modules.

Let K be a field and let $S = K[x_1, \ldots, x_n]$ be the polynomial ring over K in the indeterminates $x_1, \ldots, x_n$. A polynomial $f \in S$ is called *homogeneous* (of degree d), if all monomials in the support of f are of degree d. The polynomial ring S has a decomposition $S = \bigoplus_{i \geq 0} S_i$ where for each i, S_i is the K-vector space of homogeneous polynomials of degree i. In other words, each polynomial $f \in S$ has a unique presentation $f = \sum_i f_i$ with $f_i \in S_i$ for all i, where all f_i but finitely many are equal to 0. Notice that $S_i S_j = S_{i+j}$ for all i and j. Having this example in mind, we define

© Springer International Publishing AG, part of Springer Nature 2018   35
J. Herzog et al., *Binomial Ideals*, Graduate Texts in Mathematics 279,
https://doi.org/10.1007/978-3-319-95349-6_2

Definition 2.1 *Let K be a field. A ring R is called a* graded K-algebra, *if*

(i) $R = \bigoplus_{i \geq 0} R_i$, *where each R_i is a K-vector space;*
(ii) $R_0 = K$;
(iii) $R_i R_j \subset R_{i+j}$ *for all i,j.*

An R-module M is called a *graded R-module* if $M = \bigoplus_{i \in \mathbb{Z}} M_i$ with each M_i a K-vector space and such that $R_i M_j \subset M_{i+j}$ for all i and j. The elements of M_i are called *homogeneous* of degree i. The degree of the homogeneous element $x \in M$ will be denoted by $\deg x$.

Given a graded R-module M and an integer a, the graded R-module $M(a)$ *shifted by a* is the R-module M equipped with the new grading $M(a)_j = M_{a+j}$ for all j.

The polynomial ring $S = K[x_1, \ldots, x_n]$ can be graded by assigning to x_i the degree a_i where $a_1, \ldots, a_n$ are positive integers. Thus, if, for example, $\deg x_1 = 2$, $\deg x_2 = 3$ and $\deg x_3 = 1$, then $x_1^3 - x_2 x_3^3$ is homogeneous of degree 6. We say that a graded K-algebra R is *standard graded* if $R = K[R_1]$. Hence the polynomial ring S is standard graded, if and only if all indeterminates x_i are of degree 1.

Important examples of graded modules are graded ideals. Let R be a graded K-algebra. An ideal $I \subset R$ is called a *graded ideal*, if $I = \bigoplus_{j \in \mathbb{Z}} I_j$ where $I_j = I \cap R_j$ for all j. An ideal $I \subset R$ is graded if and only if I is generated by homogeneous elements, see Problem 2.1.

A homomorphism $\varphi: M \to N$ of graded modules is called *homogeneous* if $\varphi(M_i) \subset N_i$ for all i. Similarly, homogeneous K-algebra homomorphisms are defined. For example, if $I \subset R$ is a graded ideal, then the inclusion map $I \to R$ is a graded homomorphism. More generally, let $U \subset M$ be graded R-modules. Then U is called a *graded submodule* of M, if the inclusion map $U \to M$ is a graded homomorphism. In that case the factor module M/U is again naturally graded with grading $(M/U)_j = M_j/U_j$ for all j. In particular, if $I \subset R$ is a graded ideal, then R/I has the structure of a graded K-algebra.

Proposition 2.2 *Let R be a finitely generated graded K-algebra. Then there is a graded polynomial ring S over K and a graded ideal $I \subset S$ such that $R \cong S/I$, as graded K-algebras.*

Proof Let $r_1, \ldots, r_n$ be homogeneous generators of the K-algebra R, and let $S = K[x_1, \ldots, x_n]$ be the graded polynomial ring with $\deg x_i = \deg r_i$ for all i. There is a unique K-algebra homomorphism $\varphi: S \to R$ with $\varphi(x_i) = r_i$. This K-algebra homomorphism is homogeneous. Let $I = \operatorname{Ker} \varphi$, and let $f \in I$. We write $f = \sum_i f_i$ with f_i homogeneous of degree i. It remains to be shown that $f_i \in I$ for all i. Indeed, $0 = \varphi(f) = \sum_i \varphi(f_i)$. Since $\varphi(f_i) \in R_i$ and since $R = \bigoplus_i R_i$ it follows that $\varphi(f_i) = 0$ for all i. In other words, $f_i \in I$ for all i. $\square$

Let $S = K[x_1, \ldots, x_n]$ be the graded polynomial ring with $\deg x_i = a_i > 0$ for $i = 1, \ldots, n$. Then S_j is the K-vector space spanned by all monomial $\mathbf{x}^{\mathbf{b}}$ with $\sum_{i=1}^n a_i b_i = j$. Since there is only a finite number of vectors $\mathbf{b} \in \mathbb{Z}_{\geq 0}$ satisfying this identity, it follows that $\dim_K S_j < \infty$ for all j. More generally we have

Proposition 2.3 *Let R be a finitely generated graded K-algebra and M a finitely generated graded R-module. Then $\dim_K M_j < \infty$ for all j.*

Proof We choose a presentation S/I of R as in Proposition 2.2. Then M is a graded S-module as well, and hence we may assume that R itself is a graded polynomial ring. Let $m_1, \ldots, m_r$ be homogeneous generators of M. Then M_j is generated as a K-vector space by the homogeneous elements $\mathbf{x}^\mathbf{b} m_i$ with $\deg \mathbf{x}^\mathbf{b} + \deg m_i = j$. In particular, each monomial $\mathbf{x}^\mathbf{b}$ in such an expression is of degree $\leq j - \max\{\deg m_i : i = 1, \ldots, r\}$. Obviously there exist only finitely many such monomials. Thus the desired result follows.   □

Definition 2.4 Let R be a finitely generated graded K-algebra and M a finitely generated graded R-module. The numerical function $H(M, -) \colon \mathbb{Z} \to \mathbb{Z}_+$ with $H(M, i) = \dim_K M_i$ is called the *Hilbert function* of M. The formal Laurent series

$$\mathrm{Hilb}_M(t) = \sum_i H(M, i) t^i$$

is called the *Hilbert series* of M.

Example 2.5 Let as before $S = K[x_1, \ldots, x_n]$ be the polynomial ring. Then $H(S, i) = \dim_K S_i$ is equal to the number of monomials of degree i in S. A simple inductive argument shows that

$$H(S, i) = \binom{n + i - 1}{i}.$$

It follows that

$$\mathrm{Hilb}_S(t) = \sum_{i \geq 0} \binom{n + i - 1}{i} t^i = \frac{1}{(1 - t)^n}.$$

We will see in Section 2.2 that if R is a standard graded K-algebra, then $H_R(t)$ is always a rational function with denominator a power of $1 - t$.

As an immediate consequence of Theorem 1.19, for a graded ideal I, the computation of the Hilbert series of S/I can be reduced to the case that I is a monomial ideal.

Proposition 2.6 *Let $<$ be a monomial order on $S = K[x_1, \ldots, x_n]$, and let $I \subset S$ be a graded ideal. Then*

$$\mathrm{Hilb}_{S/I}(t) = \mathrm{Hilb}_{S/\mathrm{in}_<(I)}(t).$$

Proposition 2.6 can be improved as follows to obtain a Gröbner basis criterion.

Corollary 2.7 *Let $<$ be a monomial order on $S = K[x_1, \ldots, x_n]$, and $I \subset S$ a graded ideal. Let $\mathcal{G} = \{g_1, \ldots, g_m\}$ be a homogeneous system of generators of I and let $J = (\mathrm{in}_<(g_1), \ldots, \mathrm{in}_<(g_m))$. Then $\mathcal{G}$ is a Gröbner basis of I if and only if $\mathrm{Hilb}_{S/J}(t) = \mathrm{Hilb}_{S/\mathrm{in}_<(I)}(t)$.*

Proof Note that $J \subset \mathrm{in}_<(I)$. This together with Proposition 2.6 implies the coefficientwise inequality $\mathrm{Hilb}_{S/J}(t) \geq \mathrm{Hilb}_{S/\mathrm{in}_<(I)}(t) = \mathrm{Hilb}_{S/I}(t)$. Equality holds if and only if $J = \mathrm{in}_<(I)$. $\square$

Problems

2.1 Let R be a graded K-algebra and $I \subset R$ an ideal of R. Show that I is a graded ideal if and only if I is generated by homogeneous elements. Prove a similar result for graded R-modules.

2.2 Let $\varphi \colon M \to N$ a homomorphism of graded R-modules. Show that $\mathrm{Ker}\,\varphi$ is a graded submodule of M and $\mathrm{Im}\,\varphi$ is a graded submodule of N.

2.3

(a) Let $0 \to U \to M \to N \to 0$ be a short exact sequence of graded modules. Show that $\mathrm{Hilb}_U(t) + \mathrm{Hilb}_N(t) = \mathrm{Hilb}_M(t)$.
(b) Let $f_1, \ldots, f_m \in S = K[x_1, \ldots, x_n]$ be a regular sequence (see Definition 2.12) of homogeneous polynomials with $\deg f_i = a_i$. Use (a) to prove that

$$\mathrm{Hilb}_{S/(f_1, \ldots, f_m)}(t) = \frac{\prod_{i=1}^{m}(1 - t^{a_i})}{(1 - t)^n}.$$

(c) Let P and Q be two monomial prime ideals of $S = K[x_1, \ldots, x_n]$. Use (a) and a suitable exact sequence to compute $\mathrm{Hilb}_{S/P \cap Q}(t)$.

2.4 Let $S = K[x_1, \ldots, x_n]$ be the polynomial ring with grading given by $\deg x_i = a_i$ for $i = 1, \ldots, n$.

(a) Show that $\mathrm{Hilb}_S(t) = 1/\prod_{i=1}^{n}(1 - t^{a_i})$.
(b) Show that the following conditions are equivalent: (i) there exists an integer c such that $S_j \neq 0$ for all $j \geq c$, (ii) $\gcd(a_1, \ldots, a_n) = 1$.

2.5 Let $S = K[x_1, \ldots, x_n]$ be the polynomial ring. We define a $\mathbb{Z}^n$-grading on S by setting $S_\mathbf{a} = K\mathbf{x}^\mathbf{a}$ for $\mathbf{a} \in \mathbb{Z}^n$ with nonnegative entries. Otherwise set $S_\mathbf{a} = 0$. A finitely generated S-module M is called a $\mathbb{Z}^n$-graded S-module if $M = \bigoplus_{\mathbf{a} \in \mathbb{Z}^n} M_\mathbf{a}$ with each $M_\mathbf{a}$ a K-vector space and such that $S_\mathbf{a} M_\mathbf{b} \subset M_{\mathbf{a}+\mathbf{b}}$ for all $\mathbf{a}, \mathbf{b} \in \mathbb{Z}^n$.

(a) Show that $\dim M_\mathbf{a} < \infty$ for all $\mathbf{a} \in \mathbb{Z}^n$.

(b) We set $\mathrm{Hilb}_S(M) = \sum_{\mathbf{a}\in\mathbb{Z}^n} \dim_K M_{\mathbf{a}} \mathbf{t}^{\mathbf{a}}$ where $\mathbf{t}^{\mathbf{a}} = t_1^{a_1} \cdots t_n^{a_n}$ for $\mathbf{a} = (a_1, \ldots, a_n)$. Show that there exists $Q(\mathbf{t}) \in \mathbb{Z}[t_1^{\pm 1}, \ldots, t_n^{\pm 1}]$ and integers $d_i \geq 0$ such that $\mathrm{Hilb}_S(M) = Q(\mathbf{t})/(1 - t_1)^{d_1} \cdots (1 - t_n)^{d_n}$.

2.2 Finite Free Resolutions

Let K be a field. Throughout this section S will denote the standard graded polynomial ring $K[x_1, \ldots, x_n]$ in n indeterminates over K and $\mathfrak{m} = (x_1, \ldots, x_n)$ the graded maximal ideal of S. We will study graded free S-resolutions of graded S-modules. This will help us to better understand the nature of Hilbert functions.

We begin with a graded version of Nakayama's lemma.

Proposition 2.8 *Let M be a finitely generated graded S-module, $m_1, \ldots, m_r$ homogeneous elements of M and denote by $\bar{m}_i$ the residue class of m_i in $M/\mathfrak{m}M$. Then the elements $m_1, \ldots, m_r$ generate M if and only if their residue classes $\bar{m}_1, \ldots, \bar{m}_r$ generate $M/\mathfrak{m}M$, and $m_1, \ldots, m_r$ is a minimal system of generators of M if and only if $\bar{m}_1, \ldots, \bar{m}_r$ is a K-basis of the graded K-vector space of $M/\mathfrak{m}M$. In particular, all minimal systems of generators of M have the same length.*

Proof It is clear that if $m_1, \ldots, m_r$ generate M, then $\bar{m}_1, \ldots, \bar{m}_r$ generate $M/\mathfrak{m}M$. Conversely, let $U \subset M$ be the submodule of M generated by $m_1, \ldots, m_r$. Our hypothesis implies that $M = U + \mathfrak{m}M$, and we want to show that $U = M$. Let $m \in M$ be a homogeneous element. Since M is finitely generated, there exists an integer c such that $M_j = 0$ for all $j \leq c$. We will show by induction on $\deg m$, that $m \in U$. We may write $m = u + fn$ with homogeneous elements $u \in U$, $n \in N$ and $f \in \mathfrak{m}$ such that $\deg m = \deg u$ and $\deg n < \deg m$. If $\deg m = c$, then $n = 0$ and $m \in U$. Suppose now that $\deg m > c$. Since $\deg n < \deg m$, our induction hypothesis implies that $n \in U$, and hence $m \in U$.

If $\bar{m}_1, \ldots, \bar{m}_r$ is K-basis of $M/\mathfrak{m}M$, then no proper subset of $\{\bar{m}_1, \ldots, \bar{m}_r\}$ generates $M/\mathfrak{m}M$. Hence by the first part, no proper subset of $\{m_1, \ldots, m_r\}$ generates M. In other words, $m_1, \ldots, m_r$ is a minimal system of generators of M. The converse implication is obvious. $\square$

The least number of homogeneous generators of M is denoted by $\mu(M)$.

A finitely generated graded free S-module is a module F which admits a finite basis of homogeneous elements. If the basis elements are of degree $a_1, \ldots, a_r$, then $F \cong \bigoplus_{j=1}^r S(-a_j)$. Let M be a finitely generated graded S-module. A *homogeneous free presentation* of M is a homogeneous graded epimorphism $\epsilon \colon F \to M$ where F is a finitely generated graded S-module. The presentation is called *minimal* if rank $F = \mu(M)$. It follows from Nakayama's lemma that the free presentation $\epsilon \colon F \to M$ is minimal if and only if $\mathrm{Ker}\,\epsilon \subset \mathfrak{m}$ where $\mathfrak{m} = (x_1, \ldots, x_n)$, see Problem 2.6.

Now let M be a finitely generated S-module and let $\epsilon \colon F_0 \to M$ be a free presentation of M. By Problem 2.2, $\mathrm{Ker}\,\epsilon$ is a graded S-module for which we can again choose a homogeneous free presentation $F_1 \to \operatorname{Ker} \epsilon$, which composed with the inclusion map $\operatorname{Ker} \epsilon \to F_0$ yields the exact sequence $F_1 \to F_0 \to M \to 0$. Proceeding in this way we obtain an exact sequence of graded modules

$$\cdots \to F_i \to \cdots \to F_1 \to F_0 \to M \to 0$$

with $F_i = \bigoplus_j S(-a_{ij})$ for all i and suitable integers a_{ij}. The acyclic sequence of graded free modules

$$\mathbb{F} : \cdots \to F_i \to \cdots \to F_1 \to F_0 \to 0$$

with $H_0(\mathbb{F}) \cong M$ is called a *graded free S-resolution* of M. This sequence can be rewritten in the form

$$\mathbb{F}: \cdots \to \bigoplus_j S(-j)^{b_{ij}} \to \cdots \to \bigoplus_j S(-j)^{b_{1j}} \to \bigoplus_j S(-j)^{b_{0j}} \to 0. \quad (2.1)$$

The numbers b_{ij} are called the *graded Betti numbers* of $\mathbb{F}$.

Obviously such a resolution cannot be unique if the free presentations in the construction of $\mathbb{F}$ are not minimal. One calls $\mathbb{F}$ a *minimal graded free S-resolution* of M, if the augmentation map $F_0 \to H_0(\mathbb{F})$ is a minimal free presentation of M, and if moreover $F_i \to \operatorname{Im}(F_i \to F_{i-1})$ is a minimal free presentation for all i. By what we observed before it follows that the resolution $\mathbb{F}$ is minimal if and only if $\operatorname{Im}(F_i \to F_{i-1}) \subset \mathfrak{m} F_{i-1}$ for all $i > 0$.

An important example of a graded minimal free resolution is the resolution of $S/\mathfrak{m}$ which is provided by the Koszul complex: let R be any commutative ring (with unit) and $\mathbf{f} = f_1, \ldots, f_m$ a sequence of elements of R, and let F be a free R-module with basis $e_1, \ldots, e_m$. Then we let $K_j(\mathbf{f}; R)$ be the jth exterior power of F, that is, $K_j(\mathbf{f}; R) = \bigwedge^j F$. A basis of the free R-module $K_j(\mathbf{f}; R)$ is given by the wedge products $e_F = e_{i_1} \wedge e_{i_2} \wedge \cdots \wedge e_{i_j}$ where $F = \{i_1 < i_2 < \cdots < i_j\}$. In particular, it follows that rank $K_j(\mathbf{f}; R) = \binom{m}{j}$. The *Koszul complex* $K(\mathbf{f}; R)$ attached to the sequence $\mathbf{f}$ is given as follows: we define the differential $\partial : K_j(\mathbf{f}; R) \to K_{j-1}(\mathbf{f}; R)$ by the formula

$$\partial(e_{i_1} \wedge e_{i_2} \wedge \cdots \wedge e_{i_j}) = \sum_{k=1}^{j} (-1)^{k+1} f_{i_k} e_{i_1} \wedge e_{i_2} \wedge \cdots \wedge e_{i_{k-1}} \wedge e_{i_{k+1}} \wedge \cdots \wedge e_{i_j}.$$

One readily verifies that $\partial \circ \partial = 0$, so that $K(\mathbf{f}; R)$ is indeed a complex. Now if M is any finitely generated graded S-module we set $K(\mathbf{f}; M) = K(\mathbf{f}; R) \otimes M$ and call $K(\mathbf{f}; M)$ the Koszul complex of M with respect to the sequence $\mathbf{f}$. The ith homology of this complex is denoted $H_i(\mathbf{f}; M)$.

Some of the basic properties of Koszul complexes that we are going to use can be found in Bruns-Herzog [27]. A short introduction to this theory of complexes can also be found in the appendix of the book [94] by Herzog-Hibi.

In the particular case that $\mathbf{f}$ is the sequence $\mathbf{x} = x_1, \ldots, x_n$ and $R = K[x_1, \ldots, x_n]$, the Koszul complex is acyclic because the sequence $\mathbf{x}$ is a regular sequence (see Definition 2.12), and hence $K(\mathbf{x}; R)$ provides a graded minimal free resolution of $S/\mathfrak{m}$. In what follows we will need the following fact: let $\mathbb{G}$ be a graded free resolution of M. Then

$$H_i(\mathbf{x}; M) \cong H_i(\mathbb{G}/\mathfrak{m}\mathbb{G}) \quad \text{for all} \quad i, \tag{2.2}$$

and this is an isomorphism of finitely generated graded K-vector spaces, see, for example, [94, Corollary A.3.5].

It is known that any two graded minimal free resolutions are unique up to isomorphism. Here we show

Theorem 2.9 *Let M be a finitely generated free S-module, and let the numbers b_{ij} be the graded Betti numbers of a graded free S-resolution $\mathbb{G}$ of M. Furthermore let the numbers β_{ij} be the graded Betti numbers of a graded minimal free S-resolution $\mathbb{F}$ of M. Then*

$$\beta_{ij} \leq b_{ij} \quad \text{for all} \quad i, j.$$

In particular, the graded Betti numbers of a graded minimal free S-resolution of M depend only on M, and hence are denoted $\beta_{ij}(M)$ and are called the graded Betti numbers *of M.*

Proof Let $\mathbb{G}: \cdots \to \bigoplus_j S(-j)^{b_{1j}} \to \bigoplus_j S(-j)^{b_{0j}} \to 0$. Then $H_i(\mathbb{G}/\mathfrak{m}\mathbb{G})$ is a graded subquotient of $G_i/\mathfrak{m}G_i \cong \bigoplus_j K(-j)^{b_{ij}}$. Hence it follows that $\dim_K H_i(\mathbf{x}; M)_j = \dim_K H_i(\mathbb{G}/\mathfrak{m}\mathbb{G})_j \leq b_{ij}$.

On the other hand, since $\mathrm{Im}(F_{i+1} \to F_i)$ is contained in $\mathfrak{m}F_i$ for all i, it follows that $H_i(\mathbf{x}; M) \cong H_i(\mathbb{F}/\mathfrak{m}\mathbb{F}) \cong \mathbb{F}/\mathfrak{m}\mathbb{F}$. This implies that $\dim_K H_i(\mathbf{x}; M)_j = \beta_{ij}$. Thus the desired inequality follows. $\qquad\square$

In the proof we have seen that

$$\beta_{ij}(M) = \dim_K H_i(\mathbf{x}; M)_j \quad \text{for all } i \text{ and } j.$$

Thus, since the Koszul complex for the sequence $\mathbf{x}$ has length n, we obtain

Corollary 2.10 *Let M be a finitely generated graded S-module. Then $\beta_{ij}(M) = 0$ for all i and j with $i > n$.*

The corollary implies that there are only finitely many pairs (i, j) for which $\beta_{ij}(M) \neq 0$. One defines

$$\mathrm{proj\,dim}\, M = \max\{i : \ \beta_{ij}(M) \neq 0 \text{ for some } j\},$$

Fig. 2.1 Betti diagram

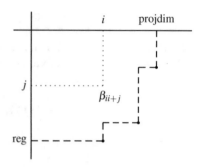

and

$$\mathrm{reg}(M) = \max\{j - i: \ \beta_{ij}(M) \neq 0 \text{ for some } i\}.$$

The number $\mathrm{proj\,dim}\, M$ is called the *projective dimension* of M, and the number $\mathrm{reg}(M)$ is called the *Castelnuovo-Mumford regularity* of M.

Figure 2.1 displays the Betti diagram of a graded S-module. The corner points of the dotted line are called the *extremal Betti numbers* of M and they represent nonzero Betti numbers.

The set of Betti numbers in the jth row of the Betti diagram is called the *jth strand* of M. Let d be the least degree of a generator of M. Then the dth strand of M is called the *linear strand* of M. The module M is said to have a *d-linear resolution* if $\beta_{i,i+j}(M) = 0$ for all i and all $j \neq d$, and M is said to have *linear relations* if M is generated in degree d and $\beta_{1,j}(M) = 0$ for all $j \neq d + 1$. Finally, M is said to have *linear quotients*, if M is generated in a single degree, and M is minimally generated by $m_1, \ldots, m_r$ such that the colon ideals

$$(m_1, \ldots, m_{i-1}) : m_i = \{f \in S: \ f m_i \in (m_1, \ldots, m_{i-1})\}$$

are generated by linear forms.

Proposition 2.11 *Suppose M has linear quotients. Then M has a linear resolution.*

Proof We proceed by induction on the number of generators of M. We may assume that M is generated in degree d. If $r = 1$, then $M \cong S(-d)/I$ where I is an ideal generated by linear forms. By Problem 2.12, I has a linear resolution. Thus M has a linear resolution. Now let $r > 1$. By induction hypothesis, the module N, generated by $m_1, \ldots, m_{r-1}$, has a d-linear resolution. Also the module M/N has a d-linear resolution, as the argument for $r = 1$ shows. Considering the long exact Tor-sequences arising from the short exact sequence $0 \to N \to M \to M/N \to 0$, we deduce that M has a linear resolution. □

Problems

2.6 Let $\epsilon: M \to N$ be a graded surjective S-module homomorphism of finitely generated graded S-modules. Show that $\mu(M) \geq \mu(N)$, and that equality holds if and only if $\operatorname{Ker} \epsilon \subset \mathfrak{m}M$.

2.7 Let $I \subset S$ be an ideal generated by a regular sequence $\mathbf{f} = f_1, \ldots, f_r$ of homogeneous elements with $\deg f_i = a_i$, see Definition 2.12. Use the fact that the Koszul complex $K(\mathbf{f}; S)$ is acyclic to show that $r \leq n$ and to compute the graded Betti numbers of S/I. What is proj dim S/I and what is reg S/I?

2.8 Let M be a finitely generated graded S-module. Show that

$$\operatorname{Hilb}_M(t) = \frac{\sum_{i=0}^{n}(-1)^{i+1}\beta_{ij}(M)t^j}{(1-t)^n}.$$

2.9 Let $I \subset S$ be a nonzero graded ideal. Show that $\sum_{i=0}^{n}(-1)^{i+1} \sum_j \beta_{ij}(S/I) = 0$.

2.10 Let $0 \to U \to M \to N \to 0$ be a short exact sequence of graded modules. Show that $\beta_{ij}(M) \leq \beta_{ij}(U) + \beta_{ij}(N)$, and give an example which shows that in general this inequality is strict.

2.11 Show that the graded Betti numbers of a module with d-linear resolution are determined by its Hilbert function.

2.12 Show that any ideal generated by linear forms has a linear resolution.

2.13 Let $I \subset S$ be a graded ideal such that $\dim_K S/I < \infty$ (in which case $\operatorname{Hilb}_{S/I}(t)$ is a polynomial). Show that I has a linear resolution if and only if I is a power of the graded maximal ideal $\mathfrak{m}$ of S.

2.14 Compute the minimal graded free resolution of $(x_1, x_2)^k$ for all k.

2.15 Let I be the ideal generated by monomials $x_i y_j$ with $1 \leq i < j \leq n$. Show that the ideal I has a linear resolution.

2.16 Let K be a field, $= K[x_1, \ldots, x_n]$ be the polynomial ring in the indeterminates $x_1, \ldots, x_n$, $T = K[y_1, \ldots, y_m]$ the polynomial in the indeterminates $y_1, \ldots, y_m$, M a finitely generated graded S-module with graded minimal free S-resolution $\mathbb{F}$, and N a finitely generated graded T-module with graded minimal free T-resolution $\mathbb{G}$. Show that the tensor product $\mathbb{F} \otimes_K \mathbb{G}$ of $\mathbb{F}$ and $\mathbb{G}$ over K is a graded minimal free $S \otimes_K T$-resolution of $M \otimes_K N$, and use this fact to show that

$$\beta_{ij}(M \otimes_K N) = \sum \beta_{i_1, j_1}(M)\beta_{i_2, j_2}(N),$$

where the sum is taken over all i_1 and i_2 with $i_1 + i_2 = i$, and over all j_1 and j_2 with $j_1 + j_2 = j$.

2.17 Let $S = K[x_1, \ldots, x_n]$ be the polynomial ring over the field K with the natural $\mathbb{Z}^n$-grading, as defined in Problem 2.5, and let M be a finitely generated $\mathbb{Z}^n$-graded S-module.

(a) Show that M admits a minimal $\mathbb{Z}^n$-graded free S-resolution $\mathbb{F}$ with each
$F_i = \bigoplus_{\mathbf{a} \in \mathbb{Z}^n} S(-\mathbf{a})^{\beta_{i,\mathbf{a}}(M)}$. The integer $\beta_{i,\mathbf{a}}(M)$ are called the *multigraded Betti numbers* of M.

(b) Show that the Koszul homology $H_i(\mathbf{x}; M)$ is a $\mathbb{Z}^n$-graded module with

$$\dim_K H_i(\mathbf{x}; M)_{\mathbf{a}} = \beta_{i,\mathbf{a}} \quad \text{for all} \quad \mathbf{a} \in \mathbb{Z}^n.$$

2.3   Dimension and Depth

We will use graded free resolutions to define dimension and depth of a graded S-module, where, as before, $S = K[x_1, \ldots, x_n]$ is the polynomial ring over K. Resolutions will also be used to introduce two other important invariants of M: the multiplicity and the a-invariant of M.

Let M be a finitely generated graded S-module. As we have seen in Problem 2.8, the Hilbert series is a rational function of the form $\mathrm{Hilb}_M(t) = P(t)/(1-t)^n$. After cancelation we obtain a presentation

$$\mathrm{Hilb}_M(t) = \frac{Q(t)}{(1-t)^d}, \quad \text{where } Q(t) \text{ a polynomial with } Q(1) \neq 0.$$

The number d is called the *Krull dimension* of M, and $Q(1)$ is called the *multiplicity* of M, denoted $e(M)$. The multiplicity is always a positive number since it is the leading coefficient of the Hilbert polynomial, see [27, Definition 4.1.5 and Proposition 4.1.9]. An equivalent definition of the Krull dimension, actually the original one, gives the Krull dimension as the maximal length of a chain of prime ideals in the support of M, see [27, Appendix]. Let $Q(t) = \sum_{i=0}^{c} h_i t^i$. The coefficient vector $(h_0, h_1, \ldots, h_c)$ is called the *h-vector* of M. Obviously, $e(M) = \sum_{i=1}^{c} h_i$. Finally, the *a-invariant* of M, denoted $a(M)$, is the degree of $\mathrm{Hilb}_M(t)$. In other words, $a(M) = \deg P(t) - n = \deg Q(t) - d$.

Definition 2.12 Let M be a finitely generated graded S-module. A sequence $\mathbf{f} = f_1, \ldots, f_m$ of homogeneous elements of positive degree of S is called a *regular sequence* on M (or an *M-sequence*), if f_1 is a nonzerodivisor on M and f_i is a nonzerodivisor on $M/(f_1, \ldots, f_{i-1})M$ for all $i > 0$. The maximal possible length of an M-sequence is called the *depth* of M, denoted depth M.

Proposition 2.13 *Let M be a finitely generated graded S-module. Then* depth $M \leq$ dim M.

Proof We proceed by induction on the depth M. The assertion is trivial if depth $M = 0$. Suppose now that depth $M = m > 0$. Then there exists a regular sequence $\mathbf{f} = f_1, \ldots, f_m$ on M. Thus $f_2, \ldots, f_m$ is a regular sequence on $M/f_1 M$. This shows that depth $M/f_1 M \geq m - 1$. Suppose depth $M/f_1 M = t > m - 1$. Then there exists a regular sequence $g_1, \ldots, g_t$ on $M/f_1 M$, and hence $f_1, g_1, \ldots, g_t$ is a regular sequence on M of length $>$ depth M, a contradiction. Thus depth $M/f_1 M = $ depth $M - 1$.

Let deg $f_1 = a$, and let $0 \to M(-a) \to M \to M/f_1 M \to 0$ be the exact sequence, where $M(-a) \to M$ is given by multiplication by f_1. Then $\mathrm{Hilb}_{M/f_1 M}(t) = (1 - t^a)\,\mathrm{Hilb}_M(t)$. Thus, if $\mathrm{Hilb}_M(t) = Q(t)/(1-t)^d$ with $d = \dim M$, then

$$\mathrm{Hilb}_{M/f_1 M}(t) = \frac{(1-t^a)Q(t)}{(1-t)^d} = \frac{Q'(t)}{(1-t)^{d-1}},$$

where $Q'(t) = Q(t)(1 + t + \cdots + t^{a-1})$. Since $Q(1) \neq 0$, we see that $Q'(1) = aQ(1) \neq 0$. Thus $\dim M/f_1 M = \dim M - 1$. By using the induction hypothesis we obtain

$$\mathrm{depth}\, M = \mathrm{depth}\, M/f_1 M + 1 \leq \dim M/f_1 M + 1 = \dim M,$$

as desired. $\square$

Definition 2.14 Let M be a finitely generated graded S-module. Then M is called a *Cohen–Macaulay module* if depth $M = \dim M$.

Let $\mathbf{f}$ be an M-sequence. The proof of Proposition 2.13 shows that M is Cohen–Macaulay if and only if $M/(\mathbf{f})M$ is Cohen–Macaulay. Another important property of a Cohen–Macaulay module is that it has no embedded prime ideal and that all minimal prime ideals have the same height. Rings with this property are called *unmixed*. Unmixedness for Cohen–Macaulay modules follows from the fact that for any finitely generated graded S-module M one has depth $M \leq \dim S/P$ for all associated prime ideals of M, see [27, Proposition 1.2.13].

On the other hand, an unmixed module need not to be Cohen–Macaulay, as the example in Problem 2.19 shows.

Theorem 2.15 (Auslander-Buchsbaum) *Let M be a finitely generated graded S-module. Then*

$$\mathrm{proj\,dim}\, M + \mathrm{depth}\, M = n.$$

Proof We proceed by induction on the depth of M. If depth $M = 0$, then $\mathfrak{m}$ is associated with M and hence there exists $m \in M$, $m \neq 0$ with $\mathfrak{m}m = 0$. It follows that $me_1 \wedge \cdots \wedge e_n \in H_n(\mathbf{x}; M)$, so that $H_n(\mathbf{x}; M) \neq 0$. Thus proj dim $M = n$, by (2.2). Suppose now that depth $M > 0$. Then there exists a homogeneous polynomial $f \in \mathfrak{m}$ which is a nonzerodivisor on M. As we noticed before, depth $M/fM = $ depth $M - 1$. Let $\mathbb{F}$ be a graded minimal free resolution of M. Multiplication with f yields a complex homomorphism $\mathbb{F} \to \mathbb{F}$ whose mapping cone $\mathbb{G}$ provides a graded minimal free resolution of M/fM, cf. [216, Section 1.5]. Note that $G_i = F_i \oplus F_{i-1}$ for all $i > 0$. This implies that proj dim $M/fM =$ proj dim $M + 1$. Applying the induction hypothesis we obtain

$$\text{proj dim } M + \text{depth } M = (\text{proj dim } M/fM + 1) + (\text{depth } M/fM - 1)$$
$$= \text{proj dim } M/fM + \text{depth } M/fM = n,$$

as desired. $\square$

Let M be a finitely generated graded Cohen–Macaulay S-module of dimension d. The Auslander–Buchsbaum theorem implies that proj dim $M = n - d$. Let $\mathbb{F}$ be the graded minimal free S-resolution of M. Then the rank of the free module F_{n-d} is called the *Cohen–Macaulay type* of M, denoted $r(M)$. It follows that $r(M) = \dim_K F_{n-d}/\mathfrak{m}F_{n-d}$, and hence by (2.2), $r(M) = \dim_K H_{n-d}(\mathbf{x}; M)$. In particular, if dim $M = 0$ it follows that $r(M) = \dim_K H_n(\mathbf{x}; M)$. Since $H_n(\mathbf{x}; M)$ is isomorphic to the *socle* $\gamma(M)$ of M, which by definition is the submodule of M whose elements are all annihilated by $\mathfrak{m}$, we see that $r(M) = \dim_K \gamma(M)$ whenever dim $M = 0$.

For later applications we need the following result and its corollaries.

Theorem 2.16 *Let M be a finitely generated graded Cohen–Macaulay S-module. Then M admits only one extremal Betti number.*

Proof Let $\mathbb{F}$ be the graded minimal free resolution of M. Suppose that proj dim $M = p$, and that M has more than one extremal Betti number. Since one of the extremal Betti numbers is always in homological degree p, there exists another extremal Betti number in homological degree $i < p$. Let $\beta_{i,i+j}(M)$ be this extremal Betti number, and let $e_1, \ldots, e_r$ be a homogeneous basis of F_i. We may assume that deg $e_1 = i+j$. Let $\partial_{i+1} : F_{i+1} \to F_i$ be the $(i + 1)$-differential in $\mathbb{F}$. Since $\beta_{i,i+j}(M)$ is an extremal Betti number of M, it follows that deg $f \leq$ deg e_1 for all homogeneous basis elements f in F_{i+1}. Thus, since ∂_{i+1} is a graded map and since $\text{Im}(\partial_{i+1}) \subset \mathfrak{m}F_i$, it follows that for all basis elements f of F_{i+1} we have

$$\partial_{i+1}(f) = \sum_{e_l \neq e_1} a_l e_l \quad \text{with } a_l \in S. \tag{2.3}$$

Dualizing the resolution of M with respect to S and using the fact that M is Cohen–Macaulay, we get the acyclic complex $\mathbb{F}^*$, since $\text{Ext}_S^i(M, S) = 0$ for $i < \text{proj dim}(M)$, see [27, Proposition 3.3.3]. On the other hand, (2.3) implies that $\partial_{i+1}^*(e_1^*) = 0$, while $e_1^* \notin \text{Im}(\partial_i^*)$ because $\text{Im}(\partial_i^*) \subset \mathfrak{m}F_i^*$. This contradicts the acyclicity of $\mathbb{F}^*$. $\square$

As an immediate consequence we have

Corollary 2.17 *Let M be a finitely generated graded Cohen–Macaulay S-module of projective dimension p. Then reg $M = \max\{j: \beta_{p,p+j}(M) \neq 0\}$ and $\beta_{p,p+\text{reg } M}(M)$ is the unique extremal Betti number of M.*

Corollary 2.18 *Let M be a finitely generated graded Cohen–Macaulay S-module of dimension d and let $H_M(t) = Q(t)/(1-t)^d$ be its Hilbert series. Then*

$$\operatorname{reg}(M) = \deg Q(t).$$

Proof Let β_{ij} be the graded Betti numbers of M. Since M is Cohen–Macaulay, it follows from the Auslander–Buchsbaum formula that proj dim $M = n - d$. By using the additivity of Hilbert series, we deduce that $H_M(t) = P(t)/(1-t)^n$, where

$$P(t) = \sum_{i=0}^{n-d} (-1)^i \sum_j \beta_{i,i+j} t^{i+j}.$$

By Corollary 2.17, $\beta_{n-d,n-d+\operatorname{reg} M}$ is the unique extremal Betti number of M, and hence $\deg P(t) = n - d + \operatorname{reg} M$. Since $P(t) = (1-t)^{n-d} Q(t)$, the assertion follows.

A graded ring $R = S/I$ is called a *Cohen–Macaulay ring*, if R as an S-module is Cohen–Macaulay. Sometimes an ideal I is called a *Cohen–Macaulay ideal* if S/I is a Cohen–Macaulay ring. A Cohen–Macaulay ring R with $r(R) = 1$ is called a *Gorenstein ring*. Gorenstein rings are a very distinguished class of Cohen–Macaulay rings. By a famous theorem of Bass [10], Gorenstein rings are characterized by the property that they are of finite injective dimension considered as modules over themselves.

The following result provides a comparison between S/I and $S/\operatorname{in}_<(I)$.

Theorem 2.19 *Let $I \subset S$ be a graded ideal, and let $<$ be a monomial order on S. Then the following holds:*

(a) $\beta_{ij}(S/I) \le \beta_{ij}(S/\operatorname{in}_<(I))$ *for all i and j;*
(b) $\dim S/I = \dim S/\operatorname{in}_<(I)$, $\operatorname{depth} S/\operatorname{in}_<(I) \le \operatorname{depth} S/I$ *and* $\operatorname{reg} S/I \le \operatorname{reg} S/\operatorname{in}_<(I)$;
(c) *if $S/\operatorname{in}_<(I)$ is Cohen–Macaulay, then S/I is Cohen–Macaulay, and $r(S/I) \le r(S/\operatorname{in}_<(I))$;*
(d) *if $S/\operatorname{in}_<(I)$ is Gorenstein, then S/I is Gorenstein;*
(e) *if $S/\operatorname{in}_<(I)$ has a linear resolution, then S/I has a linear resolution.*

Proof The proof of statement (a) can be found in [94, Corollary 3.3.3].

(b) The equality $\dim S/I = \dim S/\operatorname{in}_<(I)$ follows from Proposition 2.6 and the fact that the dimension of a graded S-module is the pole order of its Hilbert series at $t = 1$. The inequality $\operatorname{depth} S/\operatorname{in}_<(I) \le \operatorname{depth} S/I$ follows from (a) and the Auslander–Buchsbaum theorem, while the $\operatorname{reg} S/I \le \operatorname{reg} S/\operatorname{in}_<(I)$ is an immediate consequence of (a).

(c) If $S/\operatorname{in}_<(I)$ is Cohen–Macaulay, then $\dim S/\operatorname{in}_<(I) = \operatorname{depth} S/\operatorname{in}_<(I)$. Thus it follows from (b) that $\dim S/I \ge \operatorname{depth} S/I$. By Proposition 2.13, the opposite inequality always holds, and this implies that S/I is Cohen–Macaulay. Since in

this case S/I and $S/\operatorname{in}_<(I)$ have the same projective dimension, it follows from
(a) and the definition of the Cohen–Macaulay type that $r(S/I) \leq r(S/\operatorname{in}(I))$.

(d) If $S/\operatorname{in}_<(I)$ is Gorenstein, then $r(S/\operatorname{in}_<(I)) = 1$. Thus (c) implies that
$r(S/I) = 1$, and hence S/I is Gorenstein.

(e) Let a be the least degree of a generator of I, then this is also the least degree
of a generator of $\operatorname{in}_<(I)$. Since $\operatorname{in}_<(I)$ has a linear resolution it follows that
$\operatorname{reg}\operatorname{in}_<(I) = a$. Thus by (b), $\operatorname{reg}(I) \leq a$. However $\operatorname{reg}(I) \geq a$, always. Thus
$\operatorname{reg} I = a$, and this implies that I has a linear resolution.   □

Problems

2.18 Let M be a graded Cohen–Macaulay S-module of dimension d, and let $\mathbf{f} = f_1, \ldots, f_d$ be an M-sequence with $\deg f_i = a_i$ for $i = 1, \ldots, d$. Show that $M/(\mathbf{f})M$ has finite length and that $e(M) = \ell(M/(\mathbf{f})M)/\prod_{i=1}^d a_i$, where $\ell(M/(\mathbf{f})M)$ denotes the length of $M/(\mathbf{f})M$.

2.19 Let $S = K[x_1, x_2, x_3, x_4]$. Show that $S/(x_1, x_2)\cap(x_3, x_4)$ is unmixed but not Cohen–Macaulay.

2.20 A graded ideal generated by a regular sequence of homogeneous polynomials is called a complete intersection ideal. Show that if $I \subset S$ is a complete intersection ideal, then S/I is Gorenstein.

2.21 Let I be the ideal in the polynomial ring $K[x_1, \ldots, x_n, y_1, \ldots, y_n]$ generated by the binomials $x_i y_j - x_j y_i$ with $1 \leq i < j \leq n$.

(a) Show that the binomials generating I form a reduced Gröbner basis with respect to the reverse lexicographic order induced by $x_1 > x_2 > \cdots > x_n > y_1 > y_2 > \cdots > y_n$.

(b) Use (a) and Theorem 2.19 to show that I is a Cohen–Macaulay ideal with linear resolution, and compute the type and the a-invariant of S/I.

2.22 Give examples of graded ideals for which the inequalities in Theorem 2.19 (b) are strict.

2.23

(a) Show that the ideal $I = (xy - z^2, x^2)$ is a complete intersection ideal.

(b) Let $<$ be the lexicographic monomial order induced by $x > y > z$. Show that $\operatorname{in}_<(I)$ is not a complete intersection ideal, and not even a Gorenstein ideal.

2.24 Let K be a field, $S = K[x_1, \ldots, x_n]$ the polynomial ring over K in the indeterminates $x_1, \ldots, x_n$, and $I \subset S$ an ideal. Let $<$ be a monomial order on S and suppose that x_1 is a nonzerodivisor on $S/\operatorname{in}_<(I)$. Then x_1 is a nonzerodivisor on S/I.

2.4   Infinite Free Resolutions and Koszul Algebras

As in the previous sections, $S = K[x_1 \ldots, x_n]$ denotes the polynomial ring in the variables $x_1, \ldots, x_n$. Any standard graded K-algebra R of embedding dimension n is isomorphic to S/I where I is a graded ideal with $I \subset (x_1, \ldots, x_n)^2$. Let $\mathfrak{m}$ be the graded maximal ideal of R. In general and in contrast to finitely generated graded S-modules, the finitely generated graded R-modules do not have a finite projective dimension. Indeed, Serre showed that if $(R, \mathfrak{m})$ is a Noetherian local ring, then $\operatorname{proj} \dim R/\mathfrak{m} < \infty$ if and only if R is regular. The same holds true in the standard graded case considered here. Thus in our setting, the graded minimal free R-resolution of $R/\mathfrak{m}$ is infinite if and only if $I \neq 0$.

The *Poincaré series* of R is defined to be the formal power series

$$P_R(t) = \sum_{i \geq 0} \operatorname{Tor}_i^R(R/\mathfrak{m}, R/\mathfrak{m}) t^i \in \mathbb{Z}[[t]].$$

Let $\mathbb{X}$ be the graded minimal free R-resolution of $R/\mathfrak{m}$. Then $\operatorname{Tor}_i^R(R/\mathfrak{m}, R/\mathfrak{m})$ and $X_i/\mathfrak{m}X_i$ are isomorphic as graded K-vector spaces. In particular, the vector space dimension of $\operatorname{Tor}_i^R(R/\mathfrak{m}, R/\mathfrak{m})$ is equal to the rank of the free R-module X_i.

It has been an open question for several years whether $P_R(t)$ is always a rational function. A first counterexample was found by D. Anick [2]. However, there is a class of standard graded K-algebras for which $P_R(t)$ is a rational function by rather simple reasons.

Definition 2.20 A standard graded K-algebra is called *Koszul* if $R/\mathfrak{m}$ has a linear resolution, in other words, if $\operatorname{Tor}_i^R(R/\mathfrak{m}, R/\mathfrak{m})_j = 0$ for all i and all $j \neq i$.

Koszul algebras were introduced by Priddy [171]. The simplest example of a Koszul algebra is the polynomial ring S, since the Koszul complex $K(\mathbf{x}; S)$ provides a linear resolution of $S/\mathfrak{m}$.

Proposition 2.21 *Let R be a Koszul algebra. Then $P_R(t) \operatorname{Hilb}_R(-t) = 1$. In particular, $P_R(t)$ is a rational function.*

Proof Let $\mathbb{X}$ be the graded minimal free resolution of $R/\mathfrak{m}$. By assumption $X_i = R(-i)^{\beta_i}$ for all i. Hence

$$1 = \operatorname{Hilb}_{R/\mathfrak{m}}(t) = \sum_{i \geq 0}(-1)^i \operatorname{Hilb}_{X_i}(t) = \sum_{i \geq 0}(-1)^i \beta_i t^i \operatorname{Hilb}_R(t) = P_R(-t) \operatorname{Hilb}_R(t).$$

Thus the assertion follows.   □

Proposition 2.22 *Let R be a standard graded K-algebra and $\ell \in R_1$ a nonzerodivisor. Then R is Koszul if and only if $R/\ell R$ is Koszul.*

Proof Since ℓ is a nonzerodivisor on R it follows that $\operatorname{Hilb}_{R/\ell R}(t) = (1 - t)\operatorname{Hilb}_R(t)$. On the one hand, it is known [6, Proposition 3.3.5] that $P_{R/\ell R}(t) =$

$P_R(t)/(1+t)$. Thus, $P_R(t) \operatorname{Hilb}_R(-t) = 1$ if and only if $P_{R/\ell R}(t) \operatorname{Hilb}_{R/\ell R}(-t) = 1$,. Now we use the fact, proved by C. Löfwall [140], that the statement of Proposition 2.21 has a converse. In other words, a standard graded K-algebra A is Koszul if and only if $P_A(t) \operatorname{Hilb}_A(-t) = 1$. This yields the desired conclusion.

□

In general it is hard, and often impossible, to decide whether an algebra is Koszul or not. However there are necessary and also sufficient conditions for Koszulness. Let us begin with a necessary condition.

Proposition 2.23 Let $R = S/I$ be a Koszul algebra with $I \subset (x_1, \ldots, x_n)^2$. Then I is generated by polynomials of degree 2.

Proof We denote by $\bar{f}$ the residue class modulo I of a polynomial $f \in S$. Then X_1 is a free module with basis $e_1, \ldots, e_n$ and $\partial \colon X_1 \to X_0 = R$ is given by $\partial(e_i) = \bar{x}_i$ for $i = 1, \ldots, n$.

Let $f_1, \ldots, f_m$ be a minimal homogeneous system of generators of I, and write $f_i = \sum_{j=1}^{n} f_{ij} x_j$ with homogeneous polynomials f_{ij}. Then obviously the elements $u_i = \sum_{j=1}^{n} \bar{f}_{ij} e_j$ in X_1 belong to $\operatorname{Ker} \partial$, and $\deg u_i = \deg f_i$ for all i. We claim that the relations $u_1, \ldots, u_m$ together with the relations $r_{ij} = \bar{x}_i e_j - \bar{x}_j e_i$, $i < j$ form a minimal system of generators of $\operatorname{Ker} \partial$. From this it then follows that I must be generated in degree 2, if R is Koszul.

In order to prove the claim let $\sum_{j=1}^{n} \bar{g}_j e_j$ be an arbitrary element in $\operatorname{Ker} \partial$. Then $\sum_{j=1}^{n} \bar{g}_j \bar{x}_j = 0$, and so $\sum_{j=1}^{n} g_j x_j \in I$. Hence there exist $h_i \in S$ such that $\sum_{j=1}^{n} g_j x_j = \sum_{i=1}^{m} h_i f_i$. It follows that

$$\sum_{j=1}^{n} g_j x_j = \sum_{i=1}^{m} h_i \left(\sum_{j=1}^{n} f_{ij} x_j \right) = \sum_{j=1}^{n} \left(\sum_{i=1}^{m} h_i f_{ij} \right) x_j.$$

Consequently, $\sum_{j=1}^{n} (g_j - \sum_{i=1}^{m} h_i f_{ij}) x_j = 0$. This implies that $\sum_{j=1}^{n} (g_j - \sum_{i=1}^{m} h_i f_{ij}) e_j$ is an element of the kernel of the map $\bigoplus_{j=1}^{n} S e_j \to (x_1, \ldots, x_n)$ with $e_j \mapsto x_j$ for $j = 1, \ldots, n$. Since the Koszul complex $K(\mathbf{x}; S)$ is acyclic this kernel is generated by the elements $s_{kl} = x_k e_l - x_l e_k$, $k < l$. Thus there exist polynomials p_{kl} such that

$$\sum_{j=1}^{n} (g_j - \sum_{i=1}^{m} h_i f_{ij}) e_j = \sum_{k<l} p_{kl} s_{kl}.$$

Therefore, $\sum_{j=1}^{n} \bar{g}_j e_j = \sum_{i=1}^{m} \bar{h}_i u_i + \sum_{k<l} \bar{p}_{kl} r_{kl}$. This shows that the elements u_i and r_{kl} generate $\operatorname{Ker} \partial$.

Suppose that one of the elements u_i, say u_1, can be omitted in the above generating set. Then there exist polynomials q_i and g_{kl} in S such that $u_1 = \sum_{i=2}^{m} \bar{q}_i u_i + \sum_{k<l} \bar{g}_{kl} r_{kl}$, and hence

$$\sum_{j=1}^{n} f_{1j}e_j - \sum_{i=2}^{m} q_i \left(\sum_{j=1}^{n} f_{ij}e_j \right) - \sum_{k<l} g_{kl}s_{kl} \in \bigoplus_{j=1}^{n} I e_j.$$

Substituting the e_j by the x_j we obtain that $f_1 - \sum_{i=2}^{m} q_i f_i \in (x_1, \ldots, x_n)I$, which by Nakayama's lemma is impossible since $f_1, \ldots, f_m$ is a minimal system of generators of I.   □

In Chapter 4 an example is given which shows that this necessary condition is not sufficient. Now we will give a sufficient condition.

According to Conca, Trung, and Valla [39], a *Koszul filtration* of R is a finite set $\mathscr{F}$ of ideals generated by linear forms such that

(i) $\mathfrak{m} \in \mathscr{F}$;
(ii) for any $I \in \mathscr{F}$ with $I \neq 0$, there exists $J \in \mathscr{F}$ with $J \subset I$ such that I/J is cyclic and $J : I \in \mathscr{F}$.

The next result illustrates the usefulness of Koszul filtrations.

Proposition 2.24 *Assume R admits a Koszul filtration $\mathscr{F}$. Then each ideal $I \in \mathscr{F}$ admits a linear resolution. In particular, R is Koszul.*

Proof We prove by induction on i and by the number of generators of I that $\operatorname{Tor}_i^R(R/\mathfrak{m}, I)_j = 0$ for all $I \in \mathscr{F}$ and $j \neq i + 1$. Then, this implies that each $I \in \mathscr{F}$ has a linear resolution. By (i), $\mathfrak{m} \in \mathscr{F}$. Therefore it will then follow that R is Koszul.

For $i = 0$, the assertion is trivial since all $I \in \mathscr{F}$ are generated by linear forms. Now let $i > 0$. Condition (ii) implies that $I/J \cong (R/L)(-1)$ for some $L \in \mathscr{F}$. Thus we obtain a short exact sequence

$$0 \to J \to I \to (R/L)(-1) \to 0.$$

By using the fact that $\operatorname{Tor}_i^R(R/\mathfrak{m}, (R/L)(-1))_j \cong \operatorname{Tor}_{i-1}^R(R/\mathfrak{m}, L)_{j-1}$, for all j we obtain the exact sequence

$$\operatorname{Tor}_i^R(R/\mathfrak{m}, J)_j \to \operatorname{Tor}_i^R(R/\mathfrak{m}, I)_j \to \operatorname{Tor}_{i-1}^R(R/\mathfrak{m}, L)_{j-1}$$

Now $\operatorname{Tor}_{i-1}^R(R/\mathfrak{m}, L)_{j-1} = 0$ for $j \neq i + 1$, by induction on i, and $\operatorname{Tor}_i^R(R/\mathfrak{m}, J)_j = 0$ for $j \neq i + 1$ by induction on the number of generators of J. Thus the exact sequence yields that $\operatorname{Tor}_i^R(R/\mathfrak{m}, I)_j = 0$ for $j \neq i + 1$, as desired.   □

Obviously, if $\mathscr{F}$ is a Koszul filtration, then $\mathscr{F}$ contains a flag of ideals

$$0 = I_0 \subset I_1 \subset I_2 \subset \cdots \subset I_n = \mathfrak{m},$$

where $I_j \in \mathscr{F}$ for all j (and I_j/I_{j-1} is cyclic for all j). If it happens that for all j there exists k such that $I_{j+1} : I_j = I_k$, then $\{I_0, I_1, \ldots, I_n\}$ is a Koszul filtration.

Such Koszul filtrations are called *Koszul flags*. Conca, Rossi, and Valla showed [38, Theorem 2.4] that if S/I has a Koszul flag, then I has a quadratic Gröbner basis. The following theorem is a partial converse of this result.

Theorem 2.25 *Let $I \subset S$ be a graded ideal which has a quadratic Gröbner basis with respect to the reverse lexicographic order induced by $x_1 > \cdots > x_n$. Then, for all i, the colon ideals*

$$(I, x_{i+1}, \ldots, x_n) : x_i$$

are generated, modulo I, by linear forms.

For the proof of the theorem we need to recall the following result.

Lemma 2.26 *Let $\mathscr{G}$ be the reduced Gröbner basis of the graded ideal $I \subset S$ with respect to the reverse lexicographic order induced by $x_1 > \cdots > x_n$. Then*

$$\mathscr{G}' = \{f \in \mathscr{G} : x_n \nmid f\} \cup \{f/x_n : f \in \mathscr{G} \text{ and } x_n | f\}$$

is a Gröbner basis of $I : x_n$.

The proof is similar to that of Proposition 1.40.

Proof (Proof of Theorem 2.25) Let $\mathscr{G} = \{g_1, \ldots, g_m\}$ be the reduced Gröbner basis of I with respect to the reverse lexicographic order and fix $i \leq n$. Let $f_j = g_j \bmod(x_{i+1}, \ldots, x_n)$, where $f_j \in K[x_1, \ldots, x_i]$ for all j. We may assume that $\text{in}_<(g_1) > \cdots > \text{in}_<(g_m)$, and therefore, there exists an $s \leq m$ such that $f_s \neq 0$ and $f_{s+1} = \cdots = f_m = 0$. In addition, we have $\text{in}_<(f_j) = \text{in}_<(g_j)$ for $1 \leq j \leq s$. It then follows that $(I, x_{i+1}, \ldots, x_n) = (f_1, \ldots, f_s, x_{i+1}, \ldots, x_n)$ and the set $\mathscr{F} = \{f_1, \ldots, f_s, x_{i+1}, \ldots, x_n\}$ is a Gröbner basis, since

$$\text{in}_<(I, x_{i+1}, \ldots, x_n) = (\text{in}_<(I), x_{i+1}, \ldots, x_n),$$

see Problem 1.6. Moreover, $\mathscr{F}$ is reduced, since $\mathscr{G}$ is reduced. Let $J = (f_1, \ldots, f_s)$. Then

$$(I, x_{i+1}, \ldots, x_n) : x_i = (J, x_{i+1}, \ldots, x_n) : x_i = (J : x_i) + (x_{i+1}, \ldots, x_n).$$

By applying Lemma 2.26 for $J \cap K[x_1, \ldots, x_i]$, it follows that, modulo J, $(J : x_i)$ is generated by linear forms in $K[x_1, \ldots, x_i]$ which implies that $(I, x_{i+1}, \ldots, x_n)$ is also generated by linear forms modulo I. □

A particular class of Koszul filtrations which naturally occur in combinatorial contexts are the following: let R be a standard graded K-algebra and let the graded maximal ideal m be minimally generated by the homogeneous elements $u_1, \ldots, u_m$. We let $\mathscr{F}$ be the set of all ideals generated by the subsequences $u_{i_1}, \ldots, u_{i_j}$ of

$u_1, \ldots, u_m$. Suppose that for each such subsequence, $(u_{i_1}, \ldots, u_{i_{j-1}}) : u_{i_j}$ is generated by a subset of $\{u_1, \ldots, u_m\}$. Then, obviously, $\mathscr{F}$ is a Koszul filtration.

A standard graded K-algebra, whose graded maximal ideal possesses a system of generators satisfying these conditions, is called *strongly Koszul*. Of course, any strongly Koszul algebra is also Koszul.

The simplest example of a strongly Koszul algebra is the polynomial ring itself with generators $x_1, \ldots, x_n$ for the graded maximal ideal of S.

The following example of a strongly Koszul algebra is of great importance for the further theory.

Proposition 2.27 *Let $I \subset S$ be generated by monomials of degree 2. Then $R = S/I$ is strongly Koszul.*

Proof We denote by $\bar{f}$ the residue class modulo I of a polynomial $f \in S$, and will show that $J = (\bar{x}_{i_1}, \ldots, \bar{x}_{i_{k-1}}) : \bar{x}_{i_k}$ is generated by a subset of $\{\bar{x}_1, \ldots, \bar{x}_n\}$.

Let I be generated by the degree 2 monomials $u_1, \ldots, u_m$. Since I is a monomial ideal, it follows that $(\bar{x}_{i_1}, \ldots, \bar{x}_{i_{k-1}}) : \bar{x}_{i_k}$ is generated by residue classes of monomials. Let $\bar{u} \neq 0$ be such a monomial. Then $\bar{u}\bar{x}_{i_k} \in (\bar{x}_{i_1}, \ldots, \bar{x}_{i_{k-1}})$.

Suppose first that $\bar{u}\bar{x}_{i_k} = 0$. Then there exists a monomial $v \in S$ such that $ux_{i_k} = vu_i$ for some i. Since $\bar{u} \neq 0$, it follows that x_{i_k} divides u_i, say, $u_i = x_{i_k}x_l$. Then x_l divides u and $\bar{x}_l \in J$.

Next suppose $\bar{u}\bar{x}_{i_k} \neq 0$. Then there exists a monomial v such that $\bar{u}\bar{x}_{i_k} = \bar{v}\bar{x}_{i_j}$ for some $j < k$. Thus $ux_{i_k} - vx_{i_j} \in I$. If $ux_{i_k} - vx_{i_j} = 0$, then x_{i_j} divides u, and if $ux_{i_k} - vx_{i_j} \neq 0$, then, since I is a monomial ideal, it follows that $ux_{i_k} \in I$ and we are in the first case. □

This result has an important consequence.

Theorem 2.28 *Let $I \subset S$ be a graded ideal and suppose that there exists a monomial order $<$ on S such that $\mathrm{in}_<(I)$ is generated by monomials of degree 2. Then $R = S/I$ is Koszul.*

Proof We use the fact that the graded Betti numbers of K viewed as an S/I-module are less than or equal to the corresponding graded Betti numbers of K viewed as an $S/\mathrm{in}_<(I)$-module, see, for example, [60, Theorem 6.8]. Obviously this fact implies that $R = S/I$ is Koszul, if $S/\mathrm{in}_<(I)$ is Koszul. Thus the desired conclusion follows immediately from Proposition 2.27. □

Our discussions show that the following implications hold:

I has a quadratic Gröbner basis $\Rightarrow$ S/I is Koszul $\Rightarrow$ I is generated by quadrics.

None of these implication can be reversed.

Example 2.29 Let I be an ideal generated by quadrics given in Example 1.18. Then $K[x_1, x_2, \ldots, x_{10}]/I$ is not Koszul. By using a specialized software (e.g., Macaulay2) one can check that $\beta_{34}(K) = 1 \neq 0$.

Example 2.30 Let $n = 8$ and I be the ideal of $K[x_1, x_2, \ldots, x_8]$ generated by

$$f_1 = x_2x_8 - x_4x_7, \quad f_2 = x_1x_6 - x_3x_5, \quad f_3 = x_1x_3 - x_2x_4.$$

Then I has no quadratic Gröbner basis and $K[x_1, x_2, \ldots, x_8]/I$ is Koszul. The details will be explained in Example 4.28.

Let A be a standard graded K-algebra, and $B \subset A$ a K-subalgebra generated by elements of A of degree 1. Then B is standard graded as well. The algebra B is said to be an *algebra retract* of A, if there exists a surjective K-algebra homomorphism $\epsilon : A \to B$ such that the inclusion map $B \hookrightarrow A$ composed with ϵ yields a K-algebra isomorphism $B \to B$. The map ϵ is called the *retraction map* of the algebra retract $B \subset A$.

The following result can often be used as an inductive argument to prove Koszulness.

Theorem 2.31 *Let $B \subset A$ be an algebra retract of standard graded K-algebras with retraction map ϵ. Then the following conditions are equivalent:*

(i) *A is Koszul;*
(ii) *B is Koszul, and B viewed as an A-module via ϵ admits an A-linear resolution.*

Proof Let R be a standard graded K-algebra with graded maximal ideal $\mathfrak{m}$, and M a finitely generated graded R-module generated in nonnegative degree. The formal power series

$$P_R^M(s, t) = \sum_{i,j} \dim_K \operatorname{Tor}_i^R(R/\mathfrak{m}, M)_j s^j t^i$$

in the variables s and t is called the *graded Poincaré series* of M. Since for each i there exist only finitely many j with $\operatorname{Tor}_i^R(R/\mathfrak{m}, M)_j \neq 0$, we can write

$$P_R^M(s, t) = \sum_{i \geq 0} p_i^M(s) t^i,$$

where each $p_i^M(s)$ is a polynomial in s.

It has been shown in [91] that, since $B \subset A$ is an algebra retract, the following identity of formal power series holds:

$$P_A^K(s, t) = P_A^B(s, t) P_B^K(s, t).$$

Write $P_A^K(s, t) = \sum_{i \geq 0} p_i(s) t^i$, $P_A^B(s, t) = \sum_{i \geq 0} q_i(s) t^i$ and $P_B^K(s, t) = \sum_{i \geq 0} r_i(s) t^i$. Then

$$p_i(i) = \sum_{j=0}^{i} q_j(t) r_{i-j}(t) \quad \text{for all} \quad i.$$

Since the coefficients of the polynomials q_i and r_i are all nonnegative integers, it follows that

$$\deg p_i(t) = \max\{\deg q_j(t) + \deg r_{i-j}(t) ; \ j = 0, \ldots, i\}.$$

From this equation both assertions of the theorem follow at once. $\quad\square$

We end our discussions on Koszulness in this chapter by relating the Koszul property of S/I to the finite graded free S-resolution of S/I.

Theorem 2.32 *Let $I \subset S$ be a graded ideal. Then*

(a) *Suppose I has a 2-linear resolution. Then S/I is Koszul.*
(b) *Suppose that I is generated by quadrics and that $\beta_{2j}(S/I) \neq 0$ for some $j > 4$. Then S/I is not Koszul.*

Proof (a) The condition in (a) implies that all Massey operations vanish, so that S/I is a Golod ring. Therefore,

$$P^K_{S/I}(s, t) = \frac{(1 + st)^n}{1 + t - t P^{S/I}_S(s, t)}.$$

For the details of this argument we refer to the survey article on infinite free resolutions by Avramov in [6].

We use again that I has a 2-linear resolution, and deduce that

$$P^{S/I}_S(s, t) = 1 + \sum_{i \geq 1} \beta_i(S/I) s^{i+1} t^i,$$

so that $1 + t - t P^{S/I}_S(s, t) = 1 - \sum_{i \geq 1} \beta_i(S/I) s^{i+1} t^{i+1}$.

Now expanding the fraction which gives us $P^K_{S/I}(s, t)$, we see that $P^K_{S/I}(s, t)$ is a power series in the product st of the variables s and t, and this means that S/I is Koszul.

The proof of (b) needs some preparation and will be postponed. $\quad\square$

Let $(R, \mathfrak{m}, K)$ be a Noetherian local ring or a standard graded K-algebra (in which case we assume that $\mathfrak{m}$ is the graded maximal ideal of R). Tate in his famous paper [210] constructed an R-free resolution

$$X: \cdots \longrightarrow X_i \longrightarrow \cdots \longrightarrow X_2 \longrightarrow X_1 \longrightarrow X_0 \longrightarrow 0,$$

of the residue class field $R/\mathfrak{m} = K$, that is, an acyclic complex of finitely generated free R-modules X_i with $H_0(X) = K$, admitting an additional structure, namely the structure of a differential graded R-algebra. It was Gulliksen [86] and independently Schoeller [187] who proved that if Tate's construction is minimally done, as explained below, then X is indeed a minimal free R-resolution of K. For details we refer to the original paper of Tate and to a modern treatment of the theory as given in [6].

Here we sketch Tate's construction as much as is needed to prove Theorem 2.32(b). In Tate's theory X is a DG-algebra, that is, a graded skew-symmetric R-algebra with free R-modules X_i as graded components and $X_0 = R$, equipped with a differential d of degree -1 such that

$$d(ab) = d(a)b + (-1)^i a d(b) \tag{2.4}$$

for $a \in X_i$ and $b \in X$. Moreover, (X, d) is an acyclic complex with $H_0(X) = K$.

The algebra X is constructed by adjunction of variables: given any DG-algebra Y and a cycle $z \in Y_i$, then the DG-algebra $Y' = Y\langle T : dT = z \rangle$ is obtained by adjoining the variable T of degree $i + 1$ to Y in order to kill the cycle z.

If i is even we let

$$Y'_j = Y_j \oplus Y_{j-i-1} T \quad \text{with } T^2 = 0 \text{ and } d(T) = z.$$

If i is odd we let

$$Y'_j = X_j \oplus X_{j-(i+1)} T^{(1)} \oplus X_{j-2(i+1)} T^{(2)} \oplus \cdots$$

with $T^{(0)} = 1$, $T^{(1)} = T$, $T^{(i)} T^{(j)} = ((i + j)!/i!j!) T^{(i+j)}$ and $d(T^{(i)}) = z T^{(i-1)}$. The $T^{(j)}$ are called the divided powers of T. The degree of $T^{(j)}$ is defined to be $j \deg T$.

The construction of X proceeds as follows: Say, $\mathfrak{m}$ is minimally generated by $x_1, \ldots, x_n$. Then we adjoin to R (which is a DG-algebra concentrated in homological degree 0) the variables $T_{11}, \ldots, T_{1n}$ of degree 1 with $d(T_{1i}) = x_i$. The DG-algebra $X^{(1)} = R\langle T_{11}, \ldots, T_{1n} \rangle$ so obtained is nothing but the Koszul complex of the sequence $x_1, \ldots, x_n$ with values in R. If $X^{(1)}$ is acyclic, then R is regular and $X = X^{(1)}$ is the Tate resolution of K. Otherwise $H_1(X^{(1)}) \neq 0$ and we choose cycles $z_1, \ldots, z_m$ whose homology classes form a K-basis of $H_1(X^{(1)})$, and we adjoin variables $T_{21}, \ldots, T_{2m}$ of degree 2 to $X^{(1)}$ with $d(T_{2i}) = z_i$ to obtain $X^{(2)}$. It is then clear that $H_j(X^{(2)}) = 0$ for $j = 1$. Suppose $X^{(k)}$ has been already constructed with $H_j(X^{(k)}) = 0$ for $j = 1, \ldots, k-1$. We first observe that $H_k(X^{(k)})$ is annihilated by $\mathfrak{m}$. Indeed, let z be a cycle of $X^{(k)}$, then $x_i z = d(T_{1i} z)$, due to the product rule (2.4). Now one chooses a K-basis of cycles representing the homology classes of $H_k(X^{(k)})$ and adjoins variables in degree $k+1$ to kill these cycles, thereby obtaining $X^{(k+1)}$. In this way one obtains a chain of DG-algebras

$$R = X^{(0)} \subset X^{(1)} \subset X^{(2)} \subset \cdots \subset X^{(k)} \subset \cdots,$$

which in the limit yields the Tate resolution X of K. It is clear that if R is standard graded then in each step the representing cycles that need to be killed can be chosen to be homogeneous, so that X becomes a graded minimal free R-resolution of K if we assign to the variables T_{ij} inductively the degree of the cycles they do kill and apply the following rule: denote the internal degree (different from the homological degree) of a homogeneous element a of X by $\text{Deg}(a)$. Then we require that $\text{Deg } T^{(i)} = i \text{ Deg } T$ for any variable of even homological degree and furthermore $\text{Deg}(ab) = \text{Deg}(a) + \text{Deg}(b)$ for any two homogeneous elements in X.

Proof (of Theorem 2.32(b)) The Koszul complex $X^{(1)}$ as a DG-algebra over S/I is generated by the variable T_{1i} with $d(T_{1i}) = x_i$ for $i = 1, \ldots, n$. Thus $\text{Deg } T_{1i} = 1$ for all i. Let $f_1, \ldots, f_m$ be quadrics which minimally generate I, and write $f_i = \sum_{j=1}^{m} f_{ij} x_j$ with suitable linear forms f_{ij}. Then $H_1(X^{(1)})$ is minimally generated by the homology classes of the cycles $z_i = \sum_{j=1}^{m} f_{ij} T_{1j}$. Let $T_{2i} \in X^{(2)}$ be the variables of homological degree 2 with $d(T_{2i}) = z_i$ for $i = 1, \ldots, m$. Then $\text{Deg } T_{2i} = \text{Deg } z_i = 2$ for all i. To proceed in the construction of X we have to kill the cycles $w_1, \ldots, w_r$ whose homology classes form a K-basis of $H_2(X^{(2)})$. Since $\text{Tor}_i(K, S/I) \cong H_i(X^{(1)})$, our hypothesis implies that there is a cycle $z \in (X^{(1)})_2$ with $\text{Deg } z = j > 4$ which is not a boundary. Of course z is also a cycle in $X^{(2)}$ because $X^{(1)}$ is a subcomplex of $X^{(2)}$. We claim that z is not a boundary in $X^{(2)}$. To see this we consider the exact sequence of complexes

$$0 \longrightarrow X^{(1)} \longrightarrow X^{(2)} \longrightarrow X^{(2)}/X^{(1)} \longrightarrow 0,$$

which induces the long exact sequence

$$\cdots \longrightarrow H_3(X^{(2)}/X^{(1)}) \xrightarrow{\ \delta\ } H_2(X^{(1)}) \longrightarrow H_2(X^{(2)}) \longrightarrow \cdots.$$

Thus it suffices to show that the homology class $[z]$ of the cycle z is not in the image of δ. Notice that the elements $T_{1i} T_{2j}$ form a basis of the free S/I-module $(X^{(2)}/X^{(1)})_3$ and that the differential on $X^{(2)}/X^{(1)}$ maps $T_{1i} T_{2j}$ to $x_i T_{2j}$, so that $w \in (X^{(2)}/X^{(1)})_3$ is a cycle if and only if $w = \sum_{j=1}^{m} w_j T_{2j}$ where each $w_j \in X_1^{(1)}$ is a cycle. Now the connecting homomorphism δ maps $[w]$ to $[-\sum_{j=1}^{m} w_j z_j]$. It follows that $\text{Im }\delta = H_1(X^{(1)})^2$. Since $H_1(X^{(1)})$ is generated in degree 2 we conclude that the subspace $H_1(X^{(1)})^2$ of $H_2(X^{(1)})$ is generated in degree 4. Hence our element $[z] \in H_2(X^{(1)})$ which is of degree > 4 cannot be in the image of δ, as desired.

Thus the homology class of z, viewed as an element of $H_2(X^{(2)})$ has to be killed by adjoining a variable of degree $j > 4$. This shows that $\beta_{3j}^{S/I}(S/\mathfrak{m}) \neq 0$, and hence S/I is not Koszul. $\qquad\qquad\square$

Problems

2.25 Let $I \subset K[x_1, \ldots, x_5, y_1, \ldots, y_5]$ be the ideal generated by $x_1 y_2 - x_2 y_1, x_2 y_3 - x_3 y_2, x_3 y_4 - x_4 y_3, x_4 y_5 - x_5 y_4, x_1 y_5 - x_5 y_1$. By using Theorem 2.32, show that S/I is not Koszul.

2.26 Let $I \subset S$ be an ideal generated by a regular sequence of quadrics. Show that S/I is Koszul.

2.27 Let I be the ideal given in Problem 2.21. Show that $R = S/I$ is Koszul and compute the Poincaré series $P_R(t)$ of R.

2.28 Let R be the toric ring generated over K by all monomials of S of degree 2. Show that R is strongly Koszul.

2.29 Let R_1 and R_2 be standard graded K-algebras, and set $R = R_1 \otimes_K R_2$. Then $P_R(t) = P_{R_1}(t) P_{R_2}(t)$, and R is Koszul if and only if R_1 and R_2 are Koszul.

Notes

The books of Bruns–Herzog [27] and Eisenbud [57] may help to deepen the understanding of the concepts and results discussed in this chapter. A systematic introduction to Cohen–Macaulay and Gorenstein rings is given in [27]. There, one can also find (see [27, Definition 3.1.18 and Theorem 3.2.10]) the proof of the theorem of Bass [10] according to which a Cohen–Macaulay ring R is Gorenstein if and only if R, viewed as module over itself, has finite injective dimension.

Koszul algebras were first introduced by Priddy [171]. Fröberg [75] showed that S/I is Koszul if I is generated by monomials of degree 2. This result leads to the important conclusion that algebras, whose defining ideal admits a quadratic Gröbner basis, are Koszul. In [98] strongly Koszul algebras were introduced. This concept inspired Conca, Trung, and Valla [39] to introduce the more flexible notion of Koszul filtrations. From that paper, Proposition 2.24 is adopted. It provides a sufficient condition of Koszulness in terms of Koszul filtrations. The short proof for Proposition 2.22 is due to Backelin and Fröberg [7]. The proof of Proposition 2.23, in which it is shown that the defining ideal of a Koszul algebra is generated by quadrics, reproduces the proof given in [60, Proposition 6.3]. Theorem 2.31 is taken from [156, Proposition 1.4], while Theorem 2.32(b) is Lemma 1.2 of [61].

A nice survey on Koszul algebras is given in [76].

Part II
Binomial Ideals and Convex Polytopes

Chapter 3
Introduction to Binomial Ideals

Abstract In this chapter we introduce the main topic of this book: binomials and binomial ideals. Special attention is given to toric ideals. These are binomial ideals arising from an integer matrix which represents the exponent vectors of the monomial generators of a toric ring. It will be shown that the toric ideal I_A attached to the matrix A is graded if and only if A is a configuration matrix. Furthermore, it will be shown that an arbitrary binomial ideal is a toric ideal if and only if it is a prime ideal. Then we study the Gröbner basis of a binomial ideal and show that its reduced Gröbner basis consists of binomials. We introduce Graver bases and show that the reduced Gröbner basis of a binomial ideal is contained in its Graver basis. Naturally attached to a lattice $L \subset \mathbb{Z}^n$ (i.e. a subgroup of the abelian group $\mathbb{Z}^n$) there is a binomial ideal I_L, called the lattice ideal of L. It will be shown that the saturation of any binomial ideal is a lattice ideal, and that the lattice ideals are exactly those which are saturated. The ideal generated by the binomials corresponding to the basis vectors of a basis of the lattice L is called a lattice basis ideal. Its saturation is the lattice ideal I_L. The chapter closes with an introduction to Lawrence ideals and to squarefree divisor complexes.

3.1 Toric Ideals and Binomial Ideals

Gröbner bases of toric ideals play an important role in algebraic statistics and in the study of convex polytopes. In this section we introduce toric and binomial ideals and discuss some of their basic properties.

Let K be a field. We denote by $S = K[x_1, \ldots, x_n]$ the polynomial ring in the variables $x_1, \ldots, x_n$. A *binomial* belonging to S is a polynomial of the form $u - v$, where u and v are monomials in S. A *binomial ideal* is an ideal of S generated by binomials. Any binomial ideal is generated by a finite number of binomials. Indeed, let I be a binomial ideal. Since S is Noetherian, I admits a finite number of generators. Each of these generators is a linear combination of a finite number of binomials. Thus the finitely many binomials appearing in these linear combinations generate I. More generally, some authors call an expression $u - \lambda v$ a binomial, with $\lambda \in K$ and u and v monomials. For this definition monomials become also

© Springer International Publishing AG, part of Springer Nature 2018
J. Herzog et al., *Binomial Ideals*, Graduate Texts in Mathematics 279,
https://doi.org/10.1007/978-3-319-95349-6_3

binomials, since $\lambda = 0$ is not excluded. This more general concept of binomials is required to get a satisfactory theory for primary decompositions of binomial ideals, see Eisenbud and Sturmfels [58]. Since primary decompositions of a binomial ideal may depend on the base field, Kahle and Miller [123] developed the theory of mesoprimary decompositions of binomial ideals.

An important class of binomial ideals are the so-called *toric ideals*. In order to define toric ideals we let $A = (a_{ij})_{\substack{1 \le i \le d \\ 1 \le j \le n}}$ be a $d \times n$-matrix of integers and let

$$
\mathbf{a}_j = \begin{pmatrix} a_{1j} \\ a_{2j} \\ \vdots \\ a_{dj} \end{pmatrix}, \quad 1 \le j \le n
$$

be the column vectors of A. We write $\mathbb{Z}^{d \times n}$ for the set of $d \times n$-matrices $A = (a_{ij})_{\substack{1 \le i \le d \\ 1 \le j \le n}}$ with each $a_{ij} \in \mathbb{Z}$.

As usual $\mathbf{a} \cdot \mathbf{b} = \sum_{i=1}^{d} a_i b_i$ denotes the inner product of the vectors $\mathbf{a} = (a_1, \ldots, a_d)^t$ and $\mathbf{b} = (b_1, \ldots, b_d)^t$. Here $\mathbf{c}^t$ denotes transpose of a vector $\mathbf{c}$.

A matrix $A = (a_{ij})_{\substack{1 \le i \le d \\ 1 \le j \le n}} \in \mathbb{Z}^{d \times n}$ is called a *configuration matrix* (or simply a *configuration*) if there exists $\mathbf{c} \in \mathbb{Q}^d$ such that

$$
\mathbf{a}_j \cdot \mathbf{c} = 1, \quad 1 \le j \le n.
$$

For example, $A = \begin{pmatrix} 1 & 3 & 2 \\ 0 & 2 & 1 \end{pmatrix}$ is a configuration matrix, while $(a_1, \ldots, a_n) \in \mathbb{Z}^{1 \times n}$ is a configuration matrix if and only if $a_1 = a_2 = \cdots = a_n \ne 0$.

Now let $T = K[t_1^{\pm 1}, \ldots, t_d^{\pm 1}]$ be the *Laurent polynomial ring* over K in the variables $t_1, \ldots, t_d$, and let $A \in \mathbb{Z}^{d \times n}$ with column vectors $\mathbf{a}_j$. We define a K-algebra homomorphism

$$
\pi: S \to T \quad \text{with} \quad x_j \mapsto \mathbf{t}^{\mathbf{a}_j}. \tag{3.1}
$$

The image of π is the K-subalgebra $K[\mathbf{t}^{\mathbf{a}_1}, \ldots, \mathbf{t}^{\mathbf{a}_n}]$ of T, denoted $K[A]$. We call $K[A]$ the *toric ring* of A. For the configuration matrix A of the above example we have $K[A] = K[t_1, t_1^3 t_2^2, t_1^2 t_2]$.

The kernel of π is denoted by I_A and is called the *toric ideal* of A. In our example, we have $I_A = (x_1 x_2 - x_3^2)$.

Proposition 3.1 *Let $A \in \mathbb{Z}^{d \times n}$. Then $\dim K[A] = \operatorname{rank} A$.*

Proof Let $K(A)$ be the quotient field of $K[A]$. Then the Krull dimension of $K[A]$ is equal to the transcendence degree $\operatorname{tr deg}(K(A)/K)$ of $K(A)$ over K, see [27, Theorem A.16]. Let $G \subset \mathbb{Z}^d$ be the subgroup of $\mathbb{Z}^d$ generated by the column vectors of A. Then G is a free abelian group with $\operatorname{rank} A = \operatorname{rank} G$. Let $\mathbf{b}_1, \ldots, \mathbf{b}_m$ be a $\mathbb{Z}$-basis of integer vectors of G. Then $m = \operatorname{rank} A$ and $K(A) = K(\mathbf{t}^{\mathbf{b}_1}, \ldots, \mathbf{t}^{\mathbf{b}_m})$.

The desired result will follow once we have shown that the elements $\mathbf{t}^{\mathbf{b}_1}, \ldots, \mathbf{t}^{\mathbf{b}_m}$ are algebraically independent over K. To see this, let $F \in K[y_1, \ldots, y_m]$ be a polynomial with $F(\mathbf{t}^{\mathbf{b}_1}, \ldots, \mathbf{t}^{\mathbf{b}_m}) = 0$. Say, $F = \sum_{\mathbf{c}} a_{\mathbf{c}} \mathbf{y}^{\mathbf{c}}$ with $a_{\mathbf{c}} \in K$. Then

$$0 = \sum_{\mathbf{c}} a_{\mathbf{c}} \mathbf{t}^{c_1 \mathbf{b}_1 + \cdots + c_m \mathbf{b}_m}.$$

Since the vectors $\mathbf{b}_1, \ldots, \mathbf{b}_m$ are linearly independent it follows that the monomials $\mathbf{t}^{c_1 \mathbf{b}_1 + \cdots c_m \mathbf{b}_m}$ are pairwise distinct. This implies that $F = 0$. □

Given a column vector

$$\mathbf{b} = \begin{pmatrix} b_1 \\ b_2 \\ \vdots \\ b_n \end{pmatrix}$$

belonging to $\mathbb{Z}^n$, we introduce the binomial $f_{\mathbf{b}} \in S$ defined by

$$f_{\mathbf{b}} = \prod_{b_i > 0} x_i^{b_i} - \prod_{b_j < 0} x_j^{-b_j}.$$

Note that $f_{\mathbf{b}} = \mathbf{x}^{\mathbf{b}^+} - \mathbf{x}^{\mathbf{b}^-}$, where $\mathbf{b}^+$ and $\mathbf{b}^-$ are vectors in $\mathbb{Z}^n$ with entries

$$b_i^+ = \begin{cases} b_i, & \text{if } b_i \geq 0, \\ 0, & \text{if } b_i < 0, \end{cases} \quad \text{and} \quad b_i^- = \begin{cases} 0, & \text{if } b_i > 0, \\ -b_i, & \text{if } b_i \leq 0. \end{cases}$$

Note that if f is any binomial in S, then $f = u f_{\mathbf{b}}$ for a unique $\mathbf{b} \in \mathbb{Z}^n$ and a unique monomial u.

For example, if $\mathbf{b} = (1, -1, 0, 2)$, then $f_{\mathbf{b}} = x_1 x_4^2 - x_2$ and if $\mathbf{b} = (1, 2, 3, 1)$, then $f_{\mathbf{b}} = x_1 x_2^2 x_3^3 x_4 - 1$, while if $f = x_1^2 x_2 - x_1 x_2^2 x_3^3 x_4$, then $f = x_1 x_2 f_{\mathbf{b}}$ with $\mathbf{b} = (1, -1, -3, -1)$.

Theorem 3.2 *Any toric ideal is a binomial ideal. More precisely, let $A \in \mathbb{Z}^{d \times n}$. Then I_A is generated by the binomials $f_{\mathbf{b}}$ with $\mathbf{b} \in \mathbb{Z}^n$ and $A\mathbf{b} = 0$.*

Proof We first show that I_A is a binomial ideal. Let $f \in \operatorname{Ker} \pi$ with $f = \sum_u \lambda_u u$, $\lambda_u \in K$ and each u a monomial in S. We write $f = \sum_{\mathbf{c}} f^{(\mathbf{c})}$, where $f^{(\mathbf{c})} = \sum_{u, \pi(u)=\mathbf{t}^{\mathbf{c}}} \lambda_u u$—the sum taken over those monomials u which appear in f.

It follows that

$$0 = \pi(f) = \sum_{\mathbf{c}} \pi(f^{(\mathbf{c})}) = \sum_{\mathbf{c}} \Big(\sum_{u, \pi(u)=\mathbf{t}^{\mathbf{c}}} \lambda_u \Big) \mathbf{t}^{\mathbf{c}},$$

and hence $\sum_{u, \pi(u)=\mathbf{t}^{\mathbf{c}}} \lambda_u = 0$ for all $\mathbf{c}$. Thus if $f^{(\mathbf{c})} \neq 0$ and $u \in \operatorname{supp}(f^{(\mathbf{c})})$, then $f^{(\mathbf{c})} = \sum_{v \in \operatorname{supp}(f^{(\mathbf{c})})} \lambda_v (v - u)$.

Finally, let $f_{\mathbf{b}} \in S$. Then $\pi(f_{\mathbf{b}}) = \mathbf{t}^{A\mathbf{b}^+} - \mathbf{t}^{A\mathbf{b}^-}$. Hence $f_{\mathbf{b}} \in \operatorname{Ker} \pi$ if and only if $A\mathbf{b}^+ = A\mathbf{b}^-$, and this is the case if and only if $A\mathbf{b} = 0$. $\qquad\qquad\square$

Proposition 3.3 *Let $A \in \mathbb{Z}^{d \times n}$. The following conditions are equivalent:*

(i) *A is a configuration matrix;*
(ii) *for all $\mathbf{b} = (b_1, \ldots, b_n)^t \in \mathbb{Z}^n$ with $A\mathbf{b} = 0$ we have $\sum_{i=1}^n b_i = 0$;*
(iii) *I_A is a graded ideal.*

Proof (i) $\Rightarrow$ (ii): There exists $\mathbf{c} \in \mathbb{Q}^d$ such that $\mathbf{c}^t A = (1, \ldots, 1)$. Now let $\mathbf{b} = (b_1, \ldots, b_n)^t \in \mathbb{Z}^n$ with $A\mathbf{b} = 0$. Then

$$0 = \mathbf{c}^t(A\mathbf{b}) = (\mathbf{c}^t A)\mathbf{b} = (1, \ldots, 1)(b_1, \ldots, b_n)^t = \sum_{i=1}^n b_i.$$

(ii) $\Rightarrow$ (i): Let $U \subset \mathbb{Q}^n$ be the $\mathbb{Q}$-subspace of $\mathbb{Q}^n$ generated by the row vectors of A, and let $V \subset \mathbb{Q}^n$ the $\mathbb{Q}$-subspace generated by U and $(1, \ldots, 1)$. Then $U \subset V$ and (ii) implies that $U^\perp = V^\perp$, where for a $\mathbb{Q}$-subspace W of $\mathbb{Q}^n$ we denote by $W^\perp$ the $\mathbb{Q}$-subspace of $\mathbb{Q}^n$ consisting of all vectors $\mathbf{v} \in \mathbb{Q}^n$ with $\mathbf{w} \cdot \mathbf{v} = 0$ for all $\mathbf{w} \in W$. It follows that

$$U = (U^\perp)^\perp = (V^\perp)^\perp = V,$$

since U and V are finitely generated vector spaces. Hence $(1, \ldots, 1)$ is a linear combination of the row vectors of A. This implies (i).

(ii) $\Longleftrightarrow$ (iii): By Theorem 3.2, the binomials $f_{\mathbf{b}}$ with $A\mathbf{b} = 0$ generate I_A. Thus I_A is graded if and only if all $f_{\mathbf{b}}$ are homogeneous. This is the case if and only if $\sum_{i=1}^n b_i = 0$ for all $\mathbf{b}$ with $A\mathbf{b} = 0$. $\qquad\qquad\square$

It is clear that any toric ideal is a prime ideal. Theorem 3.2 has the following converse.

Theorem 3.4 *Let $I \subset S$ be a binomial prime ideal. Then I is a toric ideal.*

Proof Let $f_{\mathbf{b}}$ and $f_{\mathbf{c}}$ be two binomials. Then

$$f_{\mathbf{b}} f_{\mathbf{c}} = u f_{\mathbf{b}+\mathbf{c}} - \mathbf{x}^{\mathbf{b}^-} f_{\mathbf{c}} - \mathbf{x}^{\mathbf{c}^-} f_{\mathbf{b}} \tag{3.2}$$

for some monomial u. By using that I is a prime ideal, it follows from (3.2) that if $f_{\mathbf{b}}, f_{\mathbf{c}} \in I$, then $f_{\mathbf{b}+\mathbf{c}} \in I$, and of course also $f_{-\mathbf{b}} \in I$, since $f_{-\mathbf{b}} = -f_{\mathbf{b}}$. Thus if $L \subset \mathbb{Z}^n$ consists of all $\mathbf{b} \in \mathbb{Z}^n$ with $f_{\mathbf{b}} \in I$, then $L \subset \mathbb{Z}^n$ is a subgroup of $\mathbb{Z}^n$.

We claim that $\mathbb{Z}^n/L$ is torsionfree. Indeed, let $\mathbf{b} \in L$ with $m\mathbf{b} \in L$ for some integer $m > 1$. We have to show that $\mathbf{b} \in L$. We have that $f_{m\mathbf{b}} \in I$. If $\operatorname{char}(K) = 0$, then we decompose $f_{m\mathbf{b}} = f_{\mathbf{b}}g$, where $g = \mathbf{x}^{(m-1)\mathbf{b}^+} + \mathbf{x}^{(m-2)\mathbf{b}^+}\mathbf{x}^{\mathbf{b}^-} + \cdots + \mathbf{x}^{\mathbf{b}^+}\mathbf{x}^{(m-2)\mathbf{b}^-} + \mathbf{x}^{(m-1)\mathbf{b}^-} \in S$. By using the substitutions $x_i \mapsto 1$ for $i = 1, \ldots, n$, we easily see that $g \notin I$ since all binomials vanish on this substitution. Therefore, $f_{\mathbf{b}} \in I$, since I is a prime ideal. This implies that $\mathbf{b} \in L$.

If $\mathrm{char}(K) = p > 0$, then we write $m = p^e m'$ where $e \geq 0$, $m' \geq 1$ are integers such that p does not divide m'. In this case we decompose f_{mb} as $f_{\mathbf{b}}^{p^e} g'$ where $g' = (\mathbf{x}^{p^e \mathbf{b}^+})^{m'-1} + \cdots + (\mathbf{x}^{p^e \mathbf{b}^-})^{m'-1} \in S$. By using again the substitutions $x_i \mapsto 1$ for $i = 1, \ldots, n$, we get $g' \notin I$. Since I is a prime ideal, it follows that $f_{\mathbf{b}}^{p^e} \in I$, whence $f_{\mathbf{b}} \in I$. This implies that $\mathbf{b} \in L$.

The desired conclusion follows from Theorem 3.17. $\qquad\square$

Let I be the binomial ideal generated by $x^2 - y^2$ in $K[x, y]$. Then I is not a prime ideal, because $x^2 - y^2 = (x + y)(x - y)$, and if $\mathrm{char}(K) = 2$ it is not even a radical ideal, because in this case $x^2 - y^2 = (x - y)^2$.

The next result shows how the toric ideal of a matrix $A \in \mathbb{Z}^{d \times n}$ can be computed by elimination theory.

Let $A = (\mathbf{a}_1, \mathbf{a}_2, \ldots, \mathbf{a}_n) \in \mathbb{Z}^{d \times n}$, and let

$$S[\mathbf{t}^{\pm 1}] = K[x_1, x_2, \ldots, x_n, t_1^{\pm 1}, t_2^{\pm 1}, \ldots, t_d^{\pm 1}]$$

be the polynomial ring in $n + d$ variables and define the ideal J_A of $S[\mathbf{t}^{\pm 1}]$ by

$$J_A = (x_1 - \mathbf{t}^{\mathbf{a}_1}, x_2 - \mathbf{t}^{\mathbf{a}_2}, \ldots, x_n - \mathbf{t}^{\mathbf{a}_n}).$$

Proposition 3.5 *The toric ideal $I_A \subset S$ of A is equal to the intersection of the ideal $J_A \subset S[\mathbf{t}^{\pm 1}]$ with S, i.e.,*

$$I_A = J_A \cap S.$$

Proof If a polynomial $f = f(x_1, x_2, \ldots, x_n) \in S$ belongs to I_A, then $\pi(f) = 0$. Thus $f(\mathbf{t}^{\mathbf{a}_1}, \mathbf{t}^{\mathbf{a}_2}, \ldots, \mathbf{t}^{\mathbf{a}_n}) = 0$. Therefore the Taylor expansion of

$$f((x_1 - \mathbf{t}^{\mathbf{a}_1}) + \mathbf{t}^{\mathbf{a}_1}, (x_2 - \mathbf{t}^{\mathbf{a}_2}) + \mathbf{t}^{\mathbf{a}_2}, \ldots, (x_n - \mathbf{t}^{\mathbf{a}_n}) + \mathbf{t}^{\mathbf{a}_n})$$

with respect to $y_i = x_i - \mathbf{t}^{\mathbf{a}_i}$ for $i = 1, \ldots, n$ yields that $f \in J_A \cap S$. Hence $I_A \subset J_A \cap S$.

On the other hand, if a polynomial $f = f(x_1, x_2, \ldots, x_n) \in S$ belongs to J_A, then there exist elements $g_1, g_2, \ldots, g_n$ belonging to $S[\mathbf{t}^{\pm 1}]$ such that

$$f(\mathbf{x}) = g_1(\mathbf{x}, \mathbf{t})(x_1 - \mathbf{t}^{\mathbf{a}_1}) + \cdots + g_n(\mathbf{x}, \mathbf{t})(x_n - \mathbf{t}^{\mathbf{a}_n}).$$

Then $\pi(f) = f(\mathbf{t}^{\mathbf{a}_1}, \mathbf{t}^{\mathbf{a}_2}, \ldots, \mathbf{t}^{\mathbf{a}_n}) = 0$. Thus $f \in I_A$. Hence $J_A \cap S \subset I_A$. $\qquad\square$

By using Gröbner bases we can compute elimination as described in Section 1.4. Applied to the present case and assuming that all entries of A are nonnegative integers we proceed as follows: let $<_{\mathrm{purelex}}$ denote the pure lexicographic order on $S[\mathbf{t}]$ induced by

$$t_1 > t_2 > \cdots > t_d > x_1 > x_2 > \cdots > x_n,$$

and compute the reduced Gröbner basis $\mathscr{G}$ of J_A with respect to $<_{\text{purelex}}$. Corollary 1.35 then guarantees that $\mathscr{G} \cap S$ is the reduced Gröbner basis of I_A with respect to $<_{\text{purelex}}$. In particular, $\mathscr{G} \cap S$ is a system of generators of I_A.

We come back to our example $A = \begin{pmatrix} 1 & 3 & 2 \\ 0 & 2 & 1 \end{pmatrix}$. Then

$$J_A = (x_1 - t_1, x_2 - t_1^3 t_2^2, x_3 - t_1^2 t_2).$$

Computing the Gröbner basis of J_A with respect to the lexicographic order induced by $t_1 > t_2 > x_1 > x_2 > x_3$ we obtain: $x_1 x_2 - x_3^2, t_2 x_3^3 - x_2^2, t_2 x_1 x_3 - x_2, t_2 x_1^2 - x_3, t_1 - x_1$. Thus $I_A = (x_1 x_2 - x_3^2)$, as observed before.

Problems

3.1 Show that

$$A = \begin{pmatrix} 2 & 0 & 3 & 4 \\ 1 & -2 & 1 & -1 \\ 3 & 0 & 5 & 1 \\ 7 & -1 & 12 & 5 \end{pmatrix}$$

is a configuration matrix.

3.2 Let $A \in \mathbb{Z}^{d \times n}$. Then I_A is a principal ideal if and only if rank $A = n - 1$.

3.3 Let $A = (3, 4, 5) \in \mathbb{Z}^{1 \times 3}$. Compute I_A.

3.4 Let $I \subset K[x_1, \ldots, x_n, y_1, \ldots, y_n]$ be the ideal generated by a set $\mathscr{S}$ of 2-minors of the $2 \times n$-matrix $X = \begin{pmatrix} x_1 & \cdots & x_n \\ y_1 & \cdots & y_n \end{pmatrix}$. Show that I is a prime ideal if and only if $\mathscr{S}$ is the set of all 2-minors of X.

3.5 Let $\text{char}(K) = 0$ and let $\mathbf{b} \in \mathbb{Z}^n$. Then $I = (f_{\mathbf{b}}) \subset S$ is a radical ideal. In other words, if $g \in S$ and $g^k \in I$ for some k, then $g \in I$.

3.6 Let $\mathbf{b}_1, \ldots, \mathbf{b}_r \in \mathbb{Z}^n$ be $\mathbb{Q}$-linearly independent vectors. Then $f_{\mathbf{b}_1}, \ldots, f_{\mathbf{b}_r}$ is a regular sequence.

3.2 Gröbner Bases of Binomial Ideals

Gröbner bases of binomial ideals have many applications in combinatorics and algebraic statistics. One of the nice properties is that reduced Gröbner bases of binomial ideals consist again of binomials.

Theorem 3.6 *Let I be a binomial ideal of S. Then the reduced Gröbner basis of I with respect to an arbitrary monomial order on S consists of binomials.*

Proof In general, if f and g are binomials, then their S-polynomial $S(f, g)$ is again a binomial. It then follows from the argument done in the proof of Theorem 1.20 that a remainder of a binomial with respect to a set of binomials can be chosen as a binomial, see Problem 3.7. Thus, applying Buchberger's algorithm to a system of generators of a binomial ideal I consisting of a finite number of binomials, we obtain a minimal Gröbner basis $\mathcal{G} = \{g_1, g_2, \ldots, g_s\}$ of I, where each g_i is a binomial.

Let $g_i = u_i - v_i$, where u_i and v_i are monomials with $u_i = \text{in}_<(g_i)$. Recall that $\mathcal{G}$ is reduced if v_i cannot be divided by u_j for $i \neq j$. Suppose that $\mathcal{G}$ is not reduced and, say, v_2 is divided by u_1. Let $v_2 = wu_1$, where w is a monomial. We then replace g_2 with $g_2' = g_2 + wg_1 = u_2 - v_2'$, where $v_2' = wv_1$. Then $\{g_1, g_2', g_3, \ldots, g_s\}$ is a minimal Gröbner basis of I consisting of binomials with $v_2' < v_2$, since $v_1 < u_1$. Thus, after a finite number of steps, we obtain a reduced Gröbner basis of I consisting of binomials. $\square$

There is no analogue of Theorem 3.6 for ideals generated by polynomials with more than two terms. For example, the ideal $I = (x_1 + x_2 + x_3, x_1 + x_4 + x_5)$ has the reduced Gröbner basis $x_1 + x_4 + x_5, x_2 + x_3 - x_4 - x_5$ with respect to the reverse lexicographic order induced by $x_1 > x_2 > \cdots > x_5$. Though all generators of I only admit three terms, a polynomial with four terms belongs to the reduced Gröbner basis of I.

Binomials in a binomial ideal can be written as linear combinations of the binomial generators with coefficients which are monomials with scalars belonging to $\mathbb{Z}1_K$. Indeed, we have

Lemma 3.7 *Let $I \subset K[x_1, \ldots, x_n]$ be an ideal generated by the binomials $f_1, \ldots, f_r$. Let $\mathbf{x^u} - \mathbf{x^v}$ be a binomial belonging to I. Then there exists an expression*

$$\mathbf{x^u} - \mathbf{x^v} = \sum_{k=1}^{s} z_k \mathbf{x^{w_k}} f_{i_k},$$

where $z_k \in \mathbb{Z}1_K$, $\mathbf{w}_k \in \mathbb{Z}_{\geq 0}^n$, and $1 \leq i_k \leq r$ for $k = 1, 2, \ldots, s$.

Proof Let $\mathcal{G} = \{f_1, \ldots, f_{r'}\}$ $(r \leq r')$ be a Gröbner basis of I with respect to a given monomial order $<$, obtained by applying Buchberger's algorithm to $\{f_1, \ldots, f_r\}$. The Gröbner basis $\mathcal{G}$ may not be reduced. By the argument in the proof of Theorem 3.6, $\mathcal{G}$ consists of binomials. Let $f_i = \mathbf{x^{u_i}} - \mathbf{x^{v_i}}$ $(1 \leq i \leq r')$. We may assume that $\text{in}_<(f_i) = \mathbf{x^{u_i}}$, since we can replace f_i with $-f_i$ if needed. We will show that f_i $(r < i \leq r')$ has a presentation as stated in the lemma. Let $r < j \leq r'$. Suppose that $f_1, \ldots, f_{j-1}$ have such a presentation, and that f_j is a remainder of the S-polynomial $g = S(f_\mu, f_\nu)$ $(1 \leq \mu < \nu \leq j - 1)$ with respect to $f_1, \ldots, f_{j-1}$. Then, $g = \mathbf{x^a} f_\mu - \mathbf{x^b} f_\nu$ for some $\mathbf{a}, \mathbf{b} \in \mathbb{Z}_{\geq 0}^n$. Moreover, since $f_1, \ldots, f_{j-1}$, and g are binomials, it follows from the proof of Theorem 1.20, that

$$\mathbf{x}^{\mathbf{a}} f_\mu - \mathbf{x}^{\mathbf{b}} f_\nu = \sum_{\ell=1}^{t} z'_\ell \mathbf{x}^{\mathbf{w}'_\ell} f_{j_\ell} + f_j,$$

where $1 \leq j_\ell \leq j - 1$, $z'_\ell \in \mathbb{Z}$ and $\mathbf{w}'_\ell \in \mathbb{Z}_{\geq 0}^n$ for $\ell = 1, 2, \ldots, t$. Thus,

$$f_j = \mathbf{x}^{\mathbf{a}} f_\mu - \mathbf{x}^{\mathbf{b}} f_\nu - \sum_{\ell=1}^{t} z'_\ell \mathbf{x}^{\mathbf{w}'_\ell} f_{j_\ell}$$

has the desired presentation.

Now let $f = \mathbf{x}^{\mathbf{u}} - \mathbf{x}^{\mathbf{v}} \in I$ be arbitrary and suppose that f does not have the desired presentation, and that $\mathrm{in}_<(\mathbf{x}^{\mathbf{u}} - \mathbf{x}^{\mathbf{v}}) = \mathbf{x}^{\mathbf{u}}$ is minimal among such binomials. Since $\mathscr{G}$ is a Gröbner basis, there exists a binomial f_i ($1 \leq i \leq r$) such that $\mathrm{in}_<(f_i)$ divides $\mathrm{in}_<(f)$. Then, $f = \mathbf{x}^{\mathbf{w}} f_i + \mathbf{x}^{\mathbf{u}'} - \mathbf{x}^{\mathbf{v}}$ for some $\mathbf{w}, \mathbf{u}' \in \mathbb{Z}_{\geq 0}^n$ with $\mathbf{x}^{\mathbf{u}'} < \mathbf{x}^{\mathbf{u}}$. We may assume that $\mathbf{x}^{\mathbf{u}'} - \mathbf{x}^{\mathbf{v}} \neq 0$. Since $\mathbf{x}^{\mathbf{u}'} - \mathbf{x}^{\mathbf{v}} \in I$ satisfies $\mathrm{in}_<(\mathbf{x}^{\mathbf{u}'} - \mathbf{x}^{\mathbf{v}}) < \mathbf{x}^{\mathbf{u}}$, by the assumption for f, the binomial $\mathbf{x}^{\mathbf{u}'} - \mathbf{x}^{\mathbf{v}}$ has the desired presentation. Thus, f has the desired presentation as well, a contradiction.   □

The preceding lemma can be improved as follows.

Lemma 3.8 *Let $I \subset K[x_1, \ldots, x_n]$ be an ideal generated by the binomials $f_1, \ldots, f_r$. Let $\mathbf{x}^{\mathbf{u}} - \mathbf{x}^{\mathbf{v}}$ be a binomial belonging to I. Then, there exists an expression*

$$\mathbf{x}^{\mathbf{u}} - \mathbf{x}^{\mathbf{v}} = \sum_{k=1}^{s} \epsilon_k \mathbf{x}^{\mathbf{w}_k} f_{i_k},$$

where $\epsilon_k \in \{\pm 1\}$, $\mathbf{w}_k \in \mathbb{Z}_{\geq 0}^n$, and $1 \leq i_k \leq r$ for $k = 1, 2, \ldots, s$, and where $\mathbf{x}^{\mathbf{w}_p} f_{i_p} \neq \mathbf{x}^{\mathbf{w}_q} f_{i_q}$ for all $1 \leq p < q \leq s$.

Proof By Lemma 3.7, there exists an expression

$$\mathbf{x}^{\mathbf{u}} - \mathbf{x}^{\mathbf{v}} = \sum_{k=1}^{s} z_k \mathbf{x}^{\mathbf{w}_k} f_{i_k},$$

where $z_k \in \mathbb{Z}$, $\mathbf{w}_k \in \mathbb{Z}_{\geq 0}^n$, and $1 \leq i_k \leq r$ for $k = 1, 2, \ldots, s$. Then we can rewrite it as

$$\mathbf{x}^{\mathbf{u}} - \mathbf{x}^{\mathbf{v}} = \sum_{j=1}^{t} (\mathbf{x}^{\mathbf{a}_j} - \mathbf{x}^{\mathbf{b}_j}),$$

where each $\mathbf{x}^{\mathbf{a}_j} - \mathbf{x}^{\mathbf{b}_j}$ coincides with $\epsilon \mathbf{x}^{\mathbf{w}_k} f_{i_k}$ for some $\epsilon \in \{\pm 1\}$ and $1 \leq k \leq s$ such that

$$\mathbf{u} = \mathbf{a}_1, \ \mathbf{b}_1 = \mathbf{a}_2, \ \mathbf{b}_2 = \mathbf{a}_3, \ \ldots, \ \mathbf{b}_{t-1} = \mathbf{a}_t, \ \mathbf{b}_t = \mathbf{v}.$$

Suppose that $\mathbf{x}^{\mathbf{a}_p} - \mathbf{x}^{\mathbf{b}_p} = \mathbf{x}^{\mathbf{a}_q} - \mathbf{x}^{\mathbf{b}_q}$ for some $1 \leq p < q \leq t$. Then we have a sequence

$$\mathbf{b}_{p+1} = \mathbf{a}_{p+2}, \ \mathbf{b}_{p+2} = \mathbf{a}_{p+3}, \ \ldots, \ \mathbf{b}_{q-1} = \mathbf{a}_q, \ \mathbf{b}_q = \mathbf{b}_p = \mathbf{a}_{p+1}.$$

Hence we have

$$\sum_{j=p+1}^{q} (\mathbf{x}^{\mathbf{a}_j} - \mathbf{x}^{\mathbf{b}_j}) = 0.$$

Thus we have another expression

$$\mathbf{x}^{\mathbf{u}} - \mathbf{x}^{\mathbf{v}} = \sum_{j=1}^{p} (\mathbf{x}^{\mathbf{a}_j} - \mathbf{x}^{\mathbf{b}_j}) + \sum_{j=q+1}^{t} (\mathbf{x}^{\mathbf{a}_j} - \mathbf{x}^{\mathbf{b}_j}).$$

By repeating the same argument, we obtain the desired presentation.   □

Corollary 3.9 *Let K be a field, and let $f_1, \ldots, f_s$ be any set of binomial generators of the binomial ideal $I \subset K[x_1, \ldots, x_n]$. Let L be any other field, and let $J \subset L[x_1, \ldots, x_n]$ be the ideal generated by all binomials of I. Then $f_1, \ldots, f_s$ is as well a system of generators of J.*

Given a subset of vectors $\mathscr{B} \subset \mathbb{Z}^m$, one defines $G_{\mathscr{B}}$ to be the graph with the vertex set $\mathbb{Z}^m_{\geq 0}$ such that two vertices $\mathbf{a}$ and $\mathbf{c}$ are adjacent in $G_{\mathscr{B}}$ if $\mathbf{a} - \mathbf{c} \in \pm\mathscr{B}$. The vectors $\mathbf{a}$ and $\mathbf{c}$ are said to be *connected via $\mathscr{B}$* if they belong to the same connected component of $G_{\mathscr{B}}$. This is the case if and only if there exist $\mathbf{u}_1, \ldots, \mathbf{u}_k \in \mathscr{B}$ such that $\mathbf{a} + \mathbf{u}_1 + \cdots + \mathbf{u}_i \in \mathbb{Z}^m_{\geq 0}$ for $i = 1, \ldots, k$ and $\mathbf{c} = \mathbf{a} + \mathbf{u}_1 + \cdots + \mathbf{u}_k$.

The binomial ideal $I_{\mathscr{B}}$ in the polynomial ring $S = K[x_1, \ldots, x_m]$ is defined to be the ideal

$$I_{\mathscr{B}} = (\mathbf{x}^{\mathbf{b}^+} - \mathbf{x}^{\mathbf{b}^-} : \mathbf{b} \in \mathscr{B}).$$

Corollary 3.10 *Let $\mathbf{a}, \mathbf{b} \in \mathbb{Z}^m_{\geq 0}$. Then $\mathbf{a}$ and $\mathbf{b}$ are connected via $\mathscr{B}$, if and only if $\mathbf{x}^{\mathbf{a}} - \mathbf{x}^{\mathbf{b}} \in I_{\mathscr{B}}$.*

Proof Suppose first that $\mathbf{a}$ and $\mathbf{b}$ are connected via $\mathscr{B}$. Then there exist $\mathbf{u}_1, \ldots, \mathbf{u}_k \in \pm\mathscr{B}$ such that

$$\mathbf{a} + \mathbf{u}_1 + \mathbf{u}_2 + \cdots + \mathbf{u}_i \in \mathbb{Z}^m_{\geq 0} \quad \text{for all} \quad i = 1, \ldots, k,$$

and

$$\mathbf{b} = \mathbf{a} + \mathbf{u}_1 + \mathbf{u}_2 + \cdots + \mathbf{u}_k.$$

We show by induction on k that $\mathbf{x}^{\mathbf{a}} - \mathbf{x}^{\mathbf{b}} \in I_{\mathscr{B}}$. If $k = 1$, then $\mathbf{b} = \mathbf{a} + \mathbf{u}_1 \in \mathbb{Z}_{\geq 0}^m$. This implies that $\mathbf{u}_1 \leq \mathbf{b}$, componentwise. Let $\mathbf{c} = \mathbf{a} - \mathbf{u}_1^-$. Then $\mathbf{c} = \mathbf{b} - \mathbf{u}_1^+ \in \mathbb{Z}_{\geq 0}^m$, and hence

$$\mathbf{x}^{\mathbf{a}} - \mathbf{x}^{\mathbf{b}} = \mathbf{x}^{\mathbf{c}}(\mathbf{x}^{\mathbf{u}_1^-} - \mathbf{x}^{\mathbf{u}_1^+}) \in I_{\mathscr{B}}.$$

Now suppose $k > 1$ and that the assertion is true for $i < k$. Since $\mathbf{a} + \mathbf{u}_1$ and $\mathbf{b}$ are connected via $k-1$ edges of $G_{\mathscr{B}}$, our induction hypothesis implies that $\mathbf{x}^{\mathbf{a}+\mathbf{u}_1} - \mathbf{x}^{\mathbf{b}} \in I_{\mathscr{B}}$, and this implies that

$$\mathbf{x}^{\mathbf{a}} - \mathbf{x}^{\mathbf{b}} = (\mathbf{x}^{\mathbf{a}} - \mathbf{x}^{\mathbf{a}+\mathbf{u}_1}) + (\mathbf{x}^{\mathbf{a}+\mathbf{u}_1} - \mathbf{x}^{\mathbf{b}}) \in I_{\mathscr{B}}.$$

Conversely, suppose that $\mathbf{x}^{\mathbf{a}} - \mathbf{x}^{\mathbf{b}} \in I_{\mathscr{B}}$. Then Lemma 3.7 implies that there exist $\mathbf{u}_1, \ldots, \mathbf{u}_t \in \pm\mathscr{B}$ and monomials $\mathbf{x}^{\mathbf{c}_i}$ such that

$$\mathbf{x}^{\mathbf{a}} - \mathbf{x}^{\mathbf{b}} = \sum_{i=1}^{t} \mathbf{x}^{\mathbf{c}_i}(\mathbf{x}^{\mathbf{u}_i^+} - \mathbf{x}^{\mathbf{u}_i^-}).$$

We show by induction on t that $\mathbf{a}$ is connected to $\mathbf{b}$ via $\mathscr{B}$. If $t = 1$, then $\mathbf{x}^{\mathbf{a}} - \mathbf{x}^{\mathbf{b}} = \mathbf{x}^{\mathbf{c}_1}(\mathbf{x}^{\mathbf{u}_1^+} - \mathbf{x}^{\mathbf{u}_1^-})$. Therefore, $\mathbf{a} = \mathbf{c}_1 + \mathbf{u}_1^+$ and $\mathbf{b} = \mathbf{c}_1 + \mathbf{u}_1^-$, so that

$$\mathbf{a} - \mathbf{u}_1 = \mathbf{c}_1 + \mathbf{u}_1^- = \mathbf{b},$$

which means that $\mathbf{a}$ and $\mathbf{b}$ are connected via $\mathscr{B}$. Now let $t > 1$. Then there exists an integer i, say $i = 1$, such that $\mathbf{x}^{\mathbf{a}} = \mathbf{x}^{\mathbf{c}_1 + \mathbf{u}_1^+}$. It follows that

$$\mathbf{x}^{\mathbf{c}_1 + \mathbf{u}_1^-} - \mathbf{x}^{\mathbf{b}} = \sum_{i=2}^{t} \mathbf{x}^{\mathbf{c}_i}(\mathbf{x}^{\mathbf{u}_i^+} - \mathbf{x}^{\mathbf{u}_i^-}).$$

Hence, our induction hypothesis implies that $\mathbf{c}_1 + \mathbf{u}_1^-$ and $\mathbf{b}$ are connected via $\mathscr{B}$. Since $\mathbf{a}$ and $\mathbf{c}_1 + \mathbf{u}_1^-$ are connected via $\mathscr{B}$, the desired conclusion follows. $\square$

Theorem 3.11 *Let I be a binomial ideal of S and $\{g_1, \ldots, g_s\}$ a set of nonzero binomials in I. Then, $\{g_1, \ldots, g_s\}$ is a Gröbner basis of I with respect to a monomial order $<$, if and only if for all binomials $0 \neq u - v \in I$ either u or v belongs to $(\mathrm{in}_<(g_1), \ldots, \mathrm{in}_<(g_s))$.*

Proof Let $f = u - v \in I$. Then $\mathrm{in}_<(f) \in \mathrm{in}_<(I)$. Since $\mathrm{in}_<(f)$ is equal to u or to v, it follows that either u or v belongs to $\mathrm{in}_<(I)$. Thus, if both u and v do not belong to $(\mathrm{in}_<(g_1), \ldots, \mathrm{in}_<(g_s))$, then this ideal cannot be equal to $\mathrm{in}_<(I)$.

On the other hand, suppose that $\{g_1, \ldots, g_s\}$ is not a Gröbner basis of I. Let $\{g_1', \ldots, g_s'\}$ be the reduced Gröbner basis of I with respect to $<$. Since $\{g_1, \ldots, g_s\}$ is not a Gröbner basis of I, we have

$$\mathrm{in}_<(I) = (\mathrm{in}_<(g_1'), \ldots, \mathrm{in}_<(g_t')) \supsetneq (\mathrm{in}_<(g_1), \ldots, \mathrm{in}_<(g_s)).$$

Hence there exists $1 \leq i \leq t$ such that $\mathrm{in}_<(g_i')$ does not belong to $(\mathrm{in}_<(g_1), \ldots, \mathrm{in}_<(g_s))$. By Theorem 3.6, g_i' is a binomial. Let $g_i' = u - v$ with $\mathrm{in}_<(g_i') = u$. Since $\{g_1', \ldots, g_t'\}$ is reduced, the monomial v does not belong to $\mathrm{in}_<(I) \supset (\mathrm{in}_<(g_1), \ldots, \mathrm{in}_<(g_s))$. Thus none of the monomials u and v in the binomial $g_i' \in I$ belong to $(\mathrm{in}_<(g_1), \ldots, \mathrm{in}_<(g_s))$. $\square$

Let I be a binomial ideal of $K[\mathbf{x}] = K[x_1, x_2, \ldots, x_n]$. A nonzero binomial $f = u - v \in I$ is called *primitive*, if there is no nonzero binomial $g = u' - v' \in I$ with $g \neq f$ such that $u'|u$ and $v'|v$. The set of all primitive binomials of I is called the *Graver basis* of I.

Proposition 3.12 *Let I be a binomial ideal. Then the Graver basis of I is finite.*

Proof Let $\mathscr{S}$ be the set of all monomials $\mathbf{x^a y^b}$ and $\mathbf{x^b y^a}$ such that $\mathbf{x^a} - \mathbf{x^b}$ belongs to the Graver basis of I. By the definition of primitive binomials, there are no divisibility relations among distinct elements of $\mathscr{S}$. Hence, by Dickson's Lemma, $\mathscr{S}$ is finite. Thus the Graver basis is finite. $\square$

Theorem 3.13 *Let I be a binomial ideal and $\mathscr{G}$ its reduced Gröbner basis with respect to a given monomial order. Then any binomial $f \in \mathscr{G}$ is a primitive binomial.*

Proof Suppose that the binomial $f = u - v$ belongs to the reduced Gröbner basis $\mathscr{G}$ of I with respect to the given monomial order $<$, and that the initial monomial of f is u. Suppose that f is not primitive. Then there exists a binomial $g = u' - v' \in I$ with $g \neq f$ such that $u'|u$ and $v'|v$. If the initial term of g is v', then it contradicts the hypothesis that $\mathscr{G}$ is reduced. Hence, the initial term of g is u'. Since the binomial f belongs to the reduced Gröbner basis, its initial monomial u belongs to a minimal set of generators of $\mathrm{in}_<(I)$. Thus, we have $u = u'$. Then, $g - f = v - v'$ is a binomial belonging to I. Since v' divides v and $v \neq v'$, it follows that $\mathrm{in}_<(f - g) = v$, contradicting the assumption that $\mathscr{G}$ is a reduced Gröbner basis. $\square$

Consider the binomial ideal $I = (x^2 - yz, x - y)$. The binomial $x^2 - yz$ is a minimal generator of I but is not primitive. Hence by the previous theorem it cannot belong to any reduced Gröbner basis of I. For example, if we compute the reduced Gröbner basis of I with respect to the lexicographic order induced by $x > y > z$, we obtain $y^2 - yz, x - y$ (which in this case is also a minimal set of generators).

Corollary 3.14 *The reduced Gröbner basis of a binomial ideal is contained in its Graver basis.*

Recall that the union of the reduced Gröbner bases with respect to all possible monomial orders is finite (Corollary 1.42) and called the *universal Gröbner basis* of I. Corollary 3.14 says that the universal Gröbner basis of I is a subset of the Graver basis of I. Since the Graver basis of a binomial ideal is finite, we have another proof for the fact that the universal Gröbner basis is finite.

For the case of graded toric ideals, the Graver basis contains another important set of binomials.

Proposition 3.15 Let $A \in \mathbb{Z}^{d \times n}$ be a configuration. Then any minimal set of binomial generators of I_A is contained in its Graver basis.

Proof By Proposition 3.3, I_A is a graded ideal. Suppose that a binomial generator $f = u - v$ of I_A is not primitive. Then there exists a nonzero binomial $g = u' - v' \in I_A$ with $g \neq f$ such that $u'|u$ and $v'|v$. Let $u'' = u/u'$ and $v'' = v/v'$, and set $h = u'' - v''$. Then $f = u''g + v'h$. Since f and g belong to I_A it follows that $v'h \in I_A$, and since I_A is a prime ideal and $v' \notin I_A$ we see that $h \in I_A$. Since $\deg(g), \deg(h) < \deg(f)$, we conclude that f is not a minimal generator of I_A, a contradiction.   □

In Proposition 3.15, we cannot omit the hypothesis that I_A is graded. For example, if $A = (1, -1) \in \mathbb{Z}^{1 \times 2}$, then $I_A = (x_1 x_2 - 1) = (x_1^2 x_2^2 - 1, x_1^3 x_2^3 - 1)$. However, this minimal set of binomial generators $\{x_1^2 x_2^2 - 1, x_1^3 x_2^3 - 1\}$ consists of nonprimitive binomials.

Problems

3.7 Let f be a binomial and $\{g_1, \ldots, g_s\}$ a set of binomials.

(a) Show that a remainder of f with respect to $\{g_1, \ldots, g_s\}$ is a binomial if it is obtained by the procedure given in the proof of Theorem 1.19.
(b) Give an example for which a remainder of f with respect to $\{g_1, \ldots, g_s\}$ is not a binomial. (Hint: Consider for example the binomials $f = x_1 x_2 - x_3 x_4$, $g_1 = x_1 x_2 - x_5 x_6$, $g_2 = x_1 x_2 - x_7 x_8$ and consider a lexicographic order induced by $x_1 > \cdots > x_8$.)

3.8 Let I be a binomial ideal.

(a) Does any system of binomial generators of I contain at least one primitive binomial?
(b) Show that I can be (minimally) generated by primitive binomials.

3.9

(a) Let $I_n \subset K[x_1, \ldots, x_n, y_1, \ldots, y_n]$ be the ideal generated by the set $\mathscr{S}$ of all 2-minors of the $2 \times n$-matrix $X = \begin{pmatrix} x_1 & \cdots & x_n \\ y_1 & \cdots & y_n \end{pmatrix}$. Show that $\mathscr{S}$ is a reduced Gröbner basis of I_n with respect to $x_1 > \cdots > x_n > y_1 > \cdots > y_n$.
(b) Show that the minors generating I_3 form a Graver basis of I_3.

3.10 Let $A = (3, 4, 5) \in \mathbb{Z}^{1 \times 3}$. Compute the Graver basis of I_A.

3.3 Lattice Ideals and Lattice Basis Ideals

In this section we give another interpretation of toric ideals. A subgroup L of $\mathbb{Z}^n$ is called a *lattice*. Recall from basic algebra that L is a free abelian group of rank $m \leq n$. The binomial ideal $I_L \subset S$ generated by the binomials $f_\mathbf{b}$ with $\mathbf{b} \in L$ is called the *lattice ideal* of L.

Consider for example, the lattice $L \subset \mathbb{Z}^3$ with basis $(1, 1, 1)^t$, $(1, 0, -1)^t$. Then $\mathbf{b} \in L$ if and only if $A\mathbf{b} = 0$, where $A = (1, -2, 1)$. Thus in this case we have that I_L is a toric ideal, namely I_A. On the other hand, any toric ideal is a lattice ideal. Indeed, we have

Proposition 3.16 *Let $A \in \mathbb{Z}^{d \times n}$. Then the toric ideal I_A is equal to the lattice ideal I_L, where $L = \{\mathbf{b}: A\mathbf{b} = 0\}$.*

Proof By Theorem 3.2 we know that I_A is generated by the binomials $f_\mathbf{b}$ with $A\mathbf{b} = 0$. $\square$

Not all lattice ideals are toric ideals. The simplest such example is the ideal I_L for $L = 2\mathbb{Z} \subset \mathbb{Z}$. Here $I_L = (x^2 - 1)$. If I_L would be a toric ideal it would be a prime ideal. But $x^2 - 1 = (x + 1)(x - 1)$, and so I_L is not a prime ideal.

We have the following general result:

Theorem 3.17 *Let $L \subset \mathbb{Z}^n$ be a lattice. The following conditions are equivalent:*

 (i) *the abelian group $\mathbb{Z}^n/L$ is torsionfree;*
(ii) *I_L is a prime ideal;*

The equivalent conditions hold, if and only if I_L is a toric ideal.

Proof (i) $\Rightarrow$ (ii): Since $\mathbb{Z}^n/L$ is torsionfree, there exists an embedding $\mathbb{Z}^n/L \subset \mathbb{Z}^d$ for some d. Let $\mathbf{e}_1, \ldots, \mathbf{e}_n$ be the canonical basis of $\mathbb{Z}^n$. Then for $i = 1, \ldots, n$, $\mathbf{e}_i + L$ is mapped to $\mathbf{a}_i \in \mathbb{Z}^d$ via this embedding. It follows that $\sum_{i=1}^n b_i \mathbf{a}_i = 0$ if and only if $\mathbf{b} = (b_1, \ldots, b_n)^t \in L$. In other words, $\mathbf{b} \in L$ if and only if $A\mathbf{b} = 0$, where A is the matrix whose column vectors are $\mathbf{a}_1, \ldots, \mathbf{a}_n$. Therefore, Theorem 3.2 implies that I_L is the toric ideal of A, and hence a prime ideal.

(ii) $\Rightarrow$ (i) is already shown in the proof of Theorem 3.4.

In the proof of (i) $\Rightarrow$ (ii) we have seen that I_L is a toric ideal if $\mathbb{Z}^n/L$ is torsionfree. $\square$

Let I and J be two ideals. The *saturation* of I with respect to J is the ideal $I : J^\infty$, where by definition $I : J^\infty = \bigcup_k (I : J^k)$.

Proposition 3.18 *Let $I \subset S$ be a binomial ideal. Then $I : (\prod_{i=1}^n x_i)^\infty$ is also a binomial ideal.*

Proof We set $x = \prod_{i=1}^{n} x_i$. Then

$$I : (\prod_{i=1}^{n} x_i)^{\infty} = I S_x \cap S, \tag{3.3}$$

see Problem 3.15. Here S_x denotes the localization with respect to the multiplicatively closed set $\{1, x, x^2, \ldots\}$ consisting of the powers of x.

Consider the polynomial ring $T = K[x_1, \ldots, x_n, y_1, \ldots, y_n]$ over K in the variables $x_1, \ldots, x_n, y_1, \ldots, y_n$. Then $T/(x_1 y_1 - 1, \ldots, x_n y_n - 1) \cong S_x$, and hence $T/(I, x_1 y_1 - 1, \ldots, x_n y_n - 1)T \cong S_x/I S_x$. Therefore, $I S_x \cap S = (I, x_1 y_1 - 1, \ldots, x_n y_n - 1)T \cap S$. By Corollary 1.35, Theorem 3.2, and Theorem 3.6 it follows that $I : (\prod_{i=1}^{n} x_i)^{\infty}$ is a binomial ideal. $\square$

Theorem 3.19 *Let $I \subset S$ be a binomial ideal. Then $I : (\prod_{i=1}^{n} x_i)^{\infty}$ is a lattice ideal.*

Proof Let

$$L = \{\mathbf{b} \in \mathbb{Z}^n : \ u f_{\mathbf{b}} \in I \text{ for some monomial } u\}.$$

We claim that $L \subset \mathbb{Z}^n$ is a lattice. Indeed, if $\mathbf{b} \in L$, then $u f_{\mathbf{b}} \in I$ for some monomial u and hence $u f_{-\mathbf{b}} = -u f_{\mathbf{b}} \in I$. This shows that $-\mathbf{b} \in L$. Now let $\mathbf{c} \in L$ be another vector. Then there exists a monomial v such that $v f_{\mathbf{c}} \in I$. By using Formula (3.2) we get

$$(u f_{\mathbf{b}})(v f_{\mathbf{c}}) = uv(w f_{\mathbf{b}+\mathbf{c}} - \mathbf{x}^{\mathbf{b}^-} f_{\mathbf{c}} - \mathbf{x}^{\mathbf{c}^-} f_{\mathbf{b}}) = uvw f_{\mathbf{b}+\mathbf{c}} - \mathbf{x}^{\mathbf{b}^-} u(v f_{\mathbf{c}}) - \mathbf{x}^{\mathbf{c}^-} v(u f_{\mathbf{b}}).$$

It follows from this equation that $\mathbf{b} + \mathbf{c} \in L$. This proves the claim.

Next we claim that $I : (\prod_{i=1}^{n} x_i)^{\infty} = I_L$. By the definition of L it follows that $I_L \subset I : (\prod_{i=1}^{n} x_i)^{\infty}$. This implies that $I_L : (\prod_{i=1}^{n} x_i)^{\infty} \subset I : (\prod_{i=1}^{n} x_i)^{\infty}$. On the other hand, since $I \subset I_L$ it follows that $I : (\prod_{i=1}^{n} x_i)^{\infty} \subset I_L : (\prod_{i=1}^{n} x_i)^{\infty}$, and hence we conclude that $I : (\prod_{i=1}^{n} x_i)^{\infty} = I_L : (\prod_{i=1}^{n} x_i)^{\infty}$. Thus it suffices to show that $I_L : (\prod_{i=1}^{n} x_i)^{\infty} = I_L$. But this follows from Theorem 3.20. $\square$

Theorem 3.20 *Let $L \subset \mathbb{Z}^n$ be a lattice. Then $I_L : (\prod_{i=1}^{n} x_i)^{\infty} = I_L$.*

Proof We only need to show that $I_L : (\prod_{i=1}^{n} x_i)^{\infty} \subset I_L$. Let $f \in I_L : (\prod_{i=1}^{n} x_i)^{\infty}$. By Proposition 3.18 we may assume that f is a binomial, and we may further assume that $f = f_{\mathbf{b}}$ for some $\mathbf{b} \in \mathbb{Z}^n$. We want to show that $\mathbf{b} \in L$. Since $f_{\mathbf{b}} \in I_L : (\prod_{i=1}^{n} x_i)^{\infty}$, it follows that $1 - \mathbf{x}^{\mathbf{b}} \in I_L S_x$, where $x = \prod_{i=1}^{n} x_i$. Observe that $I_L S_x$ is generated by the binomials $1 - \mathbf{x}^{\mathbf{c}}$ with $\mathbf{c} \in L$. Therefore, $S_x/I_L S_x$ is isomorphic to the group ring $K[\mathbb{Z}^n/L]$ which admits the K-basis consisting of the elements of the group $G = \mathbb{Z}^n/L$ with group operations in multiplicative notation. In particular, the unit element 1_G of G is equal to $0 + L$. Multiplication of elements of $K[G]$ is defined by linear extension of the multiplication on G. The isomorphism $S_x/I_L S_x \to K[G]$ is given as follows: let $\mathbf{x}^{\mathbf{c}} \in S_x$ with $\mathbf{c} \in \mathbb{Z}^n$. Then $\mathbf{x}^{\mathbf{c}} + I_L S_x$ is mapped to $g = \mathbf{c} + L$ in $K[G]$.

Let $g = \mathbf{b} + L$. Then $1_G - g = 0$ in $K[\mathbb{Z}^n/L]$ because $1 - \mathbf{x}^{\mathbf{b}} \in I_L S_x$. Due to the above isomorphism, this implies that $\mathbf{b} + L = 0 + L$, and hence $\mathbf{b} \in L$, as desired. $\square$

Corollary 3.21 *Let $I \subset S$ be a binomial ideal. Then I is a lattice ideal if and only if $I : (\prod_{i=1}^n x_i)^\infty = I$.*

Let $L \subset \mathbb{Z}^n$ be a lattice and let $\mathscr{B} = \mathbf{b}_1, \ldots, \mathbf{b}_m$ be a basis of the free abelian group L. The ideal $I_{\mathscr{B}}$ is called a *lattice basis ideal* of L. In general, $I_{\mathscr{B}} \neq I_L$. Consider for example, $A = (3, 4, 5) \in \mathbb{Z}^{1 \times 3}$. The toric ideal I_A is the lattice ideal of the lattice L with basis $\mathscr{B} = (2, 1, -2), (1, -2, 1)$. Then $I_{\mathscr{B}} = (x^2 y - z^2, xz - y^2)$, while I_L contains the binomial $x^3 - yz$ which does not belong to $I_{\mathscr{B}}$.

However one has

Corollary 3.22 *Let $\mathscr{B}$ be a basis of the lattice L. Then $I_{\mathscr{B}} : (\prod_{i=1}^n x_i)^\infty = I_L$.*

Proof By Theorem 3.19 there exists a lattice $L' \subset \mathbb{Z}^n$ such that $I_{\mathscr{B}} : (\prod_{i=1}^n x_i)^\infty = I_{L'}$ and Theorem 3.20 implies that $I_{L'} = I_{\mathscr{B}} : (\prod_{i=1}^n x_i)^\infty \subset I_L : (\prod_{i=1}^n x_i)^\infty = I_L$.

It remains to be shown that $I_L \subset I_{L'}$. Let $v \in L$. We will show that $f_v \in I_{L'}$. Let $\mathscr{B} = \mathbf{b}_1, \ldots, \mathbf{b}_r$. Then $\mathbf{v} = \sum_{i=1}^r z_i \mathbf{b}_i$ with $z_i \in \mathbb{Z}$. We set $c(\mathbf{v}) = \sum_{i=1}^r |z_i|$ and show by induction on $c(\mathbf{v})$ that $f_v \in I_{L'}$. If $c(\mathbf{v}) = 1$, then $\mathbf{v} = \pm \mathbf{b}_i$ for some i, and hence $f_v = \pm f_{\mathbf{b}_i}$. Since $I_{\mathscr{B}} \subset I_{L'}$ it follows that $f_v \in I_{L'}$.

Now let $c(\mathbf{v}) > 1$, then there exist $\mathbf{w} \in \mathbb{Z}^n$ with $c(\mathbf{w}) < c(\mathbf{v})$ such that $\mathbf{v} = \mathbf{w} \pm \mathbf{b}_i$. By induction hypothesis, $f_{\mathbf{w}} \in I_{L'}$, and further $f_{\mathbf{b}_i} \in I'_L$, as shown before. Thus formula (3.2) implies that there exists a monomial u such that $u f_v \in I_{L'}$. Since $I_{L'} = I_{\mathscr{B}} : (\prod_{i=1}^n x_i)^\infty$, it follows that $f_v \in I_{L'}$. $\square$

We have seen above that I_L is not always a prime ideal. The lattice ideal I_L is not even a radical ideal if $\mathrm{char}(K) = p > 0$. Indeed, if $L = (p, -p) \subset \mathbb{Z}^2$, then $I_L = (x^p - y^p)$, and we have $f = x - y \notin I_L$ but $f^p \in I_L$.

However, if $\mathrm{char}(K) = 0$ or $\mathrm{char}(K) = p > 0$ and p is big enough, then I_L is a radical ideal. More precisely, we have

Theorem 3.23 *Let $L \subset \mathbb{Z}^n$ be a lattice and let t be the maximal order of a torsion element of $\mathbb{Z}^n/L$. If $\mathrm{char}(K) = 0$ or $\mathrm{char}(K) > t$, then I_L is a radical ideal.*

Proof Let $f \in S$ with $f^k \in I_L$. We want to show that $f \in I_L$. We have $f^k \in I_L S_x$. Suppose we have shown that $I_L S_x$ is a radical ideal. Then it follows that $f \in I_L S_x$, and hence $f \in I_L S_x \cap S$. Therefore, (3.3) and Theorem 3.20 yield that $f \in I_L$.

It remains to be shown that the group ring $K[G]$ isomorphic to $S_x/I_L S_x$ is reduced, where $G = \mathbb{Z}^n/L$ (see the proof of Theorem 3.20). Since G is of the form $\mathbb{Z}^r \oplus \bigoplus_{i=1}^{n-r} \mathbb{Z}/(m_i)\mathbb{Z}$ for some r and suitable integers $m_i > 0$, it follows that $K[G] \cong S_x/(x_1^{m_1} - 1, \ldots, x_{n-r}^{m_{n-r}} - 1)S_x$. Let $\bar{K}$ be the algebraic closure of K. If $\bar{K}[G]$ is reduced, then $K[G]$ is reduced. Thus we may assume that K is algebraically closed. Since $\mathrm{char}(K) > m_i$ for $i = 1, \ldots, n-r$, it follows that all the polynomials

$x_i^{m_i} - 1$ are separable. Hence $x_i^{m_i} - 1 = \prod_{j=1}^{m_i}(x_i - u_{ij})$ with pairwise distinct $u_{ij} \in K$. It follows that

$$(x_1^{m_1} - 1, \ldots, x_{n-r}^{m_{n-r}} - 1) = \bigcap (x_1 - u_{1j_1}, x_2 - u_{2j_2}, \ldots, x_{n-r} - u_{n-r,j_{n-r}}),$$

where the intersection is taken over all $j_i = 1, \ldots, m_i$ for $i = 1 \ldots, n - r$. This shows that $K[G]$ is indeed reduced.   □

Problems

3.11 Let k and l be positive integers such that $\gcd(k, l) = 1$. Show that $(x^k - y^k, x^l - y^l) : (xy)^\infty = (x - y)$. Which is the smallest integer m with the property that $(x^k - y^k, x^l - y^l) : (xy)^m = (x - y)$?

3.12 Let $L \subset \mathbb{Z}^n$ be a lattice. Prove that height $I_L = \operatorname{rank} L$.

3.13 Let $X = (x_{ij})$ be an $m \times n$-matrix of variables, and let $I_2(X) \subset K[X]$ be the ideal of 2-minors of $K[X]$, where $K[X]$ is the polynomial ring over the field K in the variables x_{ij}.

(a) Show that $I_2(X)$ is a lattice ideal I_L. What is the rank of L?
(b) Let $J_2(X)$ be the ideal generated by the adjacent minors $x_{ij}x_{i+1,j+1} - x_{i,j+1}x_{i+1,j}$ with $i = 1, \ldots, m - 1$ and $j = 1, \ldots, n - 1$. Show that $J_2(X)$ is a lattice basis ideal of L.

3.14 We maintain the notation of Problem 3.13. Show that the lattice basis ideal $J_2(X)$ is a radical ideal if and only if $m \leq 2$ or $n \leq 2$.

3.15 Let K be a field, and I be an ideal in the polynomial ring $S = K[x_1, \ldots, x_n]$. Set $x = \prod_{i=1}^n x_i$. Then

$$I : \left(\prod_{i=1}^n x_i\right)^\infty = IS_x \cap S.$$

Here S_x denotes the localization with respect to the multiplicatively closed set $\{1, x, x^2, \ldots\}$ consisting of the powers of x.

3.4 Lawrence Ideals

We introduce the notion of the Lawrence ideal $\Lambda(I)$ of a binomial ideal. This will help us to better understand the primitive binomials of I.

Let $K[\mathbf{x}, \mathbf{y}]$ denote the polynomial ring

$$K[\mathbf{x}, \mathbf{y}] = K[x_1, x_2, \ldots, x_n, y_1, y_2, \ldots, y_n].$$

If $f = \mathbf{x}^{\mathbf{a}} - \mathbf{x}^{\mathbf{b}}$ is a binomial belonging to $K[\mathbf{x}]$, then we introduce the binomial $f^{\sharp}$ belonging to $K[\mathbf{x}, \mathbf{y}]$ by

$$f^{\sharp} = \mathbf{x}^{\mathbf{a}} \mathbf{y}^{\mathbf{b}} - \mathbf{x}^{\mathbf{b}} \mathbf{y}^{\mathbf{a}}.$$

Given a binomial ideal I of $K[\mathbf{x}]$, the *Lawrence ideal* of I is defined to be the ideal

$$\Lambda(I) = (f^{\sharp} : f = \mathbf{x}^{\mathbf{a}} - \mathbf{x}^{\mathbf{b}} \in I). \tag{3.4}$$

Lemma 3.24 *Let I be a binomial ideal of $K[\mathbf{x}]$ and let $F = u - v$ be a binomial in $\Lambda(I)$ such that u and v are relatively prime. Then there exists a binomial $f \in I$ such that $F = f^{\sharp}$.*

Proof Let $F = \mathbf{x}^{\mathbf{a}} \mathbf{y}^{\mathbf{b}'} - \mathbf{x}^{\mathbf{b}} \mathbf{y}^{\mathbf{a}'} \in \Lambda(I)$.
Then

$$F = \sum_{i=1}^{q} h_i(x_1, \ldots, x_n, y_1, \ldots, y_n)(\mathbf{x}^{\mathbf{a}_i} \mathbf{y}^{\mathbf{b}_i} - \mathbf{x}^{\mathbf{b}_i} \mathbf{y}^{\mathbf{a}_i}), \tag{3.5}$$

where $h_i \in K[\mathbf{x}, \mathbf{y}]$ and $\mathbf{x}^{\mathbf{a}_i} - \mathbf{x}^{\mathbf{b}_i} \in I$. By substituting $y_1 = \cdots = y_n = 1$ in (3.5), one has

$$\mathbf{x}^{\mathbf{a}} - \mathbf{x}^{\mathbf{b}} = \sum_{i=1}^{q} h_i(x_1, \ldots, x_n, 1, \ldots, 1)(\mathbf{x}^{\mathbf{a}_i} - \mathbf{x}^{\mathbf{b}_i}).$$

Thus, $\mathbf{x}^{\mathbf{a}} - \mathbf{x}^{\mathbf{b}}$ belongs to I. Furthermore, by replacing in (3.5) y_i by x_i for each $i = 1, 2, \ldots, n$, we obtain

$$\mathbf{x}^{\mathbf{a}+\mathbf{b}'} - \mathbf{x}^{\mathbf{b}+\mathbf{a}'} = 0,$$

and hence $\mathbf{a} + \mathbf{b}' = \mathbf{b} + \mathbf{a}'$. Since $\mathbf{x}^{\mathbf{a}} \mathbf{y}^{\mathbf{b}'}$ and $\mathbf{x}^{\mathbf{b}} \mathbf{y}^{\mathbf{a}'}$ are relatively prime, it follows that $\mathbf{x}^{\mathbf{a}}$ and $\mathbf{x}^{\mathbf{b}}$ are relatively prime. Hence there exist nonnegative integer vectors $\mathbf{a}''$ and $\mathbf{b}''$ belonging to $\mathbb{Z}^n$ such that $\mathbf{a}' = \mathbf{a} + \mathbf{a}''$ and $\mathbf{b}' = \mathbf{b} + \mathbf{b}''$. Since $\mathbf{a} + \mathbf{b}' = \mathbf{b} + \mathbf{a}'$ and since $\mathbf{y}^{\mathbf{a}'}$ and $\mathbf{y}^{\mathbf{b}'}$ are relatively prime, one has $\mathbf{a}'' = \mathbf{b}'' = \mathbf{0}$. It then follows that $F = f^{\sharp}$, where $f = \mathbf{x}^{\mathbf{a}} - \mathbf{x}^{\mathbf{b}}$. $\qquad\square$

Lemma 3.25 *Let $L \subset \mathbb{Z}^n$ be a lattice. Then we have:*

(a) *A binomial $f \in I_L$ is primitive if and only if $f^{\sharp} \in \Lambda(I_L)$ is primitive.*
(b) *Every primitive binomial belonging to $\Lambda(I_L)$ is of the form $f^{\sharp}$, where f is a primitive binomial belonging to I_L.*

Proof

(a) Suppose that the binomial $f = \mathbf{x}^{\mathbf{a}} - \mathbf{x}^{\mathbf{b}} \in I_L$ is not primitive. Then there is a nonzero binomial $g = \mathbf{x}^{\mathbf{a}'} - \mathbf{x}^{\mathbf{b}'} \in I_L$ with $f \neq g$ for which $\mathbf{x}^{\mathbf{a}'} | \mathbf{x}^{\mathbf{a}}$ and $\mathbf{x}^{\mathbf{b}'} | \mathbf{x}^{\mathbf{b}}$. Then $\mathbf{x}^{\mathbf{a}'} \mathbf{y}^{\mathbf{b}'} | \mathbf{x}^{\mathbf{a}} \mathbf{y}^{\mathbf{b}}$ and $\mathbf{x}^{\mathbf{b}'} \mathbf{y}^{\mathbf{a}'} | \mathbf{x}^{\mathbf{b}} \mathbf{y}^{\mathbf{a}}$. Since $g^{\sharp} \neq f^{\sharp}$, it follows that $f^{\sharp}$ is not primitive.

Conversely, let $f = \mathbf{x}^{\mathbf{a}} - \mathbf{x}^{\mathbf{b}} \in I_L$, and suppose that $f^{\sharp} = \mathbf{x}^{\mathbf{a}}\mathbf{y}^{\mathbf{b}} - \mathbf{x}^{\mathbf{b}}\mathbf{y}^{\mathbf{a}} \in \Lambda(I_L)$ is not primitive. Then there is a nonzero binomial $G = \mathbf{x}^{\mathbf{a}'}\mathbf{y}^{\mathbf{b}''} - \mathbf{x}^{\mathbf{b}'}\mathbf{y}^{\mathbf{a}''} \in \Lambda(I_L)$ with $G \neq f^{\sharp}$ such that $\mathbf{x}^{\mathbf{a}'}\mathbf{y}^{\mathbf{b}''}|\mathbf{x}^{\mathbf{a}}\mathbf{y}^{\mathbf{b}}$ and $\mathbf{x}^{\mathbf{b}'}\mathbf{y}^{\mathbf{a}''}|\mathbf{x}^{\mathbf{b}}\mathbf{y}^{\mathbf{a}}$. If $\mathbf{x}^{\mathbf{a}'}\mathbf{y}^{\mathbf{b}''}$ and $\mathbf{x}^{\mathbf{b}'}\mathbf{y}^{\mathbf{a}''}$ are not relatively prime, then $\mathbf{x}^{\mathbf{a}}$ and $\mathbf{x}^{\mathbf{b}}$ are not relatively prime. Since I_L is a lattice ideal, Theorem 3.20 implies that f is not primitive. Suppose that $\mathbf{x}^{\mathbf{a}'}\mathbf{y}^{\mathbf{b}''}$ and $\mathbf{x}^{\mathbf{b}'}\mathbf{y}^{\mathbf{a}''}$ are relatively prime. By Lemma 3.24, we have $G = g^{\sharp}$, where $g = \mathbf{x}^{\mathbf{a}'} - \mathbf{x}^{\mathbf{b}'} \in I_L$. One has $\mathbf{x}^{\mathbf{a}'}|\mathbf{x}^{\mathbf{a}}$ and $\mathbf{x}^{\mathbf{b}'}|\mathbf{x}^{\mathbf{b}}$. Since $G \neq f^{\#}$ it follows that $g \neq f$. Thus f is not primitive.

(b) We first observe that $\Lambda(I_L)$ is again a lattice ideal, namely $\Lambda(I_L) = I_{L'}$, where

$$L' = \{(\mathbf{v}, -\mathbf{v}) \subset \mathbb{Z}^n \times \mathbb{Z}^n : \mathbf{v} \in L\}.$$

Since in a lattice ideal, any primitive binomial $u - v$ has the property that u and v are relatively prime, we may apply Lemma 3.24, and deduce that every primitive binomial of $\Lambda(I_L)$ is of the form $f^{\sharp}$, where $f = \mathbf{x}^{\mathbf{a}} - \mathbf{x}^{\mathbf{b}} \in I_L$. By (a), f is primitive.   □

Theorem 3.26

(a) *Let I be a binomial ideal of $K[\mathbf{x}]$ and let $\{g_1^{\sharp}, g_2^{\sharp}, \ldots, g_m^{\sharp}\}$ be a minimal set of generators of $\Lambda(I)$ with each $g_i \in I$. Then the Graver basis of I is contained in $\{g_1, g_2, \ldots, g_m\}$.*
(b) *Let I be a lattice ideal. Then the following conditions are equivalent:*

   (i) *$\{g_1, \ldots, g_m\}$ is the Graver basis of I;*
   (ii) *$\{g_1^{\sharp}, \ldots, g_m^{\sharp}\}$ is the Graver basis of $\Lambda(I)$;*
   (iii) *$\{g_1^{\sharp}, \ldots, g_m^{\sharp}\}$ is the universal Gröbner basis of $\Lambda(I)$;*
   (iv) *$\{g_1^{\sharp}, \ldots, g_m^{\sharp}\}$ is the reduced Gröbner basis of $\Lambda(I)$ with respect to any monomial order;*
   (v) *$\{g_1^{\sharp}, \ldots, g_m^{\sharp}\}$ is a minimal set of generators of $\Lambda(I)$.*

If the equivalent conditions hold, then $\{g_1^{\sharp}, \ldots, g_m^{\sharp}\}$ is the unique minimal set of generators of $\Lambda(I)$.

Proof

(a) Suppose that $f = \mathbf{x}^{\mathbf{a}} - \mathbf{x}^{\mathbf{b}} \in I$ is primitive. Since $f^{\sharp}$ belongs to $\Lambda(I) = (g_1^{\sharp}, g_2^{\sharp}, \ldots, g_m^{\sharp})$, there exist polynomials $h_1, \ldots, h_m$ belonging to $K[\mathbf{x}, \mathbf{y}]$ such that $f^{\sharp} = \mathbf{x}^{\mathbf{a}}\mathbf{y}^{\mathbf{b}} - \mathbf{x}^{\mathbf{b}}\mathbf{y}^{\mathbf{a}} = h_1 g_1^{\sharp} + \cdots + h_m g_m^{\sharp}$. Then a monomial, say $\mathbf{x}^{\mathbf{a}'}\mathbf{y}^{\mathbf{b}'}$, appearing in one of $g_1^{\sharp}, \ldots, g_m^{\sharp}$ divides $\mathbf{x}^{\mathbf{a}}\mathbf{y}^{\mathbf{b}}$. Let $g_i^{\sharp} = \mathbf{x}^{\mathbf{a}'}\mathbf{y}^{\mathbf{b}'} - \mathbf{x}^{\mathbf{b}'}\mathbf{y}^{\mathbf{a}'}$. Then $g_i = \mathbf{x}^{\mathbf{a}'} - \mathbf{x}^{\mathbf{b}'} \in I$ is a nonzero binomial such that $\mathbf{x}^{\mathbf{a}'}$ divides $\mathbf{x}^{\mathbf{a}}$ and $\mathbf{x}^{\mathbf{b}'}$ divides $\mathbf{x}^{\mathbf{b}}$. Since f is primitive, we have $f = g_i$, as desired.
(b) By Lemma 3.25, we have (i) $\iff$ (ii). Every reduced Gröbner basis is a subset of the universal Gröbner basis. Furthermore, it follows from Corollary 3.14 that the universal Gröbner basis is a subset of the Graver basis. Since every reduced Gröbner basis of $\Lambda(I)$ is a system of generators of $\Lambda(I)$, the equivalence of

(ii)–(v) follows once we have shown that the Graver basis of $\Lambda(I)$ is the unique minimal system of generators of $\Lambda(I)$ consisting of binomials.

Suppose that $\{g_1^\sharp, g_2^\sharp, \ldots, g_m^\sharp\}$ is a minimal set of generators of $\Lambda(I)$. It follows from (a) that the Graver basis of I is contained in $\{g_1, g_2, \ldots, g_m\}$. Let $\{g_{i_1}, g_{i_2}, \ldots, g_{i_k}\}$ be the Graver basis of I. By Lemma 3.25, $\{g_{i_1}^\sharp, g_{i_2}^\sharp, \ldots, g_{i_k}^\sharp\}$ is the Graver basis of $\Lambda(I)$. Since the Graver basis is a system of generators, we have $\{g_1^\sharp, g_2^\sharp, \ldots, g_m^\sharp\} = \{g_{i_1}^\sharp, g_{i_2}^\sharp, \ldots, g_{i_k}^\sharp\}$. Therefore, $\{g_1^\sharp, g_2^\sharp, \ldots, g_m^\sharp\}$ is the Graver basis of $\Lambda(I)$. In particular, a minimal set of generators of $\Lambda(I)$ is uniquely determined. □

Theorem 3.26 can be used to compute the Graver basis of a lattice ideal. For this purpose it suffices to show that for a given lattice $L \subset \mathbb{Z}^n$ a system of generators of $\Lambda(I_L)$ can be determined. We describe a method to do this.

Let $\mathscr{B}$ be a basis of L. Since $\Lambda(I_L) = I_{L'}$, where

$$L' = \{(\mathbf{v}, -\mathbf{v}) \subset \mathbb{Z}^n \times \mathbb{Z}^n : \mathbf{v} \in L\},$$

a lattice basis for L' is $\mathscr{B}' = \{(\mathbf{v}, -\mathbf{v}) : \mathbf{v} \in \mathscr{B}\}$. The lattice basis ideal of L' for the basis $\mathscr{B}'$ is the ideal

$$I_{\mathscr{B}'} = (\mathbf{x}^{\mathbf{v}_+}\mathbf{y}^{\mathbf{v}_-} - \mathbf{x}^{\mathbf{v}_-}\mathbf{y}^{\mathbf{v}_+} : \mathbf{v} \in \mathscr{B}) \quad \text{in} \quad K[\mathbf{x}, \mathbf{y}].$$

Now it follows from Corollary 3.22 that

$$\Lambda(I_L) = I_{\mathscr{B}'} : \left(\prod_{i=1}^n x_i\right)^\infty.$$

This colon ideal can be computed by using Proposition 1.39 or Proposition 1.40.

For an integer matrix $A \in \mathbb{Z}^{d \times n}$, the Lawrence ideal $\Lambda(I_A)$ of the toric ideal I_A is a toric ideal of a configuration. In the rest of the present section, we study how to construct the corresponding configuration.

The *Lawrence lifting* of an integer matrix $A \in \mathbb{Z}^{d \times n}$ is the configuration

$$\Lambda(A) = \begin{pmatrix} A & 0 \\ I_n & I_n \end{pmatrix} \in \mathbb{Z}^{(d+n) \times 2n},$$

where I_n is the $n \times n$ identity matrix. For example, if

$$A = \begin{pmatrix} 1 & 0 & -1 & 0 \\ 0 & 1 & -1 & 0 \\ 1 & 1 & 1 & 1 \end{pmatrix} \in \mathbb{Z}^{3 \times 4},$$

then its Lawrence lifting is

$$
\Lambda(A) = \begin{pmatrix}
1 & 0 & -1 & 0 & 0 & 0 & 0 & 0 \\
0 & 1 & -1 & 0 & 0 & 0 & 0 & 0 \\
1 & 1 & 1 & 1 & 0 & 0 & 0 & 0 \\
1 & 0 & 0 & 0 & 1 & 0 & 0 & 0 \\
0 & 1 & 0 & 0 & 0 & 1 & 0 & 0 \\
0 & 0 & 1 & 0 & 0 & 0 & 1 & 0 \\
0 & 0 & 0 & 1 & 0 & 0 & 0 & 1
\end{pmatrix} \in \mathbb{Z}^{7 \times 8}. \tag{3.6}
$$

We will show that $\Lambda(I_A) = I_{\Lambda(A)}$.

Let $K[\mathbf{t}, \mathbf{t}^{-1}, \mathbf{z}]$ denote the Laurent polynomial ring

$$
K[\mathbf{t}, \mathbf{t}^{-1}, \mathbf{z}] = K[t_1, t_1^{-1}, t_2, t_2^{-1}, \ldots, t_d, t_d^{-1}, z_1, z_2, \ldots, z_n].
$$

The toric ring of the Lawrence lifting $\Lambda(A)$ is

$$
K[\Lambda(A)] = K[\mathbf{t}^{\mathbf{a}_1} z_1, \mathbf{t}^{\mathbf{a}_2} z_2, \ldots, \mathbf{t}^{\mathbf{a}_n} z_n, z_1, z_2, \ldots, z_n].
$$

Let $K[\mathbf{x}, \mathbf{y}]$ denote the polynomial ring

$$
K[\mathbf{x}, \mathbf{y}] = K[x_1, x_2, \ldots, x_n, y_1, y_2, \ldots, y_n]
$$

and define the ring homomorphism

$$
\pi : K[\mathbf{x}, \mathbf{y}] \to K[\Lambda(A)]
$$

by setting $\pi(x_i) = \mathbf{t}^{\mathbf{a}_i} z_i$ for $1 \le i \le n$ and $\pi(y_j) = z_j$ for $1 \le j \le n$. The toric ideal $I_{\Lambda(A)}$ of $\Lambda(A)$ is the kernel of π.

For example, the toric ideal of the Lawrence lifting (3.6) is

$$
I_{\Lambda(A)} = (x_1 x_2 x_3 y_4^3 - x_4^3 y_1 y_2 y_3).
$$

Note that $\Lambda(A)\mathbf{w} = 0$ for $\mathbf{w} = \begin{pmatrix} \mathbf{u} \\ \mathbf{v} \end{pmatrix} \in \mathbb{Z}^{2n}$ if and only if $A\mathbf{u} = 0$ and $\mathbf{v} = -\mathbf{u}$.

Thus we have the following proposition.

Proposition 3.27 *Let $A \in \mathbb{Z}^{d \times n}$ be an integer matrix. Then $\Lambda(I_A) = I_{\Lambda(A)}$.*

Problems

3.16 Let I be a principal binomial ideal. Show that $\Lambda(I)$ is also a principal ideal.

3.17 In Proposition 3.15 it is shown that if A is a configuration, then each minimal set of generators of I_A is contained in the Graver basis of I_A. Show this is no longer true, if the binomial ideal I is not a graded ideal.

3.18 Let L be a lattice. Show that I_L is a prime ideal if and only if $\Lambda(I_L)$ is a prime ideal. Is this statement true for any other binomial ideal?

3.19 Give an example of a binomial ideal I and a binomial $f \in I$ such that f is not primitive but $f^\sharp \in \Lambda(I)$ is primitive.

3.5 The Squarefree Divisor Complex

Let K be a field, and let $T = K[t_1, \ldots, t_d]$ be the polynomial ring over K in the variables $t_1, \ldots, t_d$. For a matrix $A \in \mathbb{Z}_{\geq 0}^{d \times n}$ with column vectors $\mathbf{a}_j$, we consider the toric ring $R = K[A] = K[u_1, \ldots, u_n]$ with $u_j = \mathbf{t}^{\mathbf{a}_j}$ for $j = 1, \ldots, n$ and the K-algebra homomorphism

$$\pi : S = K[x_1, \ldots, x_n] \to T \quad \text{with} \quad x_j \mapsto \mathbf{t}^{\mathbf{a}_j}. \tag{3.7}$$

with kernel I_A.

In this section we want to analyze the Betti numbers of the free S-resolution of I_A. To this end we introduce some concepts and terminology: a *simplicial complex* Δ on $V = \{v_1, v_2, \ldots, v_n\}$ is a collection Δ of subsets of V with the property that for any $F \in \Delta$ and any $G \subset F$, it follows that $G \in \Delta$. The elements of Δ are called *faces*, and the maximal faces of Δ (maximal with respect to inclusion) are called the *facets* of Δ. The dimension $\dim F$ of a face is given as $\dim F = |F| - 1$. Finally we set $\dim \Delta$ to be the maximal dimension of a facet of Δ. Faces of Δ of dimension i are called i-*faces*. Observe that the empty set is a face of Δ whose dimension is defined to be -1.

Let $d = \dim \Delta$. Fix a field K. The *augmented oriented chain complex* of Δ (with coefficients in K) is the complex $\widetilde{\mathscr{C}}(\Delta; K)$:

$$0 \longrightarrow \mathscr{C}_d(\Delta; K) \longrightarrow \mathscr{C}_{d-1}(\Delta; K) \longrightarrow \cdots \longrightarrow \mathscr{C}_0(\Delta; K) \longrightarrow \mathscr{C}_{-1}(\Delta; K) \longrightarrow 0,$$

where

$$\mathscr{C}_i(\Delta; K) = \bigoplus_{F \in \Delta, \dim F = i} K e_F \quad \text{and} \quad \partial e_F = \sum_{k=0}^{i} (-1)^k e_{F_k},$$

and where F_k is defined as follows: let $F = \{v_{j_0}, \ldots, v_{j_i}\}$ with $j_0 < j_1 < \cdots < j_i$. Then $F_k = \{v_{j_0}, \ldots, \widehat{v}_{j_k}, \ldots, v_{j_i}\}$. We set

$$\widetilde{H}_i(\Delta; K) = H_i(\widetilde{\mathscr{C}}(\Delta; K)), \qquad i = -1, \ldots, d - 1,$$

and call $\widetilde{H}_i(\Delta; K)$ the i-th reduced simplicial homology of Δ. Similarly one defines the i-th reduced simplicial cohomology of Δ, as

$$\widetilde{H}^i(\Delta; K) = H^i(\mathrm{Hom}_K(\widetilde{\mathscr{C}}(\Delta; K), K)), \qquad i = -1, \ldots, d - 1.$$

It can be easily shown that the reduced simplicial homology of Δ does not depend on the labeling of V. On the other hand, it can be shown by examples that for $i > 1$ the vanishing or non-vanishing of $\widetilde{H}_i(\Delta; K)$ may depend on the field K. A fundamental theorem of topology (see for example [150, Theorem 34.3]) says that the reduced singular homology $\widetilde{H}_i(X; K)$ of a topological space X with triangulation Δ can be computed by means of the reduced simplicial homology. Indeed one has

$$\widetilde{H}_i(X; K) \simeq \widetilde{H}_i(\Delta; K).$$

One calls a subset $H \subset \mathbb{Z}^n$ an *affine semigroup*, if there are finitely many elements $\mathbf{a}_1, \ldots, \mathbf{a}_n \in H$ such that each element of H is a linear combination $\lambda_1 \mathbf{a}_1 + \cdots + \lambda_n \mathbf{a}_n$ with $\lambda_i \in \mathbb{Z}_{\geq 0}$ for all i. The elements $\mathbf{a}_1, \ldots, \mathbf{a}_n$ are called *generators* of H. The affine semigroup is called *positive*, if whenever we have $\mathbf{a}, -\mathbf{a} \in H$, then $\mathbf{a} = 0$.

Coming back to our algebra $R = K[A]$, we notice that it has a K-basis consisting of monomials $\mathbf{t}^{\mathbf{a}}$. We denote this monomial basis by $\mathcal{M}$. The set of exponents $\mathbf{a}$ appearing as exponents of the basis elements of $K[A]$ together with addition form a *positive affine semigroup* $H \subset \mathbb{Z}_{\geq 0}^n$ which is generated by $\mathbf{a}_1, \ldots, \mathbf{a}_n$. We will assume that these elements form a minimal system of generators of H.

Given an element $\mathbf{a} \in H$, we define the simplicial complex

$$\Delta_{\mathbf{a}} = \{F \subset [n]: u^F \text{ divides } \mathbf{t}^{\mathbf{a}} \text{ in } R\}.$$

where $u^F = \prod_{j \in F} u_j$.

The simplicial complex $\Delta_{\mathbf{a}}$ is called the *squarefree divisor complex* of H (or of A).

Let $I \subset R$ be an ideal generated by elements of $\mathcal{M}$. Then R/I is an H-graded K-algebra and the canonical residue class map $S \to R/I$ is an H-graded K-algebra homomorphism if we set $\deg x_j = \mathbf{a}_j$ for $j = 1, \ldots, n$. It follows that R/I becomes an H-graded S-module. Consequently, the K-vector spaces $\mathrm{Tor}_i^S(K, R/I)$ are H-graded. We set $\beta_{i,\mathbf{a}}(R/I) = \mathrm{Tor}_i^S(K, R/I)_{\mathbf{a}}$, and call these numbers the H-graded Betti numbers of R/I. They can be computed from the Koszul complex. Indeed, $\mathrm{Tor}_i^S(K, R/I) \cong H_i(\mathbf{x}; R/I)$, as an H-graded K-vector space, cf. (2.2).

We will describe the H-graded Betti numbers of R/I in terms of certain reduced simplicial homologies. Let Γ be a simplicial subcomplex of Δ. Then $\widetilde{\mathscr{C}}(\Gamma; K)$ is a subcomplex of $\widetilde{\mathscr{C}}(\Delta; K)$. The homology of the complex $\widetilde{\mathscr{C}}(\Delta; K)/\widetilde{\mathscr{C}}(\Gamma; K)$ is called the *relative simplicial homology*, and is denoted $\widetilde{H}(\Delta, \Gamma; K)$.

Theorem 3.28 *For $\mathbf{a} \in H$ we let $\Gamma_{\mathbf{a}} = \{F \in \Delta_{\mathbf{a}} \colon \mathbf{t}^{\mathbf{a}}/\mathbf{u}^F \in I\}$. Then*

(a) $K(\mathbf{x}; R/I)_{\mathbf{a}} \cong (\widetilde{\mathscr{C}}(\Delta_{\mathbf{a}}; K)/\widetilde{\mathscr{C}}(\Gamma_{\mathbf{a}}; K))(-1)$;

(b) $\beta_{i,\mathbf{a}}(R/I) = \dim_K \tilde{H}_{i-1}(\Delta_{\mathbf{a}}, \Gamma_{\mathbf{a}}; K)$.

Proof

(a) The free R-module $K_i(\mathbf{x}; R)$ has the multigraded decomposition

$$K_i(\mathbf{x}; R) = \bigoplus_{F \subset [n], |F| = i} R(-\deg u^F),$$

where the differentiation $K_i(\mathbf{x}; R) \rightarrow K_{i-1}(\mathbf{x}; R)$ on the component $R(-\deg u^F) \rightarrow R(-\deg u^{F'})$ is given as multiplication by $\epsilon(F, F') u_{j_k}$. Here $\epsilon(F, F') = 0$, if $F' \not\subset F$, and $\epsilon(F, F') = (-1)^{k-1}$, if $F' = F \setminus \{j_k\}\}$, $F = \{j_1 < j_2 < \cdots < j_i\}$.

Let us fix $\mathbf{a} \in H$. In order to have $R(-\deg u^F)_{\mathbf{a}} \neq 0$, we must have $\mathbf{a} - \deg u^F \in H$, which is equivalent to saying that $u^F | \mathbf{t}^{\mathbf{a}}$. If this is the case, then $R(-\deg u^F)_{\mathbf{a}}$ is a 1-dimensional K-vector space with basis element $\mathbf{t}^{\mathbf{a}}/u^F$. Thus we see that

$$K_i(\mathbf{x}; R)_{\mathbf{a}} = \bigoplus_{F \in \Delta_{\mathbf{a}}, |F| = i} K \mathbf{t}^{\mathbf{a}}/u^F.$$

With respect to these K-bases of the $K_i(\mathbf{x}; R)_{\mathbf{a}}$, the maps in $K(\mathbf{x}; R)_{\mathbf{a}}$ are the same as those in $\widetilde{\mathscr{C}}(\Delta_{\mathbf{a}}; K)(-1)$, since $\widetilde{\mathscr{C}}(\Delta_{\mathbf{a}}; K)$ is the complex of K-vector spaces with

$$\widetilde{\mathscr{C}}_{i-1}(\Delta_{\mathbf{a}}; K) = \bigoplus_{F \in \Delta_{\mathbf{a}}, |F| = i} K e_F$$

and with differentiation on the component $K e_F \rightarrow K e_{F'}$ which maps e_F to $\epsilon(F, F') e_{F'}$, as explained in Section 2.2.

Similar arguments apply to $K(\mathbf{x}; I)_{\mathbf{a}}$. Thus the short exact sequence of complexes

$$0 \longrightarrow K(\mathbf{x}; I)_{\mathbf{a}} \longrightarrow K(\mathbf{x}; R)_{\mathbf{a}} \longrightarrow K(\mathbf{x}; R/I)_{\mathbf{a}} \longrightarrow 0$$

yields the desired isomorphism (a).

(b) is an immediate consequence of (a), since we have $\operatorname{Tor}_i^R(K, R/I)_{\mathbf{a}} \cong H_i(\mathbf{x}; R/I)_{\mathbf{a}}$, see (2.2). □

For the applications to follow we need a duality statement for relative simplicial homology. Let Δ be an arbitrary simplicial complex on $[n]$. The *Alexander dual* $\Delta^{\vee}$ of Δ is defined by

$$\Delta^{\vee} = \{F \in [n] \colon \bar{F} \notin \Delta\},$$

where $\bar{F} = [n] \setminus F$.

Lemma 3.29 *Let* $\Gamma \subset \Delta$ *be simplicial complexes on* $[n]$. *Then*

$$\widetilde{H}_i(\Delta, \Gamma; K) \cong \widetilde{H}^{n-2-i}(\Gamma^\vee, \Delta^\vee; K) \cong \widetilde{H}_{n-2-i}(\Gamma^\vee, \Delta^\vee; K).$$

Proof Let $e_1, \ldots, e_n$ be a basis of the K-vector space E. Then the elements $e_F = e_{j_1} \wedge \cdots \wedge e_{j_i}$ for all $F = \{j_1 < j_2 < \cdots < j_i\}$ of cardinality i form a K-basis of $\bigwedge^i E$, and we have $\mathscr{C}_i(\Gamma; K) \subset \mathscr{C}_i(\Delta; K) \subset \bigwedge^i E$.

For all i we define an isomorphism of K-vector spaces $\bigwedge^i E \to (\bigwedge^{n-i} E)^*$, where $(\bigwedge^{n-i} E)^*$ denotes the K-dual of $\bigwedge^{n-i} E$. This isomorphism assigns to e_F the element $\epsilon(F, \bar{F})(e_{\bar{F}})^*$, where $(e_{\bar{F}})^*$ is the basis element of $(\bigwedge^{n-i} E)^*$ with $(e_{\bar{F}})^*(e_G) = 1$ if $G = \bar{F}$ and $(e_{\bar{F}})^*(e_G) = 0$, otherwise, and where the sign $\epsilon(F, \bar{F})$ is defined by the equation $e_F \wedge e_{\bar{F}} = \epsilon(F, \bar{F})e_1 \wedge \cdots \wedge e_n$. This isomorphism induces the first isomorphism between reduced simplicial homology and cohomology. For the second isomorphism see Problem 3.20. □

In general the H-graded Betti numbers $\beta_{i,\mathbf{a}}$ of $K[A]$ may depend on the base field. But nevertheless one has

Theorem 3.30 *With the notation introduced, the H-graded Betti numbers $\beta_{i,\mathbf{a}}$ are independent of K in the following cases:*

(i) $i = 0, 1, n - 1, n$;
(ii) $i = 2$ *if* $R \cong K[x_1, \ldots, x_n]$;
(iii) $i = n - 2$ *if* $I = 0$.

Proof The assertions of (i) are obvious for $i = 0$ and $i = n$. In fact, $\beta_{0,\mathbf{a}} = 1$ for $\mathbf{a} = \mathbf{0}$ and $\beta_{0,\mathbf{a}} = 0$ for $\mathbf{a} \neq \mathbf{0}$, while $\beta_{n,\mathbf{a}}$ is an H-graded component of the socle of R/I.

By Problem 3.21, $\dim_K \widetilde{H}_0(\Delta, \Gamma; K)$ is independent of K for all simplicial complexes $\Gamma \subset \Delta$. The same holds for the dimension of $\widetilde{H}_{n-2}(\Delta, \Gamma; K)$, since by Lemma 3.29, this vector space is isomorphic to $\widetilde{H}_0(\Gamma^\vee, \Delta^\vee; K)$. Thus the assertion for $i = 1$ and $i = n - 1$ follows from Theorem 3.28(b).

Under the assumptions of (ii), $\Delta_\mathbf{a}$ is a simplex on the set $\{i : a_i \neq 0\}$. Therefore, $\mathscr{C}(\Delta_\mathbf{a}; K)$ is acyclic, and hence from the long exact homology sequence arising from the short exact sequence

$$0 \to \widetilde{\mathscr{C}}(\Gamma_\mathbf{a}; K) \to \widetilde{\mathscr{C}}(\Delta_\mathbf{a}; K) \to \widetilde{\mathscr{C}}(\Delta_\mathbf{a}; K)/\widetilde{\mathscr{C}}(\Gamma_\mathbf{a}; K) \to 0.$$

we obtain $\dim_K \widetilde{H}_1(\Delta_\mathbf{a}, \Gamma_\mathbf{a}; K) = \dim_K \widetilde{H}_0(\Gamma_\mathbf{a}; K)$. This proves (ii).

Finally, if $I = 0$, then $\Gamma_\mathbf{a} = \emptyset$ for all $\mathbf{a} \in H$. Therefore, in this case, if we denote by Σ the simplex on $[n]$, Lemma 3.29 implies that

$$\dim_K \widetilde{H}_{n-3}(\Delta_\mathbf{a}, \Gamma_\mathbf{a}; K) = \dim_K \widetilde{H}_1(\Sigma, \Delta_\mathbf{a}^\vee; K) = \dim_K \widetilde{H}_0(\Delta_\mathbf{a}^\vee; K).$$

Thus (iii) follows. □

Recall that a monomial $\mathbf{x}^{\mathbf{a}}$ in S is called squarefree if the exponent vector $\mathbf{a} = (a_1, \ldots, a_n)$ is *squarefree*, which means that $0 \le a_i \le 1$ for $i = 1, \ldots, n$. Lemma 1.13 says that a monomial ideal is a squarefree monomial ideal if it is generated by squarefree monomials.

As another application of Theorem 3.28 we will prove a theorem of Hochster [116] which describes the $\mathbb{Z}^n$-graded Betti numbers of S/I when I is a squarefree monomial ideal.

Let K be a field, $S = K[x_1, \ldots, x_n]$ be the polynomial ring over K in the variables $x_1, \ldots, x_n$, and let Σ be a simplicial complex of $[n]$. The *Stanley–Reisner ideal* of Σ, denoted I_Σ, is the squarefree monomial ideal generated by the monomial $x_F = \prod_{i \in F} x_i$ with $F \subset [n]$ and $F \notin \Sigma$. Note that for any squarefree monomial ideal $I \subset S$, there exists a unique simplicial complex Σ such that $I = I_\Sigma$. The *Stanley–Reisner ring* of Σ over K is defined to be the K-algebra $K[\Sigma] = S/I_\Sigma$.

Let $\mathbf{a} = (a_1, \ldots, a_n) \in \mathbb{Z}^n$. We set $\mathrm{supp}(\mathbf{a}) = \{i \in [n]: a_i \ne 0\}$.

Theorem 3.31 (Hochster) *Let Σ be a simplicial complex on $[n]$ and $\mathbf{a} \in \mathbb{Z}^n$. Then the following holds:*

(a) $\beta_{i,\mathbf{a}}(K[\Sigma]) = 0$, *if $\mathbf{a}$ is not squarefree.*
(b) *If $\mathbf{a}$ is squarefree, then $\beta_{i,\mathbf{a}}(K[\Sigma]) = \dim_K \widetilde{H}_{|W|-i-1}(\Sigma_W; K)$ for all i, where $W = \mathrm{supp}(\mathbf{a})$ and $\Sigma_W = \{F \in \Sigma: F \subset W\}$.*

Proof In the situation of the theorem, $H = \mathbb{Z}^n_{\ge 0}$, $R = S$, and $I = I_\Sigma$. Let $\mathbf{a} \in H$. Then $\Delta_{\mathbf{a}}$ consists of all subset of $W = \mathrm{supp}(\mathbf{a})$, and hence is a simplex on W. Furthermore,

$$\Gamma_{\mathbf{a}} = \{F \in \Delta_{\mathbf{a}}: \ \mathrm{supp}(\mathbf{a} - \epsilon_F) \notin \Sigma\},$$

where ϵ_F is the unique squarefree vector with $\mathrm{supp}(\epsilon_F) = F$.

Proof of (a): Let $\mathbf{a} \in \mathbb{Z}^n_{\ge 0}$ be not squarefree, and choose j such that $a_j \ge 2$. Let $\mathscr{A}$ be the set of vectors $\mathbf{a}' \in \mathbb{Z}^n_{\ge 0}$ with $a'_j \ge 2$ and $a'_i = a_i$ for all $i \ne j$. Then $\Delta_{\mathbf{a}} = \Delta_{\mathbf{a}'}$ and $\Gamma_{\mathbf{a}} = \Gamma_{\mathbf{a}'}$ for all $\mathbf{a}' \in \mathscr{A}$. Thus Theorem 3.28(b) implies that $\beta_{i,\mathbf{a}}(K[\Sigma]) = \beta_{i,\mathbf{a}'}(K[\Sigma])$ for all $\mathbf{a}' \in \mathscr{A}$. Suppose that $\beta_{i,\mathbf{a}}(K[\Sigma]) \ne 0$. Then $\beta_{i,\mathbf{a}'}(K[\Sigma]) \ne 0$ for all $\mathbf{a}' \in \mathscr{A}$. Since $|\mathscr{A}| = \infty$, it would follow that $K[\Sigma]$ has infinitely many non-vanishing Betti numbers, a contradiction.

Proof of (b): Let $\mathbf{a}$ be squarefree and let $W = \mathrm{supp}(\mathbf{a})$. Then $F \in \Gamma_{\mathbf{a}}$ if and only if $W \setminus F \notin \Sigma_W$. This implies that $\Gamma_{\mathbf{a}}^\vee = \Sigma_W$, where the Alexander dual of $\Gamma_{\mathbf{a}}$ is taken with respect to the vertex set W. Since the Alexander dual $\Delta_{\mathbf{a}}^\vee$ with respect to W is the empty set, Theorem 3.28(b) together with Lemma 3.29 implies that

$$\beta_{i,\mathbf{a}}(K[\Sigma]) = \dim_K \widetilde{H}_{i-1}(\Delta_{\mathbf{a}}, \Gamma_{\mathbf{a}}; K) = \dim_K \widetilde{H}_{|W|-i-1}(\Sigma_W; K);$$

as desired. $\qquad\square$

Problems

3.20 Let K be a field and $C : 0 \rightarrow C_d \rightarrow \cdots \rightarrow C_i \rightarrow \cdots \rightarrow C_0 \rightarrow 0$ be a complex of finite dimensional K-vector spaces, and let $C^* = \operatorname{Hom}_K(C, K)$ be the dual complex of C. Show that $H_i(C) \cong H^i(C^*)$ for all i.

3.21 Let $\Gamma \subset \Delta$ be simplicial complexes, and let K be a field. Show that the dimension of $\widetilde{H}_0(\Gamma, \Delta; K)$ is independent of K.

3.22 Let $R = K[A]$ and $I \subset R$ be as in Theorem 3.28. Show that all H-graded Betti numbers of R/I are independent of K if $n \leq 4$, or $n = 5$ and either (i) R is the polynomial ring or else (ii) $I = 0$.

Notes

One of the first articles where binomial ideals appeared is [90]. In that paper the relation ideals of semigroup rings were identified as binomial ideals. A first systematic treatment of binomial ideals and toric rings is given in the Sturmfels' book [202] with applications to convex polytopes and integer programming. That treatment also includes, for the case of toric ideals, the basic facts presented in Sections 3.1 and 3.2. Hoşten and Shapiro [118] introduced lattice basis ideals and discussed their primary decomposition in some special cases. In the fundamental article [58], Eisenbud and Sturmfels develop a general theory of binomial ideals and their primary decomposition. In their terminology, a binomial is a polynomial with at most two terms. In that paper a more general version of Theorem 3.12 can be found. A similar result for lattice basis ideals has been shown by Fischer and Shapiro [73], cf. Corollary 3.22. In Section 3.4, Lawrence ideals attached to binomial ideals are introduced. It is shown in Proposition 3.27 that the Lawrence ideal of the toric ideal of a matrix is the toric ideal of the Lawrence lifting of this matrix. A theorem analogue to Theorem 3.26, but stated for Lawrence liftings appeared first in [204] and can also be found in [202]. Higher Lawrence liftings were introduced in [185], and have been further generalized and studied in [31]. A different definition of Lawrence ideals is given in [32] which however coincides with the definition given here, in the case that the given binomial ideal is a lattice ideal. The content of Section 3.5 is taken from [28].

For further reading we recommend the book [146] by Miller and Sturmfels, the article [58] by Eisenbud and Sturmfels, as well the papers [123, 124, 170] for newer developments.

Chapter 4
Convex Polytopes and Unimodular Triangulations

Abstract The triangulation of a convex polytope is one of the most important topics in the classical theory of convex polytopes. In this chapter the modern treatment of triangulations of convex polytopes is systematically developed. In Section 4.1 we recall fundamental materials on convex polytopes and summarize basic facts without their proofs. The highlight of Chapter 4 is Section 4.2, where unimodular triangulations of convex polytopes are introduced and studied in the frame of initial ideals of toric ideals of convex polytopes. Furthermore, the normality of convex polytopes is discussed. Finally, in Section 4.3, we study the Lawrence lifting of a configuration, which is a powerful tool for computing the Graver basis of a toric ideal. Furthermore, unimodular polytopes, which form a distinguished subclass of the class of normal polytopes, are discussed.

4.1 Foundations on Convex Polytopes

We collect fundamental material on convex polytopes and summarize basic facts on their classical theory. A detailed proof of each fact, which will be omitted, can be found in [19, 85, 221].

4.1.1 Convex Sets

A nonempty subset $C \subset \mathbb{R}^d$ is called *convex* if, for any two points $\mathbf{a}$ and $\mathbf{b}$ belonging to C, the segment

$$\{\, t\mathbf{a} + (1-t)\mathbf{b} \; : \; 0 \leq t \leq 1 \,\}$$

is contained in C. Clearly, $\mathbb{R}^d$ is a convex set. Furthermore, if $\{C_\lambda\}_{\lambda \in \Lambda}$ is a family of convex sets of $\mathbb{R}^d$ with $\cap_{\lambda \in \Lambda} C_\lambda \neq \emptyset$, then $\cap_{\lambda \in \Lambda} C_\lambda$ is again a convex set of $\mathbb{R}^d$. It then follows easily that, given a nonempty subset $X \subset \mathbb{R}^d$, there exists a unique

© Springer International Publishing AG, part of Springer Nature 2018   87
J. Herzog et al., *Binomial Ideals*, Graduate Texts in Mathematics 279,
https://doi.org/10.1007/978-3-319-95349-6_4

convex set $\operatorname{conv}(X) \subset \mathbb{R}^d$ with $X \subset \operatorname{conv}(X)$ such that, if $C \subset \mathbb{R}^d$ is a convex set with $X \subset C$, then $\operatorname{conv}(X) \subset C$. We say that $\operatorname{conv}(X)$ is the *convex hull* of X.

If X is a finite subset $\{\mathbf{a}_1, \ldots, \mathbf{a}_s\}$ of $\mathbb{R}^d$, then one has

$$\operatorname{conv}(X) = \left\{ \sum_{i=1}^{s} r_i \mathbf{a}_i \in \mathbb{R}^d \ : \ 0 \leq r_i \in \mathbb{R}, \ i = 1, \ldots, s, \ \sum_{i=1}^{s} r_i = 1 \right\}. \quad (4.1)$$

4.1.2 Convex Polytopes

A *convex polytope* of $\mathbb{R}^d$ is a convex hull of a nonempty finite set of $\mathbb{R}^d$. For example, the tetrahedron of $\mathbb{R}^3$ consisting of those points $(x, y, z) \in \mathbb{R}^3$ satisfying

$$x \geq 0, \ y \geq 0, \ z \geq 0, \ 2x + 3y + 5z \leq 1$$

is the convex hull of $\{(0, 0, 0), (1/2, 0, 0), (0, 1/3, 0), (0, 0, 1/5)\}$ and is a convex polytope of $\mathbb{R}^3$.

4.1.3 Faces

A *hyperplane* of $\mathbb{R}^d$ is a subset of $\mathbb{R}^d$ of the form

$$\mathcal{H} = \{ (z_1, \ldots, z_d) \in \mathbb{R}^d \ : \ a_1 z_1 + \cdots + a_d z_d = b \},$$

where each $a_i \in \mathbb{R}$ and $b \in \mathbb{R}$. Given a hyperplane $\mathcal{H} \subset \mathbb{R}^d$ as above, the closed half-spaces $\mathcal{H}^{(+)}$ and $\mathcal{H}^{(-)}$ of $\mathbb{R}^d$ are defined as follows:

$$\mathcal{H}^{(+)} = \{ (z_1, \ldots, z_d) \in \mathbb{R}^d \ : \ a_1 z_1 + \cdots + a_d z_d \geq b \},$$
$$\mathcal{H}^{(-)} = \{ (z_1, \ldots, z_d) \in \mathbb{R}^d \ : \ a_1 z_1 + \cdots + a_d z_d \leq b \}.$$

Let $\mathcal{P} \subset \mathbb{R}^d$ be a convex polytope. A *supporting hyperplane* of $\mathcal{P}$ is a hyperplane $\mathcal{H} \subset \mathbb{R}^d$ such that $\mathcal{H} \cap \mathcal{P} \neq \emptyset$, $\mathcal{H} \cap \mathcal{P} \neq \mathcal{P}$ and that either $\mathcal{P} \subset \mathcal{H}^{(+)}$ or $\mathcal{P} \subset \mathcal{H}^{(-)}$. A *face* of $\mathcal{P}$ is a subset of $\mathcal{P}$ of the form $\mathcal{H} \cap \mathcal{P}$, where $\mathcal{H}$ is a supporting hyperplane of $\mathcal{P}$.

We say that $\mathbf{v} \in \mathcal{P}$ is a *vertex* of $\mathcal{P}$ if $\{\mathbf{v}\}$ is a face of $\mathcal{P}$. It follows that $\mathbf{v} \in \mathcal{P}$ is a vertex of $\mathcal{P}$ if and only if the following condition is satisfied: If $\mathbf{v} = (\mathbf{v}' + \mathbf{v}'')/2$ with $\mathbf{v}', \mathbf{v}'' \in \mathcal{P}$, then $\mathbf{v}' = \mathbf{v}'' = \mathbf{v}$.

Theorem 4.1 *The number of vertices of a convex polytope is finite.*

Let $V(\mathscr{P})$ denote the set of vertices of $\mathscr{P}$. Then

$$\mathscr{P} = \mathrm{conv}(V(\mathscr{P})).$$

Furthermore, if $\mathscr{P} = \mathrm{conv}(X)$ with $X \subset \mathbb{R}^d$, then $V(\mathscr{P}) \subset X$.

Let M denote the matrix whose columns are those vectors $(\mathbf{v}, 1)^t$ with $\mathbf{v} \in V(\mathscr{P})$. Here $(\mathbf{v}, 1)^t$ is the transpose of $(\mathbf{v}, 1) \in \mathbb{R}^{d+1}$. The dimension $\dim \mathscr{P}$ of $\mathscr{P}$ is defined to be $\mathrm{rank}(M) - 1$, where $\mathrm{rank}(M)$ is the rank of M.

Let $\mathscr{F}$ be a face of $\mathscr{P}$. Then $\mathscr{F} = \mathrm{conv}(\mathscr{F} \cap V(\mathscr{P}))$. In particular, every face of $\mathscr{P}$ is again a convex polytope of $\mathbb{R}^d$. It follows from Theorem 4.1 that

Corollary 4.2 *The number of faces of a convex polytope is finite.*

The dimension $\dim \mathscr{F}$ of a face $\mathscr{F}$ is the dimension of $\mathscr{F}$ as a convex polytope of $\mathbb{R}^d$. A face of $\mathscr{P}$ of dimension 0 is a subset of $\mathscr{P}$ of the form $\{\mathbf{v}\}$ with $\mathbf{v} \in V(\mathscr{P})$. An *edge* of $\mathscr{P}$ is a face of $\mathscr{P}$ of dimension 1. A *facet* of $\mathscr{P}$ is a face $\mathscr{F}$ of $\mathscr{P}$ with $\dim \mathscr{F} = \dim \mathscr{P} - 1$. Given a face $\mathscr{F}$ of $\mathscr{P}$, there is a facet $\mathscr{F}'$ of $\mathscr{P}$ such that $\mathscr{F}$ is a face of $\mathscr{F}'$.

Let $\mathscr{F}$ be a face of $\mathscr{P}$ and $\mathscr{F}'$ a face of $\mathscr{F}$. Then $\mathscr{F}'$ is a face of $\mathscr{P}$. If $\mathscr{F}$ and $\mathscr{F}'$ are faces of $\mathscr{P}$ with $\mathscr{F} \cap \mathscr{F}' \neq \emptyset$, then $\mathscr{F} \cap \mathscr{F}'$ is a face of $\mathscr{P}$.

Let $\mathscr{F}_1, \ldots, \mathscr{F}_t$ be the facets of $\mathscr{P}$ and $\mathscr{F}_i = \mathscr{H}_i \cap \mathscr{P}$ and $\mathscr{P} \subset \mathscr{H}_i^{(+)}$, where $\mathscr{H}_i$ is a supporting hyperplane of $\mathscr{P}$, for each $1 \leq i \leq t$. Then $\mathscr{P} = \bigcap_{i=1}^{t} \mathscr{H}_i^{(+)}$. Conversely, if $\mathscr{H}_1, \ldots, \mathscr{H}_t$ are hyperplanes of $\mathbb{R}^d$ for which $\bigcap_{i=1}^{t} \mathscr{H}_i^{(+)}$ is nonempty and bounded, then $\bigcap_{i=1}^{t} \mathscr{H}_i^{(+)}$ is a convex polytope of $\mathbb{R}^d$.

4.1.4 *f*-Vectors

Let $\mathscr{P} \subset \mathbb{R}^d$ be a convex polytope with $\dim \mathscr{P} = \delta$. Write $f_i = f_i(\mathscr{P})$ for the number of faces $\mathscr{F}$ of $\mathscr{P}$ with $\dim \mathscr{F} = i$. In particular f_0 is the number of vertices of $\mathscr{P}$ and $f_{\delta-1}$ is the number of facets of $\mathscr{P}$. We say that the vector $f(\mathscr{P}) = (f_0, f_1, \ldots, f_{\delta-1})$ is the *f-vector* of $\mathscr{P}$.

4.1.5 Simplicial Polytopes

A *simplex* of $\mathbb{R}^d$ of dimension q is a convex polytope $\mathscr{Q} \subset \mathbb{R}^d$ with $\dim \mathscr{Q} = q$ such that $|V(\mathscr{Q})| = q + 1$. Every face of a simplex is a simplex. A *simplicial* polytope is a convex polytope any of whose faces is a simplex. Equivalently, a convex polytope $\mathscr{P}$ is simplicial if each of its facets is a simplex.

Problems

4.1 Compute the f-vector of a simplex of $\mathbb{R}^d$ of dimension q.

4.2 Compute the f-vector of the convex polytope $\mathcal{P} \subset \mathbb{R}^3$ which is the convex hull of $\{(1, 1, 1), (0, 1, 1), (1, 0, 1), (1, 1, 0), (-1, -1, -1)\}$.

4.3

(a) Show that $(6, 9, 5)$ is the f-vector of a convex polytope of dimension 3.
(b) Find $(v, e, f) \in \mathbb{Z}^3$ with $v > 0, e > 0, f > 0$ and $v - e + f = 2$ for which (v, e, f) cannot be the f-vector of any convex polytope of dimension 3.
(c) Find all the f-vectors of convex polytopes of dimension 3 with at most 6 vertices.

4.4 Show that a convex polytope $\mathcal{Q} \subset \mathbb{R}^d$ with $V(\mathcal{Q}) = \{\mathbf{a}_1, \ldots, \mathbf{a}_{q+1}\}$ is a simplex of dimension q if and only if the vectors $(\mathbf{a}_1, 1), \ldots, (\mathbf{a}_{q+1}, 1)$ belonging to $\mathbb{R}^{d+1}$ are linearly independent.

4.2 Normal Polytopes and Unimodular Triangulations

In algebraic combinatorics on convex polytopes the normality of convex polytopes and the unimodularity of triangulations play important roles. The systematic study of triangulations in the frame of initial ideals of toric ideals of convex polytopes will be achieved.

4.2.1 Integral Polytopes

A convex polytope $\mathcal{P} \subset \mathbb{R}^d$ is said to be *integral* if each vertex of $\mathcal{P}$ belongs to $\mathbb{Z}^d$. We often use the terminology an *integral polytope* instead of an integral convex polytopes. A $(0, 1)$-*polytope* is a convex polytope with the property that any of its vertices is a $(0, 1)$-vector.

Let $\mathcal{P} \subset \mathbb{R}^d$ be an integral convex polytope with $\mathcal{P} \cap \mathbb{Z}^d = \{\mathbf{a}_1, \ldots, \mathbf{a}_n\}$. We then introduce the configuration $A(\mathcal{P}) \in \mathbb{Z}^{(d+1) \times n}$ whose column vectors are

$$(\mathbf{a}_1, 1)^t, \ldots, (\mathbf{a}_n, 1)^t.$$

Here, as before, $(\mathbf{a}_i, 1)^t$ is the transpose of $(\mathbf{a}_i, 1) \in \mathbb{Z}^{d+1}$. For example, if $\mathcal{P} \subset \mathbb{R}^2$ is the polygon with the vertices $(0, 0)$, $(2, 0)$, and $(0, 3)$, then $\mathcal{P} \cap \mathbb{Z}^d$ consists of 7 integer vectors and

$$A(\mathcal{P}) = \begin{pmatrix} 0 & 1 & 2 & 0 & 0 & 0 & 1 \\ 0 & 0 & 0 & 1 & 2 & 3 & 1 \\ 1 & 1 & 1 & 1 & 1 & 1 & 1 \end{pmatrix}.$$

4.2.2   Integer Decomposition Property

Let $\mathscr{P} \subset \mathbb{R}^d$ be an integral polytope. Given an integer $N > 0$, the *dilated* polytope $N\mathscr{P}$ is defined as follows:

$$N\mathscr{P} = \{ N\mathbf{a} \in \mathbb{R}^d \; : \; \mathbf{a} \in \mathscr{P} \}.$$

In particular if the set of vertices of $\mathscr{P}$ is $V(\mathscr{P}) = \{\mathbf{v}_1, \ldots, \mathbf{v}_s\}$, then $V(N\mathscr{P}) = \{N\mathbf{v}_1, \ldots, N\mathbf{v}_s\}$.

Definition 4.3 We say that an integral polytope $\mathscr{P} \subset \mathbb{R}^d$ possesses the *integer decomposition property* if, for each $N > 0$ and for each $\mathbf{a} \in N\mathscr{P} \cap \mathbb{Z}^d$, there exist $\mathbf{a}_1, \ldots, \mathbf{a}_N$ belonging to $\mathscr{P} \cap \mathbb{Z}^d$, possibly $\mathbf{a}_i = \mathbf{a}_j$ for $i \neq j$, such that $\mathbf{a} = \mathbf{a}_1 + \cdots + \mathbf{a}_N$.

4.2.3   Normal Polytopes

Recall that a configuration is a matrix $A \in \mathbb{Z}^{d \times n}$ for which there exists $\mathbf{c} \in \mathbb{Q}^d$ with $\mathbf{a}_j \cdot \mathbf{c} = 1$ for $1 \leq j \leq n$. If $\mathbf{a}_1, \ldots, \mathbf{a}_n$ are the columns of A, then we define

$$\mathbb{Z}_{\geq 0} A = \left\{ \sum_{i=1}^n q_i \mathbf{a}_i \; : \; q_i \in \mathbb{Z}_{\geq 0} \right\},$$

$$\mathbb{Z} A = \left\{ \sum_{i=1}^n q_i \mathbf{a}_i \; : \; q_i \in \mathbb{Z} \right\},$$

$$\mathbb{Q}_{\geq 0} A = \left\{ \sum_{i=1}^n q_i \mathbf{a}_i \; : \; q_i \in \mathbb{Q}_{\geq 0} \right\}.$$

Definition 4.4 A configuration $A \in \mathbb{Z}^{d \times n}$ is called *normal* if

$$\mathbb{Z}_{\geq 0} A = \mathbb{Z} A \cap \mathbb{Q}_{\geq 0} A. \tag{4.2}$$

Furthermore, we say that an integral convex polytope $\mathscr{P} \subset \mathbb{R}^d$ is normal if the configuration $A(\mathscr{P}) \in \mathbb{Z}^{(d+1) \times n}$ is normal. A configuration $A \in \mathbb{Z}^{d \times n}$ is called *very ample* if

$$\mathbb{Z} A \cap \mathbb{Q}_{\geq 0} A \setminus \mathbb{Z}_{\geq 0} A \tag{4.3}$$

is a finite set. In particular, a normal configuration is very ample.

In the language of commutative algebra, it can be shown that a configuration $A \in \mathbb{Z}^{d \times n}$ is normal if and only if the toric ring $K[A]$ is normal, i.e., integrally closed in its quotient field (Problem 4.5).

Theorem 4.5 *If an integral convex polytope $\mathscr{P} \subset \mathbb{R}^d$ possesses the integer decomposition property, then $\mathscr{P}$ is normal.*

Proof In general, in the equality (4.2), the left-hand side is contained in the right-hand side. Let $\mathscr{P} \cap \mathbb{Z}^d = \{\mathbf{a}_1, \ldots, \mathbf{a}_n\}$ and $\alpha \neq 0$ belong to $\mathbb{Z} A(\mathscr{P}) \cap \mathbb{Q}_{\geq 0} A(\mathscr{P})$ with

$$\alpha = \frac{1}{q}(q_1(\mathbf{a}_1, 1) + \cdots + q_n(\mathbf{a}_n, 1)),$$

where $q > 0$ is an integer and each $q_i \in \mathbb{Z}_{\geq 0}$. Let $N = q_1 + \cdots + q_n$. Since α belongs to $\mathbb{Z} A(\mathscr{P})$, it follows that the $(d+1)$-th coordinate of α must be an integer. In other words, $(1/q)N$ must be a positive integer. Hence, by virtue of (4.1), it follows that α belongs to $(1/q)N \mathscr{P}'$, where $\mathscr{P}' \subset \mathbb{R}^{d+1}$ is the convex polytope which is the convex hull of $\{(\mathbf{a}_1, 1), \ldots, (\mathbf{a}_n, 1)\}$. Since $\mathscr{P}$ possesses the integer decomposition property, it follows that $\mathscr{P}'$ also possesses the integer decomposition property. Hence there exist nonnegative integers $q_1', \ldots, q_n'$ with $(1/q)N = q_1' + \cdots + q_n'$ for which

$$\alpha = q_1'(\mathbf{a}_1, 1) + \cdots + q_n'(\mathbf{a}_n, 1).$$

Thus $\alpha \in \mathbb{Z}_{\geq 0} A(\mathscr{P})$, as desired.   □

However, the converse of Theorem 4.5 is false.

Example 4.6 Let $\mathscr{P} \subset \mathbb{R}^3$ be the tetrahedron with the vertices

$$(0, 0, 0), (0, 1, 1), (1, 0, 1), (1, 1, 0).$$

Then $\mathscr{P}$ is normal, but cannot possess the integer decomposition property.

In fact, $\mathbb{Z}_{\geq 0} A(\mathscr{P})$ consists of those integer points $(x, y, z, w) \in \mathbb{Z}_{\geq 0}^4$ such that $x + y + z = 2w$. Furthermore, $\mathbb{Z} A(\mathscr{P})$ consists of those integer points $(x, y, z, w) \in \mathbb{Z}^4$ such that $x + y + z = 2w$. Hence $\mathbb{Z}_{\geq 0} A(\mathscr{P}) = \mathbb{Z} A(\mathscr{P}) \cap \mathbb{Q}_{\geq 0} A(\mathscr{P})$ and $\mathscr{P}$ is normal. On the other hand, even though $(1, 1, 1)$ belongs to $2\mathscr{P}$, it is impossible to write $(1, 1, 1) = \alpha + \beta$, where α and β belong to $\{(0, 0, 0), (0, 1, 1), (1, 0, 1), (1, 1, 0)\}$. Thus $\mathscr{P}$ cannot possess the integer decomposition property.

Theorem 4.7 *Let $\mathscr{P} \subset \mathbb{R}^d$ be an integral convex polytope and suppose that $\mathbb{Z} A(\mathscr{P})$ coincides with $\mathbb{Z}^{d+1}$. Then $\mathscr{P}$ is normal if and only if $\mathscr{P}$ possesses the integer decomposition property.*

Proof Work with the same notation as in the proof of Theorem 4.5. The "if" part follows from Theorem 4.5. We show that "only if" part. Let $\mathscr{P}$ be normal. Since

$\mathbb{Z}A(\mathscr{P}) = \mathbb{Z}^{d+1}$, it follows that

$$\mathbb{Z}_{\geq 0}A(\mathscr{P}) = \mathbb{Z}^{d+1} \cap \mathbb{Q}_{\geq 0}A(\mathscr{P}).$$

Let $\beta \in N\mathscr{P}$. Again, by virtue of (4.1), one has

$$\beta = q_1 \mathbf{a}_1 + \cdots + q_n \mathbf{a}_n,$$

where each $q_i \in \mathbb{Q}_{\geq 0}$ and $q_1 + \cdots + q_n = N$. Hence

$$(\beta, N) = q_1(\mathbf{a}_1, 1) + \cdots + q_n(\mathbf{a}_n, 1).$$

Thus $(\beta, N) \in \mathbb{Z}^{d+1} \cap \mathbb{Q}_{\geq 0}A(\mathscr{P})$. It then follows that $(\beta, N) \in \mathbb{Z}_{\geq 0}A(\mathscr{P})$. In other words,

$$(\beta, N) = q_1'(\mathbf{a}_1, 1) + \cdots + q_n'(\mathbf{a}_n, 1),$$

where each $q_i' \in \mathbb{Z}_{\geq 0}$ and $q_1' + \cdots + q_n' = N$. As a result,

$$\beta = q_1' \mathbf{a}_1 + \cdots + q_n' \mathbf{a}_n.$$

Thus $\mathscr{P}$ possesses the integer decomposition property, as required. □

4.2.4 Triangulations and Coverings

Let, as before, $\mathscr{P} \subset \mathbb{R}^d$ be an integral convex polytope of dimension $\dim \mathscr{P} = \delta$ and $\mathscr{P} \cap \mathbb{Z}^d = \{\mathbf{a}_1, \ldots, \mathbf{a}_n\}$. Let $A(\mathscr{P}) \in \mathbb{Z}^{(d+1)\times n}$ be a configuration whose column vectors are $(\mathbf{a}_1, 1)^t, \ldots, (\mathbf{a}_n, 1)^t$. It then follows that $\dim \mathscr{P} = \mathrm{rank}(A(\mathscr{P})) - 1$. A simplex belonging to $\mathscr{P}$ is a subset F of $\mathscr{P} \cap \mathbb{Z}^d$ for which $\mathscr{Q} = \mathrm{P}(F)$ is a simplex of $\mathbb{R}^d$, i.e., $\dim \mathscr{Q} = |F| - 1$. Thus in particular the empty set is a simplex belonging to $\mathscr{P}$ of dimension -1. Every subset of a simplex belonging to $\mathscr{P}$ is again a simplex belonging to $\mathscr{P}$. A maximal simplex belonging to $\mathscr{P}$ is a simplex belonging to $\mathscr{P}$ of dimension δ. Every simplex belonging to $\mathscr{P}$ is a subset of a maximal simplex belonging to $\mathscr{P}$ (Problem 4.8). A maximal simplex belonging to $\mathscr{P}$ is called *fundamental* if $\mathbb{Z}A(\mathscr{P}) = \mathbb{Z}A(F)$, where $A(F) \subset \mathbb{Z}^{(d+1)\times(\delta+1)}$ is the configuration whose column vectors are those $(\mathbf{a}_i, 1)^t$ with $\mathbf{a}_i \in F$.

Definition 4.8 A collection Δ of simplices belonging to $\mathscr{P}$ is called a *triangulation* of $\mathscr{P}$ if the following conditions are satisfied:

- If $F \in \Delta$ and $F' \subset F$, then $F' \in \Delta$;
- If F and G belong to Δ, then $\mathrm{P}(F) \cap \mathrm{P}(G) = \mathrm{P}(F \cap G)$;
- $\mathscr{P} = \cup_{F \in \Delta} \mathrm{P}(F)$.

Each simplex of a triangulation Δ of $\mathscr{P}$ is called a *face* of Δ. A *facet* of Δ is a face of Δ which is a maximal simplex belonging to $\mathscr{P}$. Every face of Δ is a subset of a facet of Δ. A triangulation Δ of $\mathscr{P}$ is called *unimodular* if every facet of Δ is fundamental.

Example 4.9 Let $\mathscr{P} \subset \mathbb{R}^3$ be a convex polytope whose vertices are

$$(0, 0, 0), (0, 1, 1), (1, 0, 1), (1, 1, 0), (1, 1, 1).$$

Then

$$\mathscr{P} \cap \mathbb{Z}^3 = \{(0, 0, 0), (0, 1, 1), (1, 0, 1), (1, 1, 0), (1, 1, 1)\}$$

and $\mathbb{Z}A(\mathscr{P}) = \mathbb{Z}^4$. Let

$$F_1 = \{(0, 0, 0), (0, 1, 1), (1, 0, 1), (1, 1, 0)\},$$

$$F_2 = \{(0, 1, 1), (1, 0, 1), (1, 1, 0), (1, 1, 1)\},$$

$$F_3 = \{(0, 0, 0), (0, 1, 1), (1, 0, 1), (1, 1, 1)\},$$

$$F_4 = \{(0, 0, 0), (0, 1, 1), (1, 1, 0), (1, 1, 1)\},$$

$$F_5 = \{(0, 0, 0), (1, 0, 1), (1, 1, 0), (1, 1, 1)\}.$$

Since $(1, 1, 1, 1) \notin \mathbb{Z}A(F_1)$, it follows that F_1 cannot be fundamental. Each of F_2, F_3, F_4, and F_5 is fundamental. Let Δ be a set consisting of F_3, F_4, F_5, and their subsets and Δ' a set consisting of F_1, F_2, and their subsets. Then each of Δ and Δ' is a triangulation of $\mathscr{P}$. Furthermore, Δ is unimodular and Δ' is not unimodular.

A collection Ω of maximal simplices belonging to $\mathscr{P}$ is called a *covering* of $\mathscr{P}$ if $\mathscr{P} = \bigcup_{F \in \Omega} \mathrm{P}(F)$. Every triangulation of $\mathscr{P}$ is a covering of $\mathscr{P}$. A covering Ω of $\mathscr{P}$ is called *unimodular* if every $F \in \Omega$ is fundamental.

Lemma 4.10 *Let Ω denote the set of maximal simplices belonging to $\mathscr{P}$. Then Ω is a covering of $\mathscr{P}$. Thus in particular every integral convex polytope possesses a covering.*

Proof Let $\alpha \in \mathscr{P}$ and, by using (4.1), write $\alpha = \sum_{i=1}^{n} r_i \mathbf{a}_i$, where each $r_i \in \mathbb{Q}_{\geq 0}$ and $\sum_{i=1}^{n} r_i = 1$. Among such expressions, we choose an expression for which $\{i : r_i \neq 0\}$ is minimal with respect to inclusion. Then $F = \{\mathbf{a}_i : r_i \neq 0\}$ is a simplex belonging to $\mathscr{P}$. To see why this is true, suppose that $\mathrm{P}(F)$ is not a simplex of $\mathbb{R}^d$. Let, say, $F = \{1, 2, \ldots, q\}$. Then $(\mathbf{a}_1, 1), (\mathbf{a}_2, 1), \ldots, (\mathbf{a}_q, 1)$ cannot be linearly independent. Let, say, $(\mathbf{a}_q, 1) = \sum_{i=1}^{q-1} r_i'(\mathbf{a}_i, 1)$ with each $r_i' \in \mathbb{Q}_{\geq 0}$. Then one has $\sum_{i=1}^{q-1} r_i' = 1$. Since $\alpha = \sum_{i=1}^{n} r_i \mathbf{a}_i$, where $0 \leq r_i \in \mathbb{Q}_{\geq 0}$ and $\sum_{i=1}^{q} r_i = 1$, it follows that

$$\alpha = \sum_{i=1}^{q-1} r_i \mathbf{a}_i + r_q (\sum_{i=1}^{q-1} r'_i (\mathbf{a}_i, 1)),$$

where

$$\sum_{i=1}^{q-1} r_i + r_q (\sum_{i=1}^{q-1} r'_i) = 1.$$

Thus α belongs to $P(\{\mathbf{a}_1, \mathbf{a}_2, \ldots, \mathbf{a}_{q-1}\})$, which contradicts the minimality of F. Hence F is a simplex belonging to $\mathscr{P}$. Let F' be a maximal simplex belonging to $\mathscr{P}$ with $F \subset F'$. Then $\alpha \in F'$. Hence Ω is a covering of $\mathscr{P}$, as desired. $\square$

Theorem 4.11 *An integral convex polytope which possesses a unimodular covering is normal.*

Proof Let Ω be a unimodular covering of an integral polytope $\mathscr{P} \subset \mathbb{R}^d$. What we must prove is the equality $\mathbb{Z}_{\geq 0} A(\mathscr{P}) = \mathbb{Z} A(\mathscr{P}) \cap \mathbb{Q}_{\geq 0} A(\mathscr{P})$. In general, the left-hand side is contained in the right-hand side. Let $\alpha \in \mathbb{Z} A(\mathscr{P}) \cap \mathbb{Q}_{\geq 0} A(\mathscr{P})$ and $\alpha = \sum_{i=1}^n r_i (\mathbf{a}_i, 1)$ with each $r_i \in \mathbb{Q}_{\geq 0}$. Let $r = \sum_{i=1}^n r_i > 0$ and $\alpha = (\alpha', r)$. Then, again by using (4.1), one has $(1/r)\alpha' \in \mathscr{P}$. Since Ω is a covering, it follows that there is $F \in \Omega$ with $(1/r)\alpha' \in P(F)$. Let, say, $F = \{\mathbf{a}_1, \mathbf{a}_2, \ldots, \mathbf{a}_\delta\}$, where $\delta = \dim \mathscr{P}$. Then $(1/r)\alpha' = \sum_{i=1}^\delta r'_i \mathbf{a}_i$, where each $r'_i \in \mathbb{Q}_{\geq 0}$ and $\sum_{i=1}^\delta r'_i = 1$. In particular $((1/r)\alpha', 1) \in \mathbb{Q}_{\geq 0} A(F)$, where $A(F) \subset \mathbb{Z}^{(d+1) \times (\delta+1)}$ is the configuration whose column vectors are $(\mathbf{a}_1, 1)^t, (\mathbf{a}_2, 1)^t, \ldots, (\mathbf{a}_\delta, 1)^t$. It then follows that $\alpha = r((1/r)\alpha', 1) \in \mathbb{Q}_{\geq 0} A(F)$. Since F is fundamental, one has $\mathbb{Z} A(\mathscr{P}) = \mathbb{Z} A(F)$. Hence $\alpha \in \mathbb{Z} A(F) \cap \mathbb{Q}_{\geq 0} A(F)$. Thus

$$\alpha = \sum_{i=1}^\delta q_i (\mathbf{a}_i, 1) = \sum_{i=1}^\delta r_i (\mathbf{a}_i, 1),$$

where each $q_i \in \mathbb{Z}$ and each $r'_i \in \mathbb{Q}_{\geq 0}$. Since F is a simplex belonging to $\mathscr{P}$, it follows that $(\mathbf{a}_1, 1), (\mathbf{a}_2, 1), \ldots, (\mathbf{a}_\delta, 1)$ are linearly independent. Thus $q_i = r_i$ for each $1 \leq i \leq \delta$. Hence $\alpha \in \mathbb{Z}_{\geq 0} A(F) \subset \mathbb{Z}_{\geq 0} A(\mathscr{P})$, as desired. $\square$

Corollary 4.12 *An integral convex polytope which possesses a unimodular triangulation is normal.*

The simplex $\mathscr{P} \subset \mathbb{R}^3$ of Example 4.6 clearly possesses a unimodular triangulation, but cannot possess the integer decomposition property.

4.2.5 Regular Triangulations

Let $\mathscr{P} \subset \mathbb{R}^d$ be an integral convex polytope with $\mathscr{P} \cap \mathbb{Z}^d = \{\mathbf{a}_1, \dots, \mathbf{a}_n\}$ and $A(\mathscr{P}) \in \mathbb{Z}^{(d+1)\times n}$ the configuration whose column vectors are $(\mathbf{a}_1, 1)^t, \dots, (\mathbf{a}_n, 1)^t$. Let $T = K[t_1^{\pm 1}, \dots, t_d^{\pm 1}, s]$ denote the Laurent polynomial ring in $(d+1)$ variables over a field K and $S = K[x_1, \dots, x_n]$ the polynomial ring in n variables over K. Given $\mathbf{a} = (a_1, \dots, a_d) \in \mathbb{Z}^d$, one can associate the Laurent monomial $\mathbf{t}^{\mathbf{a}} = t_1^{a_1} \cdots t_n^{a_d} \in T$. The toric ring $K[\mathscr{P}]$ of $\mathscr{P}$ is the toric ring $K[A(\mathscr{P})]$ of $A(\mathscr{P})$ and the toric ideal $I_\mathscr{P}$ is the toric ideal $I_{A(\mathscr{P})}$ of $A(\mathscr{P})$. In other words, $K[\mathscr{P}]$ is the subring of T generated by those Laurent monomials $\mathbf{t}^{\mathbf{a}_1}s, \dots, \mathbf{t}^{\mathbf{a}_n}s$ and $I_\mathscr{P}$ is the ideal of S which is the kernel of the ring homomorphism $\pi : S \to T$ defined by $\pi(x_i) = \mathbf{t}^{\mathbf{a}_i}s$ for $1 \leq i \leq n$.

Fix a monomial order $<$ on S and study the initial ideal $\mathrm{in}_<(I_\mathscr{P})$ of $I_\mathscr{P}$ with respect to $<$. Recall that the radical $\sqrt{\mathrm{in}_<(I_\mathscr{P})}$ of $\mathrm{in}_<(I_\mathscr{P})$ is the subset of S consisting of those polynomials $f \in S$ with $f^N \in \mathrm{in}_<(I_\mathscr{P})$ for some $N = N_f > 0$.

Lemma 4.13 *A subset F of $\mathscr{P} \cap \mathbb{Z}^d$ is a simplex belonging to $\mathscr{P}$ if*

$$\prod_{\mathbf{a}_i \in F} x_i \notin \sqrt{\mathrm{in}_<(I_\mathscr{P})}. \tag{4.4}$$

Proof Let $F = \{\mathbf{a}_{i_1}, \mathbf{a}_{i_2}, \dots, \mathbf{a}_{i_N}\} \subset \mathscr{P} \cap \mathbb{Z}^d$ satisfy (4.4). What we must prove is that the vectors $(\mathbf{a}_{i_1}, 1), (\mathbf{a}_{i_2}, 1), \dots, (\mathbf{a}_{i_N}, 1)$ belonging to $\mathbb{Q}^{d+1}$ are linearly independent. If not, then one has $(q_1, q_2, \dots, q_N) \neq (0, 0, \dots, 0)$ with each $q_i \in \mathbb{Z}$ such that

$$q_1(\mathbf{a}_{i_1}, 1) + q_2(\mathbf{a}_{i_2}, 1) + \cdots + q_N(\mathbf{a}_{i_N}, 1) = \mathbf{0}.$$

Let $U_+ = \{k : q_k > 0\}$ and $U_- = \{k : q_k < 0\}$. Then

$$\sum_{k \in U_+} q_k(\mathbf{a}_{i_k}, 1) = \sum_{k' \in U_-} -q_{k'}(\mathbf{a}_{i_{k'}}, 1).$$

Thus, in $T = K[t_1^{\pm 1}, \dots, t_d^{\pm 1}, s]$, one has

$$\prod_{k \in U_+} (\mathbf{t}^{\mathbf{a}_{i_k}}s)^{q_k} = \prod_{k' \in U_-} (\mathbf{t}^{\mathbf{a}_{i_{k'}}}s)^{-q_{k'}}.$$

Hence the binomial

$$\prod_{k \in U_+} x_{i_k}^{q_k} - \prod_{k' \in U_-} x_{i_{k'}}^{-q_{k'}}$$

belongs to $I_{\mathscr{P}}$. Thus either $u = \prod_{k \in U_+} x_{i_k}^{q_k}$ or $v = \prod_{k' \in U_-} x_{i_{k'}}^{-q_{k'}}$ belongs to $\mathrm{in}_<(I_{\mathscr{P}})$. Hence either $\sqrt{u}$ or $\sqrt{v}$ belongs to $\sqrt{\mathrm{in}_<(I_{\mathscr{P}})}$. Thus $\prod_{\mathbf{a}_i \in F} x_i \in \sqrt{\mathrm{in}_<(I_{\mathscr{P}})}$. $\qquad\square$

Now, we write $\Delta(\mathrm{in}_<(I_{\mathscr{P}}))$ for the set of those subsets $F \subset \mathscr{P} \cap \mathbb{Z}^d$ satisfying the condition (4.4). In other words,

$$\Delta(\mathrm{in}_<(I_{\mathscr{P}})) = \{ F \subset \mathscr{P} \cap \mathbb{Z}^d : \prod_{\mathbf{a}_i \in F} x_i \notin \sqrt{\mathrm{in}_<(I_{\mathscr{P}})} \}.$$

Lemma 4.13 says that $\Delta(\mathrm{in}_<(I_{\mathscr{P}}))$ consists of simplices belonging to $\mathscr{P}$.

Theorem 4.14 *The collection* $\Delta(\mathrm{in}_<(I_{\mathscr{P}}))$ *of simplices belonging to* $\mathscr{P}$ *is a triangulation of* $\mathscr{P}$.

Proof First it follows immediately from Lemma 4.13 that if $F \in \Delta(\mathrm{in}_<(I_{\mathscr{P}}))$ and $F' \subset F$, then $F' \in \Delta(\mathrm{in}_<(I_{\mathscr{P}}))$.

Second, given F and F' belonging to $\Delta(\mathrm{in}_<(I_{\mathscr{P}}))$, we show $\mathrm{conv}(F) \cap \mathrm{conv}(F') = \mathrm{conv}(F \cap F')$. One has $\mathrm{conv}(F \cap F') \subset \mathrm{conv}(F) \cap \mathrm{conv}(F')$. If $\mathrm{conv}(F) \cap \mathrm{conv}(F') \neq \mathrm{conv}(F \cap F')$, then there exist nonnegative integers q_i, q_i', q_j, q_k for which

$$\sum_{\mathbf{a}_i \in F \cap F'} q_i \mathbf{a}_i + \sum_{\mathbf{a}_j \in F \setminus F'} q_j \mathbf{a}_j = \sum_{\mathbf{a}_i \in F \cap F'} q_i' \mathbf{a}_i + \sum_{\mathbf{a}_k \in F' \setminus F} q_k \mathbf{a}_k,$$

$$\sum_{\mathbf{a}_i \in F \cap F'} q_i + \sum_{\mathbf{a}_j \in F \setminus F'} q_j = \sum_{\mathbf{a}_i \in F \cap F'} q_i' + \sum_{\mathbf{a}_k \in F' \setminus F} q_k,$$

$$\sum_{\mathbf{a}_j \in F \setminus F'} q_j \neq 0, \qquad \sum_{\mathbf{a}_k \in F' \setminus F} q_k \neq 0.$$

Then the binomial

$$\prod_{\mathbf{a}_i \in F \cap F'} x_i^{q_i} \prod_{\mathbf{a}_j \in F \setminus F'} x_j^{q_j} - \prod_{\mathbf{a}_i \in F \cap F'} x_i^{q_i'} \prod_{\mathbf{a}_k \in F' \setminus F} x_k^{q_k}$$

belongs to the toric ideal $I_{\mathscr{P}}$. Hence either $u = \prod_{\mathbf{a}_i \in F \cap F'} x_i^{q_i} \prod_{\mathbf{a}_j \in F \setminus F'} x_j^{q_j}$ or $v = \prod_{\mathbf{a}_i \in F \cap F'} x_i^{q_i'} \prod_{\mathbf{a}_k \in F' \setminus F} x_k^{q_k}$ belongs to the initial ideal $\mathrm{in}_<(I_{\mathscr{P}})$. Thus either $\sqrt{u}$ or $\sqrt{v}$ belongs to $\sqrt{\mathrm{in}_<(I_{\mathscr{P}})}$. As a result, either $\prod_{\mathbf{a}_i \in F} x_i$ or $\prod_{\mathbf{a}_i \in F'} x_i$ belongs to $\sqrt{\mathrm{in}_<(I_{\mathscr{P}})}$, which contradict the fact that each of F and F' belongs to $\Delta(\mathrm{in}_<(I_{\mathscr{P}}))$.

Third we prove $\mathscr{P} = \bigcup_{F \in \Delta(\mathrm{in}_<(I_{\mathscr{A}}))} \mathrm{conv}(F)$. It is known [94, Theorem 3.1.2] that there exists a nonzero and nonnegative integer vector $\omega = (\omega_1, \ldots, \omega_n)$ with $\mathrm{in}_<(I_{\mathscr{P}}) = \mathrm{in}_\omega(I_{\mathscr{P}}) = (\mathrm{in}_\omega(f) : 0 \neq f \in I_{\mathscr{P}})$, where $\mathrm{in}_\omega(f)$ is the sum of all terms of f such that the inner product of its exponent vector and ω is maximal.

Suppose $\text{conv}(\mathscr{P}) \neq \bigcup_{F \in \Delta(\text{in}_<(I_{\mathscr{P}}))} \text{conv}(F)$ and choose $\alpha \in \text{conv}(\mathscr{P}) \cap \mathbb{Q}^d$ with $\alpha \notin \bigcup_{F \in \Delta(\text{in}_<(I_{\mathscr{P}}))} \text{conv}(F)$. The set $\mathscr{X} \subset \mathbb{Q}^n$ of nonnegative vectors $(r_1, \ldots, r_n) \in \mathbb{Q}^n$ with $\sum_{i=1}^n r_i = 1$ for which $\alpha = \sum_{i=1}^n r_i \mathbf{a}_i$ is a bounded closed set of the distance space $\mathbb{Q}^n$ and the function $\omega_1 r_1 + \cdots + \omega_n r_n$ on $\mathscr{X}$ is continuous. Hence, by virtue of the extreme value theorem, there is $(r_1^*, \ldots, r_n^*) \in \mathscr{X}$ with

$$\omega_1 r_1^* + \cdots + \omega_n r_n^* = \min\{\omega_1 r_1 + \cdots + \omega_n r_n \ : \ (r_1, \ldots, r_n) \in \mathscr{X}\}.$$

Let $r_i^* = q_i^*/N$, where N is a positive integer and where each q_i^* is a nonnegative integer. Then $N\alpha = \sum_{i=1}^n q_i^* \mathbf{a}_i$ with $\sum_{i=1}^n q_i^* = N$. Let $u = \prod_{i=1}^n x_i^{q_i^*}$. If $u \notin \sqrt{\text{in}_<(I_{\mathscr{P}})}$, then $F = \{\mathbf{a}_i \in \mathscr{P} ; r_i^* \neq 0\} \in \Delta(\text{in}_<(I_{\mathscr{P}}))$ and $\alpha \in \bigcup_{F \in \Delta(\text{in}_<(I_{\mathscr{P}}))} \text{conv}(F)$, which contradict $\alpha \notin \bigcup_{F \in \Delta(\text{in}_<(I_{\mathscr{P}}))} \text{conv}(F)$. Hence $u \in \sqrt{\text{in}_<(I_{\mathscr{P}})}$. Thus there is an integer $m > 0$ with $u^m = \prod_{i=1}^n x_i^{mq_i^*} \in \text{in}_<(I_{\mathscr{P}})$. Macaulay's Theorem 1.19 says that there is a monomial $v = \prod_{i=1}^n x_i^{p_i}$ of degree Nm with $v \notin \text{in}_<(I_{\mathscr{P}})$ for which $u^m - v \in I_{\mathscr{P}}$.

Now, since $\text{in}_<(I_{\mathscr{P}}) = \text{in}_\omega(I_{\mathscr{P}})$, it follows that

$$\omega_1 m q_1^* + \cdots + \omega_n m q_n^* > \omega_1 p_1 + \cdots + \omega_n p_n. \tag{4.5}$$

Since $u^m - v \in I_{\mathscr{P}}$, one has $mN\alpha = \sum_{i=1}^n m q_i^* \mathbf{a}_i = \sum_{i=1}^n p_i \mathbf{a}_i$. Thus

$$\alpha = \sum_{i=1}^n (p_i/mN)\mathbf{a}_i, \qquad \sum_{i=1}^n p_i/mN = 1.$$

Hence $(p_1, \ldots, p_n)/mN \in \mathscr{X}$. However, the inequality (4.5) then contradicts the minimality of $\omega_1 r_1^* + \cdots + \omega_n r_n^*$. $\square$

A triangulation Δ of $\mathscr{P}$ is called *regular* if there is a monomial order $<$ on S with $\Delta = \Delta(\text{in}_<(I_{\mathscr{P}}))$.

Example 4.15 The integral convex polytope $\mathscr{P} \subset \mathbb{R}^3$ with the vertices

$$(0, 0, 0), \ (0, 1, 1), \ (1, 0, 1), \ (1, 1, 0), \ (1, 1, 1)$$

possesses exactly two triangulations Δ and Δ' given in Example 4.9 and each of them is regular. (Problem 4.9.)

Example 4.16 A typical nonregular triangulation is now given. Let $\mathscr{P} \subset \mathbb{R}^2$ be the integral convex polytope with the vertices

$$\mathbf{a}_1 = (0, 2), \ \mathbf{a}_2 = (4, -2), \ \mathbf{a}_3 = (-4, -2).$$

Let

$$\mathbf{a}_4 = (0, 1), \ \mathbf{a}_5 = (2, -1), \ \mathbf{a}_6 = (-2, -1).$$

Fig. 4.1 A nonregular triangulation.

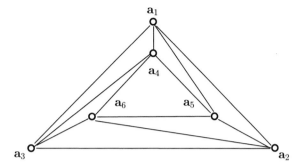

Then the triangulation of $\mathscr{P}$ consisting of

$$\{a_1, a_2, a_5\}, \quad \{a_1, a_4, a_5\}, \quad \{a_2, a_3, a_6\},$$

$$\{a_2, a_5, a_6\}, \quad \{a_1, a_3, a_4\}, \quad \{a_3, a_4, a_6\}, \quad \{a_4, a_5, a_6\}$$

and their subsets (Figure 4.1) is not regular. See Problem 4.10.

It is natural to ask when a regular triangulation $\Delta(\mathrm{in}_<(I_\mathscr{P}))$ is unimodular. Recall that $\mathrm{in}_<(I_\mathscr{P})$ is called squarefree if $\mathrm{in}_<(I_\mathscr{P}) = \sqrt{\mathrm{in}_<(I_\mathscr{P})}$. By Lemma 1.13, $\mathrm{in}_<(I_\mathscr{P})$ is squarefree if and only if $\mathrm{in}_<(I_\mathscr{P})$ is generated by squarefree monomials.

Theorem 4.17 *A regular triangulation $\Delta(\mathrm{in}_<(I_\mathscr{P}))$ is unimodular if and only if $\mathrm{in}_<(I_\mathscr{P})$ is squarefree.*

In order to prove Theorem 4.17 techniques on Hilbert functions together with Ehrhart functions will be required. Let f_i denote the number of faces F of $\Delta(\mathrm{in}_<(I_\mathscr{P}))$ with $|F| = i + 1$. We say that the sequence

$$f(\Delta(\mathrm{in}_<(I_\mathscr{P}))) = (f_0, f_1, \ldots, f_\delta),$$

where $\delta = \dim \mathscr{P}$, is the *f-vector* of $\Delta(\mathrm{in}_<(I_\mathscr{P}))$.

Lemma 4.18 *A monomial $u = x_1^{q_1} \cdots x_n^{q_n} \in S$ does not belong to $\sqrt{\mathrm{in}_<(I_\mathscr{P})}$ if and only if $W = \{a_i : q_i > 0\}$ is a face of $\Delta(\mathrm{in}_<(I_\mathscr{P}))$.*

Proof Since $\sqrt{\mathrm{in}_<(I_\mathscr{P})}$ is generated by squarefree monomials, it follows that a monomial $u = x_1^{q_1} \cdots x_n^{q_n}$ does not belong to $\sqrt{\mathrm{in}_<(I_\mathscr{P})}$ if and only if $\sqrt{u} = \prod_{q_i>0} x_i = \prod_{a_i \in W} x_i$ does not belong to $\sqrt{\mathrm{in}_<(I_\mathscr{P})}$. □

Corollary 4.19 *The number of monomials of S of degree N which do not belong to $\sqrt{\mathrm{in}_<(I_\mathscr{P})}$ is*

$$\sum_{i=0}^{\delta} f_i \binom{N-1}{i}, \qquad N = 1, 2, \ldots$$

Proof Let W be a face of $\Delta(\text{in}_<(I_{\mathscr{P}}))$ with $|W| = i + 1$. Then the number of monomials $u = x_1^{q_1} \cdots x_n^{q_n}$ of degree N with $u \notin \sqrt{\text{in}_<(I_{\mathscr{P}})}$ for which $W = \{\mathbf{a}_i : q_i > 0\}$ is

$$\binom{(i+1) + (N - i - 1) - 1}{N - i - 1} = \binom{N - 1}{i}.$$

Since the number of faces W of $\Delta(\text{in}_<(I_{\mathscr{P}}))$ with $|W| = i + 1$ is f_i, the desired result follows.   □

Let $\mathscr{P}^* \subset \mathbb{R}^{d+1}$ be the integral convex polytope which is the convex hull of $\{(\mathbf{a}, 1) \in \mathbb{R}^{d+1} : \mathbf{a} \in \mathscr{P} \cap \mathbb{Z}^d\}$ and

$$\mathbb{Z}\mathscr{P}^* = \mathbb{Z}(\mathbf{a}_1, 1) + \cdots + \mathbb{Z}(\mathbf{a}_n, 1).$$

In other words, $\mathscr{P}^* = \{(\alpha, 1) \in \mathbb{R}^{d+1} : \alpha \in \mathscr{P}\}$. Recall that the dilated polytope $N\mathscr{P}^* \subset \mathbb{R}^{d+1}$ is the convex polytope

$$N\mathscr{P}^* = \{N\alpha : \alpha \in \mathscr{P}^*\}, \qquad N = 1, 2, \ldots$$

Let $i(\mathscr{P}, N)$ denote the number of integer points $\alpha \in N\mathscr{P}^*$ which belong to $\mathbb{Z}\mathscr{P}^*$, that is to say,

$$i(\mathscr{P}, N) = |N\mathscr{P}^* \cap \mathbb{Z}\mathscr{P}^*|, \qquad N = 1, 2, \ldots$$

We say that $i(\mathscr{P}, N)$ is the *normalized Ehrhart function* of $\mathscr{P}$.

Lemma 4.20 *A maximal simplex F belonging to $\mathscr{P}$ is fundamental if and only if*

$$|N \cdot F^* \cap \mathbb{Z}_{\geq 0}\mathscr{P}^*| = \binom{\delta + N}{\delta}, \qquad N = 1, 2, \ldots, \tag{4.6}$$

where $\dim \mathscr{P} = \delta$.

Proof Let a maximal simplex $F = \{\mathbf{a}_{i_1}, \ldots, \mathbf{a}_{i_{\delta+1}}\}$ belonging to $\mathscr{P}$ be fundamental. Then

$$N \cdot F^* \cap \mathbb{Z}F^* = N \cdot F^* \cap \mathbb{Z}\mathscr{P}^*.$$

Let $\mathbf{a} \in N \cdot F^* \cap \mathbb{Z}F^*$ and write

$$\mathbf{a} = \sum_{j=1}^{\delta+1} r_j(\mathbf{a}_{i_j}, 1) = \sum_{j=1}^{\delta+1} q_j(\mathbf{a}_{i_j}, 1),$$

where each $0 \leq r_j \in \mathbb{Q}$ with $\sum_{j=1}^{\delta+1} r_j = N$ and where each $q_j \in \mathbb{Z}$. Since F is a simplex and since $\sum_{j=1}^{\delta+1} r_j = N$, Problem 4.4 guarantees that $r_j = q_j$ for $1 \leq j \leq \delta + 1$. Hence

$$N \cdot F^* \cap \mathbb{Z} \mathscr{P}^* = N \cdot F^* \cap \mathbb{Z}_{\geq 0} \mathscr{P}^*.$$

Again, by using Problem 4.4, it follows that $|N \cdot F^* \cap \mathbb{Z}_{\geq 0} \mathscr{P}^*|$ is equal to the number of sequences $(q_1, \ldots, q_{\delta+1}) \in \mathbb{Z}_{\geq 0}^{\delta+1}$ with $\sum_{j=1}^{\delta+1} N_j = N$. Hence the left-hand side of the formula (4.6) is $\binom{(\delta+1)+N-1}{N} = \binom{\delta+N}{\delta}$, as desired.

Now, suppose that a maximal simplex $F = \{\mathbf{a}_{i_1}, \ldots, \mathbf{a}_{i_{\delta+1}}\}$ belonging to $\mathscr{P}$ is not fundamental. Thus $\mathbb{Z} \mathscr{P}^* \neq \mathbb{Z} F^*$. Problem 4.12 then says that $\mathbb{Z}_{\geq 0} \mathscr{P}^* \neq \mathbb{Z} F^*$. Fix $\mathbf{a} \in \mathbb{Z} \mathscr{P}^* \setminus \mathbb{Z} F^*$. Let $\mathbb{Q} \mathscr{P}^*$ denote the vector space over $\mathbb{Q}$ spanned by $\mathbb{Z} \mathscr{P}^*$. Since F is a maximal simplex belonging to $\mathscr{P}$, it follows that $\{(\mathbf{a}_{i_1}, 1), \ldots, (\mathbf{a}_{i_{\delta+1}}, 1)\}$ is a $\mathbb{Q}$-basis of $\mathbb{Q} \mathscr{P}^*$. Thus one can write $\mathbf{a} = \sum_{j=1}^{\delta+1} r_j (\mathbf{a}_{i_j}, 1)$ with each $r_j \in \mathbb{Q}$. Since $\mathbf{a} \in \mathbb{Z}^{d+1}$, one has $\sum_{j=1}^{\delta+1} r_j \in \mathbb{Z}$. Choose $\mathbf{b} = \sum_{j=1}^{\delta+1} q_j (\mathbf{a}_{i_j}, 1) \in \mathbb{Z}_{\geq 0} F^*$ with each $q_j \in \mathbb{Z}_{\geq 0}$ for which each $r_j + q_j > 0$. Let $\sum_{j=1}^{\delta+1} (r_j + q_j) = N$. Then $\mathbf{a} + \mathbf{b} \in N \cdot F^*$. Since $\mathbf{a} + \mathbf{b} \in \mathbb{Z}_{\geq 0} \mathscr{P}^* \setminus \mathbb{Z}_{\geq 0} F^*$, it follows that

$$|N \cdot F^* \cap \mathbb{Z}_{\geq 0} \mathscr{P}^*| > |N \cdot F^* \cap \mathbb{Z}_{\geq 0} F^*| = \binom{\delta + N}{\delta}.$$

Hence F fails to satisfy the formula (4.6), as required.   □

Let $K[\mathbf{t}, \mathbf{t}^{-1}, s] = K[t_1, t_1^{-1}, \ldots, t_d, t_d^{-1}, s]$ denote the Laurent polynomial ring in $d + 1$ variables over a field K and $K[A(\mathscr{P})] \subset K[\mathbf{t}, \mathbf{t}^{-1}, s]$ the toric ring of the configuration $A(\mathscr{P}) \in \mathbb{Z}^{(d+1) \times n}$. Thus $K[A(\mathscr{P})]$ is the subring of $K[\mathbf{t}, \mathbf{t}^{-1}, s]$ generated by the monomials $\mathbf{t}^{\mathbf{a}_1} s, \ldots, \mathbf{t}^{\mathbf{a}_n} s$ with each $\deg(\mathbf{t}^{\mathbf{a}_i} s) = 1$. Let $H(K[A(\mathscr{P})], N)$ denote the number of monomials of degree N belonging to $K[A(\mathscr{P})]$, that is to say,

$$H(K[A(\mathscr{P})], N) = |\{\mathbf{t}^{\mathbf{a}} s^N : \mathbf{t}^{\mathbf{a}} s^N \in K[A(\mathscr{P})]\}|, \qquad N = 1, 2, \ldots$$

In other words,

$$H(K[A(\mathscr{P})], N) = |N \mathscr{P}^* \cap \mathbb{Z}_{\geq 0} \mathscr{P}^*|, \qquad N = 1, 2, \ldots$$

We say that $H(K[A(\mathscr{P})], N)$ is the *Hilbert function* of $K[A(\mathscr{P})]$.

It follows from Macaulay's Theorem 1.19 that

Lemma 4.21 *The number of monomials $u \in S = K[x_1, \ldots, x_n]$ of degree N not belonging to $\mathrm{in}_<(I_\mathscr{P})$ is equal to $H(K[A(\mathscr{P})], N)$.*

Lemma 4.22 *Let $f(\Delta(\mathrm{in}_<(I_\mathscr{P}))) = (f_0, f_1, \ldots, f_\delta)$ be the f-vector of the triangulation $\Delta(\mathrm{in}_<(I_\mathscr{P}))$, $i(\mathscr{P}, N)$ the normalized Ehrhart function of $\mathscr{P}$ and $H(K[A(\mathscr{P})], N)$ the Hilbert function of $K[A(\mathscr{P})]$.*

(a) *One has*

$$\sum_{i=0}^{\delta} f_i \binom{N-1}{i} \leq H(K[A(\mathscr{P})], N) \leq i(\mathscr{P}, N), \qquad N = 1, 2, \ldots$$

(b) *The integral polytope $\mathscr{P}$ is normal if and only if*

$$H(K[A(\mathscr{P})], N) = i(\mathscr{P}, N), \qquad N = 1, 2, \ldots$$

(c) *The triangulation $\Delta(\mathrm{in}_<(I_{\mathscr{P}}))$ is unimodular if and only if*

$$\sum_{i=0}^{\delta} f_i \binom{N-1}{i} = H(K[A(\mathscr{P})], N), \qquad N = 1, 2, \ldots \qquad (4.7)$$

Proof

(a) Since $\mathrm{in}_<(I_{\mathscr{P}}) \subset \sqrt{\mathrm{in}_<(I_{\mathscr{P}})}$, the left inequality follows from Corollary 4.19 and Lemma 4.21. Furthermore, if a monomial $\prod_{i=1}^{d}(\mathbf{t}^{\mathbf{a}_i}_i)^{q_i} s^N$ of degree N belongs to $K[A(\mathscr{P})]$, then $N = \sum_{i=1}^{d} q_i$ and $\sum_{i=1}^{d} q_i(\mathbf{a}_i, 1)$ belongs to $N\mathscr{P}^* \cap \mathbb{Z}\mathscr{P}^*$. Hence the right inequality follows.

(b) We claim

$$\mathbb{Q}_{\geq 0}\mathscr{P}^* \cap \mathbb{Z}\mathscr{P}^* = \{\mathbf{0}\} \cup \left(\cup_{N=1}^{\infty}(N\mathscr{P}^* \cap \mathbb{Z}\mathscr{P}^*) \right). \qquad (4.8)$$

Clearly the right-hand side of (4.8) is contained in the left-hand side of (4.8). Let $\alpha \in \mathbb{Q}_{\geq 0}\mathscr{P}^* \cap \mathbb{Z}\mathscr{P}^*$ and

$$\alpha = \sum_{i=1}^{d} r_i(\mathbf{a}_i, 1) = \sum_{i=1}^{d} q_i(\mathbf{a}_i, 1),$$

with each $0 \leq r_i \in \mathbb{Q}$ and $q_i \in \mathbb{Z}$. One has $\sum_{i=1}^{d} r_i = \sum_{i=1}^{d} q_i$. Let $\sum_{i=1}^{d} r_i = N$. Then $N \in \mathbb{Z}_{\geq 0}$. Thus $\alpha \in N\mathscr{P}^*$, as desired.

It follows from (4.8) that $\mathscr{P}$ is normal if and only if

$$\mathbb{Z}_{\geq 0}\mathscr{P}^* = \{\mathbf{0}\} \cup \left(\cup_{N=1}^{\infty}(N\mathscr{P}^* \cap \mathbb{Z}\mathscr{P}^*) \right). \qquad (4.9)$$

Let $\mathscr{H}_N \subset \mathbb{R}^{d+1}$ be the hyperplane consisting of those points $(y_1, \ldots, y_n, N) \in \mathbb{R}^{d+1}$. Then one has (4.9) if and only if

$$\mathbb{Z}_{\geq 0}\mathscr{P}^* \cap \mathscr{H}_N = N\mathscr{P}^* \cap \mathbb{Z}\mathscr{P}^*, \qquad N = 1, 2, \ldots \qquad (4.10)$$

Since $|\mathbb{Z}_{\geq 0}\mathcal{P}^* \cap \mathcal{H}_N| = H(K[A(\mathcal{P})], N)$ and $|N\mathcal{P}^* \cap \mathbb{Z}\mathcal{P}^*| = i(\mathcal{P}, N)$ and since the left-hand side of (4.10) is contained in the right-hand side of (4.10), it follows that one has (4.9) if and only if $H(K[A(\mathcal{P})], N) = i(\mathcal{P}, N)$ for $N = 1, 2, \ldots$

(c) In general, given a simplex F belonging to $\mathcal{P}$, its interior $F^{(i)}$ is defined to be

$$F^{(i)} = \left\{ \sum_{\mathbf{a}_i \in F} r_i \mathbf{a}_i \ : \ 0 < r_i \in \mathbb{Q}, \ \sum_{\mathbf{a}_i \in F} r_i = 1 \right\}.$$

Let W and W' be faces of $\Delta(\text{in}_<(I_\mathcal{P}))$ with $W \neq W'$. Then, since $P(W) \cap P(W') = P(W \cap W')$, one has $W^{(i)} \cap W'^{(i)} = \emptyset$. Thus $\mathcal{P}$ possesses the direct sum decomposition

$$\mathcal{P} = \bigcup_{W \in \Delta(\text{in}_<(I_\mathcal{P}))} W^{(i)}. \tag{4.11}$$

Hence

$$N\mathcal{P}^* \cap \mathbb{Z}_{\geq 0}\mathcal{P}^* = \bigcup_{W \in \Delta(\text{in}_<(I_\mathcal{P}))} N(W^{(i)})^* \cap \mathbb{Z}_{\geq 0}\mathcal{P}^*, \tag{4.12}$$

where

$$(W^{(i)})^* = \{(\alpha, 1) \in \mathbb{R}^{d+1} \ : \ \alpha \in W^{(i)}\}.$$

One has $|N\mathcal{P}^* \cap \mathbb{Z}_{\geq 0}\mathcal{P}^*| = H(K[A(\mathcal{P})], N)$. If $W \in \Delta(\text{in}_<(I_\mathcal{P}))$ with $|W| = i + 1$, then $|N(W^{(i)})^* \cap \mathbb{Z}_{\geq 0}\mathcal{P}^*| \geq \binom{N-1}{i}$. Thus

$$\left| \bigcup_{W \in \Delta(\text{in}_<(I_\mathcal{P}))} N(W^{(i)})^* \cap \mathbb{Z}_{\geq 0}\mathcal{P}^* \right| \geq \sum_{i=0}^{\delta} f_i \binom{N-1}{i}.$$

As a result, one has (4.7) if and only if the following condition ($\sharp$) is satisfied:

($\sharp$) Each face $W \in \Delta(\text{in}_<(I_\mathcal{P}))$ with $|W| = i + 1$ enjoys the property that

$$|N(W^{(i)})^* \cap \mathbb{Z}_{\geq 0}\mathcal{P}^*| = \binom{N-1}{i}, \qquad N = 1, 2, \ldots \tag{4.13}$$

We show that ($\sharp$) is equivalent to the condition that $\Delta(\text{in}_<(I_\mathcal{P}))$ is unimodular.

Let, in general, F be a maximal simplex belonging to $\mathcal{P}$. Since the number of simplices W belonging to $\mathcal{P}$ with $W \subset F$ and with $|W| = i + 1$ is $\binom{\delta+1}{i+1}$,

$$|N \cdot F^* \cap \mathbb{Z}_{\geq 0}\mathscr{P}^*| \geq \sum_{i=0}^{\delta} \binom{\delta+1}{i+1}\binom{N-1}{i}, \qquad N = 1, 2, \ldots$$

Furthermore, counting the number of monomials of degree N in $\delta + 1$ variables yields

$$\binom{\delta+N}{\delta} = \sum_{i=0}^{\delta} \binom{\delta+1}{i+1}\binom{N-1}{i}.$$

It then follows from Lemma 4.20 that F is fundamental if and only if each simplex $W \subset F$ belonging to $\mathscr{P}$ with $|W| = i + 1$ enjoys the property (4.13). In particular if the condition ($\sharp$) holds, then each facet $F \in \Delta(\mathrm{in}_<(I_\mathscr{P}))$ is fundamental. Hence $\Delta(\mathrm{in}_<(I_\mathscr{P}))$ is unimodular. Conversely, suppose that $\Delta(\mathrm{in}_<(I_\mathscr{P}))$ is unimodular. Then each facet $F \in \Delta(\mathrm{in}_<(I_\mathscr{P}))$ is fundamental. Since each face $W \in \Delta(\mathrm{in}_<(I_\mathscr{P}))$ is a subset of a facet $F \in \Delta(\mathrm{in}_<(I_\mathscr{P}))$, the condition ($\sharp$) is satisfied. □

Theorem 4.17 now follows from Lemma 4.22. In fact,

Proof (Proof of Theorem 4.17) It follows from Corollary 4.19 and Lemma 4.21 that $\sqrt{\mathrm{in}_<(I_\mathscr{P})} = \mathrm{in}_<(I_\mathscr{P})$ if and only if the equalities (4.7) hold. Thus the desired result follows from Lemma 4.22 (c). □

Corollary 4.23 *An integral convex polytope $\mathscr{P} \subset \mathbb{R}^d$ is normal if there is a monomial order $<$ on S with $\sqrt{\mathrm{in}_<(I_\mathscr{P})} = \mathrm{in}_<(I_\mathscr{P})$.*

Corollary 4.24 *Suppose that $\mathscr{P}$ possesses a regular unimodular triangulation. Then*

$$i(\mathscr{P}, N) = \sum_{i=0}^{\delta} f_i \binom{N-1}{i}, \qquad N = 1, 2, \ldots$$

The converse of Corollary 4.23 is false.

Example 4.25 Let $\mathscr{P} \subset \mathbb{R}^{10}$ be the integral convex polytope with $\dim \mathscr{P} = 9$ whose vertices are

$$\mathbf{e}_1 + \mathbf{e}_2, \quad \mathbf{e}_2 + \mathbf{e}_3, \quad \mathbf{e}_3 + \mathbf{e}_4, \quad \mathbf{e}_4 + \mathbf{e}_5, \quad \mathbf{e}_1 + \mathbf{e}_5,$$

$$\mathbf{e}_1 + \mathbf{e}_6, \quad \mathbf{e}_2 + \mathbf{e}_6, \quad \mathbf{e}_2 + \mathbf{e}_7, \quad \mathbf{e}_3 + \mathbf{e}_7, \quad \mathbf{e}_3 + \mathbf{e}_8,$$

$$\mathbf{e}_4 + \mathbf{e}_8, \quad \mathbf{e}_4 + \mathbf{e}_9, \quad \mathbf{e}_5 + \mathbf{e}_9, \quad \mathbf{e}_1 + \mathbf{e}_{10}, \quad \mathbf{e}_5 + \mathbf{e}_{10}.$$

Then the following five binomials appear in any minimal set of binomial generators of $I_\mathscr{P}$:

$$x_2 x_5 x_8 x_{14} - x_1^2 x_9 x_{15}, \quad x_1 x_7 x_3 x_{10} - x_2^2 x_6 x_{11}, \quad x_2 x_4 x_9 x_{12} - x_3^2 x_8 x_{13},$$

$$x_3x_5x_{11}x_{15} - x_4^2x_{10}x_{14}, \quad x_1x_4x_6x_{13} - x_5^2x_7x_{12}. \tag{4.14}$$

It is easy to see that there exists no monomial order such that the initial monomial of any binomial in (4.14) is squarefree (Problem 4.13). Hence there is no monomial order $<$ on $K[x_1, \ldots, x_{15}]$ with $\sqrt{\text{in}_<(I_{\mathscr{P}})} = \text{in}_<(I_{\mathscr{P}})$. On the other hand, we can check by using a specialized software (e.g., TOPCOM) that $\mathscr{P}$ has a nonregular unimodular triangulation and hence is normal.

Our work on initial ideals and regular triangulations has been naturally achieved in the frame of configurations arising from integral convex polytopes. On the other hand, however, it is straightforward to recognize that, in the language of commutative algebra, Corollary 4.23 can be interpreted in the following:

Corollary 4.26 Let $A \in \mathbb{Z}^{d \times n}$ be a configuration and $K[A]$ its toric ring. Let $I_A \subset S = K[x_1, \ldots, x_n]$ be the toric ideal of A. If there is a monomial order $<$ on S with $\sqrt{\text{in}_<(I_A)} = \text{in}_<(I_A)$, then $K[A]$ is normal, i.e., integrally closed in its quotient field.

Proposition 4.27 Let $A \in \mathbb{Z}^{d \times n}$ be a $(0, 1)$ configuration and $K[A]$ its toric ring. If I_A has a quadratic Gröbner basis, then $K[A]$ is normal.

Proof Let $A = (\mathbf{a}_1, \ldots, \mathbf{a}_n)$. Suppose that there exists a monomial order $<$ such that a Gröbner basis $\{g_1, \ldots, g_s\}$ of I_A is quadratic. We may assume that $\{g_1, \ldots, g_s\}$ is reduced. By Theorem 3.6, each g_i is a binomial. If $g_i = x_j^2 - x_k x_\ell$ for some $1 \leq i \leq s$, then $2\mathbf{a}_j = \mathbf{a}_k + \mathbf{a}_\ell$. Since $\mathbf{a}_j, \mathbf{a}_k$, and $\mathbf{a}_\ell$ are $(0,1)$ vectors, we have $\mathbf{a}_j = \mathbf{a}_k = \mathbf{a}_\ell$, and hence $j = k = \ell$. Thus $g_i = 0$, which is a contradiction. Hence both monomials in g_i are squarefree for each $1 \leq i \leq s$. Thus $\text{in}_<(I_A)$ is squarefree. By Corollary 4.26, $K[A]$ is normal. □

Example 4.28 (Example 2.30) Let

$$A = (\mathbf{a}_1, \ldots, \mathbf{a}_8) = \begin{pmatrix} 1 & 1 & 1 & 1 & 0 & 0 & 0 & 0 \\ 1 & 0 & 0 & 1 & 1 & 0 & 0 & 1 \\ 1 & 1 & 0 & 0 & 1 & 0 & 1 & 0 \\ 0 & 1 & 1 & 0 & 0 & 1 & 1 & 0 \\ 0 & 0 & 1 & 1 & 0 & 1 & 0 & 1 \\ 0 & 0 & 0 & 0 & 1 & 1 & 0 & 0 \\ 0 & 0 & 0 & 0 & 0 & 0 & 1 & 1 \end{pmatrix} \in \mathbb{Z}^{7 \times 8}.$$

Then the toric ideal I_A of A is generated by the quadratic binomials

$$x_2x_8 - x_4x_7, \quad x_1x_6 - x_3x_5, \quad x_1x_3 - x_2x_4.$$

Let $\alpha = (0, 1, 1, 1, 1, 1, 1)^t$. Since we have

$$\alpha = \frac{1}{2}(\mathbf{a}_5 + \mathbf{a}_6 + \mathbf{a}_7 + \mathbf{a}_8) = \mathbf{a}_4 + \mathbf{a}_5 + \mathbf{a}_7 - \mathbf{a}_1,$$

the vector α belongs to $\mathbb{Q}_{\geq 0}A \cap \mathbb{Z}A$. However, α does not belong to $\mathbb{Z}_{\geq 0}A$. Thus $K[A]$ is not normal. By Proposition 4.27, I_A has no quadratic Gröbner bases. We now show that $K[A] \cong K[x_1, \ldots, x_8]/I_A$ is Koszul. Let $B = (\mathbf{a}_1, \ldots, \mathbf{a}_7) \in \mathbb{Z}^{7 \times 7}$ be a subconfiguration of A. Then the toric ideal I_B of B has a quadratic Gröbner basis

$$\{x_1 x_6 - x_3 x_5, \quad x_1 x_3 - x_2 x_4\}$$

with respect to a reverse lexicographic order induced by $x_1 < \cdots < x_7$. Hence $K[B] \cong K[x_1, \ldots, x_7]/I_B$ is Koszul. Then $(K[x_1, \ldots, x_7]/I_B)[x_8]$ is also Koszul. Since

$$K[x_1, \ldots, x_8]/I_A = (K[x_1, \ldots, x_7]/I_B)[x_8]/(x_2 x_8 - x_4 x_7)$$

and since $x_2 x_8 - x_4 x_7$ is a nonzerodivisor on $(K[x_1, \ldots, x_7]/I_B)[x_8]$, it follows from Corollary 2.22 that $K[A]$ is Koszul.

Problems

4.5 Show that a configuration $A \in \mathbb{Z}^{d \times n}$ is normal if and only if the toric ring $K[A]$ is normal.

4.6 Let $\mathscr{P} \subset \mathbb{R}^3$ be the integral polytope with the vertices

$$(2, 0, 0), (0, 2, 0), (0, 0, 2), (1, 1, 0), (1, 0, 1), (0, 1, 1).$$

Show that $\mathscr{P}$ possesses the integer decomposition property.

4.7 Find an example of a very ample integral polytope which is nonnormal.

4.8 Let $\mathscr{P} \subset \mathbb{R}^d$ be an integral convex polytope. Show that every simplex belonging to $\mathscr{P}$ is a subset of a maximal simplex belonging to $\mathscr{P}$.

4.9 In Example 4.15, show that the integral polytope $\mathscr{P} \subset \mathbb{R}^3$ possesses exactly two triangulations and each of them is regular.

4.10 In Example 4.16, show that the triangulation is not regular.

4.11 Find the normalized Ehrhart function of the integral convex polytope $\mathscr{P} \subset \mathbb{R}^2$ with the vertices $(0, 0)$, $(3, 0)$, $(0, 2)$, $(4, 3)$.

4.12 In the proof of Lemma 4.20 show that if $\mathbb{Z}_{\geq 0}\mathscr{P}^* = \mathbb{Z}F^*$, then $\mathbb{Z}\mathscr{P}^* = \mathbb{Z}F^*$.

4.13 Show that there exists no monomial order such that the initial monomial of any binomial in (4.14) is squarefree.

4.3 Unimodular Polytopes

Unimodular polytopes, which form a distinguished subclass of the class of normal polytopes, are discussed.

An integral convex polytope $\mathscr{P} \subset \mathbb{R}^d$ is called *unimodular* if every triangulation of $\mathscr{P}$ is unimodular. For example, the integral convex polytope $\mathscr{P} \subset \mathbb{R}^3$ discussed in Example 4.9 cannot be unimodular.

Theorem 4.29 *Given an integral convex polytope $\mathscr{P} \subset \mathbb{R}^d$, the following conditions are equivalent:*

(i) *$\mathscr{P}$ is unimodular;*
(ii) *Every maximal simplex belonging to $\mathscr{P}$ is fundamental;*
(iii) *Every regular triangulation of $\mathscr{P}$ is unimodular;*
(iv) *The initial ideal $\mathrm{in}_{<_{\mathrm{lex}}}(I_{\mathscr{P}})$ is squarefree for any lexicographic order $<_{\mathrm{lex}}$.*

Proof Each of (ii) $\Rightarrow$ (i) $\Rightarrow$ (iii) $\Rightarrow$ (iv) is clear. We prove (iv) $\Rightarrow$ (ii). Let $I_{\mathscr{P}} \subset S = K[x_1, \ldots, x_n]$ be the toric ideal of $\mathscr{P}$. Let F be an arbitrary maximal simplex belonging to $\mathscr{P}$ and fix a total order $<$ on the variables of S with the property that, for $\mathbf{a}_i$ and $\mathbf{a}_j$ belonging to $\mathscr{P} \cap \mathbb{Z}^d$, if $\mathbf{a}_i \in F$ and $\mathbf{a}_j \notin F$, then $x_i < x_j$. Let $<_{\mathrm{lex}}$ be the lexicographic order on S induced by $<$. We claim that F belongs to the regular triangulation $\Delta(\mathrm{in}_{<_{\mathrm{lex}}}(I_{\mathscr{P}}))$. In fact, if $F \notin \Delta(\mathrm{in}_{<_{\mathrm{lex}}}(I_{\mathscr{P}}))$, then

$$\prod_{\mathbf{a}_i \in F} x_i \in \sqrt{\mathrm{in}_{<_{\mathrm{lex}}}(I_{\mathscr{P}})} = \mathrm{in}_{<_{\mathrm{lex}}}(I_{\mathscr{P}}).$$

Thus there exists a binomial $f = u - v \in I_{\mathscr{P}}$ with $f \neq 0$ for which $\mathrm{in}_{<_{\mathrm{lex}}}(f) = u = \prod_{\mathbf{a}_i \in F} x_i$. Since F is a simplex, it follows that $I_{\mathscr{P}} \cap K[\{x_i : \mathbf{a}_i \in F\}] = (0)$. In particular $f \notin K[\{x_i : \mathbf{a}_i \in F\}]$. Thus there is j_0 with $\mathbf{a}_{j_0} \notin F$ such that x_{j_0} divides v. Since $x_i < x_{j_0}$ for each i with $\mathbf{a}_i \in F$, one has $u <_{\mathrm{lex}} v$, which contradicts $\mathrm{in}_{<_{\mathrm{lex}}}(f) = u$. Thus $F \in \Delta(\mathrm{in}_{<_{\mathrm{lex}}}(I_{\mathscr{P}}))$. Since $\Delta(\mathrm{in}_{<_{\mathrm{lex}}}(I_{\mathscr{P}}))$ is unimodular, it follows that F is fundamental, as desired. □

In general, a configuration $A = (\mathbf{a}_1, \mathbf{a}_2, \ldots, \mathbf{a}_n) \in \mathbb{Z}^{d \times n}$ is called *unimodular* if, for an arbitrary monomial order $<$ on $S = K[x_1, \ldots, x_n]$, the initial ideal $\mathrm{in}_{<}(I_A)$ of the toric ideal I_A is squarefree. It follows from the proof of Theorem 4.29 that A is unimodular if and only if, with respect to any lexicographic order $<_{\mathrm{lex}}$ on S, the initial ideal $\mathrm{in}_{<_{\mathrm{lex}}}(I_A)$ is squarefree.

Given a monomial $u \in S$, let $\mathrm{var}(u)$ denote the set of those variables x_i which divides u. Moreover, for a binomial $f = u - v$, where u and v are monomials belonging to S with $u \neq v$, let

$$\mathrm{var}(f) = \mathrm{var}(u) \cup \mathrm{var}(v).$$

We say that a binomial $f = u - v$ is *squarefree* if each of u and v is squarefree.

An irreducible binomial f belonging to a toric ideal I_A is called a *circuit* of I_A if there is no binomial $g \in I_A$ with $g \neq 0$ such that $\mathrm{var}(g) \subset \mathrm{var}(f)$ and $\mathrm{var}(g) \neq \mathrm{var}(f)$.

First, we observe the following fact.

Lemma 4.30 *Let $f = u - v$ be a binomial, where u and v are relatively prime. Then f is reducible if and only if there exist monomials u' and v' together with an integer $p > 1$ for which $u = u'^p$ and $v = v'^p$.*

Proof Since u and v are relatively prime, there exists a vector $\mathbf{b} \in \mathbb{Z}^n$ such that $f = f_{\mathbf{b}}$. Then the ideal (f) is a lattice ideal I_L where L is a lattice generated by $\mathbf{b}$. By Theorem 3.17, I_L is prime if and only if the abelian group $\mathbb{Z}^n / L$ is torsionfree, that is, there exist no integers $p > 1$ such that $\mathbf{b} = p\mathbf{b}'$ for some $\mathbf{b}' \in \mathbb{Z}^n$. Thus f is irreducible if and only if there exist no monomials u' and v' together with an integer $p > 1$ for which $u = u'^p$ and $v = v'^p$. □

It follows from Lemma 4.30 that every primitive binomial of a toric ideal is irreducible.

Lemma 4.31 *Let $g \in I_A$ be an irreducible binomial and $f \in I_A$ a circuit. Suppose that $\mathrm{var}(g) = \mathrm{var}(f)$. Then $g = f$.*

Proof Let, say, $x_1 \in \mathrm{var}(f)$ and $f = x_1^p u - v$, where $p \geq 1$ and $x_1 \notin \mathrm{var}(u)$. Let $g = x_1^q u' - v'$ with $x_1 \notin \mathrm{var}(u')$. Since each of the binomials $(x_1^p u)^q - v^q$ and $(x_1^q u')^p - v'^p$ belongs to I_A, one has $h = u^q v'^p - u'^p v^q \in I_A$. Since f is a circuit and since $\mathrm{var}(h) \subset \mathrm{var}(f)$ with $x_1 \notin \mathrm{var}(h)$, it follows that $h = 0$ and $u^q v'^p = u'^p v^q$. Furthermore, since $\mathrm{var}(u) \cap \mathrm{var}(v) = \emptyset$ and $\mathrm{var}(u') \cap \mathrm{var}(v') = \emptyset$, one has $u^q = u'^p$ and $v^q = v'^p$. Let $p \neq q$, say, $p < q$. Then there exist a prime number $k > 1$ and an integer $\ell \geq 1$ such that k^ℓ divides q, but k^ℓ does not divide p. If $x_i^{a_i}$ divides either u' or v', then k divides a_i. Hence $g = x_1^q u' - v' = (x_1^{q'} u_0')^k - (v_0')^k$ cannot be irreducible. Similarly, if $p > q$, then f cannot be irreducible. As a result, one has $p = q$. Thus $f = g$, as desired. □

Lemma 4.32 *Given a binomial $f = u - v$ with $f \neq 0$ belonging to a toric ideal I_A, there is a circuit $g = u' - v' \in I_A$ with $\mathrm{var}(u') \subset \mathrm{var}(u)$ and $\mathrm{var}(v') \subset \mathrm{var}(v)$.*

Proof By virtue of Lemma 4.30, one can assume that f is irreducible. We work by using induction on $|\mathrm{var}(f)|$. If $f = u - v \in I_A$ is an irreducible binomial with $|\mathrm{var}(f)| = 3$, then f is a circuit. Let $\mathrm{var}(f) = \{x_{i_1}, x_{i_2}, \ldots, x_{i_q}\}$ with $q > 3$. Considering the ideal $I_A \cap K[x_{i_1}, x_{i_2}, \ldots, x_{i_q}]$, which is the toric ideal of the subconfiguration of A with the column vectors $\mathbf{a}_{i_1}, \mathbf{a}_{i_2}, \ldots, \mathbf{a}_{i_q}$, one can assume that $\mathrm{var}(f) = \{x_1, x_2, \ldots, x_n\}$. Furthermore, since f is irreducible, one has $\mathrm{var}(u) \cap \mathrm{var}(v) = \emptyset$. Let $g = u' - v' \in I_A$ be a circuit. Since $\mathrm{var}(g) \subset \mathrm{var}(f) = \{x_1, x_2, \ldots, x_n\}$, we may assume that $\mathrm{var}(u) \cap \mathrm{var}(u') \neq \emptyset$. For each $x_i \in \mathrm{var}(u) \cap \mathrm{var}(u')$, we write a_i (resp. b_i) for the maximal integer for which $x_i^{a_i} | u$ (resp. $x_i^{b_i} | u'$). Similarly, for each $x_j \in \mathrm{var}(v) \cap \mathrm{var}(v')$, we write a_j (resp. b_j) for the maximal integer for which $x_j^{a_j} | v$ (resp. $x_j^{b_j} | v'$). Let a/b be the smallest rational number in the nonempty finite set

$$\{a_i/b_i \ : \ x_i \in \text{var}(u) \cap \text{var}(u')\} \cup \{a_j/b_j \ : \ x_j \in \text{var}(v) \cap \text{var}(v')\}.$$

Let $f^* = u^b - v^b$ and $g^* = u'^a - v'^a$, each of which belongs to I_A. If $x_i \in \text{var}(u) \cap \text{var}(u')$, then $ab_i \leq ba_i$. If $x_j \in \text{var}(v) \cap \text{var}(v')$, then $ab_j \leq ba_j$. Now, write $h = u'' - v'' \in I_A$ for the binomial arising from $u^b v'^a - u'^a v^b$ by canceling those variables which appear in both $u^b v'^a$ and $u'^a v^b$. Since $\text{var}(f) = \{x_1, x_2, \ldots, x_n\}$, it follows that $\text{var}(u'') \subset \text{var}(u)$ and $\text{var}(v'') \subset \text{var}(v)$. Furthermore, no variable x_k with $a/b = a_k/b_k$ can belong to $\text{var}(h)$. Let $h = 0$. Since $\text{var}(u) \cap \text{var}(v) = \emptyset$, one has $\text{var}(u) \subset \text{var}(u')$ and $\text{var}(v) \subset \text{var}(v')$. In addition, since $\text{var}(f) = \{x_1, x_2, \ldots, x_n\}$, one has $\text{var}(u) = \text{var}(u')$ and $\text{var}(v) = \text{var}(v')$. Let $h \neq 0$. Then, by assumption of induction, it follows that there is a circuit $g_0 = u_0 - v_0 \in I_A$ for which $\text{var}(u_0) \subset \text{var}(u'')$ and $\text{var}(v_0) \subset \text{var}(v'')$. $\quad\square$

Theorem 4.33 *Every circuit of a toric ideal I_A belongs to the universal Gröbner basis of I_A.*

Proof Given a circuit $f = u - v \in I_A$, we fix a lexicographic order $<_{\text{lex}}$ such that (i) if $x_i \in \text{var}(f)$ and $x_j \notin \text{var}(f)$, then $x_i <_{\text{lex}} x_j$ and (ii) $v <_{\text{lex}} u$. Let $\mathcal{G}$ denote the reduced Gröbner basis of I_A with respect to $<_{\text{lex}}$. We claim $f \in \mathcal{G}$. Since $u = \text{in}_{<_{\text{lex}}}(f) \in \text{in}_{<_{\text{lex}}}(I_A)$, there is an irreducible binomial $g = u' - v' \in \mathcal{G}$ with $v' <_{\text{lex}} u'$ for which u' divides u. In particular $\text{var}(u') \subset \text{var}(u) \subset \text{var}(f)$. Suppose $\text{var}(v') \not\subset \text{var}(f)$. Then by using (i) one has $u' <_{\text{lex}} v'$, which contradict $v' <_{\text{lex}} u'$. Hence $\text{var}(v') \subset \text{var}(f)$. Thus $\text{var}(g) \subset \text{var}(f)$. Since f is a circuit, one has $\text{var}(g) = \text{var}(f)$. Lemma 4.31 then guarantees that $g = f$, as desired. $\quad\square$

For a configuration $A \in \mathbb{Z}^{d \times n}$, let $\mathcal{C}_A$, $\mathcal{U}_A$, and $\mathcal{G}r_A$ denote the set of all circuits, the universal Gröbner basis, and the Graver basis of I_A, respectively. By Theorems 3.13 and 4.33, we have

$$\mathcal{C}_A \subset \mathcal{U}_A \subset \mathcal{G}r_A.$$

By using the technique used in the proof of Theorem 4.33, it follows that

Lemma 4.34 *Let $f = u - v \in I_A$ be a circuit. Then there exist lexicographic orders $<_{\text{lex}}$ and $<'_{\text{lex}}$ such that*

(i) $u = \text{in}_{<_{\text{lex}}}(f)$ *and* $f \in \mathcal{G}_{<_{\text{lex}}}(I_A)$;
(ii) $v = \text{in}_{<'_{\text{lex}}}(f)$ *and* $f \in \mathcal{G}_{<'_{\text{lex}}}(I_A)$,

where, say, $\mathcal{G}_{<_{\text{lex}}}(I_A)$ is the reduced Gröbner basis of I_A with respect to $<_{\text{lex}}$.

Theorem 4.35 *A configuration $A \in \mathbb{Z}^{d \times n}$ is unimodular if and only if each circuit of the toric ideal I_A is squarefree.*

Proof First, suppose that $A \in \mathbb{Z}^{d \times n}$ is unimodular. Let $f = u - v \in I_A$ be a circuit. By using Lemma 4.34 it follows that there exist lexicographic orders $<_{\text{lex}}$ and $<'_{\text{lex}}$

such that $u = \text{in}_{<_{\text{lex}}}(f)$ with $f \in \mathcal{G}_{<_{\text{lex}}}(I_A)$ and $v = \text{in}_{<'_{\text{lex}}}(f)$ with $f \in \mathcal{G}_{<'_{\text{lex}}}(I_A)$. Now, since A is unimodular, each of $\text{in}_{<_{\text{lex}}}(I_A)$ and $\text{in}_{<'_{\text{lex}}}(I_A)$ is squarefree. Thus each of u and v is squarefree.

Second, suppose that each circuit of the toric ideal I_A is squarefree. We claim that every primitive binomial of I_A is a circuit. Let $f = u - v \in I_A$ be a primitive binomial. Lemma 4.32 says that there is a circuit $g = u' - v' \in I_A$ with $\text{var}(u') \subset \text{var}(u)$ and $\text{var}(v') \subset \text{var}(v)$. Since each of u' and v' is squarefree, one has $u'|u$ and $v'|v$. Since f is primitive, one has $f = g$. Thus g is a circuit. In particular every primitive binomial of I_A is squarefree. Now, Theorem 3.13 guarantees that each binomial belonging to the reduced Gröbner basis with respect to any monomial order is primitive. Hence every initial ideal of I_A is squarefree. Thus A is unimodular, as desired.   □

Corollary 4.36 Let a configuration $A \in \mathbb{Z}^{d \times n}$ be unimodular. Then all of the following sets (i), (ii), and (iii) coincide:

(i) *The set of circuits of* I_A;
(ii) *The universal Gröbner basis of* I_A;
(iii) *The Graver basis of* I_A.

Proof The Graver basis of I_A is the set of primitive binomials of I_A. Since A is unimodular, it follows from the proof of Theorem 4.35 that every primitive binomial belonging to I_A is a circuit. Hence the Graver basis of I_A is a subset of the set of circuits of I_A. On the other hand, Theorem 4.33 guarantees that the set of circuits of I_A is a subset of the universal Gröbner basis of I_A. Since, in general, the universal Gröbner basis of I_A is a subset of the Graver basis of I_A, it follows that all of the above sets (i), (ii), and (iii) coincide.   □

Lemma 4.37 Let $A \in \mathbb{Z}^{d \times n}$ be an integer matrix and $\Lambda(A) \in \mathbb{Z}^{(d+n) \times 2n}$ its Lawrence lifting. Then every irreducible binomial belonging to $I_{\Lambda(A)}$ is of the form $f^\sharp$, where f is an irreducible binomial belonging to I_A.

Proof Let F be an irreducible binomial belonging to $I_{\Lambda(A)}$. By Lemma 3.24, there exists a binominal $f \in I_A$ such that $F = f^\sharp$.

Now, we show that f is irreducible. If f is reducible, then by using Lemma 4.30 there exists an integer $p > 1$ together with nonnegative integer vectors $\mathbf{a}_0$ and $\mathbf{b}_0$ belonging to $\mathbb{Z}^n$ for which $\mathbf{a} = p\mathbf{a}_0$ and $\mathbf{b} = p\mathbf{b}_0$. Hence

$$f^\sharp = (\mathbf{x}^{\mathbf{a}_0}\mathbf{y}^{\mathbf{b}_0})^p - (\mathbf{x}^{\mathbf{b}_0}\mathbf{y}^{\mathbf{a}_0})^p,$$

which contradicts the fact that $f^\sharp$ is irreducible.   □

Lemma 4.38 Let $A \in \mathbb{Z}^{d \times n}$ be a configuration and $\Lambda(A) \in \mathbb{Z}^{(d+n) \times 2n}$ its Lawrence lifting.

(a) *A binomial* $f \in I_A$ *is a circuit if and only if* $f^\sharp \in I_{\Lambda(A)}$ *is a circuit.*
(b) *Furthermore, every circuit belonging to* $I_{\Lambda(A)}$ *is of the form* $f^\sharp$, *where* f *is a circuit belonging to* I_A.

Proof

(a) Let f and g be binomials belonging to I_A. Then $\mathrm{var}(g) \subset \mathrm{var}(f)$ if and only if $\mathrm{var}(g^\sharp) \subset \mathrm{var}(f^\sharp)$. Furthermore, $\mathrm{var}(g) = \mathrm{var}(f)$ if and only if $\mathrm{var}(g^\sharp) = \mathrm{var}(f^\sharp)$. Hence if f is not a circuit, then $f^\sharp$ is not a circuit. Conversely, if $f^\sharp$ is not a circuit, then there is an irreducible binomial $F \in I_{\Lambda(A)}$ for which $\mathrm{var}(F) \subset \mathrm{var}(f^\sharp)$ with $\mathrm{var}(F) \neq \mathrm{var}(f^\sharp)$. By using Lemma 4.37, one has $F = g^\sharp$, where $g \in I_A$ is an irreducible binomial. Since $\mathrm{var}(g^\sharp) \subset \mathrm{var}(f^\sharp)$ with $\mathrm{var}(g^\sharp) \neq \mathrm{var}(f^\sharp)$, it follows that f cannot be a circuit.

(b) Since every circuit is irreducible, it follows from Lemma 4.37 that every circuit belonging to $I_{\Lambda(A)}$ is of the form $f^\sharp$, where f is an irreducible binomial belonging to I_A. Now, the desired result follows from (a). □

A subconfiguration $B = (\mathbf{a}_{i_1}, \mathbf{a}_{i_2}, \ldots, \mathbf{a}_{i_m}) \in \mathbb{Z}^{d \times m}$, where $1 \leq m \leq n$, of a configuration $A = (\mathbf{a}_1, \mathbf{a}_2, \ldots, \mathbf{a}_n) \in \mathbb{Z}^{d \times n}$ is called *combinatorial pure* if there is a face F of $\mathrm{conv}(\{\mathbf{a}_1, \mathbf{a}_2, \ldots, \mathbf{a}_n\})$ such that

$$\{\mathbf{a}_1, \mathbf{a}_2, \ldots, \mathbf{a}_n\} \cap F = \{\mathbf{a}_{i_1}, \mathbf{a}_{i_2}, \ldots, \mathbf{a}_{i_m}\}.$$

We call $K[B]$ a *combinatorial pure subring* of $K[A]$ if B is a combinatorial pure subconfiguration of A.

Example 4.39 The Lawrence lifting $\Lambda(B)$ of the submatrix $B = (\mathbf{a}_1, \ldots, \mathbf{a}_m)$ of a matrix $A = (\mathbf{a}_1, \ldots, \mathbf{a}_n)$ is a combinatorial pure subconfiguration of $\Lambda(A)$.

Lemma 4.40 *Every combinatorial pure subconfiguration of a normal configuration is normal. Moreover, every combinatorial pure subconfiguration of a very ample configuration is very ample.*

Proof Let $A = (\mathbf{a}_1, \ldots, \mathbf{a}_n) \in \mathbb{Z}^{d \times n}$ be a configuration and $B = (\mathbf{a}_{i_1}, \ldots, \mathbf{a}_{i_m}) \in \mathbb{Z}^{d \times m}$ a combinatorial pure subconfiguration of A. It is enough to show that

$$\mathbb{Z}B \cap \mathbb{Q}_{\geq 0}B \setminus \mathbb{Z}_{\geq 0}B \subset \mathbb{Z}A \cap \mathbb{Q}_{\geq 0}A \setminus \mathbb{Z}_{\geq 0}A.$$

Let $\alpha \in \mathbb{Z}B \cap \mathbb{Q}_{\geq 0}B \setminus \mathbb{Z}_{\geq 0}B$. It is clear that α belongs to $\mathbb{Z}A \cap \mathbb{Q}_{\geq 0}A$. Suppose that α belongs to $\mathbb{Z}_{\geq 0}A$. Then we have

$$\alpha = \sum_{k=1}^{m} q_{i_k} \mathbf{a}_{i_k} = \sum_{j=1}^{n} z_j \mathbf{a}_j,$$

where $0 \leq q_{i_k} \in \mathbb{Q}$ $(1 \leq k \leq m)$ and $0 \leq z_j \in \mathbb{Z}$ $(1 \leq j \leq n)$. Since A is a configuration, it follows that $\sum_{k=1}^{m} q_{i_k} = \sum_{j=1}^{n} z_j$.

On the other hand, since B is a combinatorial pure subconfiguration of A, there is a face F of $\mathrm{conv}(\{\mathbf{a}_1, \mathbf{a}_2, \ldots, \mathbf{a}_n\})$ such that

$$\{\mathbf{a}_1, \mathbf{a}_2, \ldots, \mathbf{a}_n\} \cap F = \{\mathbf{a}_{i_1}, \mathbf{a}_{i_2}, \ldots, \mathbf{a}_{i_m}\}.$$

By the definition of faces, there exists a vector $\mathbf{w} \in \mathbb{R}^d$ such that

$$\mathbf{w} \cdot \mathbf{a}_k \begin{cases} = 1, & \text{if } k \in \{i_1, \ldots, i_m\}, \\ < 1, & \text{otherwise.} \end{cases}$$

Hence $\mathbf{w} \cdot \alpha = \sum_{k=1}^{m} q_{i_k} = \sum_{j=1}^{n} z_j$ and $z_j = 0$ for all $j \notin \{i_1, \ldots, i_m\}$. Thus $\alpha \in \mathbb{Z}_{\geq 0} B$, which is a contradiction. $\qquad\square$

Lemma 4.41 *Let $A \in \mathbb{Z}^{d \times n}$ be a configuration and suppose that there is a circuit $f = u - v \in I_A$ with $\mathrm{var}(f) = \{x_{i_1}, x_{i_2}, \ldots, x_{i_m}\}$ such that none of the monomials u and v is squarefree. Then the subconfiguration B of A consisting of the column vectors $\mathbf{a}_{i_1}, \mathbf{a}_{i_2}, \ldots, \mathbf{a}_{i_m}$ cannot be very ample.*

Proof Let $I = I_A \cap K[x_{i_1}, x_{i_2}, \ldots, x_{i_m}]$. Then I coincides with the toric ideal I_B of B. Since $f \in I_B$ and is a circuit of I_B, one can assume that $\mathrm{var}(f) = \{x_1, x_2, \ldots, x_n\}$ with $A = B$. It follows from Lemma 4.31 that $I_A = (f)$.

Let $u = x_1^2 u'$ and $v = x_2^2 v'$. Since f is circuit, f is irreducible, and hence u' ($\neq 1$) is not divided by x_2 and v' ($\neq 1$) is not divided by x_1. Since $\pi(x_1^2 u') = \pi(x_2^2 v')$, one has $\pi(x_1^2 u') \pi(x_2^2 v') = (\pi(x_1^2 u'))^2$. Hence $\pi(u') \pi(v') = (\pi(x_1 u')/\pi(x_2))^2$. Let x_k be a variable with $k \neq 1, 2$ and let $\mathbf{t}^{\mathbf{a}_m} = \pi(x_k^m) \pi(x_1 u')/\pi(x_2)$ be the Laurent monomial belonging to $K[t_1, t_1^{-1}, t_2, t_2^{-1}, \ldots, t_n, t_n^{-1}]$. Then $\mathbf{a}_m \in \mathbb{Q}_{\geq 0} A \cap \mathbb{Z} A$ for all positive integer m. Suppose that there exists a monomial w such that $\pi(w) = \mathbf{t}^{\mathbf{a}_m}$. It then follows that the binomial $f' = x_1 u' x_k^m - x_2 w$ belongs to I_A. Since $I_A = (f)$ and $x_1 u' x_k^m$ is divided by neither $x_1^2 u'$ nor $x_2^2 v'$, we have $f' = 0$. Hence x_2 must divide u', which is a contradiction. Thus, $\mathbf{a}_m$ does not belong to $\mathbb{Z}_{\geq 0} A$ for all $m > 0$ and hence A is not very ample. $\qquad\square$

Theorem 4.42 *Given a configuration $A \in \mathbb{Z}^{d \times n}$ and its Lawrence lifting $\Lambda(A)$, the following conditions are equivalent:*

(i)   *A is unimodular;*
(ii) *$\Lambda(A)$ is unimodular;*
(iii) *$\Lambda(A)$ is normal.*
(iv) *$\Lambda(A)$ is very ample.*

Proof First (ii) $\Rightarrow$ (iii) $\Rightarrow$ (iv) is known (Theorem 4.11). Second (i) $\Leftrightarrow$ (ii) follows from Theorem 4.35 and Lemma 4.38.

Now, in order to prove (iv) $\Rightarrow$ (i), suppose that A is not unimodular. Then there is a circuit $f = \mathbf{x}^{\mathbf{a}} - \mathbf{x}^{\mathbf{b}} \in I_A$ such that either $\mathbf{x}^{\mathbf{a}}$ or $\mathbf{x}^{\mathbf{b}}$ is not squarefree. Thus in the circuit $f^{\sharp} = \mathbf{x}^{\mathbf{a}} \mathbf{y}^{\mathbf{b}} - \mathbf{x}^{\mathbf{b}} \mathbf{y}^{\mathbf{a}}$, none of the monomials $\mathbf{x}^{\mathbf{a}} \mathbf{y}^{\mathbf{b}}$ and $\mathbf{x}^{\mathbf{b}} \mathbf{y}^{\mathbf{a}}$ is squarefree. Let, say, $\mathrm{var}(f) = \{x_1, x_2, \ldots, x_m\}$ and B the subconfiguration of A consisting of the column vectors $\mathbf{a}_1, \mathbf{a}_2, \ldots, \mathbf{a}_m$. Since

$$\mathrm{var}(f^{\sharp}) = \{x_1, x_2, \ldots, x_m, y_1, y_2, \ldots, y_m\}$$

and since

$$\Lambda(B) = \begin{pmatrix} \mathbf{a}_1 & \mathbf{a}_2 & \cdots & \mathbf{a}_m & \mathbf{0} & \mathbf{0} & \cdots & \mathbf{0} \\ \mathbf{e}_1 & \mathbf{e}_2 & \cdots & \mathbf{e}_m & \mathbf{e}_1 & \mathbf{e}_2 & \cdots & \mathbf{e}_m \end{pmatrix},$$

by using Lemma 4.41, the Lawrence lifting $\Lambda(B)$ cannot be very ample. Since $\Lambda(B)$ is a combinatorial pure subconfiguration (Example 4.39) of $\Lambda(A)$, it follows from Lemma 4.40 that $\Lambda(A)$ cannot be very ample, as desired. □

Problems

4.14 Show that every integral polytope of dimension 2 is unimodular.

4.15 Let $\mathscr{P} \subset \mathbb{R}^3$ denote the integral polytope of dimension 3 with the 8 vertices $(\epsilon_1, \epsilon_2, \epsilon_3)$, where each $\epsilon_i \in \{0, 1\}$. Show that $\mathscr{P}$ cannot be unimodular.

4.16 Show that there exist configurations A, B, and C such that $\mathscr{C}_A = \mathscr{U}_A \neq \mathscr{G}r_A$, $\mathscr{C}_B \neq \mathscr{U}_B = \mathscr{G}r_B$, and $\mathscr{C}_C \neq \mathscr{U}_C \neq \mathscr{G}r_C$.

Notes

The study of convex polytopes originated in the Euler's formula $v - e + f = 2$ of convex polytopes of dimension 3. On the other hand, the study of integral convex polytope might originate in Pick's formula, which is a formula to compute the area of an integral convex polygon by counting integer points contained in the polygon.

Grünbaum's book [85] is the fundament on classical theory of convex polytopes, where rich references contributing to the development of convex polytopes are listed. Ziegler [221] presents a wealth of material on the modern theory of convex polytopes. A quick introduction to the theory of convex polytopes is Brøndsted [19], which invites the reader to the three highlights of convex polytopes known as Dehn–Sommerville relations (1927), the upper bound theorem (McMullen, 1970), and the lower bound theorem (Barnette, 1973).

In 1975, a revolution of convex polytopes occurred. Richard Stanley [194] proved the upper bound conjecture for spheres affirmatively by using commutative algebra, viz., the Reisner's theorem [177] on Cohen–Macaulay rings. We refer the reader to Stanley [199], Bruns–Herzog [27], and Hibi [105] for further information. See also Hochster [116]. Historically the encounter of convex polytopes with Cohen–Macaulay rings was achieved by Hochster [115]. Furthermore, in 1980, Stanley [195] and Billera–Lee [16] succeeded in proving the McMullen's g-conjecture, which characterizes the f-vectors of simplicial convex polytopes. In particular, in [195] Stanley employed the theory of toric varieties [48, 152].

The topics of normal polytopes and unimodular triangulations is one of the high-lights of the modern theory of integral convex polytopes. The integer decomposition property is important in the theory of integer programming [188].

There is an integral polytope which possesses a unimodular covering, but no unimodular triangulation [25]. Regular triangulations were introduced by Gelfand–Kapranov–Zelevinsky [80] in their study on hypergeometric functions. We refer the reader to [139] for the information about the geometry of regular triangulations. Theorem 4.14, which interprets regular triangulations as Stanley–Reisner complexes by using Gröbner bases theory, is due to Sturmfels, as well as Theorem 4.17, see [201]. In his book [202], Sturmfels develops a systematic study on convex polytopes in the frame of Gröbner bases. Corollary 4.23 is a powerful tool to show that an integral polytope is normal. Example 4.25 is discovered in [158] and Example 4.28 is discovered in [159, Example 2.2].

The set of circuits for unimodular polytopes was discussed in [204]. Combi-natorial pure subrings appeared first in Ohsugi–Herzog–Hibi [156]. The definition of combinatorial pure subrings as given in this chapter is taken from [155], and differs slightly from the definition in [156]. Theorem 4.42, which characterizes unimodular Lawrence liftings, can also be found in [156] (without the statement of very ampleness). This characterization was extended in [12] to lattice ideals by Bayer-Popescu-Sturmfels. The results on very ampleness are due to [167].

Part III
Applications in Combinatorics and Statistics

Chapter 5
Edge Polytopes and Edge Rings

Abstract The convex polytopes arising from finite graphs and their toric ideals
have been studied by many authors. The present chapter is devoted to introducing the
foundation on the topics. In Section 5.1, we summarize basic terminologies on finite
graphs. A basic fact on bipartite graphs is proved. The edge polytope of a finite graph
is introduced in Section 5.2. We study the dimension, the vertices, the edges, and the
facets of edge polytopes. In Section 5.3, the edge ring of a finite graph and its toric
ideal is discussed. One of the main results is a combinatorial characterization for the
toric ideal of an edge ring to be generated by quadratic binomials (Theorem 5.14).
The problem of the normality of edge polytopes is studied in Section 5.4. It turns out
that the odd cycle condition in the classical graph theory characterizes the normality
of an edge polytope. Furthermore, it is shown that an edge polytope is normal if and
only if it possesses a unimodular covering (Theorem 5.20). Finally, in Section 5.5,
Gröbner bases of toric ideals arising from bipartite graphs will be discussed. In
particular, we show that the toric ideal of the edge ring of a bipartite graph is
generated by quadratic binomials if and only if it possesses a quadratic Gröbner
basis (Theorem 5.27).

5.1 Finite Graphs

Let $[d] = \{1, 2, \ldots, d\}$ and $\binom{[d]}{k}$ the set of k-element subsets of $[d]$, where $d \geq 1$
and $0 \leq k \leq d$. Let G be a finite simple graph on the vertex set $V(G) = [d]$, where
$d \geq 2$, and $E(G) = \{e_1, e_2, \ldots, e_n\}$, where each $e_i \in \binom{[d]}{2}$, the set of edges of G.
Recall that a finite graph is *simple* if it possesses no loops and no multiple edges.
The *degree* of a vertex $i \in V(G)$ is the number of edges $e \in E(G)$ with $i \in e$. Let
$\deg_G i$ denote the degree of a vertex i of G.

A *subgraph* of G is a finite simple graph G' on $V(G') \subset [d]$ with $E(G') \subset$
$E(G)$. Given a nonempty subset $W \subset [d]$, the *induced subgraph* of G on $W \subset [d]$
is the subgraph $G|_W$ of G with $E(G|_W) = \{e \in E(G) : e \subset W\}$. A *spanning
subgraph* of G is a subgraph H with $V(H) = V(G) = [d]$.

© Springer International Publishing AG, part of Springer Nature 2018
J. Herzog et al., *Binomial Ideals*, Graduate Texts in Mathematics 279,
https://doi.org/10.1007/978-3-319-95349-6_5

A *walk of length q* of G connecting $i \in [d]$ with $j \in [d]$ is a sequence of edges

$$\Gamma = (\{i_0, i_1\}, \{i_1, i_2\}, \dots, \{i_{q-1}, i_q\}) \tag{5.1}$$

of G with each $i_k \in [d]$ for which $i_0 = i$ and $i_q = j$. A walk may be regarded as a subgraph of G in the obvious way. An *even walk* is a walk of even length. An *odd walk* is a walk of odd length. A *closed walk* is a walk of the form (5.1) with $i_0 = i_q$.

A *cycle of length q* is a closed walk of the form:

$$C = (\{i_0, i_1\}, \{i_1, i_2\} \dots, \{i_{q-1}, i_0\}) \tag{5.2}$$

with $i_k \neq i_\ell$ for all $0 \leq k < \ell \leq q - 1$. A *chord* of a cycle (5.2) is an edge $e \in E(G)$ of the form $e = \{i_k, i_\ell\}$, where $0 \leq k < \ell \leq q - 1$, with $e \notin E(C)$. A *minimal cycle* is a cycle with no chord.

If $e = \{i_k, i_\ell\}$, where $0 \leq k < \ell \leq q - 1$, and $e' = \{i_{k'}, i_{\ell'}\}$, where $0 \leq k' < \ell' \leq q - 1$, are chords of a cycle C of (5.2), then we say that e and e' *cross* in C if either $k < k' < \ell < \ell'$ or $k' < k < \ell' < \ell$ and if either $\{i_k, i_{k'}\}, \{i_\ell, i_{\ell'}\}$ are edges of C or $\{i_k, i_{\ell'}\}, \{i_\ell, i_{k'}\}$ are edges of C.

When a cycle C of (5.2) is an even cycle, a chord $e = \{i_k, i_\ell\}$, where $0 \leq k < \ell \leq q - 1$, is called an *even-chord* if $\ell - k$ is odd and is called an *odd-chord* if $\ell - k$ is even.

Let C and C' be cycles of G with $V(C) \cap V(C') = \emptyset$, then a *bridge* between C and C' is an edge $e = \{i, j\}$ of G with $i \in V(C)$ and $j \in V(C')$.

A finite simple graph G is *connected* if, for any two vertices i and j of G, there exists a walk of G connecting i with j. The *connected components* of G are the induced subgraphs $G|_{W_1}, \dots, G|_{W_s}$ of G such that each $G|_{W_i}$ is connected with $W_1 \cup \dots \cup W_s = [d]$ and that one has $\{i, j\} \notin E(G)$ if $i \in W_k$ and $j \in W_\ell$ with $k \neq \ell$.

The *complete graph* on $[d]$ is the simple graph G on $[d]$ whose edges are those $\{i, j\}$ with $1 \leq i < j \leq d$.

A finite graph G on $[d]$ is called *bipartite* if there is a decomposition $[d] = V \cup V'$, where $V \neq \emptyset$, $V' \neq \emptyset$ and $V \cap V' = \emptyset$ such that each edge of G is of the form $\{i, j\}$ with $i \in V$ and $j \in V'$.

The *complete bipartite graph* on $[d] = V \cup V'$ is the bipartite graph whose edges are those $\{i, j\}$ with $i \in V$ and $j \in V'$.

A *forest* is a finite simple graph with no cycle. A connected forest is called a *tree*. A *spanning tree* of a finite simple graph G is a spanning subgraph of G which is a tree.

Lemma 5.1 *A finite simple graph G is bipartite if and only if every cycle of G is even. In particular, every forest is a bipartite graph.*

Proof **(Only if)** Suppose that G is a bipartite graph on $[d]$ with the decomposition $[d] = U \cup V$. Let $C = \{\{v_1, v_2\}, \{v_2, v_3\}, \dots, \{v_{q-1}, v_q\}, \{v_q, v_1\}\}$ be a cycle of length q of G with $v_1 \in U$. Then, $v_2 \in V$ and $v_3 \in U$. In general, one has $v_i \in U$ if i is odd and $v_i \in V$ if i is even. Since $v_q \in V$, it follows that q is even. This completes a proof of "Only If" part.

(**If**) one can assume that G is connected. Suppose that every cycle of G is of even length. Let u and v be vertices of G. Let W be a walk of G of length q connecting u with v and W' a walk of G of length q' connecting u with v. Since every cycle of G is even, it follows that $q + q'$ is even. In other words, either (i) both q and q' are even or (ii) both q and q' are odd.

Now, fix a vertex v_0 of G. Let U (resp., V) denote the vertices w of G such that there is a walk of even (resp., odd) length connecting v_0 with w. Then, $U \cap V = \emptyset$ with $v_0 \in U$. Let $w, w' \in U$ with $w \neq w'$ and with $\{w, w'\} \in E(G)$. Since $w \in U$, there is an even walk connecting v_0 with w. It then follows that there is a walk of odd length which connects v_0 with w', a contradiction. Thus, $\{w, w'\} \notin E(G)$ for w and w' belonging to U with $w \neq w'$. Similarly, $\{w, w'\} \notin E(G)$ for w and w' belonging to V with $w \neq w'$. Hence, every edge of G is of the form $\{u, v\}$ with $u \in U$ and $v \in V$. Thus, G is bipartite, as desired. $\square$

Let G be a finite simple graph on $[d]$ and G' on $[d']$. We say that G is isomorphic to G' if $d = d'$ and if there is a permutation σ on $[d]$ for which

$$E(G') = \{\{\sigma(i), \sigma(j)\} : \{i, j\} \in E(G)\}.$$

Let G be the finite simple graph on $[d]$. A permutation σ on $[d]$ is called an *automorphism* of G if

$$E(G) = \{\{\sigma(i), \sigma(j)\} : \{i, j\} \in E(G)\}.$$

Problems

5.1

(a) Classify all finite simple graphs on $[d]$, up to isomorphism, with $1 \leq d \leq 4$.
(b) Classify all finite connected simple graphs on $[d]$, up to isomorphism, with $1 \leq d \leq 4$.
(c) Classify all finite simple bipartite graphs on $[d]$, up to isomorphism, with $1 \leq d \leq 4$.
(d) Classify all finite connected simple bipartite graphs on $[d]$, up to isomorphism, with $1 \leq d \leq 4$.
(e) Classify all forests on $[d]$, up to isomorphism, with $1 \leq d \leq 4$.
(f) Classify all trees on $[d]$, up to isomorphism, with $1 \leq d \leq 4$.

5.2 Let G be the finite connected simple graph on $[5]$ whose edges are those $\{i, j\}$ with $2 \leq |i - j|$. How many spanning trees does G have?

5.3 Let G be the finite connected simple graph on $[6]$ whose edges are

$$\{1, 2\}, \{2, 3\}, \{1, 3\}, \{3, 4\}, \{4, 5\}, \{5, 6\}, \{4, 6\}.$$

How many automorphisms does G have?

5.2   Edge Polytopes of Finite Graphs

Let G be a finite simple graph on the vertex set $V(G) = [d]$ and $E(G) = \{e_1, \ldots, e_n\}$ the set of edges of G. Let $\mathbf{e}_1, \ldots, \mathbf{e}_d$ denote the canonical unit coordinate vectors of $\mathbb{R}^d$. If $e = \{i, j\}$ is an edge of G, then we define $\rho(e) \in \mathbb{R}^d$ by setting $\rho(e) = \mathbf{e}_i + \mathbf{e}_j$. We write $\mathscr{P}_G$ for the convex hull of the finite set $\{\rho(e) : e \in E(G)\} \subset \mathbb{R}^d$ and call $\mathscr{P}_G$ the *edge polytope* of G.

Lemma 5.2 *One has* $\mathscr{P}_G \cap \mathbb{Z}^d = \{\rho(e) : e \in E(G)\}$. *Furthermore, the set of vertices of* $\mathscr{P}_G$ *coincides with* $\mathscr{P}_G \cap \mathbb{Z}^d$.

Proof Let $\mathscr{H}$ denote the hyperplane of $\mathbb{R}^d$ defined by the equation $z_1 + \cdots + z_d = 2$. Since each $\rho(e)$ with $e \in E(G)$ belongs to $\mathscr{H}$, it follows that $\mathscr{P}_G \subset \mathscr{H}$. Let $\alpha = (b_1, \ldots, b_d) \in \mathscr{P}_G \cap \mathbb{Z}^d$ and write $\alpha = \sum_{e \in E(G)} a_e \rho(e)$ with each $0 \leq a_e \in \mathbb{R}_{\geq 0}$ and with $\sum_{e \in E(G)} a_e = 1$. If $b_i \geq 1$, then $i \in e$ for all e with $a_e > 0$. Thus, $b_i = 1$. Since $b_1 + \cdots + b_d = 2$, there is $j \neq i$ with $b_j = 1$. Then, $j \in e$ for all e with $a_e > 0$. Hence, $e = \{i, j\}$ if $a_e > 0$. Thus, $\alpha = \rho(\{i, j\}) \in \mathscr{P}_G$, as required.

Let $V(\mathscr{P}_G)$ denote the set of vertices of $\mathscr{P}_G$. In general, $V(\mathscr{P}_G) \subset \{\rho(e) : e \in E(G)\}$. If, say, $e = \{1, 2\} \in E(G)$ and $e \notin V(\mathscr{P}_G)$, then $\rho(e) = (\rho(e') + \rho(e''))/2$, where $\rho(e')$ and $\rho(e'')$ belong to $V(\mathscr{P}_G)$ with $e \neq e'$ and $e \neq e''$. Say, $1 \notin e'$ and $i \in e'$ with $i \geq 3$. Then, $(\rho(e') + \rho(e''))/2 \neq \mathbf{e}_1 + \mathbf{e}_2$. This contradiction guarantees that $e = \{1, 2\}$ belongs to $V(\mathscr{P}_G)$. Hence, $V(\mathscr{P}_G) = \mathscr{P}_G \cap \mathbb{Z}^d$, as desired.   □

Lemma 5.3 *Let* $e = \{i, j\}$ *and* $e' = \{k, \ell\}$ *be the vertices of* $\mathscr{P}_G$ *with* $e \neq f$. *Then, the line segment* $[\rho(e), \rho(e')]$ *which is the convex hull of* $\{\rho(e), \rho(e')\}$ *in* $\mathbb{R}^d$ *is an edge of* $\mathscr{P}_G$ *if and only if the induced subgraph of* G *on* $\{i, j\} \cup \{k, \ell\}$ *contains no cycle of length* 4. *In particular, if* e *and* e' *possess exactly one common vertex, then* $[\rho(e), \rho(e')]$ *is an edge of* $\mathscr{P}_G$.

Proof Let G' denote the induced subgraph of G on $\{i, j\} \cup \{k, \ell\}$ and $\mathscr{F} = \mathscr{P}_{G'}$. Since $\mathscr{F}$ is a face of $\mathscr{P}_G$, the segment $[\rho(e), \rho(e')]$ is a face of $\mathscr{P}_G$ if and only if $[\rho(e), \rho(e')]$ is a face of $\mathscr{F}$. If e and f have exactly one common vertex, then $\mathscr{F}$ is a simplex and $[\rho(e), \rho(e')]$ is a face of $\mathscr{F}$. If e and f have no common vertex, say $e = \{1, 2\}$ and $f = \{3, 4\}$, then $\mathscr{F}$ can be regarded as a subpolytope of the convex hull of $\{(1, 1, 0), (1, 0, 1), (0, 1, 1), (1, 0, 0), (0, 1, 0), (0, 0, 1)\} \subset \mathbb{R}^3$. It then follows that $[(1, 1, 0), (0, 0, 1)]$ is a face of $\mathscr{F}$ if and only if $\mathscr{F}$ is a simplex. Moreover, $\mathscr{F}$ is a simplex if and only if G' contains no cycle of length 4. Hence, the segment $[\rho(e), \rho(e')]$ is a face of $\mathscr{P}_G$ if and only if G' contains no cycle of length 4, as desired.   □

Lemma 5.4 *Suppose that* G *is connected. Then,* $\dim \mathscr{P}_G = d - 1$ *if* G *possesses at least one odd cycle. If* G *is bipartite, then* $\dim \mathscr{P}_G = d - 2$.

Proof Let G be connected with at least one odd cycle. Then, one can find a connected spanning subgraph G' of G with d edges such that G' has exactly one odd cycle and it is a unique cycle of G'. Then, $\mathscr{P}_{G'}$ is a $(d-1)$-simplex. Hence, dim $\mathscr{P}_G \geq d - 1$. Since G is lying on the hyperplane of $\mathbb{R}^d$ defined by the equation $z_1 + \cdots + z_d = 2$, one has dim $\mathscr{P}_G \leq d - 1$. Thus, dim $\mathscr{P}_G = d - 1$.

Let G be bipartite with the decomposition $[d] = U \cup V$. Let $\mathscr{H}_1$ be the hyperplane of $\mathbb{R}^d$ defined by the equation $\sum_{i \in U} z_i = 1$ and $\mathscr{H}_2$ the hyperplane of $\mathbb{R}^d$ defined by $\sum_{j \in V} z_j = 1$. Then, $\mathscr{P}_G \subset \mathscr{H}_1 \cap \mathscr{H}_2$. Thus, dim $\mathscr{P}_G \leq d - 2$. Let G'' be a spanning tree of G. Then, $\mathscr{P}_{G''}$ is a $(d-2)$-simplex. Hence, dim $\mathscr{P}_G \geq d - 2$. As a result, one has dim $\mathscr{P}_G = d - 2$, as required. □

Lemma 5.5 *Let G be a finite connected simple graph on $[d]$ with at least one odd cycle and H a subgraph of G. Then, $F(H) = \{\rho(e) : e \in E(H)\} \subset \mathscr{P}_G \cap \mathbb{Z}^d$ is a maximal simplex belonging to $\mathscr{P}_G$ if and only if H satisfies the following conditions:*

- *H is a spanning subgraph of G;*
- *H has d edges;*
- *Every cycle of H is odd;*
- *Every connected component of H possesses exactly one odd cycle.*

Proof Let a subgraph H satisfy the required conditions and $H_1, \ldots, H_q$ the connected components of H. Then, dim $\mathscr{P}_{H_k} = |V(H_k)| - 1$ for $1 \leq k \leq q$. Since dim $\mathscr{P}_H = q - 1 + \sum_{k=1}^{q} \dim \mathscr{P}_{H_k}$ and since $[d] = \cup_{k=1}^{q} V(H_k)$, one has dim $\mathscr{P}_H = d - 1$. Since $|V(H)| = d$, it follows that $\mathscr{P}_H$ is a $(d-1)$-simplex.

Now, suppose that H is a subgraph of G for which $\mathscr{P}_H$ is a $(d-1)$-simplex. Then, H must be a spanning subgraph of G with d edges. Let $H_1, \ldots, H_q$ be the connected components of H. Again, since dim $\mathscr{P}_H = q - 1 + \sum_{k=1}^{q} \dim \mathscr{P}_{H_k}$, one has dim $\mathscr{P}_{H_k} = |V(H_k)| - 1$ for $1 \leq k \leq q$. Since H has d edges, each H_k is a spanning subgraph on $V(H)$ with $|V(H_k)|$ edges. In particular, each H_k possesses exactly one cycle. Since dim $\mathscr{P}_{H_k} = |V(H_k)| - 1$, a unique cycle of each H_k must be odd. □

Lemma 5.6 *Let G be a finite connected simple bipartite graph on $[d]$ and H a subgraph of G. Then, $F(H) = \{\rho(e) : e \in E(H)\} \subset \mathscr{P}_G \cap \mathbb{Z}^d$ is a maximal simplex belonging to $\mathscr{P}_G$ if and only if H is a spanning tree of G.*

Proof If H is a spanning tree of G, then H has $(d-1)$ edges and dim $\mathscr{P}_H = d - 2$. Thus, $\mathscr{P}_H$ is a $(d-2)$-simplex. Hence, $F(H)$ is a maximal simplex belonging to $\mathscr{P}_G$.

Let H be a subgraph of G and suppose that $F(H)$ is a maximal simplex belonging to $\mathscr{P}_G$. If H is disconnected with $k \geq 2$ connected components, then dim $\mathscr{P}_H = (k-1) + (d-2k) = d - 1 - k \leq d - 3$. Hence, H must be connected. Since H is a spanning subgraph of G with $(d-1)$ edges, it follows that H is a spanning tree, as desired. □

Lemma 5.7 *Every face of an edge polytope is again an edge polytope. More precisely, if $\mathscr{P}_G$ is the edge polytope, then each face of $\mathscr{P}_G$ is of the form $\mathscr{P}_{G'}$, where G' is a subgraph of G.*

Proof Let $\mathscr{F}$ be a face of $\mathscr{P}_G$, then $\mathscr{F} = \mathrm{conv}(\{\rho(e) : e \in E(G)\} \cap \mathscr{F})$. Thus, $\mathscr{F} = \mathscr{P}_H$, where H is a subgraph of G with $E(H) = \{e \in E(G) : \rho(e) \in \mathscr{F}\}$.

□

A nonempty subset $T \subset [d]$ is called *independent* if no edge of G is of the form $e = \{i, j\}$ with $i \in T$ and $j \in T$. If $T \subset [d]$ is independent, then we write $N(G; T)$ for the set of those $i \in [d]$ for which there is $j \in T$ with $\{i, j\} \in E(G)$.

To find the facets of $\mathscr{P}_G$ is of interest. Lemma 5.8 below describes the facets of $\mathscr{P}_G$. Since Lemma 5.8 will be never quoted, we refer the reader to [157, Theorem 1.7] for its proof.

Lemma 5.8

(a) *Let G be a finite simple connected graph on $[d]$ with at least one odd cycle and G' a subgraph of G. Then, $\mathscr{P}_{G'}$ is a facet of $\mathscr{P}_G$ if and only if one of the following conditions is satisfied:*

  • $E(G') = \{e \in E(G) : i \notin e\}$, *where $i \in [d]$ for which every connected component of $G_{[d]\setminus\{i\}}$ has at least one odd cycle.*
  • $E(G') = \{e \in E(G) : e \cap T \neq \emptyset\} \cup \{e \in E(G) : e \cap (T \cup N(G; T)) = \emptyset\}$, *where $\emptyset \neq T \subset [d]$ is independent for which: (i) the bipartite graph consisting of those edges $e \in E(G)$ with $e \cap T \neq \emptyset$ is connected and (ii) either $T \cup N(G; T) = [d]$ or every connected component of the subgraph $G_{[d]\setminus(T\cup N(G;T))}$ has at least one odd cycle.*

(b) *Let G be a finite simple connected bipartite graph on $[d] = V \cup V'$ and G' a subgraph of G. Then, $\mathscr{P}_{G'}$ is a facet of $\mathscr{P}_G$ if and only if one of the following conditions is satisfied:*

  • $E(G') = \{e \in E(G) : i \notin e\}$, *where $i \in [d]$ for which $G_{[d]\setminus\{i\}}$ is connected.*
  • $E(G') = \{e \in E(G) : e \cap T \neq \emptyset\} \cup \{e \in E(G) : e \cap (T \cup N(G; T)) = \emptyset\}$, *where $\emptyset \neq T \subset V$ is independent for which: (i) the bipartite graph consisting of those edges $e \in E(G)$ with $e \cap T \neq \emptyset$ is connected and (ii) $G_{[d]\setminus(T\cup N(G;T))}$ is a connected graph with at least one edge.*

Problems

5.4 Let G be the complete graph on $[4]$. Find the edges and facets of the edge polytope $\mathscr{P}_G$ and compute the f-vector of $\mathscr{P}_G$.

5.5 Let G be the complete bipartite graph on $[5] = V \cup V'$ with $|V| = 2$ and $|V'| = 3$. Find the edges and facets of the edge polytope $\mathscr{P}_G$ and compute the f-vector of $\mathscr{P}_G$.

5.6 Let G be the finite simple graph on $[5]$ whose edges are those $\{i, j\}$ with $1 \leq i < j \leq 4$ together with $\{4, 5\}$. Find the edges and facets of the edge polytope $\mathscr{P}_G$.

5.3 Toric Ideals of Edge Rings

Let, as before, $[d] = \{1, 2, \ldots, d\}$ denote the vertex set and G a finite simple connected graph on $[d]$ and $E(G) = \{e_1, \ldots, e_n\}$ the set of edges of G.

Let $K[\mathbf{t}] = K[t_1, \ldots, t_d]$ denote the polynomial ring in d variables over a field K. If $e = \{i, j\}$ is an edge of G, then we define $u^e \in K[\mathbf{t}]$ for the quadratic monomial $t_i t_j$. We write $K[G]$ for the toric ring $K[\{u^e : e \in E(G)\}]$ and call $K[G]$ the *edge ring* of G.

Let $S = K[x_1, \ldots, x_n]$ denote the polynomial ring in n variables over K and define the surjective ring homomorphism $\pi : S \to K[G]$ by setting $\pi(x_i) = u^{e_i}$ for $1 \leq i \leq n$. The kernel of π is denoted by I_G and is called the *toric ideal* of $K[G]$.

Given an even closed walk

$$\Gamma = (e_{i_1}, e_{i_2}, \ldots, e_{i_{2q}})$$

of G with each $e_k \in E(G)$, we write f_Γ for the binomial

$$f_\Gamma = \prod_{k=1}^{q} x_{i_{2k-1}} - \prod_{k=1}^{q} x_{i_{2k}}$$

belonging to I_G. We often employ the abbreviated notation

$$f_\Gamma = f_\Gamma^{(+)} - f_\Gamma^{(-)},$$

where

$$f_\Gamma^{(+)} = \prod_{k=1}^{q} x_{i_{2k-1}}, \qquad f_\Gamma^{(-)} = \prod_{k=1}^{q} x_{i_{2k}}$$

Lemma 5.9 *The toric ideal I_G is generated by all the binomials f_Γ, where Γ is an even closed walk of G.*

Proof It follows from Theorem 3.2 that every toric ideal is generated by binomials. Let I_G' denote the binomial ideal generated by those binomial f_Γ, where Γ is an even closed walk of G. Choose a binomial $f = \prod_{k=1}^{q} x_{i_k} - \prod_{k=1}^{q} x_{j_k}$ belonging to I_G with $i_k \neq j_{k'}$ for all k and k'. Let, say, $\pi(x_{i_1}) = t_1 t_2$. Since $\pi(\prod_{k=1}^{q} x_{i_k}) = \pi(\prod_{k=1}^{q} x_{j_k})$, one has $\pi(x_{j_m}) = t_2 t_r$ for some m with $r \neq 1$. Say, $m = 1$ and $r = 3$. Thus, $\pi(x_{j_1}) = t_2 t_3$. Then, $\pi(x_{i_\ell}) = t_3 t_s$ for some ℓ with $s \neq 2$. Repeated application of these procedures yields an even closed walk $\Gamma' = (e_{i_1}, e_{j_1}, e_{i_2}, e_{j_2}, \ldots, e_{i_p}, e_{j_p})$ with $f_{\Gamma'} = \prod_{k=1}^{p} x_{i_k} - \prod_{k=1}^{p} x_{j_k} \in I_G$. Since $\pi(\prod_{k=1}^{p} x_{i_k}) = \pi(\prod_{k=1}^{p} x_{j_k})$ and since $\pi(\prod_{k=1}^{q} x_{i_k}) = \pi(\prod_{k=1}^{q} x_{j_k})$, one has $\pi(\prod_{k=p+1}^{q} x_{i_k}) = \pi(\prod_{k=p+1}^{q} x_{j_k})$. Hence, $\prod_{k=p+1}^{q} x_{i_k} - \prod_{k=p+1}^{q} x_{j_k}$ belongs to I_G. Working with induction on $q \geq 2$ enables us to assume that

$\prod_{k=p+1}^{q} x_{i_k} - \prod_{k=p+1}^{q} x_{j_k}$ belongs to I_G'. Now, one has

$$f = \prod_{k=p+1}^{q} x_{i_k} \left(\prod_{k=1}^{p} x_{i_k} - \prod_{k=1}^{p} x_{j_k} \right) + \prod_{k=1}^{p} x_{j_k} \left(\prod_{k=p+1}^{q} x_{i_k} - \prod_{k=p+1}^{q} x_{j_k} \right)$$

$$= f_{\Gamma'} \prod_{k=p+1}^{q} x_{i_k} + \prod_{k=1}^{p} x_{j_k} \left(\prod_{k=p+1}^{q} x_{i_k} - \prod_{k=p+1}^{q} x_{j_k} \right).$$

It then follows that the binomial f belongs to I_G'. Hence, $I_G = I_G'$, as desired. □

An even closed walk Γ of G is called *primitive* if there exists no even closed walk Γ' of G with $f_{\Gamma'} \neq f_{\Gamma}$ for which $f_{\Gamma'}^{(+)}$ divides $f_{\Gamma}^{(+)}$ and $f_{\Gamma'}^{(-)}$ divides $f_{\Gamma}^{(-)}$.

Lemma 5.10 *A binomial $f \in I_G$ is primitive if and only if there exists a primitive even closed walk Γ of G such that $f = f_{\Gamma}$. In particular, the toric ideal I_G is generated by those binomials f_{Γ}, where Γ is a primitive even closed walk of G.*

Proof It follows from the proof of Lemma 5.9 that, for every binomial $f = u - v \in I_G$, where u and v are monomials of $S = K[x_1, \ldots, x_n]$ with $\deg u = \deg v$, there is an even closed walk Γ of G such that $f_{\Gamma'}^{(+)}$ divides u and $f_{\Gamma'}^{(-)}$ divides v. Hence, every primitive binomial of I_G is of the form f_{Γ}, where Γ is an even closed walk of G. It then follows that f_{Γ} is primitive if and only if Γ is primitive.

In addition by Proposition 3.15, I_G is generated by those binomials f_{Γ}, where Γ is a primitive even closed walk of G. □

Lemma 5.11 *A primitive even closed walk Γ of G is one of the following:*

(i) *Γ is an even cycle of G;*

(ii) *$\Gamma = (C_1, C_2)$, where each of C_1 and C_2 is an odd cycle of G having exactly one common vertex;*

(iii) *$\Gamma = (C_1, \Gamma_1, C_2, \Gamma_2)$, where each of C_1 and C_2 is an odd cycle of G with $V(C_1) \cap V(C_2) = \emptyset$ and where Γ_1 and Γ_2 are walks of G of the forms $\Gamma_1 = (e_{i_1}, \ldots, e_{i_r})$ and $\Gamma_2 = (e_{i_1'}, \ldots, e_{i_{r'}'})$ such that Γ_1 combines $j \in e_{i_1} \cap e_{i_{r'}'} \cap V(C_1)$ with $j' \in e_{i_r} \cap e_{i_1'} \cap V(C_2)$ and Γ_2 combines j' with j. Furthermore, none of the vertices belonging to $V(C_1) \cup V(C_2)$ appears in each of $e_{i_1} \setminus \{j\}, e_{i_2}, \ldots, e_{i_{r-1}}, e_{i_r} \setminus \{j'\}, e_{i_1'} \setminus \{j\}, e_{i_2'}, \ldots, e_{i_{r'-1}'}, e_{i_{r'}'} \setminus \{j'\}$.*

Proof Let Γ be a primitive even closed walk

$$\Gamma = (e_{i_1}, e_{i_2}, \ldots, e_{i_{2q}}) = (\{j_0, j_1\}, \{j_1, j_2\}, \ldots, \{j_{2q-1}, j_0\})$$

of G of length $2q$. If $j_k \neq j_\ell$ for all $k \neq \ell$, then Γ is an even cycle of G, which is of the form required in (i).

Let $j_k \neq j_{k'}$ for all $0 \leq k < k' < r \leq 2q - 1$ and $j_{k''} = j_r$ for some $0 \leq k'' < r$. Then, $\Gamma = (C_1, \Gamma')$, where C_1 is a cycle of G with

$$C_1 = (\{j_{k''}, j_{k''+1}\}, \{j_{k''+1}, j_{k''+2}\}, \ldots, \{j_{r-1}, j_r\})$$

and where Γ' is a closed walk of G with

$$\Gamma' = (\{j_r, j_{r+1}\}, \{j_{r+1}, j_{r+2}\}, \ldots, \{j_{2q-1}, j_0\}, \{j_0, j_1\}, \ldots, \{j_{k''-1}, j_{k''}\}).$$

Since Γ is primitive, it follows that C_1 is an odd cycle and Γ' is an odd closed walk. In order to simplify the notation of $\Gamma = (C_1, \Gamma')$, one can write C_1 and Γ' for

$$C_1 = (\{j_0, j_1\}, \{j_1, j_2\}, \ldots, \{j_{r-1}, j_0\})$$

and

$$\Gamma' = (\{j_0, j_{r+1}\}, \{j_{r+1}, j_{r+2}\}, \ldots, \{j_{2q-1}, j_0\}).$$

Let $(V(C_1) \cap V(\Gamma')) \setminus \{j_0\} \neq \emptyset$. Let $j_0 \neq j_a \in (V(C_1) \cap V(\Gamma'))$ with $1 \leq a \leq r - 1$ and $j_a = j_b$ with $r + 1 \leq b \leq 2q - 1$. Since C_1 is an odd cycle, one of the walks

$$\Gamma_1 = (\{j_0, j_1\}, \{j_1, j_2\}, \ldots, \{j_{a-1}, j_a\})$$

and

$$\Gamma_2 = (\{j_a, j_{a+1}\}, \{j_{a+1}, j_{a+2}\}, \ldots, \{j_{r-1}, j_0\})$$

is odd. Furthermore, since Γ' is odd, one of the walks

$$\Gamma_3 = (\{\{j_0, j_{r+1}\}, \{j_{r+1}, j_{r+2}\}, \ldots, \{j_{b-1}, j_b\})$$

and

$$\Gamma_4 = (\{\{j_b, j_{b+1}\}, \{j_{b+1}, j_{b+2}\}, \ldots, \{j_{2q-1}, j_0\})$$

is odd. In particular, one of the closed walks $(\Gamma_1, \Gamma_3), (\Gamma_1, \Gamma_4), (\Gamma_2, \Gamma_3)$, and (Γ_2, Γ_4) must be even. This is impossible, since Γ is primitive. As a result, one has $(V(C_1) \cap V(\Gamma')) = \{j_0\}$. If Γ' is a cycle of G, then Γ is of the form required in (ii).

Let $j_0 = j_c$ for some $r + 2 \leq c \leq 2q - 2$. Since Γ' is an odd closed walk, one of the walks

$$\Gamma_5 = (\{j_0, j_{r+1}\}, \{j_{r+1}, j_{r+2}\}, \ldots, \{j_{c-1}, j_c\})$$

and

$$\Gamma_6 = (\{j_c, j_{c+1}\}, \{j_{c+1}, j_{c+2}\}, \ldots, \{j_{2q-1}, j_0\})$$

is odd. Since C_1 is an odd cycle, one of the closed walks (C_1, Γ_5) and (C_1, Γ_6) is even. Again, this is impossible, since Γ is primitive. As a result, one has $j_0 \neq j_c$ for all $r + 2 \leq c \leq 2q - 2$.

Now, suppose that Γ' is not a cycle. Then, $\Gamma' = (\Gamma_7, \Gamma_8, \Gamma_9)$, where Γ_7 is a walk of G combining j_0 with $j' \in V(\Gamma_8)$, where Γ_8 is a closed walk of G and where Γ_9 is a walk of G combining j' with j_0. Since Γ is primitive, it follows that Γ_8 must be an odd closed walk. If Γ_8 is a cycle of G, then Γ is of the form required in (iii). If Γ_8 is not a cycle of G, then repeating the above technique guarantees that Γ is of the desired form in (iii). □

Corollary 5.12 *Let G be a bipartite graph. Then, every primitive even closed walk is an even cycle. In particular, the toric ideal I_G is generated by those binomials f_C, where C is an even cycle of G.*

An even closed walk Γ of G is called *fundamental* if every even closed walk Γ' of the induced subgraph $G|_{V(\Gamma)}$ of G on $V(\Gamma)$ with $f_{\Gamma'} \neq 0$ satisfies either $f_\Gamma = f_{\Gamma'}$ or $f_\Gamma = -f_{\Gamma'}$.

Lemma 5.13 *Let Γ be a fundamental even closed walk of G and suppose that the toric ideal I_G is generated by $f_{\Gamma_1}, f_{\Gamma_2}, \ldots, f_{\Gamma_s}$, where each f_{Γ_i} is an even closed walk of G. Then, either $f_\Gamma = f_{\Gamma_i}$ or $f_\Gamma = -f_{\Gamma_i}$ for some $1 \leq i \leq s$.*

Proof Since $f_\Gamma \in I_G$, there is f_{Γ_i} for which $f_{\Gamma_i}^{(+)}$ divides either $f_\Gamma^{(+)}$ or $f_\Gamma^{(-)}$. It then follows that each vertex of Γ_i must belong to $V(\Gamma)$. Hence, Γ_i is an even closed walk of the induced subgraph $G|_{V(\Gamma)}$. Thus, f_Γ coincides with either f_{Γ_i} or $-f_{\Gamma_i}$. □

We are now in the position to state a combinatorial criterion for the toric ideal I_G to be generated by quadratic binomials.

Theorem 5.14 *Let G be a finite connected simple graph. Then, the toric ideal I_G is generated by quadratic binomials if and only if the following conditions are satisfied:*

(i) *If C is an even cycle of G of length ≥ 6, then either C has an even-chord or C has three odd-chords e, e', e'' such that e and e' cross in C;*

(ii) *If C_1 and C_2 are minimal odd cycles with exactly one common vertex, then there exists an edge $\{i, j\} \notin E(C_1) \cup E(C_2)$ with $i \in V(C_1)$ and $j \in V(C_2)$;*

(iii) *If C_1 and C_2 are minimal odd cycles with $V(C_1) \cap V(C_2) = \emptyset$, then there exist at least two bridges between C_1 and C_2.*

Proof (**Only if**) Suppose that the toric ideal I_G of G is generated by quadratic binomials. Since every primitive binomial of I_G is of the form f_Γ, where Γ is an even closed walk of G, it follows that I_G is generated by those quadratic binomials f_C, where C is a cycle of G of length 4.

(i) Let C be an even cycle of length ≥ 6. Since $f_C \in I_G$ and since I_G is generated by quadratic binomials, one can find two quadratic binomials f_{C_1} and f_{C_2}, where both C_1 and C_2 are cycles of length 4, for which $f_{C_1}^{(+)}$ divides $f_C^{(+)}$ and

$f_{C_2}^{(+)}$ divides $f_C^{(-)}$. Then, each of C_1 and C_2 yields either two even-chords of C or two odd-chords which cross in C. If one of these chords is an even-chord, then C satisfies the required condition. Suppose that each of these chords are odd-chords. The odd-chords e and e' arising from C_1 cross in C. Let e'' and e''' be the odd-chords arising from C_2. Since $C_1 \neq C_2$, it follows that either $e'' \notin \{e, e'\}$ or $e''' \notin \{e, e'\}$. Hence, C has at least three odd-chords two of which cross in C, as desired.

(ii) Let C_1 and C_2 be minimal odd cycles of G with exactly one common vertex and suppose that there exists no edge $\{i, j\} \notin E(C_1) \cup E(C_2)$ with $i \in V(C_1)$ and $j \in V(C_2)$. Since the even closed walk $\Gamma = (C_1, C_2)$ of length ≥ 6 is fundamental, it follows from Lemma 5.13 that I_G cannot be generated by quadratic binomials.

(iii) Let C_1 and C_2 be minimal odd cycles of G with $V(C_1) \cap V(C_2) = \emptyset$ and suppose that there exists no bridge between C_1 and C_2. Since G is connected, there is a walk $\Gamma_1 = (\{v_0, v_1\}, \{v_1, v_2\}, \ldots, \{v_{t-1}, v_t\})$ of length $t \geq 2$ with $v_0 \in C_1$ and $v_t \in C_2$. One can assume that t is the smallest length of those walks Γ, where Γ connects a vertex of C_1 with a vertex of C_2. Let Γ denote the even closed walk $(C_1, \Gamma_1, C_2, -\Gamma_1)$, where $-\Gamma_1 = (\{v_t, v_{t-1}\}, \ldots, \{v_2, v_1\}, \{v_1, v_0\})$. If the induced subgraph $G|_{V(\Gamma)}$ is equal to Γ, then Γ is fundamental of length $t + 3 \geq 5$. Lemma 5.13 then says that I_G cannot be generated by quadratic binomials. Thus, $G|_{V(\Gamma)} \neq \Gamma$ and there is an edge $e \in E(G|_{V(\Gamma)}) \setminus E(\Gamma)$. Since C_1 and C_2 are minimal odd cycles of G with $V(C_1) \cap V(C_2) = \emptyset$, since there is no bridge between C_1 and C_2 and since t is the minimum length, it follows that either $e = \{i, v_1\}$ with $i \in V(C_1)$ or $e = \{v_{t-1}, j\}$ with $j \in V(C_2)$, say, $e = \{i, v_1\}$ with $i \in V(C_1)$. One can find an odd cycle C_3 ($\neq C_1$) with $E(C_3) \subset E(C_1) \cup \{\{i, v_1\}, \{v_0, v_i\}\}$ and choose a minimal odd cycle C_4 with $V(C_4) \subset V(C_3)$. Since C_1 is minimal, one has $v_1 \in V(C_4)$. Let Γ' be the even closed walk $(C_4, \Gamma_2, C_2, -\Gamma_2)$, where $\Gamma_2 = (\{v_1, v_2\}, \{v_2, v_3\}, \ldots, \{v_{t-1}, v_t\})$. Since the degree of the binomial $f_{\Gamma'}$ is at least $t + 2 \geq 4$, again by Lemma 5.13, one has $G|_{V(\Gamma')} \neq \Gamma'$ and one can find an edge $e' = \{v_{t-1}, j\} \in E(G|_{V(\Gamma')}) \setminus E(\Gamma')$ with $j \in V(C_2)$. One can then find an odd cycle C_5 ($\neq C_2$) with $E(C_5) \subset E(C_2) \cup \{\{v_{t-1}, j\}, \{v_{t-1}, v_t\}\}$ and choose a minimal odd cycle C_6 with $V(C_6) \subset V(C_5)$. Since C_2 is minimal, one has $v_{t-1} \in V(C_6)$. Let Γ'' denote the even closed walk $(C_4, \Gamma_3, C_6, -\Gamma_3)$, where $\Gamma_3 = (\{v_1, v_2\}, \{v_2, v_3\}, \ldots, \{v_{t-2}, v_{t-1}\})$. (If $t = 2$, then $\Gamma = \emptyset$.) Since $G|_{V(\Gamma'')} = \Gamma''$, it follows that Γ'' is fundamental. Since the degree of $f_{\Gamma''}$ is at least $t + 1 \geq 3$, Lemma 5.13 says that I_G cannot be generated by quadratic binomials. This contradiction says that there exists a bridge between C_1 and C_2. Now, if there exists exactly one bridge $b \in E(G)$ between C_1 and C_2, then the even closed walk (C_1, b, C_2, b) is fundamental of length ≥ 8. Again, Lemma 5.13 says that I_G cannot be generated by quadratic binomials. Hence, at least two bridges between C_1 and C_2 exist.

(**If**) By virtue of Lemma 5.10, what we must prove is that, given a primitive even closed walk Γ of G of length $2q \geq 6$, the binomial f_Γ belongs to the ideal $(I_G)_{<q}$ which is generated by the binomials of degree $< q$ belonging to I_G.

If e is an edge of G, then we write x_e for the variable of $S = K[x_1, \ldots, x_n]$ with $\pi(x_e) = u^e \in K[\mathbf{t}]$.

First Step: Let Γ be a primitive even closed walk of G of length $2q \geq 6$ of the form in (i) of Lemma 5.11. Thus, Γ is an even cycle $C = (\{v_1, v_2\}, \{v_2, v_3\}, \ldots, \{v_{2q}, v_1\})$.

(a) Suppose that C has an even-chord $e = \{v_1, v_{2t}\}$ with $2 \leq t < q$. Let C_1 be the even cycle

$$(e, \{v_{2t}, v_{2t+1}\}, \{v_{2t+1}, v_{2t+2}\}, \ldots, \{v_{2q-1}, v_{2q}\}, \{v_{2q}, v_1\})$$

and C_2 the even cycle

$$(e, \{v_{2t}, v_{2t-1}\}, \{v_{2t-1}, v_{2t-2}\}, \ldots, \{v_3, v_2\}, \{v_2, v_1\}).$$

Then, $f_C = g f_{C_1} - h f_{C_2} \in (I_G)_{<q}$, where $g = f_{C_2}^{(-)}/x_e$ and $h = f_{C_1}^{(-)}/x_e$.

(b) Suppose that C has no even-chord and that C has three odd-chords e, e', and e'' such that e and e' cross in C. Let $e = \{v_1, v_t\}$ and $e' = \{v_2, v_{t+1}\}$ with $3 \leq t \leq 2q - 1$. Let

$$\Gamma = (\{v_t, v_{t-1}\}, \{v_{t-1}, v_{t-2}\}, \ldots, \{v_3, v_2\})$$

and

$$\Gamma' = (\{v_{t+1}, v_{t+2}\}, \{v_{t+2}, v_{t+3}\}, \ldots, \{v_{2q-1}, v_{2q}\}, \{v_{2q}, v_1\}).$$

Let $C_1 = (e, \Gamma, e', \Gamma')$ and $C_2 = (e, \{v_t, v_{t+1}\}, e', \{v_2, v_1\})$ be even cycles. Then, $f_C = f_{C_1} - h f_{C_2}$ with $h = f_{C_1}^{(+)}/x_e x_{e'}$. The binomial f_{C_1} is of degree q and f_{C_2} is quadratic. Let $e'' = \{v_i, v_j\}$, $S = \{v_1, v_{t+1}, v_{t+2}, \ldots, v_{2q}\}$, and $T = \{v_2, v_3, \ldots, v_t\}$. Let $v_i \in S$ and $v_j \in T$. Since e'' is an odd-chord of C, it follows that e'' is an even-chord of C_1. Hence, $f_C \in (I_G)_{<q}$.

Now, assume that each of v_i and v_j belongs to T with $2 \leq i < j \leq t$. Let C_3 be a minimal odd cycle with $V(C_3) \subset S \cup \{v_t\}$ and C_4 a minimal odd cycle with $V(C_4) \subset S \cup \{v_2\}$. For a while, suppose that C has no chord $\{v_{i'}, v_{j'}\}$ with $2 \leq i' < j' \leq t$ for which either $i' = 2$ or $j' = t$. (Since C has no even-chord, it follows that $\{v_2, v_t\}$ cannot be a chord of C.) Let C_5 be a minimal odd cycle with $V(C_5) \subset \{v_i, v_{i+1}, \ldots, v_j\}$. Since C_3 and C_5 are odd cycles with $V(C_3) \cap V(C_5) = \emptyset$, one can find a bridge $b = \{v_k, v_\ell\}$ between C_3 and C_5. The bridge must be an odd-chord of C with $v_k \in S$ and $v_\ell \in T$. Thus, b is an even-chord of C_1. On the other hand, suppose that C has a chord $\{v_2, v_{j'}\}$ with $2 < j' < t$ and that C has no chord $\{v_2, v_{j''}\}$ with $2 < j'' < j' < t$. Let C_6 be a minimal odd cycle with $V(C_6) \subset \{v_2, v_3, \dots, v_{j'}\}$. If C_4 and C_6 have exactly common vertex ($= v_2$), then one can find a bridge $b' = \{v_{k'}, v_{\ell'}\}$ between C_4 and C_6 with $v_{k'} \in S$ and $v_{\ell'} \in T$, which is an odd-chord of C. Thus, b' is an even-chord of C_1. Finally, if $V(C_4) \cap V(C_6) = \emptyset$, then there exist at least two bridges between C_4 and C_6, one of which is of the form $b'' = \{v_{k''}, v_{\ell''}\}$ with $v_{k''} \in S$ and $v_{\ell''} \in T$. Thus, b'' is an even-chord of C_1.

Second Step: Let Γ be a primitive even closed walk of length ≥ 6 of the form in
(ii) of Lemma 5.11. Thus, $L = (C_1, C_2)$, where C_1 and C_2 are odd cycles of G
with exactly one common vertex. Let

$$C_1 = (\{w, v_1\}, \{v_2, v_3\}, \dots, \{v_{2s-1}, v_{2s}\}, \{v_{2s}, w\})$$

and

$$C_2 = (\{w, v_1'\}, \{v_2', v_3'\}, \dots, \{v_{2t-1}', v_{2t}'\}, \{v_{2t}', w\}).$$

(a) Suppose that there is an edge $e = \{v_i, v_j'\}$ of G with $1 \leq i \leq 2s$ and $1 \leq j \leq 2t$. Let, say, i and j be even. Let Γ_1 be the even closed walk

$$\Gamma_1 = (\{w, v_1\}, \{v_1, v_2\}, \dots, \{v_{i-1}, v_i\}, e, \{v_j', v_{j+1}'\}, \dots, \{v_{2t-1}', v_{2t}'\}, \{v_{2t}', w\})$$

of G of length $i + 2t - j + 2$ and Γ_2 the even closed walk

$$\Gamma_2 = (\{w, v_1'\}, \{v_1', v_2'\}, \dots, \{v_{j-1}', v_j'\}, e, \{v_i, v_{i+1}\}, \dots, \{v_{2s-1}, v_{2s}\}, \{v_{2s}, w\}).$$

of G of length $j + 2s - i + 2$. Then, $f_\Gamma = g f_{\Gamma_1} - h f_{\Gamma_2}$, where $g = f_{\Gamma_2}^{(+)}/x_e$
and $h = f_{\Gamma_1}^{(+)}/x_e$.

(b) Suppose that none of the edges $\{v_i, v_j'\}$ with $1 \leq i \leq 2s$ and $1 \leq j \leq 2t$ belongs to $E(G)$. Let, say, $e = \{v_i, w\}$, where i is even, be a chord of C_1. If

$$\Gamma_1 = (\{w, v_1\}, \{v_1, v_2\}, \dots, \{v_{i-1}, v_i\}, e, C_2)$$

and

$$\Gamma_2 = (e, \{v_i, v_{i+1}\}, \{v_{i+1}, v_{i+2}\}, \dots, \{v_{2s-1}, v_{2s}\}, \{v_{2s}, w\}),$$

then $f_\Gamma = g f_{\Gamma_1} - h f_{\Gamma_2}$, where $g = f_{\Gamma_2}^{(+)}/x_e$ and $h = f_{\Gamma_1}^{(+)}/x_e$.

(c) Suppose that none of the edges $\{v_i, v_j'\}$ with $1 \leq i \leq 2s$ and $1 \leq j \leq 2t$ belongs to $E(G)$. In addition, suppose that none of the edges $\{v_i, w\}$ and $\{v_j', w\}$ with $1 < i < 2s$ and $1 < j < 2t$ belongs to $E(G)$. Then, either C_1 or C_2 cannot be minimal. If C_1 is not minimal, then there is a chord $e = \{v_i, v_j\}$ with $1 \leq i < j \leq 2s$. Let C_3 denote the odd cycle of G with $e \in E(C_3) \subset E(C_1) \cup \{e\}$ and C_4 the even cycle of G with $e \in E(C_4) \subset E(C_1) \cup \{e\}$. Let $w \notin V(C_3)$. Since

$V(C_2) \cap V(C_3) = \emptyset$, even though each of C_2 and C_3 might not be minimal, there exist at least two bridges between C_2 and C_3. Thus, in particular, since none of the edges $\{v_i, v'_j\}$ with $1 \le i \le 2s$ and $1 \le j \le 2t$ belongs to $E(G)$, one can find a chord $\{v_i, w\}$ of C_1. This contradicts our hypothesis. Let $w \in V(C_3)$. Let $\Gamma_1 = (C_2, C_3)$. Let, say, x_e divides $f_{\Gamma_1}^{(+)}$ and $f_{C_4}^{(+)}$. Let $g = f_{\Gamma_1}^{(+)}/x_e$ and $h = f_{C_4}^{(+)}/x_e$. Then, either $f_\Gamma = gf_{C_4} - hf_{\Gamma_1}$ or $f_\Gamma = -gf_{C_4} + hf_{\Gamma_1}$.

Third Step: Let Γ be a primitive even closed walk of length ≥ 6 of the form in (iii) of Lemma 5.11. Let $\Gamma = (C_1, \Gamma_1, C_2, \Gamma_2)$, where

$$C_1 = (\{v_1, v_2\}, \{v_2, v_3\}, \ldots, \{v_{2s}, v_{2s+1}\}, \{v_{2s+1}, v_1\})$$

and

$$C_2 = (\{v'_1, v'_2\}, \{v'_2, v'_3\}, \ldots, \{v'_{2t}, v'_{2t+1}\}, \{v'_{2t+1}, v'_1\})$$

are odd cycles of G with $V(C_1) \cap V(C_2) = \emptyset$ and where Γ_1 and Γ_2 are walks of G both of which connect v_1 with v'_1. Since there exist at least two bridges between C_1 and C_2, one can find a bridge $e = \{v_i, v'_j\}$, say, $j' \ne 1$. Since Γ is an even closed walk of G, the sum of the length of Γ_1 and the length of Γ_2 must be even. When both the length of Γ_1 and the length of Γ_2 are odd, one assume that both i and j are odd. When both the length of Γ_1 and the length of Γ_2 are even, one assume that i is odd and j is even. Let Γ_3 be the even closed walk

$$(e, \{v'_j, v'_{j-1}\}, \ldots, \{v'_2, v'_1\}, \Gamma_1, \{v_1, v_2\}, \ldots, \{v_{i-1}, v_i\})$$

and Γ_4 the even closed walk

$$(e, \{v'_j, v'_{j+1}\}, \ldots, \{v'_{2t+1}, v'_1\}, \Gamma_2, \{v_1, v_{2s+1}\}, \ldots, \{v_{i+1}, v_i\}).$$

Then, $f_\Gamma = gf_{\Gamma_3} - hf_{\Gamma_4}$, where $g = f_{\Gamma_4}^{(+)}/x_e$ and $h = f_{\Gamma_3}^{(+)}/x_e$.   □

Corollary 5.15 *Let G be a bipartite graph. Then, I_G is generated by quadratic binomials if and only if every cycle of G of length ≥ 6 has a chord.*

Problems

5.7 Compute the toric ideal of the finite simple graph G on [6] with the edges

$$\{1, 2\}, \{2, 3\}, \{1, 3\}, \{4, 5\}, \{5, 6\}, \{4, 6\}, \{3, 4\}.$$

5.8 Let C the cycle of length $2n$ on $[2n]$ and G the bipartite graph which is obtained by adding the edge $\{1, n + 1\}$ to C. Compute the toric ideal of G.

5.9 Find the smallest integer $d \geq 1$ such that there is a finite simple connected graph on $[d]$ whose toric ideal I_G cannot be generated by quadratic binomials.

5.4 Normality and Unimodular Coverings of Edge Polytopes

Let G be a finite connected simple graph. We say that G satisfies the *odd cycle condition* if, for any two odd cycles C_1 and C_2 of G with $V(C_1) \cap V(C_2) = \emptyset$, there is a bridge between C_1 and C_2.

Theorem 5.16 *Let G be a finite connected simple graph on $[d]$ with at least one odd cycle and suppose that G satisfies the odd cycle condition. Then, the edge polytope $\mathscr{P}_G$ possesses the integer decomposition property.*

Proof Let $\alpha \in n\mathscr{P}_G \cap \mathbb{Z}^d$. By virtue of Lemma 4.10, one can find a subgraph H of G satisfying the conditions of Lemma 5.5 for which $\alpha \in n\mathscr{P}_H \cap \mathbb{Z}^d$. Let $H_1, \ldots, H_s$ be the connected components of H and C_k a unique odd cycle of H_k for $1 \leq k \leq s$.

We write $\alpha = \sum_{e \in E(H)} a_e \rho(e)$ with each $a_e \in \mathbb{R}_{\geq 0}$ and with $\sum_{e \in E(H)} a_e = n$. Since

$$\alpha = \sum_{e \in E(H)} \lfloor a_e \rfloor \rho(e) + \sum_{e \in E(H)} (a_e - \lfloor a_e \rfloor) \rho(e)$$

belongs to $\mathbb{Z}^d$, it follows that $\sum_{e \in E(H)} (a_e - \lfloor a_e \rfloor) \rho(e)$ belongs to $\mathbb{Z}^d$. Thus, if $i \in [d]$ with $\deg_H i = 1$ and $i \in e$, then $a_e - \lfloor a_e \rfloor = 0$ and $a_e \in \mathbb{Z}_{\geq 0}$. Let H' denote the subgraph of H obtained by removing all vertices $i \in V(H)$ with $\deg_H i = 1$ and all edges $e \in E(H)$ with $i \in e$. Then, $\sum_{e \in E(H')} (a_e - \lfloor a_e \rfloor) \rho(e) \in \mathbb{Z}^d$. If $\deg_{H'} i = 1$ and if $e = \{i, j\} \in E(H')$, then $a_e - \lfloor a_e \rfloor = 0$. Thus, $a_e \in \mathbb{Z}_{\geq 0}$. Hence, repeated applications of such the technique guaranty that, for each edge $e \in E(H)$ which belongs to none of the cycles $c_1, \ldots, c_s$, one has $a(e) \in \mathbb{Z}_{\geq 0}$. Thus

$$\sum_{e \in E(C_1) \cup \cdots \cup E(C_s)} (a_e - \lfloor a_e \rfloor) \rho(e) \in \mathbb{Z}^d.$$

Since $V(C_k) \cap V(C_\ell) = \emptyset$ for $k \neq \ell$, it follows that

$$\sum_{e \in E(C_k)} (a_e - \lfloor a_e \rfloor) \rho(e) \in \mathbb{Z}^d$$

for $1 \leq k \leq s$. Now, since C_k is an odd cycle, it follows that either $\lfloor a_e \rfloor = 0$ for all $e \in E(C_k)$ or $\lfloor a_e \rfloor = 1/2$ for all $e \in E(C_k)$. Suppose that $\lfloor a_e \rfloor = 1/2$ for all $e \in E(C_k)$ and $\lfloor a_e \rfloor = 1/2$ for all $e \in E(C_\ell)$ with $k \neq \ell$. Let, say, $V(C_k) = \{1, 2, \ldots, 2p - 1\}$ and $V(C_\ell) = \{2p, 2p + 1, \ldots 2q\}$. Let $E(C_k) = \{e_1, e_2, \ldots, e_{2p-1}\}$ and $E(C_\ell) = \{e_{2p}, e_{2p+1}, \ldots, e_{2q}\}$, where $e_1 = $

$\{1, 2\}, e_2 = \{2, 3\}, \ldots, e_{2q-2} = \{2q - 2, 2q - 1\}, e_{2q-1} = \{2q - 1, 1\}$ and where $e_{2p} = \{2p, 2p + 1\}, e_{2p+1} = \{2p + 1, 2p + 2\}, \ldots, e_{2q-1} = \{2q - 1, 2q\}$, $e_{2q} = \{2q, 2p\}$. Since G satisfies the odd cycle condition, there is a bridge e' between C_k and C_ℓ. Let, say, $e' = \{1, 2q\}$. Then,

$$\frac{1}{2} \sum_{e \in E(C_k) \cup E(C_\ell)} \rho(e) = \rho(e') + \sum_{j=1}^{q-1} \rho(e_{2j}).$$

Thus, each $\alpha \in n \mathscr{P}_G \cap \mathbb{Z}^d$ can be expresses in the form $\alpha = \sum_{e \in E(G)} a_e \rho(e)$ with each $a_e \in \mathbb{Z}_{\geq 0}$, as desired.   □

Lemma 5.17 *Let G be a finite connected simple graph on $[d]$ with at least one odd cycle. Let H be a connected spanning subgraph which possesses exactly one odd cycle. Then, the subset $F(H) = \{\rho(e) : e \in E(H)\}$ of $\mathscr{P}_G \cap \mathbb{Z}^d$ is a fundamental maximal simplex belonging to $\mathscr{P}_G$.*

Proof It follows from Lemma 5.5 that $F(H)$ is a maximal simplex belonging to $\mathscr{P}_G$. Recall that $A(\mathscr{P}_G) \subset \mathbb{Z}^{d+1}$ is the configuration whose column vectors are those $(\mathbf{e}_i + \mathbf{e}_j, 1)^t$ with $\{i, j\} \in E(G)$. We show that each $\rho(e)$ with $e \in E(G)$ belongs to $\mathbb{Z}A(F(H))$. Let $e = \{i, j\}$. Since H is a connected spanning subgraph of H and since H possesses an odd cycle, one can find an odd walk $\Gamma = (e_1, e_2, \ldots, e_{2q-1})$ of H which connects $i \in e_1$ with $j \in e_{2q-1}$. Then $\rho(e) = \sum_{k=1}^{2q-1} (-1)^{k+1} \rho(e_k)$, as desired. It then follows that $\mathbb{Z}A(F(H)) = \mathbb{Z}A(\mathscr{P}_G)$.   □

Lemma 5.18 *Let $1 < s < t$ and G the finite simple graph on $V(G) = \{1, 2, \ldots, 2t\}$ which consists of two odd cycles C and C' with $V(C) = \{1, 2, \ldots, 2s - 1\}$ and $V(C') = \{2s, 2s + 1, \ldots, 2t\}$ together with the bridge $e' = \{1, 2t\}$ between C and C'. Let $\alpha \in \mathscr{P}_G$ and write $\alpha = \sum_{e \in E(G)} a_e \rho(e)$ with each $0 \leq a_e \in \mathbb{R}$ and with $\sum_{e \in E(G)} a_e = 1$. Then, one can assume that $a_e = 0$ for at least one edge $e \in E(C) \cup E(C')$.*

Proof Let $E(C) = \{e_1, e_2, \ldots, e_{2s-1}\}$ and $E(C') = \{e_{2s}, e_{2s+1}, \ldots, e_{2t}\}$, where $e_i = \{i, i + 1\}$ for $1 \leq i \leq 2s - 2$, $e_{2s-1} = \{2s - 1, 1\}$, $e_{2s} = \{2t, 2s\}$, and $e_j = \{j - 1, j\}$ for $2s + 1 \leq j \leq 2t$. Let $W = \{1, 3, \ldots, 2s - 1, 2s, 2s + 2, \ldots, 2t\}$. We then define $\delta \geq 0$ by setting

$$\delta = \min(\{a_k : k \in W\}.$$

Then, replacing a_{e_k} with $a_{e_k} - \delta$ if $k \in W$ and with $a_{e_k} + \delta$ if $k \notin W$ and replacing $a_{e'}$ with $a_{e'} + 2\delta$ in a given expression for α yields a required expression.   □

Theorem 5.19 *Let G be a finite connected simple graph on $[d]$ with at least one odd cycle and suppose that G satisfies the odd cycle condition. Let Ω denote the set of those maximal simplices $F(H)$ belonging to $\mathscr{P}_G$, where H is a connected spanning subgraph of G with exactly one odd cycle and where $F(H) = \{\rho(e) : e \in E(H)\} \subset \mathscr{P}_G \cap \mathbb{Z}^d$. Then, Ω is a unimodular covering of $\mathscr{P}_G$.*

Proof Lemma 5.17 says that each $F(H) \in \Omega$ is fundamental. Thus, it is suffices to show that $\mathscr{P}_G = \cup_{F(H) \in \Omega} \mathscr{P}_H$. Let $\alpha \in \mathscr{P}_G$. By virtue of Lemma 4.10, one can find a subgraph H' of G satisfying the conditions of Lemma 5.5 for which $\alpha \in \mathscr{P}_{H'} \cap \mathbb{Z}^d$. Let $H'_1, \ldots, H'_s$ denote the connected components of H' and C_k a unique odd cycle of H'_k for $1 \leq k \leq s$. If $s = 1$, then $F(H') \in \Omega$. Let $s > 1$. Let $\alpha = \sum_{e \in E(H')} a_e \rho(e)$ with each $a_e \in \mathbb{R}_{\geq 0}$ and with $\sum_{e \in E(H')} a_e = 1$. By using Lemma 5.18, in the above expression for α, one can replace one of the edges (say, e') belonging to either $E(C_1)$ or $E(C_2)$ with a bridge $e_{(C_1, C_2)}$ between C_1 and C_2. Let H'' denote the subgraph of G obtained from H' by removing $e' \in E(H)$ and by adding $e_{(C_1, C_2)}$. Then, H'' is a subgraph of G satisfying the conditions of Lemma 5.5 for which $\alpha \in \mathscr{P}_{H'} \cap \mathbb{Z}^d$. Furthermore, the number of connected components of H'' is $s - 1$. Now, the induction hypothesis guarantees that $\alpha \in \cup_{F(H) \in \Omega} \mathscr{P}_H$, as desired. □

Corollary 5.20 *Let G be a finite connected simple graph on $[d]$ with at least one odd cycle. Then, the following conditions are equivalent:*

(i) *The edge polytope $\mathscr{P}_G$ is normal;*
(ii) *The edge polytope $\mathscr{P}_G$ possesses the integer decomposition property;*
(iii) *The edge polytope $\mathscr{P}_G$ possesses a unimodular covering;*
(iv) *The finite graph G satisfies the odd cycle condition.*

Proof First of all, (iii) $\Rightarrow$ (i) and (ii) $\Rightarrow$ (i) follow from Theorems 4.5 and 4.11. Furthermore, (iv) $\Rightarrow$ (ii) and (iv) $\Rightarrow$ (iii) follow from Theorems 5.16 and 5.19.

To complete our proof, we must show (i) $\Rightarrow$ (iv). Suppose that G fails to satisfy the odd cycle condition. Choose two odd cycles C and C' with no bridge. Let $E(C) = \{\{1, 2\}, \{2, 3\}, \ldots, \{2s - 2, 2s - 1\}, \{2s - 1, 1\}\}$ and $E(C') = \{\{2s, 2s + 1\}, \{2s + 1, 2s + 2\}, \ldots, \{2t - 1, 2t\}, \{2t, 2s\}\}$. Since G is connected and since C is odd, one can find an odd walk

$$\Gamma = (\{1, i_1\}, \{i_1, i_2\}, \ldots, \{i_{2p-1}, i_{2p}\}, \{i_{2p}, 2t\})$$

of G connecting $1 \in V(C)$ with $2t \in V(C')$.

Again, recall that $A(\mathscr{P}_G) \subset \mathbb{Z}^{d+1}$ is the configuration whose column vectors are those $(e_i + e_j, 1)^t$ with $\{i, j\} \in E(G)$. Let $\alpha = e_1 + e_2 + \cdots + e_{2t}$. Since

$$\alpha = \frac{1}{2} \sum_{e \in E(C) \cup E(C')} \rho(e),$$

one has $\alpha \in \mathbb{Q}_{\geq 0} A(\mathscr{P}_G)$. Furthermore, $\alpha = \sum_{j=1}^{t-1} \rho(\{2j, 2j + 1\}) + \beta$, where

$$\beta = \rho(\{1, i_1\}) + \sum_{k=1}^{2p-1} (-1)^k \rho(\{i_k, i_{k+1}\}) + \rho(\{i_{2p}, 2t\}).$$

Fig. 5.1 A graph whose edge
polytope is the polytope in
Example 4.25.

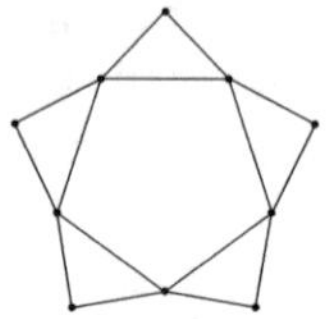

Thus, $\alpha \in \mathbb{Z}A(\mathscr{P}_G)$. Hence, $\alpha \in \mathbb{Z}A(\mathscr{P}_G) \cap \mathbb{Q}_{\geq 0}A(\mathscr{P}_G)$. Since $\alpha \notin \mathbb{Z}_{\geq 0}A(\mathscr{P}_G)$, it follows that $\mathbb{Z}_{\geq 0}A(\mathscr{P}_G) \neq \mathbb{Z}A(\mathscr{P}_G) \cap \mathbb{Q}_{\geq 0}A(\mathscr{P}_G)$. Thus, $\mathscr{P}_G$ cannot be normal, as desired.   □

Example 5.21 A normal polytope none of whose regular triangulation is unimodular given in Example 4.25 is the edge polytope of the graph in Figure 5.1. Hence, the conditions in Corollary 5.20 are not equivalent to the existence of a regular unimodular triangulation.

Recall that $\mathscr{P}_G$ is called unimodular if every triangulation of $\mathscr{P}_G$ is unimodular. In order to classify unimodular edge polytopes, we characterize circuits of I_G.

Proposition 5.22 *Let G be a finite connected simple graph. A binomial f is a circuit of I_G if and only if there exists an even closed walk Γ with $f = f_\Gamma$ satisfying one of the following:*

(i) *Γ is an even cycle of G;*
(ii) *$\Gamma = (C_1, C_2)$, where each of C_1 and C_2 is an odd cycle of G having exactly one common vertex;*
(iii) *$\Gamma = (C_1, e_{i_1}, \ldots, e_{i_r}, C_2, e_{i_r}, \ldots, e_{i_1})$, where each of C_1 and C_2 is an odd cycle of G and where $(e_{i_1}, \ldots, e_{i_r})$ is a path of G which combines $j \in V(C_1)$ with $j' \in V(C_2)$ satisfying that $V(C_1) \cap V(C_2) = \emptyset$, $V(\{e_{i_1}, \ldots, e_{i_r}\}) \cap V(C_1) = \{j\}$, and $V(\{e_{i_1}, \ldots, e_{i_r}\}) \cap V(C_2) = \{j'\}$.*

Proof Suppose that f is a circuit of I_G. Since any circuit is primitive, there exists an even closed walk Γ with $f = f_\Gamma$ satisfying one of the conditions in Proposition 5.11. If Γ satisfies none of (i), (ii), and (iii) above, then Γ is of the form $\Gamma = (C_1, \Gamma_1, C_2, \Gamma_2)$ satisfying the condition (iii) in Proposition 5.11 and does not satisfy the condition (iii) above. It then follows that there exists an even closed walk $\Gamma' = (C_1, e_{i_1}, \ldots, e_{i_r}, C_2, e_{i_r}, \ldots, e_{i_1})$ satisfying condition (iii) above such that $var(f_{\Gamma'}) \subsetneq var(f_\Gamma)$, which is a contradiction.

Let Γ be an even closed walk satisfying one of (i), (ii), and (iii) above. By Lemma 4.30, f_Γ is irreducible. Let H be the subgraph of G with the edge set $E(\Gamma)$ and H' a proper subgraph of H. Then, each connected component of H' has at most one cycle and has no even cycle. Hence, there exists no even closed walk Γ' satisfying one of the conditions above such that $var(f_{\Gamma'}) \subsetneq var(f_\Gamma)$. Thus, f_Γ is a circuit of I_G.   □

Proposition 5.23 *Let G be a finite connected simple graph. Then, $\mathscr{P}_G$ is unimodular if and only if, any two odd cycles of G have at least one common vertex.*

Proof Let Γ be an even closed walk satisfying one of the conditions in Proposition 5.22. Then, f_Γ is squarefree if and only if Γ satisfies either (i) or (ii). Thus, by Theorem 4.35, $\mathscr{P}_G$ is unimodular if and only if G has no even closed walks of type (iii) in Proposition 5.22. Since G is connected, this condition holds if and only if any two odd cycles of G have at least one common vertex.   □

Theorem 5.24 *Let G be a finite connected simple bipartite graph. Then,*

(i) *The edge polytope $\mathscr{P}_G$ is unimodular and in particular normal;*
(ii) *The edge polytope $\mathscr{P}_G$ possesses the integer decomposition property.*

Proof Since G has no odd cycles, (i) follows from Proposition 5.23. Now, we show that $\mathscr{P}_G$ possesses the integer decomposition property. Let $\alpha \in n\mathscr{P}_G \cap \mathbb{Z}^d$, where $[d]$ is the vertex set of G. By using Lemma 4.10 together with Lemma 5.6, one can find a spanning tree H of G with $\alpha \in n\mathscr{P}_H \cap \mathbb{Z}^d$. Let $\alpha = \sum_{e\in E(H)} a_e \rho(e)$ with each $0 \leq a_e \in \mathbb{R}$ and with $\sum_{e\in E(H)} a_e = 1$. Since H is a tree, there is a vertex $i \in V(H)$ with $\deg_H i = 1$ (Problem 5.10). We then employ the technique which appear in the proof of Theorem 5.16. If $e = \{i, j\} \in E(H)$ with $\deg_H i = 1$, then the subgraph H' which is obtained by removing e is again a tree. It then follows that $a_e \in \mathbb{Z}_{\geq 0}$, as desired.   □

Corollary 5.25 *Let G be a finite connected simple graph and suppose that the toric ideal I_G is generated by quadratic binomials, then the edge polytope $\mathscr{P}_G$ is normal.*

Proof If G has at least one odd cycle and if I_G is generated by quadratic binomials, then Theorem 5.14 guarantees that G satisfies the odd cycle condition. It then follows from Corollary 5.20 that $\mathscr{P}_G$ is normal. If G is bipartite, then Theorem 5.24 says that $\mathscr{P}_G$ is normal.   □

Finally, Hochster [115] says that every normal toric ring is Cohen–Macaulay. It then follows that

Corollary 5.26 *Let G be a finite simple connected graph and suppose that G satisfies the odd cycle condition. Then, the edge ring $K[G]$ is Cohen–Macaulay. In particular, the edge ring of every bipartite graph is Cohen–Macaulay.*

Problems

5.10 Show that every tree possesses a vertex of degree one.

5.11 Find the smallest integer $d \geq 1$ such that there is a finite simple connected graph on $[d]$ whose edge polytope $\mathscr{P}_G$ is not normal.

5.12 Find a unimodular covering of the edge polytope of the complete graph on $[5]$.

5.13 Find a unimodular triangulation of the edge polytope of the complete bipartite graph on $[5] = V \cup V'$ with $|V| = 2$ and $|V'| = 3$.

5.14 Let G be the bipartite graph on [5] with edges

$$\{1, 2\}, \{2, 3\}, \{3, 4\}, \{1, 4\}, \{1, 5\}.$$

Compute the normalized Ehrhart function of the edge polytope $\mathcal{P}_G$.

5.5   Koszul Bipartite Graphs

It would, of course, be of interest to classify the finite simple graphs G for which the toric ideal I_G possesses a Gröbner basis consisting of quadratic binomials. However, to find the complete classification is presumably hopeless. On the other hand, Theorem 5.27 below says that, for a bipartite graph G, its toric ideal I_G possesses a Gröbner basis consisting of quadratic binomials if and only if I_G is generated by quadratic binomials.

Theorem 5.27 *Let G be a finite connected simple bipartite graph. Then, the following conditions are equivalent:*

(i) *Every cycle of length ≥ 6 has a chord;*
(ii) *The toric ideal I_G possesses a Gröbner basis consisting of quadratic binomials;*
(iii) *The edge ring $K[G]$ is Koszul;*
(iv) *The toric ideal I_G is generated by quadratic binomials.*

Proof It follows from Theorem 2.28 and Proposition 2.23 that (ii) $\Rightarrow$ (iii) $\Rightarrow$ (iv). Furthermore, Corollary 5.15 says that (iv) $\Leftrightarrow$ (i). Thus, (i) $\Rightarrow$ (ii) remains to be proved.

Let G be a bipartite graph on $[d]$ with the partition $[d] = U \cup V$. Let $U = \{u_1, \ldots, u_s\}$ and $V = \{v_1, \ldots, v_t\}$. Let $A = (a_{ij})_{1 \leq i \leq s, 1 \leq j \leq t}$ be the *incidence matrix* of G. In other words, the rows of A are indexed by U and the columns of A are indexed by V such that $a_{ij} = 1$ if $\{u_i, v_j\} \in E(G)$ and $a_{ij} = 0$ if $\{u_i, v_j\} \notin E(G)$. In general, given integer vectors $\mathbf{a} = (a_1, \ldots, a_q)$ and $\mathbf{b} = (b_1, \ldots, b_q)$, we introduce the order $\prec$ defined by setting $\mathbf{a} \prec \mathbf{b}$ if the rightmost nonzero component of the vector $\mathbf{a} - \mathbf{b}$ is negative. Let $\delta_A = (\delta_2, \delta_3, \ldots, \delta_{s+t})$ with $\delta_k = \sum_{i+j=k} a_{ij}$. Let $\mathbf{a}_1, \ldots, \mathbf{a}_s$ denote the rows of A. Suppose that $i_1 < i_2$ and $\mathbf{a}_{i_2} \prec \mathbf{a}_{i_1}$. Let A' denote the new matrix obtained by permuting the rows $\mathbf{a}_{i_1}$ and $\mathbf{a}_{i_2}$ of A. Then, $\delta_A \prec \delta_{A'}$. Hence, repeating permutations of rows and columns of A yields the matrix A'' which maximizes $\delta_{A''}$. One can then assume that the rows and the columns of A are simultaneously arranged in the order $\prec$. Suppose that A has a submatrix

$$\begin{pmatrix} a_{i_1 j_1} & a_{i_1 j_2} \\ a_{i_2 j_1} & a_{i_2 j_2} \end{pmatrix} = \begin{pmatrix} 1 & 1 \\ 1 & 0 \end{pmatrix} = B$$

with $i_1 < i_2$ and $j_1 < j_2$. Since $\mathbf{a}_{i_1} \preceq \mathbf{a}_{i_2}$, there exists an index $j_3 > j_2$ for which $(a_{i_1 j_3}, a_{i_2 j_3}) = (0, 1)$ and $a_{i_1 k} = a_{i_2 k}$ for all $k > j_3$. Similarly, there exists an index $i_3 > i_2$ for which $(a_{i_3 j_1}, a_{i_3 j_2}) = (0, 1)$ and $a_{\ell j_1} = a_{\ell j_2}$ for all $\ell > i_3$. If $a_{i_3 j_3} = 1$, then A has the submatrix

$$
\begin{pmatrix}
a_{i_1 j_1} & a_{i_1 j_2} & a_{i_1 j_3} \\
a_{i_2 j_1} & a_{i_2 j_2} & a_{i_2 j_3} \\
a_{i_3 j_1} & a_{i_3 j_2} & a_{i_3 j_3}
\end{pmatrix}
=
\begin{pmatrix}
1 & 1 & 0 \\
1 & 0 & 1 \\
0 & 1 & 1
\end{pmatrix}.
$$

This submatrix represents the cycle of G of length 6 with no chord, which contradicts the condition (i). Hence, $a_{i_3 j_3} = 0$. Thus,

$$
\begin{pmatrix}
a_{i_1 j_1} & a_{i_1 j_2} & a_{i_1 j_3} \\
a_{i_2 j_1} & a_{i_2 j_2} & a_{i_2 j_3} \\
a_{i_3 j_1} & a_{i_3 j_2} & a_{i_3 j_3}
\end{pmatrix}
=
\begin{pmatrix}
1 & 1 & 0 \\
1 & 0 & 1 \\
0 & 1 & 0
\end{pmatrix}.
$$

Since $\mathbf{a}_{i_2} \prec \mathbf{a}_{i_3}$, there exists an index $j_4 > j_3$ for which $(a_{i_2 j_4}, a_{i_3 j_4}) = (0, 1)$ and $a_{i_2 k} = a_{i_3 k}$ for all $k > j_4$. Similarly, there exists an index $i_4 > i_3$ for which $(a_{i_4 j_2}, a_{i_4 j_3}) = (0, 1)$ and $a_{\ell j_2} = a_{\ell j_3}$ for all $\ell > i_4$. If $a_{i_4 j_4} = 1$, then A has the submatrix

$$
\begin{pmatrix}
a_{i_1 j_1} & a_{i_1 j_2} & a_{i_1 j_3} & a_{i_1 j_4} \\
a_{i_2 j_1} & a_{i_2 j_2} & a_{i_2 j_3} & a_{i_2 j_4} \\
a_{i_3 j_1} & a_{i_3 j_2} & a_{i_3 j_3} & a_{i_3 j_4} \\
a_{i_4 j_1} & a_{i_4 j_2} & a_{i_4 j_3} & a_{i_4 j_4}
\end{pmatrix}
=
\begin{pmatrix}
1 & 1 & 0 & 0 \\
1 & 0 & 1 & 0 \\
0 & 1 & 0 & 1 \\
0 & 0 & 1 & 1
\end{pmatrix}.
$$

This submatrix represents the cycle of G of length 8 with no chord, which contradicts the condition (i). Repeating this argument guarantees that B cannot be a submatrix of A.

Now, we employ the reverse lexicographic order $<_{\mathrm{rev}}$ on $S = K[\{x_e : e \in E(G)\}]$ induced by the ordering

$$
\{u_1, v_1\} < \{u_1, v_2\} < \ldots < \{u_1, v_t\} < \{u_2, v_1\} < \{u_2, v_2\} < \ldots < \{u_2, v_t\}
$$
$$
< \ldots < \{u_s, v_1\} < \{u_s, v_2\} < \ldots < \{u_s, v_t\}
$$

of edges of G. Every cycle C of G of length 4 appears in A as the submatrix

$$
\begin{pmatrix}
1 & 1 \\
1 & 1
\end{pmatrix}
$$

with the initial part

$$
\begin{pmatrix}
 & 1 \\
1 &
\end{pmatrix}.
$$

Let $C_1, \ldots, C_m$ denote the cycles of G of length 4. Since (iv) $\Leftrightarrow$ (i), the toric ideal I_G is generated by the binomials $f_{C_1}, \ldots, f_{C_m}$. In order to show that $\{f_{C_1}, \ldots, f_{C_m}\}$ is a Gröbner basis of I_G with respect to $<_{\text{rev}}$, Buchberger criterion (Theorem 1.29) can be applied. Let $S(f_{C_i}, f_{C_j})$ denote the S-polynomial of f_{C_i} and f_{C_j}. If the initial monomial $\text{in}_{<_{\text{rev}}}(f_{C_i})$ of f_i and $\text{in}_{<_{\text{rev}}}(f_{C_j})$ of f_j are relatively prime, then $S(f_{C_i}, f_{C_j})$ reduces to 0 with respect to $f_{C_1}, \ldots, f_{C_m}$. Suppose that $\text{in}_{<_{\text{rev}}}(f_{C_i})$ and $\text{in}_{<_{\text{rev}}}(f_{C_j})$ of f_j are not relatively prime.

Let $|E(C_i) \cap E(C_j)| = 2$, say, $C_i = (e_1, e_2, e_3, e_4)$ and $C_j = (e_1, e_2, e_5, e_6)$ with $e_2 < e_1 < e_3 < e_4$ and $e_2 < e_1 < e_5 < e_6$. Then, $S(f_{C_i}, f_{C_j}) = x_2 f_C$, where $C = (e_4, e_3, e_5, e_6)$ is a cycle of G of length 4. Hence, $S(f_{C_i}, f_{C_j})$ reduces to 0 with respect to f_C.

Let $|E(C_i) \cap E(C_j)| = 1$, say, $C_i = (e_2, e_3, e_4, e_1)$ and $C_j = (e_5, e_6, e_7, e_1)$ with $e_4 < e_1 < e_3 < e_2$ and $e_7 < e_1 < e_6 < e_5$. Then, $S(f_{C_i}, f_{C_j}) = f_\Gamma$, where $\Gamma = (e_2, e_3, e_4, e_7, e_6, e_5)$ is a cycle of G of length 6. Then, Γ appears in A as one of the following submatrices:

$$\begin{pmatrix} * & 1 & 1 \\ 1 & * & 1 \\ 1 & 1 & x \end{pmatrix}, \begin{pmatrix} * & 1 & 1 \\ 1 & 1 & x \\ 1 & * & 1 \end{pmatrix}, \begin{pmatrix} 1 & * & 1 \\ * & 1 & 1 \\ 1 & 1 & x \end{pmatrix}$$

$$\begin{pmatrix} 1 & * & 1 \\ 1 & 1 & x \\ * & 1 & 1 \end{pmatrix}, \begin{pmatrix} 1 & 1 & * \\ * & 1 & 1 \\ 1 & x & 1 \end{pmatrix}, \begin{pmatrix} 1 & 1 & * \\ 1 & x & 1 \\ * & 1 & 1 \end{pmatrix}$$

Each of the above six matrices contains the submatrix

$$F = \begin{pmatrix} 1 & 1 \\ 1 & a \end{pmatrix}.$$

Since B cannot be a submatrix of A, it follows that $a = 1$ and Γ has a chord. Let C' denote the cycle of G of length 4 which F represents. Then, $\text{in}_{<_{\text{rev}}}(f_{C'})$ divides $\text{in}_{<_{\text{rev}}}(f_\Gamma)$ and

$$f_\Gamma - \frac{\text{in}_{<_{\text{rev}}}(f_\Gamma)}{\text{in}_{<_{\text{rev}}}(f_{C'})} f_{C'} = x_e f_{C''},$$

where $e \in E(\Gamma)$ and C'' is a cycle of length 4 of the induced subgraph $G_{E(\Gamma)}$. Hence, f_Γ reduces to 0 with respect to $f_{C'}$ and $f_{C''}$, as desired.   □

Example 5.28 Let G be a graph in Figure 5.2. Then, the toric ideal I_G is generated by quadratic binomials and coincides with the ideal given in Example 1.18. Moreover, as stated in Example 2.29, $K[G]$ is not Koszul. Hence, conditions (iii) and (iv) in Theorem 5.27 are not equivalent for nonbipartite graphs. Note that the edge polytope $\mathscr{P}_G$ is unimodular by Proposition 5.23.

Fig. 5.2 A graph whose toric
ideal is the ideal in
Example 1.18.

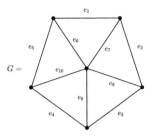

Problems

5.15 Classify all non-Koszul bipartite graphs, up to isomorphic, with at most 8 vertices.

Notes

Early references on the toric ideal of an edge polytope are [190, 213]. In particular, Villarreal [213] found a correspondence between generators of the toric ideals of the edge polytopes and even closed walks of the graphs, see Lemma 5.9. A characterization of simplices of edge polytopes in terms of graphs appeared in De Loera–Sturmfels–Thomas [51], see Lemma 5.5. These results are introduced in the lecture note [202, Chapter 9]. Inspired by such results, the systematic study on the edge polytope and toric ideal of a finite simple graph originated by Ohsugi–Hibi [157, 159].

Ohsugi–Hibi [157] proved that the following three conditions are equivalent: (i) the edge polytope of a graph is normal; (ii) it has a unimodular covering; and (iii) the graph satisfies the odd cycle condition, see Corollary 5.20. Simis–Vasconcelos–Villarreal [191] showed (i) ⇔ (iii) independently. Note that the odd cycle condition appeared in a classical paper [77] in graph theory. A normal edge polytope none of whose regular triangulations is unimodular was given in [158], see Example 5.21. Ohsugi [154] give a nontrivial infinite series of normal edge polytopes none of whose regular triangulations is unimodular.

A combinatorial criterion for the toric ideal I_G to be generated by quadratic binomials appeared in the paper [159], see Theorem 5.14. In the paper [160], it was shown that, for a bipartite graph G, its toric ideal I_G has a Gröbner basis consisting of quadratic binomials if and only if I_G is generated by quadratic binomials, see Theorem 5.27. A graph whose toric ideal is generated by quadratic binomials and whose toric ring is not Koszul is given in [159, Example 2.1], see Example 5.28. Hibi–Nishiyama–Ohsugi–Shikama [112] give a nontrivial infinite series of finite graphs with the property that their toric ideals are generated by quadratic binomials and possesses no quadratic Gröbner bases. It is known that the toric ideal of a graph has a Gröbner basis consisting of quadratic binomials if the graph is (i) a complete multipartite graph [161] (Example 9.27), and (ii) a gap-free graph [47].

Important sets of binomials in the toric ideals of edge rings were studied by many researchers. A characterization of circuits of the toric ideals of graphs was given in [213, Proposition 4.2], see Proposition 5.22. Ohsugi–Hibi [168] characterize the graphs whose toric ideals are generated by: (i) squarefree circuits, and (ii) circuits having at least one squarefree monomial. There exist several classes of graphs whose toric ideals satisfy this condition and whose toric rings are nonnormal. On the other hand, a characterization of universal Gröbner bases was given in Tatakis–Thoma [208]. Ohsugi–Hibi gave a necessary condition for a binomial to be primitive (Lemma 5.11) in [159] and discussed indispensable binomials in [162]. Reyes–Tatakis–Thoma [178] extended these results and characterized primitive binomials, minimal generators, indispensable binomials, and fundamental binomials in graph theoretical terms. Ogawa–Takemura–Hara [153] gave another characterization for primitive binomials.

Other ring-theoretical properties of edge rings were studied in [163] (Gorenstein), [81, 127, 209] (complete intersection), and [110] (strongly Koszul).

A graph-theoretical characterization of an edge of edge polytopes (Lemma 5.3) was given in [165]. Using this fact, the combinatorial structure of edge polytopes was discussed in [165] (simple edge polytopes), [111, 212] (number of edges), and [109] (separating hyperplanes of edge polytopes).

Chapter 6
Join-Meet Ideals of Finite Lattices

Abstract One of the most natural classes of binomial ideals arising from combinatorics is the class of join-meet ideals of finite lattices. The purpose of the present chapter is mainly to study Gröbner bases of join-meet ideals. In Section 6.1, we collect fundamental definitions and basic results on classical lattice theory. Especially, a complete proof of the characterization of distributive lattices due to Dedekind is supplied. The algebraic theory of join-meet ideals, which originated in the study on those ideals of finite distributive lattices, is introduced in Section 6.2. The highlight is the fact that the join-meet ideal of a finite lattice is a prime ideal if and only if the lattice is distributive. Furthermore, with respect to a certain reverse lexicographic order, it is shown that the set of binomial generators of the join-meet ideal of a finite lattice is a Gröbner basis of the ideal if and only if the lattice is distributive. We then devote Section 6.3 to the discussion of join-meet ideals of finite non-distributive modular lattices. Furthermore, in Section 6.4, join-meet ideals of planar distributive lattices will be studied. Finally, in Section 6.5, via the theory of canonical modules and the a-invariant, projective dimension together with regularity of join-meet ideals will be discussed.

6.1 Review on Classical Lattice Theory

Recall from Chapter 1 that a partial order on a set P is a binary relation $\leq$ on P such that, for all a, b, c belonging to P, one has:

- $a \leq a$ (reflexivity);
- $a \leq b$ and $b \leq a \Rightarrow a = b$ (antisymmetry);
- $a \leq b$ and $b \leq c \Rightarrow a \leq c$ (transitivity).

A set P with a partial order is called a *partially ordered set*. In combinatorics, a partially ordered set is often called a *poset* for short.

Every poset P studied in the present section is finite. A subset $C \subset P$ is called a *chain* of P if C is a totally ordered subset with respect to the induced order. In other words, a chain is a subset $C = \{a_1, a_2, \ldots, a_k\}$ of P with $a_1 < a_2 < \cdots < a_k$. The *length* of a chain C is $|C| - 1$. Let $\mathrm{rank}(P)$ denote the *rank* of P, which is the

© Springer International Publishing AG, part of Springer Nature 2018   141
J. Herzog et al., *Binomial Ideals*, Graduate Texts in Mathematics 279,
https://doi.org/10.1007/978-3-319-95349-6_6

maximal length of a chain of P. A subset P' of a poset P is called a *subposet* if, for a and b belonging to P', one has $a < b$ in P' if and only $a < b$ in P.

Let P and Q be finite posets. A map $\varphi : P \to Q$ is *order-preserving* if, for $a, b \in P$ with $a \leq b$ in P, one has $\varphi(a) \leq \varphi(b)$ in Q. We say that P is isomorphic to Q if there exists a bijection $\varphi : P \to Q$ such that both φ and its inverse φ^{-1} are order-preserving.

Let $<$ be a partial order on a set P. Then, the *dual* partial order on P is the partial order $<^*$ such that $a < b$ if and only if $b <^* a$ for all $a, b \in P$. The set P with the partial order $<^*$ is called the dual poset of P and is written as P^*. One has $(P^*)^* = P$.

A finite *lattice* is a finite poset L such that, for any two elements a and b belonging to L, there is a unique greatest lower bound $a \wedge b$, called the *meet* of a and b, and there is a unique least upper bound $a \vee b$, called the *join* of a and b. Thus, in particular a finite lattice possesses both a unique minimal element $\hat{0}$ and a unique maximal element $\hat{1}$. A subposet L' of a finite lattice L is called a *sublattice* of L if L' is a lattice and, for $a, b \in L'$, the meet of a and b in L' coincides with that in L and the join of a and b in L' coincides with that in L. The dual poset L^* of a finite lattice is again a lattice, which will be called the dual lattice of L. It follows that if $c = a \vee b$ and $c' = a \wedge b$ in L, then $c = a \wedge b$ and $c' = a \vee b$ in L^*.

Example 6.1

(a) Let $\mathscr{B}_n$ denote the set of all subsets of $[n]$, ordered by inclusion. Then, $\mathscr{B}_n$ is a lattice, called the *boolean lattice* of rank n.

(b) Let $n > 0$ be an integer and $\mathscr{D}_n$ the set of all divisors of n, ordered by divisibility. Then, $\mathscr{D}_n$ is a lattice, called the *divisor lattice* of n. Thus, in particular a boolean lattice is a divisor lattice.

Fig. 6.1 A boolean lattice and a divisor lattice.

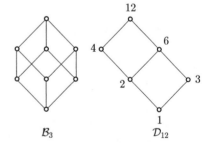

$\mathcal{B}_3$ $\qquad$ $\mathcal{D}_{12}$

A finite lattice L is called *distributive* if, for all a, b, c belonging to L, one has:

$$a \vee (b \wedge c) = (a \vee b) \wedge (a \vee c),$$

$$a \wedge (b \vee c) = (a \wedge b) \vee (a \wedge c).$$

Every divisor lattice is a distributive lattice and, in particular, every boolean lattice is a distributive lattice. Every sublattice of a finite distributive lattice is again a distributive lattice. The dual lattice of a distributive lattice is again a distributive lattice.

Lemma 6.2 *Let L be a finite lattice. Then, the following conditions are equivalent:*

(i) *For all $a, b, c \in L$, one has $a \vee (b \wedge c) = (a \vee b) \wedge (a \vee c)$;*
(ii) *For all $a, b, c \in L$, one has $a \wedge (b \vee c) = (a \wedge b) \vee (a \wedge c)$.*

Proof We prove (i) $\Rightarrow$ (ii). Then, since the dual lattice L^* is distributive, the converse (ii) $\Rightarrow$ (i) also follows. Suppose (i) and $a, b, c \in L$. One has:

$$(a \wedge b) \vee (a \wedge c) = ((a \wedge b) \vee a) \wedge ((a \wedge b) \vee c)$$

$$= a \wedge (c \vee (a \wedge b))$$

$$= a \wedge ((c \vee a) \wedge (c \vee b))$$

$$= (a \wedge (c \vee a)) \wedge (c \vee b)$$

$$= a \wedge (b \vee c),$$

as desired.   □

Lemma 6.2 does guarantee that, for individual elements $a, b, c \in L$, *neither* $a \vee (b \wedge c) = (a \vee b) \wedge (a \vee c) \Rightarrow a \wedge (b \vee c) = (a \wedge b) \vee (a \wedge c)$ *nor* $a \wedge (b \vee c) = (a \wedge b) \vee (a \wedge c) \Rightarrow a \vee (b \wedge c) = (a \vee b) \wedge (a \vee c)$. In fact,

Example 6.3 Let $L = \{\hat{0}, a, b, c, \hat{1}\}$ with $\hat{0} < a < c < \hat{1}$ and $\hat{0} < b < \hat{1}$. Then, $a \wedge (b \vee c) = a = (a \wedge b) \vee (a \wedge c)$. However, $a \vee (b \wedge c) = a$ and $(a \vee b) \wedge (a \vee c) = c$. Furthermore, in L^*, one has $a \vee (b \wedge c) = a = (a \vee b) \wedge (a \vee c)$, but $a \wedge (b \vee c) = a$ and $(a \wedge b) \vee (a \wedge c) = c$.

Let $P = \{p_1, \ldots, p_n\}$ be a finite poset with a partial order $\leq$. A *poset ideal* of P is a subset α of P with the property that, whenever $a \in \alpha$ and $b \in P$ with $b \leq a$, one has $b \in \alpha$. In particular, the empty set as well as P itself is a poset ideal. Let $\mathcal{J}(P)$ denote the set of poset ideals of P. If α and β are poset ideals of P, then each of the sets $\alpha \cap \beta$ and $\alpha \cup \beta$ is again a poset ideal. It then follows that $\mathcal{J}(P)$ is a finite lattice ordered by inclusion.

Fig. 6.2 A poset and its lattice of poset ideals.

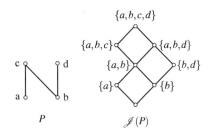

P

$\mathcal{J}(P)$

Furthermore, it follows easily that $\mathscr{J}(P)$ is a distributive lattice whose rank is equal to $|P|$. Now, Birkhoff's fundamental structure theorem for finite distributive lattices guarantees that the converse is true. In fact,

Theorem 6.4 (Birkhoff) *Given a finite distributive lattice L, there is a unique finite poset P such that L is isomorphic to $\mathscr{J}(P)$.*

Proof Let L be a finite distributive lattice. An element $a \in L$ with $a \neq \hat{0}$ is called *join-irreducible* if, whenever $a = b \vee c$ with $b, c \in L$, one has either $a = b$ or $a = c$. Let P denote the subposet of L consisting of all join-irreducible elements of L.

We claim that L is isomorphic to $\mathscr{J}(P)$. To see why this is true, we define the map $\varphi : \mathscr{J}(P) \to L$ by setting $\varphi(\alpha) = \bigvee_{a \in \alpha} a$, where $\alpha \in \mathscr{J}(P)$. In particular, $\varphi(\emptyset) = \hat{0}$. Clearly, φ is order-preserving. Since each element $a \in L$ can be the join of the join-irreducible elements b with $b \leq a$ in L, it follows that $\varphi(\alpha) = a$, where α is a poset ideal of P consisting of those $b \in P$ with $b \leq a$. Thus, φ is surjective.

The highlight of the proof is to show that φ is injective. Let α and β be poset ideals of P with $\alpha \neq \beta$, say, $\beta \not\subseteq \alpha$. Let b^* be a maximal element of β with $b^* \notin \alpha$. We show $\varphi(\alpha) \neq \varphi(\beta)$. Suppose, on the contrary, that $\varphi(\alpha) = \varphi(\beta)$. Thus,

$$\bigvee_{a \in \alpha} a = \bigvee_{b \in \beta} b. \qquad (6.1)$$

Since L is distributive, it follows that:

$$\left(\bigvee_{a \in \alpha} a \right) \wedge b^* = \bigvee_{a \in \alpha} (a \wedge b^*).$$

Since $a \wedge b^* < b^*$ and since b^* is join-irreducible, it follows that $(\bigvee_{a \in \alpha} a) \wedge b^* < b^*$. However, since $b^* \in \beta$, one has:

$$\left(\bigvee_{b \in \beta} b \right) \wedge b^* = \bigvee_{b \in \beta} (b \wedge b^*) = b^*.$$

This contradicts (6.1). Hence, φ is injective.

Now, the inverse map φ^{-1} is defined as follows: For each element $c \in L$, $\varphi^{-1}(c)$ is the set of join-irreducible elements $a \in L$ with $a \leq c$. Clearly, $\varphi^{-1}(c) \in \mathscr{J}(P)$ and φ^{-1} is order-preserving. As a result, L is isomorphic to $\mathscr{J}(P)$ with the bijective order-preserving map φ, as desired.

Finally, since P is isomorphic to the subposet consisting of all join-irreducible elements of the distributive lattice $\mathscr{J}(P)$, it follows that, for two finite posets P and Q, if $\mathscr{J}(P)$ is isomorphic to $\mathscr{J}(Q)$, then P is isomorphic to Q. In other words, the existence of a finite poset P such that L is isomorphic to $\mathscr{J}(P)$ is unique. $\square$

A finite lattice L is called *modular* if the following condition is satisfied: If, for all a, b, c belonging to L, one has:

$$a \leq c \Rightarrow a \vee (b \wedge c) = (a \vee b) \wedge c.$$

Every distributive lattice is modular. In fact, if L is distributive, then $a \vee (b \wedge c) = (a \vee b) \wedge (a \vee c)$ and $a \vee c = c$ since $a \leq c$. Every sublattice of a finite modular lattice is again a modular lattice.

The dual lattice L^* of a modular lattice L is modular. In fact, if $a \leq c$ in L^*, then $c \leq a$ in L. Thus, in L, one has $c \vee (b \wedge a) = (c \vee b) \wedge a$. Hence, in L^*, one has $c \wedge (b \vee a) = (c \wedge b) \vee a$.

Example 6.5 Let G be a finite group and $L(G)$ the poset consisting of all normal subgroups of G, ordered by inclusion. Then, $L(G)$ is a lattice. In fact, if H and H' are normal subgroups of G, then HH' and $H \cap H'$ are normal subgroups of G. Thus, $H \vee H' = HH'$ and $H \wedge H' = H \cap H'$. It is not difficult to show that $L(G)$ is a modular lattice. Furthermore, $L(G)$ is a distributive lattice if and only if G is a cyclic group.

Example 6.6 Let $\mathbb{F}_q$ denote the q-element finite field and $V_n(q)$ the vector space of dimension n over $\mathbb{F}_q$. Let $L_n(q)$ denote the poset consisting of all subspaces of $V_n(q)$, ordered by inclusion. Then, $L_n(q)$ is a lattice and is modular.

The *pentagon lattice* N_5 is the simplest non-modular lattice. The *diamond lattice* M_5 is the simplest non-distributive modular lattice.

Lemma 6.7 *Let L be a finite modular lattice and $a, b, c \in L$. Then,*

$$(a \wedge (b \vee c)) \vee (b \wedge (c \vee a)) = (a \vee b) \wedge (b \vee c) \wedge (c \vee a),$$

$$(a \vee (b \wedge c)) \wedge (b \vee (c \wedge a)) = (a \wedge b) \vee (b \wedge c) \vee (c \wedge a).$$

Proof Since $a \wedge (b \vee c) \leq a \leq c \vee a$ and since L is modular, it follows that:

$$(a \wedge (b \vee c)) \vee (b \wedge (c \vee a)) = ((a \wedge (b \vee c)) \vee b) \wedge (c \vee a)$$

$$= (b \vee (a \wedge (b \vee c))) \wedge (c \vee a)$$

$$= (b \vee a) \wedge (b \vee c) \wedge (c \vee a)$$

$$= (a \vee b) \wedge (b \vee c) \wedge (c \vee a).$$

Since the dual lattice L^* is modular, the second equality follows.   □

Lemma 6.8 *Let L be a finite modular lattice and $a, b, c \in L$. Let*

$$e = (b \wedge c) \vee (a \wedge (b \vee c)),$$

$$f = (c \wedge a) \vee (b \wedge (c \vee a)),$$

$$g = (a \wedge b) \vee (c \wedge (a \vee b)).$$

Then,

$$e \wedge f = f \wedge g = g \wedge e = (a \wedge b) \vee (b \wedge c) \vee (c \wedge a), \qquad (6.2)$$

$$e \vee f = f \vee g = g \vee e = (a \vee b) \wedge (b \vee c) \wedge (c \vee a). \qquad (6.3)$$

Proof It follows from Lemma 6.7 that

$$e \vee f = ((b \wedge c) \vee (a \wedge (b \vee c))) \vee ((c \wedge a) \vee (b \wedge (c \vee a)))$$

$$= (b \wedge c) \vee (a \wedge (b \vee c)) \vee (b \wedge (c \vee a)) \vee (c \wedge a)$$

$$= (b \wedge c) \vee ((a \vee b) \wedge (b \vee c) \wedge (c \vee a)) \vee (c \wedge a)$$

Since each of $b \wedge c$ and $c \wedge a$ is less than or equal to each of $a \vee b, b \vee c, c \vee a$, the formula (6.3) follows. Furthermore, since L^* is modular, the formula (6.2) also follows.   □

Lemma 6.9 *In Lemma 6.8, if any two of the elements e, f, g are equal, then*

$$a \vee (b \wedge c) = (a \vee b) \wedge (a \vee c), \quad a \wedge (b \vee c) = (a \wedge b) \vee (a \wedge c).$$

Proof Let, say, $e = f$. Then, $e \wedge f = e \vee f$. Hence,

$$(a \wedge b) \vee (b \wedge c) \vee (c \wedge a) = (a \vee b) \wedge (b \vee c) \wedge (c \vee a).$$

Thus,

$$a \wedge ((a \wedge b) \vee (b \wedge c) \vee (c \wedge a)) = a \wedge ((a \vee b) \wedge (b \vee c) \wedge (c \vee a)).$$

We show

$$a \wedge ((a \wedge b) \vee (b \wedge c) \vee (c \wedge a)) = (a \wedge b) \vee (a \wedge c), \qquad (6.4)$$

$$a \wedge ((a \vee b) \wedge (b \vee c) \wedge (c \vee a)) = a \wedge (b \vee c). \qquad (6.5)$$

Since $(a \wedge b) \vee (c \wedge a) \leq a$ and since L is modular, it follows that:

$$a \wedge ((a \wedge b) \vee (b \wedge c) \vee (c \wedge a))$$

$$= (((a \wedge b) \vee (c \wedge a)) \vee (b \wedge c)) \wedge a$$

$$= ((a \wedge b) \vee (c \wedge a)) \vee ((b \wedge c) \wedge a)$$

$$= ((a \wedge b) \vee (c \wedge a)) \vee (a \wedge b \wedge c)$$

$$= (a \wedge b) \vee (c \wedge a) = (a \wedge b) \vee (a \wedge c).$$

Thus, (6.4) follows. Since $a \leq a \vee b$ and $a \leq c \vee a$, the equality (6.5) follows.

It follows from (6.4) and (6.5) that $a \wedge (b \vee c) = (a \wedge b) \vee (a \wedge c)$. Since L^* is modular, $a \vee (b \wedge c) = (a \vee b) \wedge (a \vee c)$ also follows.   □

Theorem 6.10 (Dedekind)

(a) *A finite lattice L is modular if and only if no sublattice of L is isomorphic to the pentagon lattice N_5.*
(b) *A modular lattice L is distributive if and only if no sublattice of L is isomorphic to the diamond lattice M_5.*
(c) *A finite lattice L is distributive if and only if any sublattice of L is isomorphic to neither N_5 nor M_5.*

Proof Since N_5 is non-modular and M_5 is non-distributive, the "only if" part of each of (a) and (b) follows.

Let L be non-modular. Then, there exist a, b, and c belonging to L for which $a < c$ and $a \vee (b \wedge c) < (a \vee b) \wedge c$. Let

$$L' = \{b, \; a \vee b, \; b \wedge c, \; a \vee (b \wedge c), \; (a \vee b) \wedge c\}.$$

We claim that L' is a sublattice of L which is isomorphic to N_5. Clearly,

$$b \wedge c \leq a \vee (b \wedge c) < (a \vee b) \wedge c \leq a \vee b.$$

and

$$(a \vee (b \wedge c)) \vee b = a \vee b, \quad ((a \vee b) \wedge c) \wedge b = b \wedge c. \tag{6.6}$$

Let $b \wedge c = a \vee (b \wedge c)$. Then, $a \leq b \wedge c$. Hence, $a \vee (b \wedge c) = (a \vee b) \wedge c$, which contradicts our hypothesis. Similarly, if $(a \vee b) \wedge c = a \vee b$, then $a \vee b \leq c$ and, again, $a \vee (b \wedge c) = (a \vee b) \wedge c$. As a result, one has:

$$b \wedge c < a \vee (b \wedge c) < (a \vee b) \wedge c < a \vee b. \tag{6.7}$$

It follows from (6.6) and (6.7) that $b \notin L' \setminus \{b\}$. Hence, L' is isomorphic to N_5.

Let a modular lattice L be non-distributive. By using Lemma 6.2, there exist $a, b, c \in L$ for which $a \wedge (b \vee c) \neq (a \wedge b) \vee (a \wedge c)$. Let $e, f, g \in L$ be defined as in Lemma 6.8. Then, $e \wedge f = f \wedge g = g \wedge e$ and $e \vee f = f \vee g = g \vee e$. Furthermore, Lemma 6.9 guarantees that $e \neq f, f \neq g$, and $g \neq e$. Hence, the five-element sublattice $L' = \{e \wedge f, e, f, g, e \vee f\}$ is isomorphic to M_5.

Finally, (c) follows from (a) and (b).   □

In general, we say that an element a of a finite poset P *covers* $b \in P$ if $b < a$ and there is no $c \in P$ with $b < c < a$. A finite lattice L is called *semimodular* if the following condition is satisfied: If a and b belonging to L cover $a \wedge b$, then $a \vee b$ covers both a and b.

Lemma 6.11 *Every modular lattice is semimodular.*

Proof Let L be a modular lattice and $a, b \in L$. Suppose that both a and b cover $a \wedge b$. If $a \vee b$ does not cover, say, a, then there is $c \in L$ with $a < c < a \vee b$. Since L is modular and since $a < c$, one has $a \vee (b \wedge c) = (a \vee b) \wedge c$. However, since $b \wedge c = a \wedge b$, one has $a \vee (b \wedge c) = a$. In addition, $(a \vee b) \wedge c = c$. Thus, $a \vee (b \wedge c) < (a \vee b) \wedge c$, a contradiction. □

The *centered hexagon lattice* D_2 is semimodular but not modular. Furthermore, the dual lattice of D_2 cannot be semimodular.

Since every non-modular semimodular lattice possesses the pentagon lattice N_5 as a sublattice and since N_5 cannot be semimodular, it follows that a sublattice of a semimodular lattice might not be semimodular.

A *rank function* of a finite poset P is a map $\rho : P \to \mathbb{Z}_{\geq 0}$ such that $\rho(a) = 0$ if a is a minimal element of P and that $\rho(a) = \rho(b) + 1$ if a covers b. A rank function is unique if it exists. For example, the boolean lattice $\mathscr{B}_n$ of rank n possesses a rank function ρ satisfying $\rho(\alpha) = |\alpha|$ for each $\alpha \subset [n]$. Furthermore, in general, every finite distributive lattice $L = \mathscr{J}(P)$ possesses a rank function ρ satisfying $\rho(\alpha) = |\alpha|$ for each poset ideal $\alpha \subset P$.

A finite poset P is called *pure* if all maximal chains of P have the same length $(= \text{rank}(P))$. If $a, b \in P$ with $a < b$, then an *interval* $[a, b]$ of P is the subposet of P consisting of those $c \in P$ with $a \leq c \leq b$. Every interval of a pure poset is pure. Every interval of a finite lattice is again a lattice. Every interval of a distributive (resp., modular, semimodular) lattice is distributive (resp., modular, semimodular).

Lemma 6.12 *Every pure poset possesses a rank function.*

Proof Let P be pure. Given $a \in P$, we write $P_{\leq a}$ for the subposet $\{b \in P : b \leq a\}$ of P. Since P is pure, it follows that $P_{\leq a}$ is pure. Let $\rho(a) = \text{rank}(P_{\leq a})$. We claim ρ is a rank function of P. Clearly, one has $\rho(a) = 0$ if a is a minimal element. Let $a, b \in P$ for which a covers b. Then, $\text{rank}(P_{\leq b}) \leq \text{rank}(P_{\leq a}) - 1$. Since $P_{\leq a}$ is pure, there is a chain of P of the form $a_0 < a_1 < \cdots < a_{r-1} = b < a$, where $r = \text{rank}(P_{\leq a})$. Thus, $\text{rank}(P_{\leq b}) \geq \text{rank}(P_{\leq a}) - 1$. Hence, $\text{rank}(P_{\leq b}) = \text{rank}(P_{\leq a}) - 1$. In other words, one has $\rho(a) = \rho(b) + 1$. □

Lemma 6.13 *Ever semimodular lattice is pure.*

Proof Let L be a finite semimodular lattice. We show that L is pure by using induction on $\text{rank}(L)$. Let

$$\hat{0} < a_1 < a_2 < \cdots < a_k < \hat{1}, \quad \hat{0} < b_1 < b_2 < \cdots < b_{k'} < \hat{1}$$

be maximal chains of L. We claim $k = k'$. Since both a_1 and b_1 cover $\hat{0}$, it follows that $a_1 \vee b_1$ covers both a_1 and b_1. Let $\ell = \text{rank}([a_1 \vee b_1, \hat{1}])$. Each of the intervals $[a_1, \hat{1}]$ and $[b_1, \hat{1}]$ is a modular lattice containing $a_1 \vee b_1$ whose rank is less than $\text{rank}(L)$. Hence, each of $[a_1, \hat{1}]$ and $[b_1, \hat{1}]$ is pure of rank $\ell + 1$. Thus, one has $k = k' = \ell + 2$, as desired. □

Theorem 6.14 *A finite lattice L is modular if and only if L possesses a rank function satisfying*

$$\rho(a) + \rho(b) = \rho(a \wedge b) + \rho(a \vee b) \quad \text{for all} \quad a, b \in L \tag{6.8}$$

Proof Let L be a finite modular lattice and ρ its rank function. By using induction on $\text{rank}(L)$, we show that ρ satisfies (6.8). Let $a, b \in L$ with $\rho(a) + \rho(b) \neq \rho(a \wedge b) + \rho(a \vee b)$. Since every interval of a modular lattice is modular, one can assume that $a \wedge b = \hat{0}$ and $a \vee b = \hat{1}$. In particular, $\rho(a \wedge b) = 0$. Let, say, $\rho(a) + \rho(b) < \rho(\hat{1})$. Since L is pure, there is $c \in L$ with $a < c$ for which $\rho(c) + \rho(b) = \rho(\hat{1})$. Since $a < c$ and since L is modular, it follows that $a \vee (b \wedge c) = (a \vee b) \wedge c$. Since $a \vee b = \hat{1}$, one has $(a \vee b) \wedge c = c$. Thus, $a \vee (b \wedge c) = c$. Hence, $b \wedge c \neq \hat{0}$. Since $\text{rank}([b \wedge c, \hat{1}]) < \text{rank}(L)$ and since $[b \wedge c, \hat{1}]$ is modular, it follows that $\rho(c) + \rho(b) = \rho(b \wedge c) + \rho(\hat{1}) > \rho(\hat{1})$. This contradicts $\rho(c) + \rho(b) = \rho(\hat{1})$.

Suppose that a finite lattice L is pure and its rank function satisfies (6.8). Let $a, b, c \in L$ with $a \leq c$. In general, one has $a \vee (b \wedge c) \leq (a \vee b) \wedge c$. Now,

$$\rho(a \vee (b \wedge c)) = \rho(a) + \rho(b \wedge c) - \rho(a \wedge b \wedge c)$$
$$= \rho(a) + \rho(b) + \rho(c) - \rho(b \vee c) - \rho(a \wedge b \wedge c)$$
$$= \rho(a) + \rho(b) + \rho(c) - \rho(b \vee c) - \rho(a \wedge b),$$
$$\rho((a \vee b) \wedge c) = \rho(a \vee b) + \rho(c) - \rho(a \vee b \vee c)$$
$$= \rho(a) + \rho(b) - \rho(a \wedge b) + \rho(c) - \rho(a \vee b \vee c)$$
$$= \rho(a) + \rho(b) - \rho(a \wedge b) + \rho(c) - \rho(b \vee c).$$

Hence, $a \vee (b \wedge c) = (a \vee b) \wedge c$, as required.   □

Problems

6.1 Let $1 < p_1 < p_2 < \cdots < p_s$ be prime numbers and $n = p_1 p_2 \cdots p_s$. Show that the boolean lattice of rank s is isomorphic to the divisor lattice of n.

6.2

(a) Show that every divisor lattice is distributive.
(b) Find a distributive lattice which is isomorphic to no divisor lattice.
(c) Is every sublattice of a boolean lattice again a boolean lattice?
(d) Is every sublattice of a divisor lattice again a divisor lattice?

6.3 Let $P = \{a, b, c, d, e\}$ be a finite poset with $b < d, b < e, c < d, c < e$. Find the distributive lattice $\mathscr{J}(P)$.

6.4

(a) Find a finite poset P with $\mathscr{J}(P) = \mathscr{B}_4$.
(b) Find a finite poset P with $\mathscr{J}(P) = \mathscr{D}_{24}$.

6.5 Let G be a finite group and $L(G)$ the finite lattice consisting of all normal subgroups of G, ordered by inclusion.

(i) Show that $L(G)$ is a modular lattice.
(ii) Show that if $L(G)$ is a distributive lattice, then G is abelian.
(iii) Using (ii) show that $L(G)$ is a distributive lattice, if and only if G is a cyclic group.

6.6 Let L be a finite lattice. An element $a \in L$ is called an *atom* if a covers $\hat{0}$, i.e., $\hat{0} < a$ and $\hat{0} < b < a$ for no $b \in L$. A finite lattice L is called *atomic* if every element is a join of atoms. A *geometric lattice* is a finite semimodular atomic lattice.

(i) Show that every boolean lattice is a geometric lattice.
(ii) Show that the finite lattice $L_n(q)$ of Example 6.6 is a geometric lattice.
(iii) Show that a finite distributive L lattice is atomic if and only if L is boolean.

6.7

(a) Find a rank function of a boolean lattice.
(b) Find a rank function of $L_n(q)$.
(c) Find a rank function of a distributive lattice.

6.2   Gröbner Bases of Join-Meet Ideals

Let L be a finite lattice and K be a field. Let $K[L] = K[\{x_a : a \in L\}]$ denote the polynomial ring in $|L|$ variables over K. Given a and b belonging to L, we introduce the binomial ideal $f_{a,b} \in K[L]$ by setting:

$$f_{a,b} = x_a x_b - x_{a \wedge b} x_{a \vee b}.$$

In particular, $f_{a,b} = 0$ if and only if a and b are comparable in L. The *join-meet ideal* of L is the ideal $I_L \subset K[L]$ which is generated by those binomials $f_{a,b}$ with $a, b \in L$.

Example 6.15 The join-meet ideal of the pentagon lattice N_5 of Figure 6.3 is generated by $f_{a,b} = x_a x_b - x_{\hat{0}} x_{\hat{1}}$ and $f_{c,b} = x_c x_b - x_{\hat{0}} x_{\hat{1}}$. The join-meet ideal of the diamond lattice M_5 of Figure 6.4 is generated by $f_{a,b} = x_a x_b - x_{\hat{0}} x_{\hat{1}}$, $f_{b,c} = x_b x_c - x_{\hat{0}} x_{\hat{1}}$, and $f_{c,a} = x_c x_a - x_{\hat{0}} x_{\hat{1}}$.

A monomial order $<$ on $K[L]$ is called *compatible* if, for all $a, b \in L$ for which a and b are incomparable, one has $\mathrm{in}_< f_{a,b} = x_a x_b$.

Example 6.16 Let $<$ be a total order on the variables of $K[L]$ with the property that one has $x_a < x_b$ if $a < b$ in L. In combinatorics, such a total order is called a *linear extension* of L. Let $<_{\mathrm{rev}}$ denote the reverse lexicographic order induced by the ordering $<$. It then follows that $<_{rev}$ is a compatible monomial order on $K[L]$. We call $<_{rev}$ a *rank reverse lexicographic order* on $K[L]$.

Theorem 6.17 *Let L be a finite lattice and fix a compatible monomial order $<$ on $K[L]$. Let $\mathcal{G}_L$ denote the set of binomials $f_{a,b} \in I_L$ for which a and b belonging to L are incomparable. Then, the following conditions are equivalent:*

(i) *$\mathcal{G}_L$ is a Gröbner basis of I_L with respect to $<$.*
(ii) *L is a distributive lattice.*

Proof Suppose that a finite lattice L is not a distributive lattice. By virtue of Theorem 6.10, it follows that L contains either the pentagon lattice N_5 or the diamond lattice M_5. Work with the same notation a, b, and c as in Figures 6.3 and 6.4. It then follows from Example 6.15 that the initial monomial of the S-polynomial $S(f_{a,b}, f_{b,c}) \in I_L$ is of the form $x_e x_f x_g$, where $\{e, f, g\}$ is a chain of L of length 2. Thus, none of the monomials $x_e x_f$, $x_f x_g$, $x_g x_e$ can belong to the monomial ideal $(\{in_<(f) : f \in \mathcal{G}_L\})$. Hence, $\mathcal{G}_L$ cannot be a Gröbner basis of I_L with respect to $<$.

Now, suppose that L is a distributive lattice. We claim that Buchberger's criterion guarantees that $\mathcal{G}_L$ is a Gröbner basis of I_L with respect to $<$. Let $a, b, c \in L$ with $b \neq c$, where a and b are incomparable in L and a and c are incomparable in L. Then,

$$S(f_{a,b}, f_{a,c}) = x_c f_{a,b} - x_b f_{a,c}$$
$$= x_b(x_{a \wedge c} x_{a \vee c}) - x_c(x_{a \wedge b} x_{a \vee b})$$
$$= (f_{b,a \wedge c} + x_{a \wedge b \wedge c} x_{b \vee (a \wedge c)}) x_{a \vee c} - (f_{c,a \wedge b} + x_{a \wedge b \wedge c} x_{c \vee (a \wedge b)}) x_{a \vee b}$$
$$= x_{a \vee c} f_{b,a \wedge c} - x_{a \vee b} f_{c,a \wedge b} + x_{a \wedge b \wedge c}(x_{b \vee (a \wedge c)} x_{a \vee c} - x_{c \vee (a \wedge b)} x_{a \vee b})$$

Furthermore,

$$x_{b \vee (a \wedge c)} x_{a \vee c} - x_{c \vee (a \wedge b)} x_{a \vee b}$$
$$= (f_{b \vee (a \wedge c), a \vee c} + x_{(b \vee (a \wedge c)) \wedge (a \vee c)} x_{(b \vee (a \wedge c)) \vee (a \vee c)})$$
$$\quad - (f_{c \vee (a \wedge b), a \vee b} + x_{(c \vee (a \wedge b)) \wedge (a \vee b)} x_{(c \vee (a \wedge b)) \vee (a \vee b)}).$$

Now,

$$x_{(b \vee (a \wedge c)) \wedge (a \vee c)} x_{(b \vee (a \wedge c)) \vee (a \vee c)} - x_{(c \vee (a \wedge b)) \wedge (a \vee b)} x_{(c \vee (a \wedge b)) \vee (a \vee b)}$$
$$= x_{(b \vee (a \wedge c)) \wedge (a \vee c)} x_{a \vee b \vee c} - x_{(c \vee (a \wedge b)) \wedge (a \vee b)} x_{a \vee b \vee c}$$
$$= x_{a \vee b \vee c}(x_{(b \vee (a \wedge c)) \wedge (a \vee c)} - x_{(c \vee (a \wedge b)) \wedge (a \vee b)}).$$

Since L is distributive, it follows that

$$(b \vee (a \wedge c)) \wedge (a \vee c) = (a \vee b) \wedge (b \vee c) \wedge (c \vee a) = (c \vee (a \wedge b)) \wedge (a \vee b).$$

Hence, the binomial

$$x_{b \vee (a \wedge c)} x_{a \vee c} - x_{c \vee (a \wedge b)} x_{a \vee b}$$

reduces to 0. Thus, $S(f_{a,b}, f_{a,c})$ reduces to 0, as desired.   □

Let $P = \{p_1, \ldots, p_n\}$ be a finite poset and $L = \mathscr{J}(P)$ the finite distributive lattice which consists of all poset ideals of P, ordered by inclusion. Let $S = K[x_1, \ldots, x_n, t]$ the polynomial ring in $n+1$ variables over a field K with $\deg t = 1$. Given a poset ideal $\alpha \subset P$, we introduce the monomial $u_\alpha \in S$ by setting:

$$u_\alpha = (\prod_{p_i \in \alpha} x_i) t.$$

In particular, $u_\emptyset = t$ and $u_P = x_1 x_2 \cdots x_n t$. Let $\mathscr{R}_K[L] \subset S$ denote the toric ring which is generated by those monomials u_α with $\alpha \in \mathscr{J}(P)$. We then define the surjective ring homomorphism $\pi : K[L] \to \mathscr{R}_K[L]$ by setting $\pi(x_\alpha) = u_\alpha$ for all $\alpha \in L = \mathscr{J}(P)$. Let $\mathrm{Ker}(\pi)$ denote the kernel of π.

Lemma 6.18 *One has $I_L \subset \mathrm{Ker}(\pi)$.*

Proof Let $\alpha, \beta \in L = \mathscr{J}(P)$. Then, $\alpha \wedge \beta = \alpha \cap \beta$ and $\alpha \vee \beta = \alpha \cup \beta$. Thus,

$$\begin{aligned}
u_{\alpha \wedge \beta} u_{\alpha \vee \beta} &= (\prod_{p_i \in \alpha \cap \beta} x_i)(\prod_{p_i \in \alpha \cup \beta} x_i) t^2 \\
&= (\prod_{p_i \in \alpha} x_i)(\prod_{p_i \in \beta} x_i) t^2 \\
&= u_\alpha u_\beta.
\end{aligned}$$

Thus, $\pi(x_\alpha x_\beta) = \pi(x_{\alpha \wedge \beta} x_{\alpha \vee \beta})$. Hence, $I_L \subset \mathrm{Ker}(\pi)$, as required.   □

Theorem 6.19 *The set $\mathscr{G}_L$ of binomials is a Gröbner basis of $\mathrm{Ker}(\pi)$ with respect to a compatible monomial order $<$.*

Proof A basic technique by using Theorem 1.19 can be applied. Let $\mathrm{in}_<(\mathscr{G}_L)$ denote the set of initial monomials $\mathrm{in}_<(f_{\alpha,\beta})$ with $f_{\alpha,\beta} \in \mathscr{G}_L$. Thus, $\mathrm{in}_<(\mathscr{G}_L)$ consists of those quadratic monomials $x_\alpha x_\beta$ with $\alpha, \beta \in L$ such that α and β are incomparable in L. Let $\mathrm{in}_<(\mathrm{Ker}(\pi))$ denote the initial ideal of $\mathrm{Ker}(\pi)$ with respect to $<$. It then follows from Lemma 6.18 that $(\mathrm{in}_<(\mathscr{G}_L)) \subset \mathrm{in}_<(\mathrm{Ker}(\pi))$.

Let $\mathscr{B}$ denote the set of those monomials $w \in K[L]$ with $w \notin (\mathrm{in}_<(\mathscr{G}_L))$ and $\mathscr{B}'$ that of those monomials $w \in K[L]$ with $w \notin \mathrm{in}_<(\mathrm{Ker}(\pi))$. Theorem 1.19 guarantees that $\mathscr{B}'$ is a K-basis of $\mathscr{R}_K[L] = K[L]/\mathrm{Ker}(\pi)$. Since $\mathscr{B}' \subset \mathscr{B}$, in order to show $(\mathrm{in}_<(\mathscr{G}_L)) = \mathrm{in}_<(\mathrm{Ker}(\pi))$, it suffices to prove that $\mathscr{B}$ is linearly independent in $\mathscr{R}_K[L] = K[L]/\mathrm{Ker}(\pi)$.

Now, what we must prove is that, for $w, w' \in \mathscr{B}$ with $w \neq w'$, one has $\pi(w) \neq \pi(w')$. It follows that:

$$w = x_{\alpha_1} x_{\alpha_2} \cdots x_{\alpha_p}, \quad w' = x_{\beta_1} x_{\beta_2} \cdots x_{\beta_q},$$

where

$$\alpha_1 \leq \alpha_2 \leq \cdots \leq \alpha_p, \quad \beta_1 \leq \beta_2 \leq \cdots \leq \beta_q$$

in L. In order to show $\pi(w) \neq \pi(w')$, one can assume that $p = q$ and $\alpha_i \neq \beta_j$ for all i and j. Let $\alpha_1 \not\subseteq \beta_1$. Then, there is $p_\xi \in P$ for which $p_\xi \in \alpha_1$ and $p_\xi \not\in \beta_1$. Since each α_i and β_j is a poset ideal of P and since

$$\alpha_1 \subset \alpha_2 \subset \cdots \subset \alpha_p, \quad \beta_1 \subset \beta_2 \subset \cdots \subset \beta_q$$

as subsets of P, it follows that $p_\xi \in \alpha_i$ for all $1 \leq i \leq p$. Hence, x_ξ^p appears in $\pi(w)$. However, since $p_\xi \not\in \beta_1$, the power r for which x_ξ^r appears in $\pi(w')$ is at most $p - 1$. Hence, $\pi(w) \neq \pi(w')$, as desired.   □

The proof of Theorem 6.19, which is based on Theorem 1.19, supplies a somewhat surprising proof of (ii) $\Rightarrow$ (i) of Theorem 6.17 without using of Buchberger's criterion.

Since, in general, a Gröbner basis of an ideal is a set of generators of the ideal (Corollary 1.16), it follows from Theorem 6.19 that:

Corollary 6.20 *Let L be a finite distributive lattice. Then, the set $\mathscr{G}_L$ of binomials is a system of generators of* $\mathrm{Ker}(\pi)$. *In particular,* $I_L = \mathrm{Ker}(\pi)$.

Theorem 6.21 *Given a finite lattice L, the following conditions are equivalent:*

(i) *I_L is a prime ideal;*
(ii) *L is a distributive lattice.*

Proof Since every toric ideal is a prime ideal, it follows from Corollary 6.20 that I_L is prime if L is distributive.

Let L be finite lattice which is not distributive. By virtue of Theorem 6.10, that L contains either the pentagon lattice N_5 or the diamond lattice M_5. Work with the same notation a, b, and c as in Figures 6.3 and 6.4. It then follows from Example 6.15 that, even though $x_b \not\in I_L$ and $x_a - x_c \not\in I_L$, one has $x_b(x_a - x_c) \in I_L$. Thus, I_L cannot be a prime ideal.   □

Fig. 6.3 The pentagon lattice.

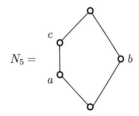

$$N_5 =$$

Fig. 6.4 The diamond
lattice.

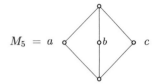

$$M_5 = a$$

Problems

6.8 Compute the join-meet ideal of the centered hexagon lattice D_2 of Figure 6.5.

Fig. 6.5 The centered
hexagon lattice.

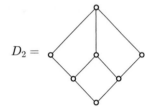

$$D_2 =$$

6.9 Work with the same notation as in the paragraph just before Lemma 6.18. Let $\hat{P} = P \cup \{\hat{0}, \hat{1}\}$, where $\hat{0} < p_i < \hat{1}$ for $1 \leq i \leq n$. An *order-reversion* map on $\hat{P}$ is a map $\sigma : \hat{P} \to \{0, 1, 2, \ldots\}$ for which $\sigma(\hat{1}) = 0$ and $\sigma(a) \leq \sigma(b)$ if $a > b$ in $\hat{P}$. Let $\Omega(\hat{P})$ denote the set of order-preserving maps on $\hat{P}$. Given an order-reversion map σ on $\hat{P}$, we introduce the monomial $w_\sigma \in S$ by setting:

$$w_\sigma = \prod_{i=1}^{n} x_i^{\sigma(p_i)} t^{\sigma(\hat{0})}.$$

Show that the set of monomials $\{w_\sigma : \sigma \in \Omega(\hat{P})\}$ is a K-basis of $\mathcal{R}_K[L]$.

6.10 By using the result of Problem 6.9, show that $\mathcal{R}_K[L]$ is normal.

6.3   Join-Meet Ideals of Modular Non-distributive Lattices

We now turn to the problem of characterizing modular non-distributive lattices in terms of initial ideals of join-meet ideals.

Lemma 6.22 *Let L be a finite modular non-distributive lattice. Then, L possesses a sublattice $L' = \{x, a, b, c, y\}$ which is isomorphic to the diamond lattice M_5, where $x < a, b, c < y$, for which $\rho(y) - \rho(x) = 2$, where ρ is the unique rank function of L.*

Proof Since L is a modular non-distributive lattice, it follows from Theorem 6.10 that there exists a sublattice $L_1 = \{x, a, b, c, y\}$ of L with $x < a, b, c < y$ which is isomorphic to the diamond lattice M_5. Suppose that $\rho(y) - \rho(x) > 2$. One can then assume that there is $e \in L$ with $x < e < c < y$. Clearly, $a \wedge e = a \wedge c = x$ and $a \vee e \le a \vee c = y$. If $a \vee e = y$, then L possesses a sublattice $\{x, a, c, e, y\}$ which is isomorphic to the pentagon lattice N_5. However, since L is modular, Theorem 6.10 says that L cannot possess a sublattice which is isomorphic to N_5. Hence, $a \vee e < y$. Let $f = a \vee e$ and L_2 the sublattice $\{x, a, b, f, y\}$ of L. Then, $b \vee f = y$. Again, since L cannot possess a sublattice which is isomorphic to N_5, one has $b \wedge f > x$. Let $g = b \wedge f$. Let L_3 be the sublattice $\{x, a, e, g, f\}$. One has $a \wedge g = a \wedge e = e \wedge g = x$ and, since L is modular with $a \le f$, it follows that:

$$a \vee g = a \vee (b \wedge f) = (a \vee b) \wedge f = f.$$

Furthermore, $a \wedge e = f$. If $e \vee g = f$, then L_3 is a sublattice of L which is isomorphic to the diamond lattice M_5 with $\rho(f) - \rho(x) < \rho(y) - \rho(x)$.

   Let $h = e \vee g < f$ and L_4 the sublattice $\{x, a, g, h, f\}$ of L. Since $a \le f$, one has:

$$a \vee g = a \vee (b \wedge f) = (a \vee b) \wedge f = f.$$

If $a \wedge h = x$, then L possesses a sublattice which is isomorphic to N_5. Thus, $a \wedge h > x$. Let $k = a \wedge h$ and L_5 the sublattice $\{x, e, g, k, h\}$ of L. One has $e \wedge g = e \wedge k = g \wedge k = x$. Again, since L is modular, it follows that:

$$e \vee k = e \vee (a \wedge h) = (e \vee a) \wedge h = h,$$

$$g \vee k = g \vee (a \wedge h) = (g \vee a) \wedge h$$

$$= (a \vee (b \wedge f)) \wedge h = ((a \vee b) \wedge f) \wedge h = f \wedge h = h.$$

Hence, L_5 is a sublattice of L which is isomorphic to the diamond lattice M_5 with $\rho(h) - \rho(x) < \rho(y) - \rho(x)$.

   Continuing these constructions yields a desired sublattice of L which is isomorphic to the diamond lattice M_5.   □

Theorem 6.23 *Let L be a finite non-distributive modular lattice. Then, for an arbitrary monomial order $<$ on $K[L]$, the initial ideal $\mathrm{in}_<(I_L)$ of the join-meet ideal I_L of L cannot be squarefree.*

Proof Let $L' = \{\xi, a, b, c, \zeta\}$ be a sublattice of L with $\xi < a, b, c < \zeta$ and with $\rho(\zeta) - \rho(\xi) = 2$ such that L' is isomorphic to the diamond lattice M_5 (Lemma 6.22). Let $a_1, \ldots, a_k$ be the elements of L, where $k \ge 3$, such that, for each $1 \le i < j \le k$, one has $a_i \wedge a_j = \xi$ and $a_i \vee a_j = \zeta$. Hence, in $K[L]/I_L$, one has $x_{a_i} x_{a_j} = x_\xi x_\zeta$ for $1 \le i < j \le k$. Let $<$ be an arbitrary monomial order on $K[L]$ with $x_{a_1} < x_{a_2} < \cdots < x_{a_k}$.

(First Step)   Suppose that $x_\xi x_\zeta < x_{a_i} x_{a_j}$ for all $1 \le i < j \le k$. Let $a = a_k$ and $f = x_\xi x_a^2 x_\zeta - x_\xi^2 x_\zeta^2$. We claim $f \in I_L$. In fact,

$$f = x_a(x_a(x_\xi x_\zeta - x_{a_1} x_{a_2}) + x_{a_2}(x_{a_1} x_a - x_\xi x_\zeta)) + x_\xi x_\zeta (x_{a_2} x_a - x_\xi x_\zeta).$$

Let $\mathrm{in}_<(I_L)$ be squarefree. Since $f \in I_L$, one has $x_\xi x_a x_\zeta \in \mathrm{in}_<(I_L)$. Thus, there is a binomial h belonging to the reduced Gröbner basis of I_L with respect to $<$ for which $\mathrm{in}_<(h)$ divides $x_\xi x_a x_\zeta$. Hence, there is a binomial $g = x_\xi x_a x_\zeta - u \in I_L$ with $g \ne 0$, where u is a monomial of degree 3 with $\mathrm{in}_<(g) = x_\xi x_a x_\zeta$. Let $g = \sum_{q=1}^N x_{b_q} g_q$, where $g_q = v_q - w_q$ with $v_q = x_{c_q} x_{c_q'}$ and $w_q = x_{c_q \wedge c_q'} x_{c_q \vee c_q'}$. Let $x_{b_1} v_1 = x_\xi x_a x_\zeta$ and $x_{b_q} w_q = x_{b_{q+1}} v_{q+1}$ for $1 \le q < N$. A crucial fact is that, for each variable x_δ appearing in $x_{b_q} g_q$, one has $\delta \in [\xi, \zeta]$. We observe that if c_q and c_q' belong to $[\xi, \zeta]$, then $c_q \wedge c_q'$ and $c_q \vee c_q'$ belong to $[\xi, \zeta]$. Since $x_{b_1} v_1 = x_\xi x_a x_\zeta$ and $x_{b_q} w_q = x_{b_{q+1}} v_{q+1}$ for $1 \le q < N$, the observation guarantees that, for each variable x_δ appearing in $x_{b_q} g_q$, one has $\delta \in [\xi, \zeta]$. In particular, $u = x_{q_N} w_N$ is a monomial consisting of those variables x_δ with $\delta \in [\xi, \zeta]$, say, $u = x_\ell x_m x_n$. Since $u = x_{b_N} x_{c_N \wedge c_N'} x_{c_N \vee c_N'}$, it follows that $\ell = m = n$ cannot occur. Furthermore, since $x_{b_q} w_q = x_{b_{q+1}} v_{q+1}$ for $1 \le q < N$, by using (6.8) one has:

$$\rho(\xi) + \rho(a) + \rho(\zeta) = \rho(\ell) + \rho(m) + \rho(n).$$

Since $\rho(\zeta) - \rho(\xi) = 2$, it follows that:

$$\rho(\ell) + \rho(m) + \rho(n) = 3\rho(\xi) + 3. \tag{6.9}$$

Let $\rho(\ell) \ge \rho(m) \ge \rho(n)$. We then claim that $\rho(n) = \rho(\xi)$. Let $\rho(n) > \rho(\xi)$. Then, by using (6.9) one has $\rho(\ell) = \rho(m) = \rho(n) = \rho(\xi) + 1$. Hence, each of ℓ, m, n belongs to $\{a_1, a_2, \ldots, a_k\}$. It then follows that $g = x_\xi x_a x_\zeta - x_{a_p} x_{a_{p'}} x_{a_{p''}}$. Since $\ell = m = n$ cannot occur, one has, say, $p \ne p'$. Since $x_\xi x_\zeta < x_{a_i} x_{a_j}$ for all $1 \le i < j \le k$ and since $x_{p''} \le x_a$, it follows that $x_\xi x_a x_\zeta < x_{a_p} x_{a_{p'}} x_{a_{p''}}$, which contradicts $\mathrm{in}_<(g) = x_\xi x_a x_\zeta$. This shows $\rho(n) = \rho(\xi)$.
Since $\rho(n) = \rho(\xi)$, it follows from (6.9) that $\rho(\ell) + \rho(m) = 2\rho(\xi) + 3$. Since $\rho(\xi) + 2 \ge \rho(\ell) \ge \rho(m) \ge \rho(\xi)$, one has $\rho(\ell) = \rho(\xi) + 2$ and $\rho(m) = \rho(\xi) + 1$. We then have $g = x_\xi x_a x_\zeta - x_\zeta x_{a_{i_0}} x_\xi$ with $1 \le i_0 < k$. Since $a_{i_0} < a_k = a$, it follows that $x_\xi x_a x_\zeta < x_\zeta x_{a_{i_0}} x_\xi$, which again contradicts $\mathrm{in}_<(g) = x_\xi x_a x_\zeta$.
(Second Step)   Suppose that there exist $1 \le i < j \le k$ with $x_\xi x_\zeta > x_{a_i} x_{a_j}$. Let $x_{a'} x_{a''}$ be the smallest monomial with respect to $<$ among those monomials $x_{a_i} x_{a_j}$ with $1 \le i < j \le k$. In particular, one has $x_\xi x_\zeta > x_{a'} x_{a''}$. We claim $x_{a'}^2 x_{a''} - x_{a'} x_{a''}^2 \in I_L$. In fact,

$$x_{a'}^2 x_{a''} - x_{a'} x_{a''}^2$$

$$= (x_{a'} - x_{a''})(x_{a'} x_{a''} - x_\xi x_\zeta) - x_{a'}(x_{a'''} x_{a''} - x_\xi x_\zeta) + x_{a''}(x_{a'} x_{a'''} - x_\xi x_\zeta),$$

where $a''' \in \{a_1, \ldots, a_k\} \setminus \{a', a''\}$ is arbitrary. Let $\mathrm{in}_<(I_L)$ be squarefree. Then, $x_{a'} x_{a''} \in \mathrm{in}_<(I_L)$. Hence, there is a binomial $g = x_{a'} x_{a''} - x_\ell x_m \in I_L$ with $\mathrm{in}_<(g) = x_{a'} x_{a''}$. Since $x_\xi x_\zeta > x_{a'} x_{a''}$, one has $x_\ell x_m \neq x_\xi x_\zeta$. It then follows that $x_\ell x_m = x_{a_i} x_{a_j}$ for some $1 \leq i < j \leq k$. However, the choice of $x_{a'} x_{a''}$ says that $x_{a_i} x_{a_j} > x_{a'} x_{a''}$, which contradicts $\mathrm{in}_<(g) = x_{a'} x_{a''}$. $\qquad \square$

Example 6.24 Let $M_5 = \{\xi, a, b, c, \zeta\}$ be the diamond lattice with $\xi < a, b, c < \zeta$. Let $<_{\text{purelex}}$ denote the pure lexicographic order induced by the ordering $x_a > x_\xi > x_b > x_c > x_\zeta$ of the variables. Then, the reduced Gröbner basis of the join-meet ideal I_{M_5} with respect to $<_{\text{purelex}}$ consists of $f_{a,b}$, $f_{b,c}$, $f_{c,a}$ together with $x_b^2 x_c - x_b x_c^2$.

Example 6.25 Let $N_5 = \{\xi, a, b, c, \zeta\}$ be the pentagon lattice with $\xi < a < b < \zeta$ and $\xi < c < \zeta$, which is a non-modular lattice. We claim that, for an arbitrary monomial order $<$, the initial ideal $\mathrm{in}_<(I_{N_5})$ is squarefree. Let $I = I_{N_5} = (f, g)$, where $f = x_a x_c - x_\xi x_\zeta$ and $g = x_b x_c - x_\xi x_\zeta$. Since the dual lattice N_5^* is again N_5, the following three cases arises:

(i) $x_a x_c < x_b x_c < x_\xi x_\zeta$,
(ii) $x_a x_c < x_\xi x_\zeta < x_b x_c$,
(iii) $x_\xi x_\zeta < x_a x_c < x_b x_c$.

In (i), the S-polynomial $S(f, g)$ is $f - g = -x_b x_c + x_a x_c$. Since $\mathrm{in}_<(S(f, g)) < \mathrm{in}_<(f)$ and $\mathrm{in}_<(S(f, g)) < \mathrm{in}_<(g)$, the S-polynomial $S(f, g)$ cannot reduce to 0 with respect to $\{f, g\}$. Let $h = -S(f, g)$. Then, $I_{N_5} = (f, h)$. Since $x_\xi x_\zeta$ and $x_b x_c$ are relatively prime, it follows that $\{f, h\}$ is the reduced Gröbner basis of I. In (ii), the initial monomials of f and g are relatively prime, it follows that $\{f, g\}$ is a Gröbner basis of I. In (iii), the S-polynomial $S(f, g)$ is $h = x_a x_\xi x_\zeta - x_b x_\xi x_\zeta$. Since $x_a x_c < x_b x_c$, one has $x_a < x_b$. Thus, $\mathrm{in}_<(h) = x_b x_\xi x_\zeta$. Since $S(g, h) = x_\xi x_\zeta f$ and since the initial monomials of f and g are relatively prime, it follows that $\{f, g, h\}$ is a Gröbner basis of I with respect to $<$.

Recall from Problem 1.8 that if there is a monomial order $<$ on $K[L]$ such that the initial ideal $\mathrm{in}_<(I_L)$ is squarefree, then I_L is a radical ideal.

Example 6.26 Since the diamond lattice $M_5 = \{\xi, a, b, c, \zeta\}$ with $\xi < a, b, c < \zeta$ is a non-distributive modular lattice, it follows from Theorem 6.23 that, for an arbitrary monomial order $<$, the initial ideal $\mathrm{in}_<(I_{M_5})$ cannot be squarefree. However, I_{M_5} is a radical ideal. In fact, the primary decomposition of I_{M_5} is

$$I_{M_5} = (x_a - x_c, x_b - x_c, -x_c^2 + x_\xi x_\zeta)$$
$$\cap (x_a, x_b, x_\zeta) \cap (x_b, x_c, x_\zeta) \cap (x_c, x_a, x_\zeta)$$
$$\cap (x_\xi, x_a, x_b) \cap (x_\xi, x_b, x_c) \cap (x_\xi, x_c, x_a)$$

Each of the ideals appearing in the right-hand side of the above primary decomposition of I_{M_5} is a prime ideal. It then follows that I_{M_5} is a radical ideal, as required.

Example 6.27 Let L be the non-distributive modular lattice of Figure 6.6. We show that the join-meet ideal I_L of L is not radical.

Fig. 6.6 A non-radical modular lattice.

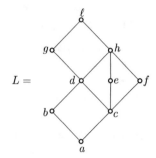

$$L =$$

In $K[L]/I_L$, one has:

$$
\begin{aligned}
x_a x_\ell x_g (x_d - x_f)^2 &= x_a x_\ell x_g x_d^2 - 2 x_a x_\ell x_g x_d x_f + x_a x_\ell x_g x_f^2 \\
&= x_a x_g^2 x_h x_d - x_a x_g^2 x_h x_f - x_a x_g x_f (x_g x_h - x_\ell x_f) \\
&= x_a x_g^2 x_h (x_d - x_f) - x_a x_g x_f x_\ell (x_d - x_f) \\
&= x_a x_g^2 x_h (x_d - x_f) - x_a x_\ell^2 x_c (x_d - x_f).
\end{aligned}
$$

Furthermore,

$$
x_a x_h (x_d - x_f) = x_b (x_d x_e - x_c x_h) + (x_f - x_d)(x_b x_e - x_a x_h) - x_b (x_e x_f - x_c x_h)
$$

and $x_\ell x_c (x_d - x_f)$ belong to I_L. Hence,

$$
(x_a x_\ell x_g (x_d - x_f))^2 = (x_a x_\ell x_g)((x_a x_\ell x_g)(x_d - x_f)^2) \in I_L.
$$

It then follows that $x_a x_\ell x_g (x_d - x_f) \in \sqrt{I_L}$. Now, with respect to the reverse lexicographic order $<_{\text{rev}}$ induced by $x_a > x_b > \cdots > x_\ell$ the reduced Gröbner basis of I_L consists of the binomials

$$
x_g x_h - x_d x_\ell, \quad x_f x_g - x_c x_\ell, \quad x_e x_g - x_c x_\ell, \quad x_e x_f - x_c x_h, \quad x_d x_f - x_c x_h,
$$

$$
x_b x_f - x_a x_h, \quad x_d x_e - x_c x_h, \quad x_b x_e - x_a x_h, \quad x_b x_c - x_a x_d,
$$

$$
x_c x_e x_\ell - x_c x_f x_\ell, \quad x_c x_d x_\ell - x_c x_f x_\ell, \quad x_c x_e x_h - x_c x_f x_h,
$$

$$
x_a x_e x_h - x_a x_f x_h, \quad x_c x_d x_h - x_c x_f x_h, \quad x_a x_d x_h - x_a x_f x_h,
$$

$$
x_c x_f^2 x_\ell - x_c^2 x_h x_\ell, \quad x_a x_d^2 x_\ell - x_a x_c x_h x_\ell, \quad x_c x_f^2 x_h - x_c^2 x_h^2, \quad x_a x_f^2 x_h - x_a x_c x_h^2.
$$

Let $I_L = \sqrt{I_L}$. Then, $x_a x_\ell x_g (x_d - x_f) \in I_L$. Hence, its initial monomial $x_a x_\ell x_g x_d$ belongs to $\text{in}_{<_{\text{rev}}}(I_L)$. However, the initial monomial of none of the above binomials divides $x_a x_\ell x_g x_d$. Thus, $x_a x_\ell x_g x_d$ cannot belong to $\text{in}_{<_{\text{rev}}}(I_L)$. Hence, $I_L \neq \sqrt{I_L}$ and I_L cannot be radical, as desired.

Problems

6.11

(a) Find all possible initial ideals of the join-meet ideal of M_5, the diamond lattice.
(b) Find the universal Gröbner basis of the join-meet ideal of N_5, the pentagon lattice.

6.12 In Example 6.26, show that the ideal $(x_a - x_c, x_b - x_c, -x_c^2 + x_\xi x_\zeta)$ is a prime ideal.

6.13 Let $N = \{\xi, a, b, c, d, \zeta\}$ be a non-modular lattice with $\xi < a < b < \zeta$ and $\xi < c < d < \zeta$. Is the join-meet ideal of N radical?

6.4   Join-Meet Ideals of Planar Distributive Lattices

A finite distributive lattice $L = \mathscr{J}(P)$ is called *planar* if P can be decomposed into a disjoint union

$$P = \{p_1, \ldots, p_n\} \cup \{q_1, \ldots, q_m\} \tag{6.10}$$

such that each of $\{p_1, \ldots, p_n\}$ and $\{q_1, \ldots, q_m\}$ is a chain of P with

$$p_1 < \cdots < p_n, \quad q_1 < \cdots < q_m,$$

where $n \geq 0$, $m \geq 0$ and $|P| = n + m$.

Example 6.28 Let P be the finite poset of Figure 6.7. Then, P can be decomposed into the disjoint union $\{p_1, p_2, p_3\} \cup \{q_1, q_2, q_3, q_4\}$ with $p_1 < p_2 < p_3$ and $q_1 < q_2 < q_3 < q_4$. It turns out that, since $q_1 < p_3$ and $p_2 < q_3$, the finite distributive lattice $\mathscr{J}(P)$ coincides with the planar distributive lattice L of Figure 6.7.

A *clutter* of a finite poset P is a subset A of P for which any two elements belonging to A are incomparable in P. Thus, in particular the empty set as well as a single-element subset of P is a clutter of P.

Lemma 6.29 *A finite distributive lattice $L = \mathscr{J}(P)$ is planar if and only if P possesses no three-element clutter.*

Fig. 6.7 A planar distributive
lattice.

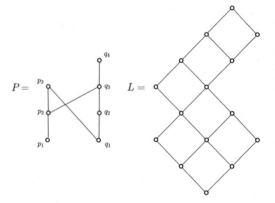

$$P = \qquad \qquad \qquad L =$$

Proof Let a finite distributive lattice $L = \mathcal{J}(P)$ be planar and $P = C \cup C'$ with
$C \cap C'$, where each of $C = \{p_1, \ldots, p_n\}$ and $C' = \{q_1, \ldots, q_m\}$ is a chain with
$p_1 < \cdots < p_n$ and $q_1 < \cdots < q_m$. Let A be a subset of P with $|A| = 3$. One has
either $|A \cap C| \geq 2$ or $|A \cap C'| \geq 2$. Thus, A cannot be a clutter.

Now, suppose that P possesses no three-element clutter. Let $a \in P$ be a maximal
element of P. By using induction on $|P|$, it follows that $P \setminus \{a\}$ can be decomposed
into the disjoint union $\{p_1, \ldots, p_n\} \cup \{q_1, \ldots, q_m\}$ with $p_1 < \cdots < p_n$ and $q_1 <
\cdots < q_m$, where $|P| = n + m + 1$. Let $n \geq 1$ and $m \geq 1$. Since P possesses no
three-element clutter, it follows that $\{p_n, q_m, a\}$ cannot be a clutter of P.

If p_n and q_m are incomparable, then one has either $p_n < a$ or $q_m < a$. Let, say,
$p_n < a$. Then, $\{p_1, \ldots, p_n, a\} \cup \{q_1, \ldots, q_m\}$ is a desired decomposition of P.

If, say, $p_n < q_m$, then $b < q_m$ for all $b \in P \setminus \{a, q_m\}$. Again, by using
induction on $|P|$, it follows that $P \setminus \{q_m\}$ can be decomposed into the disjoint union
$\{p'_1, \ldots, p'_{n'}\} \cup \{q'_1, \ldots, q'_{m'}\}$ with $p'_1 < \cdots < p'_{n'}$ and $q'_1 < \cdots < q'_{m'}$. Since a is a
maximal element of P, one has either $p'_{n'} = a$ or $q'_{m'} = a$. Let, say, $p'_{n'} = a$. Then,
$q'_{m'} < q_m$. Hence, $\{p'_1, \ldots, p'_{n'}\} \cup \{q'_1, \ldots, q'_{m'}, q_m\}$ is a desired decomposition of
P.

As a result, $L = \mathcal{J}(P)$ is a planar distributive lattice. $\square$

Corollary 6.30 *A finite distributive lattice L is planar if and only if L possesses no
sublattice which is isomorphic to $\mathcal{B}_3$, the Boolean lattice of rank 3.*

Proof If $L = \mathcal{J}(P)$ is not planar, then P possesses a three-element clutter $\{a, b, c\}$.
Let I denote the smallest poset ideal containing a, b, and c. Let $I' = I \setminus \{a, b, c\}$.
Then, in $\mathcal{J}(P)$, the interval $[I', I]$ is isomorphic to $\mathcal{B}_3$.

Suppose that $L = \mathcal{J}(P)$ possesses a sublattice L' which is isomorphic to $\mathcal{B}_3$.
Thus, L consists of 8 elements

$$\alpha \wedge \beta \wedge \gamma, \quad \alpha, \quad \beta, \quad \gamma, \quad \alpha \vee \beta, \quad \beta \vee \gamma, \quad \gamma \vee \alpha, \quad \alpha \vee \beta \vee \gamma,$$

where each of α, β, and γ is a poset ideal of P. Since, say, $\beta \vee \gamma < \alpha \vee \beta \vee \gamma$, one has $\alpha \not\subset \beta \cup \gamma$. Let $a \in \alpha \setminus (\beta \cup \gamma)$, $b \in \beta \setminus (\gamma \cup \alpha)$, and $c \in \gamma \setminus (\alpha \cup \beta)$. If, say, $a < b$, then $a \in \beta$, a contradiction. Thus, $\{a, b, c\}$ must be a clutter of P. $\square$

Let $L = \mathscr{J}(P)$ be a finite planar distributive lattice and suppose that P possesses a decomposition (6.10). Let $K_{n,m}$ denote the complete bipartite graph on the vertex set $(\{0\} \cup [n]) \cup (\{0\} \cup [m])$. Given a poset ideal β of P, we write $a(\beta)$ for the biggest integer i with $p_i \in \beta$ and $b(\beta)$ for the biggest integer j with $q_j \in \beta$. Let $e(\beta)$ denote the edge $\{a(\beta), b(\beta)\}$ of $K_{n,m}$ and write $G(P)$ for the bipartite subgraph of $K_{n,m}$ consisting of those edges $e(\beta)$ for which β is a poset ideal of P.

Example 6.31 The bipartite graph arising from the finite planar distributive lattice $L = \mathscr{J}(P)$ of Figure 6.7 is

Fig. 6.8 The bipartite graph
arising from a planar
distributive lattice.

$$G(P) =$$

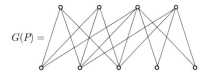

Lemma 6.32 *Every cycle of $G(P)$ of length ≥ 6 has a chord.*

Proof Let $C = (e_1, e_2, \ldots, e_{2\ell})$ be a cycle of $G(P)$ of length 2ℓ with $\ell \geq 3$, where each e_i is an edge of $G(P)$. It then follows that there exist $1 \leq k < k' \leq \ell$ with $e_k = \{i, j\}$ and $e_{k'} = \{i', j'\}$, where $i, i' \in [n]$ and $j, j' \in [m]$, such that $i < i'$ and $j > j'$. Since e_k and $e_{k'}$ are edges of $G(P)$, each of the subsets

$$\beta = \{p_1, \ldots, p_i\} \cup \{q_1, \ldots, q_j\}, \quad \beta' = \{p_1, \ldots, p_{i'}\} \cup \{q_1, \ldots, q_{j'}\}$$

of P is a poset ideal of P. Thus, in particular each of $\beta \cap \beta'$ and $\beta \cup \beta'$ is again a poset ideal of P. Hence, $e'' = \{i', j\}$ and $e''' = \{i, j'\}$ are edges of $G(P)$. Since C is of length ≥ 6, it follows that either e'' or e''' cannot be an edge belonging to C. Hence, either e'' or e''' can be a chord of C. $\square$

Let $T = K[t_1, \ldots, t_n, s_1, \ldots, s_m]$ be the polynomial ring in $(n + m)$ variables over a field K and $K[G(P)] \subset T$ the toric ring of $G(P)$. Recall that $K[G(P)]$ is generated by those monomials $t_i s_j$ with $i \in [n]$ and $j \in [m]$ for which $\{i, j\}$ is an edge of $G(P)$. We define the ring homomorphism $\pi : K[L] \to K[G(P)]$ by setting $\psi(x_\beta) = t_{a(\beta)} s_{b(\beta)}$ for $\beta \in \mathscr{J}(P)$.

Lemma 6.33 *The ring homomorphism π is surjective.*

Proof Let $t_i s_j \in K[G(P)]$. Then, $\{i, j\}$ is an edge of $G(P)$. Hence, there is a poset ideal β of P with $a(\beta) = i$ and $b(\beta) = j$. Thus, one has $\pi(x_\beta) = t_i s_j$. $\square$

Lemma 6.34 *The kernel of π coincides with the join-meet ideal I_L.*

Proof Lemma 6.32 says that every cycle of $G(P)$ of length ≥ 6 has a chord. It then follows from Theorem 5.27 that the kernel of π is generated by those quadratic binomials arising from cycles of $G(P)$ of length 4.

Let $C = (e_1, e_2, e_3, e_4)$ be a cycle of G of length 4 with $e_1 = \{i, j\}, e_2 = \{i', j\}, e_3 = \{i', j'\}, e_4 = \{i, j'\}$, where $i, i' \in [n]$ and $j, j' \in [m]$ with $i < i'$ and $j < j'$. Let $\alpha_1, \alpha_2, \alpha_3, \alpha_4$ be poset ideals of P with $e(\alpha_k) = e_k$ for $1 \leq k \leq 4$. Then, $\alpha_2 \cap \alpha_4 = \alpha_1$ and $\alpha_2 \cup \alpha_4 = \alpha_3$. Hence, the binomial arising from C is equal to $x_{\alpha_2} x_{\alpha_4} - x_{\alpha_1} x_{\alpha_4}$, which belongs to $\mathcal{G}_L$.

Let $\alpha, \beta \in L = \mathcal{J}(P)$ with $e(\alpha) = \{i, j\}$ and $e(\beta) = \{i', j'\}$, where $i, i' \in [n]$ and $j, j' \in [m]$, and where α and β are incomparable in L. One has, say, $i < i'$ and $j > j'$. Then, the binomial $f_{\alpha,\beta} = x_\alpha x_\beta - x_{\alpha \wedge \beta} x_{\alpha \vee \beta} \in \mathcal{G}_L$ coincides with the binomial arising from the cycle $C = (e_1, e_2, e_3, e_4)$, where $e_1 = \{i, j\}, e_2 = \{i', j\}, e_3 = \{i', j'\}, e_4 = \{i, j'\}$, of G of length 4.

Hence, the kernel of π coincides with the join-meet ideal I_L, as desired.   □

Theorem 6.35 *Given a finite modular lattice L, the following conditions are equivalent:*

(i) *L is a planar distributive lattice;*
(ii) *$\mathrm{in}_<(I_L)$ is squarefree with respect to an arbitrary pure lexicographic order;*
(iii) *$\mathrm{in}_<(I_L)$ is squarefree with respect to an arbitrary monomial order.*

Proof ((i) $\Rightarrow$ (iii)) Let L be a planar distributive lattice. It follows from Lemma 6.34 that I_L can be identified with the toric ideal of a finite bipartite graph. By Theorem 5.24, $\mathrm{in}_<(I_L)$ is squarefree with respect to an arbitrary monomial order.

((ii) $\Rightarrow$ (i)) Theorem 6.23 guarantees that every finite non-distributive modular lattice fails to satisfy the condition (ii). Thus, L must be a distributive lattice.

Let L be a finite non-planar distributive lattice. Corollary 6.30 says that L possesses a sublattice L' which is isomorphic to $\mathcal{B}_3$. Let $a, b, c \in L$ and L' consist 8 elements

$$\xi = a \wedge b \wedge c, \quad a, \quad b, \quad c, \quad e = a \vee b, \quad f = b \vee c, \quad g = c \vee a, \quad \zeta = a \vee b \vee c.$$

Let $<_{\mathrm{purelex}}$ denote the pure lexicographic order induced by the ordering of the variables as follows:

- $x_a < x_b < x_c < x_\xi < x_\zeta < x_e < x_g < x_f$;
- $x_f < x_h$ for all $h \in L \setminus \{\xi, a, b, c, e, f, g, \zeta\}$.

It follows that the minimal system of monomial generators of the initial ideal $\mathrm{in}_{<_{\mathrm{purelex}}}(I_{\mathcal{B}_3})$ contains $\xi \zeta^2$. Since $<_{\mathrm{purelex}}$ is an elimination order, it follows from Corollary 1.35 that the minimal system of monomial generators of $\mathrm{in}_{<_{\mathrm{purelex}}}(I_L)$ must contain $\xi \zeta^2$. Hence, $\mathrm{in}_{<_{\mathrm{purelex}}}(I_L)$ cannot be squarefree.

Finally, (iii) $\Rightarrow$ (ii) is trivial.   □

Recall that the divisor lattice of an integer $n \geq 1$ is the finite lattice $\mathcal{D}_n$ consisting of all divisors of n ordered by divisibility. Every Boolean lattice is a divisor lattice. Every divisor lattice is a distributive lattice.

Let, in general, L be a finite pure lattice with its rank function ρ. A *cut edge* of L is a pair (a, b) of elements of L with $\rho(b) = \rho(a) + 1$ such that

$$|\{c \in L : \rho(c) = \rho(a)\}| = |\{c \in L : \rho(c) = \rho(b)\}| = 1.$$

Lemma 6.36 *Let L be a planar distributive lattice with no cut edge. Then, L is the divisor lattice $\mathcal{D}_{2 \cdot 3^r}$ with $r \geq 1$ if and only if no sublattice of L is isomorphic to the lattice $\mathcal{J}(C_4)$ of Figure 6.9.*

Fig. 6.9 The cycle of length 4 and its distributive lattice.

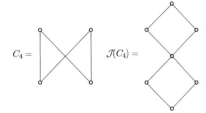

$$C_4 = \qquad \mathcal{J}(C_4) =$$

Proof "Only If" follows easily. Now, "If" is proved. Let P be decomposed into a disjoint union (6.10) with $n \geq m \geq 1$. What we must prove is that $m = 1$ and that q_1 and p_i are incomparable in P for $1 \leq i \leq n$. Let $n \geq m \geq 2$. Since L has no cut edge, there is no element of P which is comparable with any other element of P. In particular, p_1 and q_1 are incomparable in P. In order to prove the existence of a sublattice of L which is isomorphic to $\mathcal{J}(C_4)$, we must show that there exist $1 \leq i < m$ and $1 \leq j < n$ such that p_i and q_j are incomparable in P and that p_{i+1} and q_{j+1} are incomparable in P.

If p_2 and q_2 are incomparable in P, then we are done. Suppose that, say, $p_2 > q_2$ and write $j_0 \geq 2$ for the biggest integer with $p_2 > q_{j_0}$. If $j_0 = m$, then p_2 is comparable with any other element of P. Thus, $j_0 < m$. Then, p_2 and q_{j_0+1} are incomparable in P. In fact, if $p_2 < q_{j_0+1}$, then again p_2 is comparable with any other element of P. Furthermore, p_1 and q_{j_0} are incomparable in P. In fact, if $p_1 < q_{j_0}$, then q_{i_0} is comparable with any other element of P. Hence, p_2 and q_{j_0+1} are incomparable in P, and p_1 and q_{j_0} are incomparable in P, as required. $\square$

Theorem 6.37 *Let L be a finite lattice with no cut edge. Then, the following conditions are equivalent:*

(i) *L is the divisor lattice of $2 \cdot 3^r$ with $r \geq 1$;*
(ii) *I_L possesses a quadratic Gröbner basis with respect to an arbitrary monomial order.*

Proof ((i) $\Rightarrow$ (ii)) Let $L = \mathscr{D}_{2 \cdot 3^r}$ be the divisor lattice of $2 \cdot 3^r$ with $r \geq 1$. Let P be the finite poset with $L = \mathscr{J}(P)$. Then, $P = \{a, b_1, \ldots, b_r\}$, where $b_1 < \cdots < b_r$ and where a is incomparable with each of b_i. Hence, as was seen in the proof of Theorem 6.35, the join-meet ideal I_L can be identified with the toric ideal of the complete bipartite graph $K_{2,r}$ on $[2] \cup [r]$. Since every cycle of $K_{2,r}$ is of length 4, it follows from Corollary 5.12 that each primitive binomial of the toric ideal of $K_{2,r}$ is quadratic. Hence, I_L possesses a quadratic Gröbner basis with respect to an arbitrary monomial order, as desired.

((ii) $\Rightarrow$ (i)) Example 6.25 says that L cannot possess the pentagon lattice N_5 as a sublattice. Furthermore, Example 6.24 says that L cannot possess the diamond lattice M_5 as a sublattice. In addition, it follows from the proof of Theorem 6.35 that L cannot possess the Boolean lattice $\mathscr{B}_3$ of rank 3 as a sublattice. Hence, L must be a planar distributive lattice. Let $L = \mathscr{J}(P)$. Suppose that P is not the divisor lattice of $2 \cdot 3^r$ with $r \geq 1$. Lemma 6.36 says that L contains a sublattice L' which is isomorphic to $\mathscr{J}(C_4)$. Let L' consist of

$$\xi = a \wedge b, \quad a, \quad b, \quad c = a \vee b = e \wedge f, \quad e, \quad f, \quad \zeta = e \vee f.$$

Let $<_{\text{purelex}}$ denote the pure lexicographic order induced by the ordering

$$x_c > x_\zeta > x_e > x_f > x_a > x_b > x_\xi$$

of the variables. Then, the monomial $x_a x_b x_\zeta$ is contained in the minimal system of monomial generators of the initial ideal $\text{in}_{<_{\text{purelex}}}$. Hence, I_L fails to satisfy the condition (ii) and L must be the divisor lattice of $2 \cdot 3^r$ with $r \geq 1$, as desired. $\square$

Problems

6.14 Let $\mathscr{B}_3$ be the Boolean lattice of rank 3.

 (i) Find an initial ideal of the join-meet ideal of $\mathscr{B}_3$ which is not squarefree.
(ii) Find a Gröbner basis of the join-meet ideal of $\mathscr{B}_3$ which is not quadratic.

6.15 Find a Gröbner basis of the join-meet ideal of $\mathscr{D}_{36}$, the divisor lattice of $36 = 2^2 \cdot 3^2$, which is not quadratic.

6.5 Projective Dimension and Regularity of Join-Meet Ideals

Let L be a finite distributive lattice. In this section, we determine the regularity and projective dimension of the join-meet ideal I_L of L.

We fix a field K. The residue class ring $K[L]/I_L$ can be identified with the toric ring $\mathcal{R}_K[L]$, as explained in Section 6.2. Nowadays, the toric ring $\mathcal{R}_K[L]$ is called the *Hibi ring* of L (with respect to K).

By Birkhoff's theorem, $L = \mathcal{J}(P)$, where P is the set of join-irreducible elements of L and where $\mathcal{J}(P)$ is the set of poset ideals of P. Let S be the polynomial ring over K in the variables t and x_p with $p \in P$. For each $\alpha \in \mathcal{J}(P)$, we set

$$u_\alpha = \prod_{p \in \alpha} x_p t. \tag{6.11}$$

Then, by Theorem 6.19,

$$\mathcal{R}_K[L] \cong K[u_\alpha : \alpha \in L].$$

Let $\hat{P} = P \cup \{\hat{0}, \hat{1}\}$, where $\hat{0} < p < \hat{1}$ for all $p \in P$. An *order-reversion* map on $\hat{P}$ is a map $\sigma : \hat{P} \to \mathbb{Z}_{\geq 0}$ for which $\sigma(\hat{1}) = 0$ and $\sigma(p) \leq \sigma(q)$ if $p > q$ in $\hat{P}$. The set of order-reversing maps on $\hat{P}$ is denoted $\Omega(\hat{P})$. Given $\sigma \in \Omega(\hat{P})$, we set

$$w_\sigma = \prod_{p \in P} x_p^{\sigma(p)} t^{\sigma(\hat{0})}.$$

By Problem 6.9, the set of monomials $\{w_\sigma : \sigma \in \Omega(\hat{P})\}$ is the monomial K-basis of $\mathcal{R}_K[L]$. We set $\deg w_\sigma = \sigma(\hat{0})$. With this definition given, $\mathcal{R}_K(L)$ is a standard graded K-algebra. By Problem 6.10, $\mathcal{R}_K[L]$ is a normal domain. According to a theorem of Hochster [115], a normal toric ring is Cohen–Macaulay.

We will use the information regarding the monomial K-basis of $\mathcal{R}_K[L]$ to compute its Krull dimension.

Theorem 6.38 *Let L be a finite distributive lattice, and let P be the poset of join-irreducible elements of L. Then,*

$$\dim \mathcal{R}_K[L] = |P| + 1.$$

Proof Since $\mathcal{R}_K[L]$ is an affine domain, it follows that $\dim \mathcal{R}_K[L]$ is equal to the transcendence degree over K of the quotient field $Q(\mathcal{R}_K[L])$ of $\mathcal{R}_K[L]$.

We claim that $Q(\mathcal{R}_K[L]) = Q(S)$, where as above, $S = K[t, \{x_p : p \in P\}]$. Since $\operatorname{tr}\deg(Q(S)/K) = |P|+1$, the theorem will follow. Obviously, $Q(\mathcal{R}_K[L]) \subset Q(S)$. Thus, in order to show that the two quotient fields are the same, it suffices to show that the variable t and as well as the variables x_p belong to $Q(\mathcal{R}_K[L])$. This is clear for t, because $t = u_\emptyset$. Now, let $p \in P$, and let $\alpha = \{q \in P : q \leq p\}$ and $\beta = \{q \in P : q < p\}$. Then, both, α and β are poset ideals of P, and $u_\alpha/u_\beta = x_p$. Thus, $x_p \in Q(\mathcal{R}_K[L])$. □

Corollary 6.39 *Let L and P be as in Theorem 6.38. Then,*

$$\text{proj dim } I_L = |L| - |P| - 2.$$

Proof By the Auslander–Buchsbaum formula (Theorem 2.15), we have $\text{proj dim } \mathcal{R}_K[L] + \text{depth } \mathcal{R}_K[L] = |L|$. Since $\mathcal{R}_K[L]$ is Cohen–Macaulay, we have $\text{depth } \mathcal{R}_K[L] = \dim \mathcal{R}_K[L]$. Thus, together with Theorem 6.38, it follows that

$$\text{proj dim } I_L = \text{proj dim } \mathcal{R}_K[L] - 1 = |L| - \dim \mathcal{R}_K[L] - 1 = |L| - |P| - 2.$$

$\square$

Next, we will study the regularity of I_L. For this purpose, we have to recall a few facts about the canonical module of a Cohen–Macaulay ring. The results that we are quoting can all be found in [27].

Let R be a Cohen–Macaulay standard graded K-algebra of dimension d with graded maximal ideal m, and let $H_R(t) = Q(t)/(1-t)^d$ be the Hilbert series of R. By Corollary 2.18, one has $\text{reg } R = \deg Q(t)$. The a-invariant $a(R)$ of R is defined to be the degree of the Hilbert series of R, which by definition is equal to $\deg Q(t) - d$.

Thus, we see that in combination with Theorem 6.38 we obtain

$$\text{reg } I_L = \text{reg } \mathcal{R}_K[L] + 1 = a(\mathcal{R}_K[L]) + |P| + 2. \tag{6.12}$$

The a-*invariant* of R can be expressed in terms of the *canonical module* ω_R, which, up to isomorphisms, is uniquely determined by the property that

$$\text{Ext}_R^i(R/\mathfrak{m}, \omega_R) = \begin{cases} R/\mathfrak{m} & \text{if } i = d, \\ 0 & \text{if } i \neq d. \end{cases}$$

The canonical module is a graded R-module and following Goto and Watanabe [84], who introduced the a-invariant, we have:

$$a(R) = -\min\{i : (\omega_R)_i \neq 0\}. \tag{6.13}$$

By (6.12), it remains to compute the a-invariant of $\mathcal{R}_K[L]$ in order to determine the regularity of I_L. For this purpose, we use formula (6.13). The canonical module of a normal toric ring has the following interpretation: Let A be a configuration matrix. The set $C = \mathbb{Z}_{\geq 0}A$ is an *affine semigroup* . We let $\mathbb{R}_{\geq 0}C$ be the cone spanned by C, and define the *relative interior* of C as:

$$\text{relint}(C) = C \cap \text{relint}(\mathbb{R}_{\geq 0}C).$$

Here, $\text{relint}(\mathbb{R}_{\geq 0}C)$ is the interior of $\mathbb{R}_{\geq 0}C$ with respect to its affine hull.

Theorem 6.40 (Danilov, Stanley) *Let $A \in \mathbb{Z}^{m \times n}$ be a configuration matrix, and assume that the affine semigroup $C = \mathbb{Z}_{\geq} A$ generated by A is normal. Then, $K[C] \subset K[t_1^{\pm}, \ldots, t_m^{\pm}]$, and $\omega_{K[C]}$ has the monomial K-basis consisting of all $\mathbf{t}^{\mathbf{c}}$ with $\mathbf{c} \in \mathrm{relint}(C)$.*

By using the theorem of Danilov and Stanley, we get

Theorem 6.41 *Let L and P be as in Theorem 6.38, and let $\Omega^s(\hat{P})$ be the set of strictly order reversing maps $\sigma : \hat{P} \to \mathbb{Z}_{\geq 0}$, that is, maps with $\sigma(\hat{1}) = 0$ and $\sigma(p) < \sigma(q)$ if $p > q$ in $\hat{P}$. Then, the monomial K-basis of $\omega_{\mathscr{R}_K[L]}$ consists of the set of monomials $w_\sigma = \prod_{p \in P} x_p^{\sigma(p)} t^{\sigma(\hat{0})}$ with $\sigma \in \Omega^s(\hat{P})$.*

Proof Let $A \subset \mathbb{Z}^{m \times n}$ be the configuration matrix corresponding to the generators (6.11) of $\mathscr{R}_K[L]$. Here, $m = |P| + 1$ and $n = |L|$. Let C be the affine semigroup of generated by A. According to Theorem 6.40, we have to show that the exponent vector of $w_\sigma = \prod_{p \in P} x_p^{\sigma(p)} t^{\sigma(\hat{0})}$ with $\sigma \in \Omega(\hat{P})$ belongs to relint C if and only if $\sigma \in \Omega^s(\hat{P})$.

Let $\hat{P} = \{p_1, \ldots, p_{m-1}\} \cup \{\hat{0}, \hat{1}\}$. Let U denote the set of those $i \in [m-1]$ for which $\hat{1}$ covers p_i and V the set of those $j \in [m-1]$ for which p_j covers $\hat{0}$. Let W be the set of pairs $(k, \ell) \in [m-1] \times [m-1]$ for which p_k covers p_ℓ. For each $i \in U$, write $H_i^* \subset \mathbb{R}^m$ for the closed half-space of $\mathbb{R}^m$ defined by $x_i \geq 0$. For each $j \in V$, write $H_j^{**} \subset \mathbb{R}^m$ for the closed half-space of $\mathbb{R}^m$ defined by $x_m \geq x_j$. Furthermore, for each $(k, \ell) \in W$, write $H_{(k,\ell)} \subset \mathbb{R}^m$ for the closed half-space of $\mathbb{R}^m$ defined by $x_\ell \geq x_k$. We then claim

$$\mathbb{R}_{\geq 0} C = \left(\bigcap_{i \in U} H_i^* \right) \bigcap \left(\bigcap_{j \in V} H_j^{**} \right) \bigcap \left(\bigcap_{(k,\ell) \in W} H_{(k,\ell)} \right). \tag{6.14}$$

Clearly, the left-hand side of (6.14) is contained in the right-hand side of (6.14). Let $\mathbf{a} = (a_1, \ldots, a_m) \in \mathbb{R}^m$ belong to the right-hand side of (6.14). Our work is to show that $\mathbf{a} \in \mathbb{R}_{\geq 0} C$. Let $q \geq 0$ denote the number of nonzero components of $\mathbf{a}$. If $q = 0$, then $\mathbf{a}$ is the origin of $\mathbb{R}^m$ and the origin belongs to $\mathbb{R}_{\geq 0} C$. Let $q > 0$ and $\alpha = \{p_i \in P : a_i > 0\}$. It then follows that α is a poset ideal of P. Let $\rho(\alpha) = \sum_{i \in \alpha} \mathbf{e}_i + \mathbf{e}_m$, where $\mathbf{e}_1, \ldots, \mathbf{e}_m$ are the canonical unit coordinate vectors of $\mathbb{R}^m$. Let $r = \min\{a_i : a_i > 0\}$ and $\mathbf{b} = r\rho(\alpha)$. Since the number of nonzero components of $\mathbf{a} - r\mathbf{b}$ is less than q and since $\mathbf{a} - r\mathbf{b}$ belongs to the right-hand side of (6.14), it follows that $\mathbf{a} - r\mathbf{b} \in \mathbb{R}_{\geq 0} C$. Since $r\mathbf{b} \in \mathbb{R}_{\geq 0} C$, one has $\mathbf{a} = (\mathbf{a} - r\mathbf{b}) + r\mathbf{b} \in \mathbb{R}_{\geq 0} C$. This completes the proof of (6.14).

We now claim that the right-hand side of (6.14) is irredundant. Let $\mathscr{H}$ denote the set of closed half-spaces in the right-hand side of (6.14). If $i \in U$, then $-\mathbf{e}_i$ belongs to $(\bigcap_{H_i^* \neq H \in \mathscr{H}} H) \setminus \mathbb{R}_{\geq 0} C$. If $j \in V$, then $\mathbf{e}_j$ belongs to $(\bigcap_{H_j^{**} \neq H \in \mathscr{H}} H) \setminus \mathbb{R}_{\geq 0} C$. If $(k, \ell) \in W$, then $(\sum_{p_i \leq p_k} \mathbf{e}_i) - \mathbf{e}_\ell$ belongs to $(\bigcap_{H_{(k,\ell)} \neq H \in \mathscr{H}} H) \setminus \mathbb{R}_{\geq 0} C$.

Since the right-hand side of (6.14) is irredundant and since the affine hull of $\mathbb{R}_{\geq 0} C$ is $\mathbb{R}^m$, it follows that the facets of $\mathbb{R}_{\geq 0} C$ are those $[H] \cap \mathbb{R}_{\geq 0} C$ with $H \in \mathscr{H}$, where $[H]$ is the hyperplane of $\mathbb{R}^m$ which is the boundary of H. It then follows that

$$\text{relint}(\mathbb{R}_{\geq 0} C) = \mathbb{R}_{\geq 0} C \setminus \bigcup_{H \in \mathcal{H}} [H].$$

Hence, $\text{relint}(C)$ consists of those $(a_1, \ldots, a_m) \in \mathbb{Z}_{\geq 0}$ such that: (i) $a_i > 0$ for $i \in U$, (ii) $a_m > a_j$ for $j \in V$, and (iii) $a_\ell > a_k$ for $(k, \ell) \in W$. As a result, for $\sigma \in \Omega(\hat{P})$, the vector $(\sigma(p_1), \ldots, \sigma(p_{m-1}), \sigma(\hat{0}))$ belongs to $\text{relint } C$ if and only if $\sigma \in \Omega^s(\hat{P})$, as desired. $\qquad\square$

We use the convention to set $\text{reg}(I) = 1$ if I is the zero ideal. Now, we have all the tools available to prove:

Theorem 6.42 *Let L be a finite distributive lattice and P its poset of join-irreducible elements. Then,*

$$\text{reg } I_L = |P| - \text{rank } P.$$

Proof By formula (6.12), it remains to be shown that $a(\mathcal{R}_K[L]) = -\text{rank } P - 2$. Since $\text{rank } \hat{P} = \text{rank } P + 2$, this equation for the a-invariant will follow from (6.13), once we have shown that $\min\{i : (\omega_R)_i \neq 0\} = \text{rank } \hat{P}$.

Let $\sigma \in \Omega^s(\hat{P})$ and let $\hat{0} < p_1 < \cdots < p_r < \hat{1}$ be a maximal chain in $\hat{P}$ with $r = \text{rank } P + 1$. Then,

$$0 < \sigma(p_r) < \sigma(p_{r-1}) < \cdots < \sigma(p_1) < \sigma(\hat{0})).$$

It follows that $\sigma(\hat{0}) \geq \text{rank } \hat{P}$, and hence Theorem 6.41 implies that $\min\{i : (\omega_L)_i \neq 0\} \geq \text{rank } \hat{P}$.

In order to prove equality, we consider the depth function $\delta \colon \hat{P} \to \mathbb{Z}_{\geq 0}$ which for $p \in \hat{P}$ is defined to be the supremum of the lengths of chains ascending from p. Obviously, $\delta \in \Omega^s(\hat{P})$ and $\delta(\hat{0}) = \text{rank } \hat{P}$. This concludes the proof of the theorem. $\qquad\square$

Problems

6.16 Determine all finite posets P for which I_L with $L = \mathscr{I}(P)$ has a linear resolution.

6.17 Compute the regularity of I_L when L is the Boolean lattice $\mathscr{B}_n$

6.18 Give the complete list of posets P for which I_L with $L = \mathscr{I}(P)$ has regularity three.

6.19 Given integers $r \leq d - 2$, show that there exists a finite distributive lattice L such that $\dim \mathscr{R}_K[L] = d$ and $\text{reg } I_L = r$.

Notes

Birkhoff [17] is the basic source of classical lattice theory. A quick discussion on the lattice theory can be found in, e.g., Stanley [200, Chapter 3]. The highlights of Section 6.1 are Theorem 6.4, Birkhoff's fundamental structure theorem for finite distributive lattices, and Theorem 6.10 due to Dedekind, which characterizes distributive lattices and modular lattices. Except for Theorem 6.10, the topics discussed in Section 1 appear in [200, Chapter 3].

The join-meet ideal of a finite distributive lattice together with the toric ring $\mathscr{R}_K[L]$ is introduced by [104]. Theorem 6.21, Problem 6.9, Problem 6.10 as well as Theorem 6.41 are discussed in [104]. Furthermore, it is proved in [104] that $\mathscr{R}_K[L]$ with $L = \mathscr{J}(P)$ is Gorenstein if and only if P is pure. In the monograph [146], the toric ring $\mathscr{R}_K[L]$ is called *Hibi ring*. The study of join-meet ideals of arbitrary lattices originated in [95] and [69]. Section 6.3 is due to [69] and Section 6.4 is due to [95].

The Hibi ring of $L = \mathscr{J}(P)$ coincides with the Ehrhart ring [105, p. 97] of the *order polytope* [197] of P. Thus, its Hilbert series can be computed explicitly by using the theory of P-partitions, developed in Stanley's dissertation [198]. See also [72] and [176]. In particular, the formula $a(\mathscr{R}_K[L]) = -\text{rank}\,P - 2$ in the proof of Theorem 6.42 follows.

In [197], together with the order polytope $\mathcal{O}(P)$, the *chain polytope* $\mathscr{C}(P)$ of a finite poset P is also studied. A basic question when $\mathcal{O}(P)$ and $\mathscr{C}(P)$ are unimodularly equivalent is solved in [108]. The Hibi ring $\mathscr{R}_K[L]$ is an *algebra with straightening laws* [105, Chapter XIII] on $L = \mathscr{J}(P)$. Furthermore, it turns out [107] that the Ehrhart ring of $\mathscr{C}(P)$ is again an algebra with straightening laws on $L = \mathscr{J}(P)$.

In the frame of combinatorics and commutative algebra, the Hibi ring has been studied in many articles. For example, the articles [5, 35, 59, 64, 66, 98], and [99] have contributed to the development of the theory of Hibi rings. In [64], the question when the join-meet ideal of a finite distributive lattice is an extremal Gorenstein ideal is solved. Furthermore, in [61] a characterization for the join-meet ideal of a finite planar distributive lattice to be linearly related is given. It would, of course, be of interest to find a characterization for the join-meet ideal of a finite distributive lattice to be linearly related. In [66], the pseudo-Gorenstein Hibi ring is completely classified. In [99], the nearly Gorenstein Hibi ring is completely classified. In [98], the strongly Koszul Hibi ring is completely classified. The study on Gröbner bases of join-meet ideals of finite distributive lattices with respect to lexicographic orders is partially done in [5].

The Hibi ring is naturally related with determinantal rings and ideals [23, 65]. Let X be an $m \times n$-matrix of indeterminates with $2 \le m < n$ and Δ a pure simplicial complex on $[n]$ of dimension $m - 1$. Given a facet $F = \{a_1, \ldots, a_m\}$ with $1 \le a_1 < \cdots < a_d \le n$, we write $\mu_F = [a_1, \ldots, a_m]$ for the maximal minor of X with columns $a_1, \ldots, a_m$. We then introduce the ideal $J_\Delta \subset K[X]$, where $K[X]$ is the polynomial ring in mn variables, which is generated by those μ_F with $F \in \mathscr{F}(\Delta)$, where $\mathscr{F}(\Delta)$ is the set of facets of Δ. The ideal J_Δ is called the *determinant facet ideal* of Δ. In [65], the problem when J_Δ is a prime ideal as well as that when the generators of J_Δ form a Gröbner basis is studied.

To introduce a generalization of Hibi rings and join-meet ideals of finite distributive lattices is, of course, of interest. Such work has been done by, for instance, [68, 71] and [15]. Let P and Q be finite posets and $K[\{x_p^q : p \in P, q \in Q\}]$ the polynomial ring in $|P||Q|$ variables over a field K. Let $\mathrm{Hom}(P, Q)$ denote the set of order-preserving maps $\varphi : P \to Q$. Then, $\mathrm{Hom}(P, Q)$ is a finite poset by setting $\varphi \leq \psi$ if $\varphi(p) \leq \psi(p)$ for all $p \in P$. Given $\varphi \in \mathrm{Hom}(P, Q)$, we associate the monomial $u_\varphi = \prod_{p \in P} x_p^{\phi(p)}$. The toric ring $K[P, Q]$ which is generated by those monomials u_φ with $\varphi \in \mathrm{Hom}(P, Q)$ is called the *isotonian algebra* of (P, Q). When Q is a chain $C_1 : q_1 < q_2$ of length 1, then $K[P, C_1]$ is isomorphic to the Hibi ring $\mathscr{R}_K[L]$ with $L = \mathscr{J}(P)$. In [15], it is conjectured that $K[P, Q]$ is always normal and a partial answer of the conjecture is obtained. In addition, the problem when $K[P, Q]$ possesses a quadratic Gröbner basis is discussed.

On the other hand, the monomial ideal generated by the monomials u_α, where α is a poset ideal of P, is also deeply studied by, for instance, [92] and [93]. We refer the reader to the monograph [94] for the detailed information.

Furthermore, the Hibi ring appears in representation theory [121, 122, 130–134, 215] and in algebraic geometry [20–22, 50, 83, 136, 137, 179, 180, 192, 193, 211], See also [14] for a topic in statistics.

Chapter 7
Binomial Edge Ideals and Related Ideals

Abstract In this chapter we consider classes of binomial ideals which are naturally attached to finite simple graphs. The first of these classes are the binomial edge ideals. These ideals may also be viewed as ideals generated by a subset of 2-minors of a $(2 \times n)$-matrix of indeterminates. Their Gröbner bases will be computed. Graphs whose binomial edge ideals have a quadratic Gröbner basis are called closed graphs. A full classification of closed graphs is given. For an arbitrary graph the initial ideal of the binomial edge ideal (for a suitable monomial order) is a squarefree monomial ideal. This has the pleasant consequence that the binomial edge ideal itself is a radical ideal. Its minimal prime ideals are determined in terms of cut point properties of the underlying graph. Based on this information, the closed graphs whose binomial edge ideal is Cohen–Macaulay are classified. In the subsequent sections, the resolution of binomial edge ideals is considered and a bound for the Castelnuovo–Mumford regularity of these ideals is given. Finally, the Koszul property of binomial edge ideals is studied. Intimately related to binomial edge ideals are permanental edge ideals and Lovász, Saks, and Schrijver edge ideals. Their primary decomposition will be studied.

7.1 Binomial Edge Ideals and Their Gröbner Bases

Let G be a finite simple graph, that is, G has no loops and no multiple edges. Unless otherwise stated, G will always be a finite simple graph without isolated vertices. We denote by $V(G)$ the set of vertices of G and by $E(G)$ the set of edges of G. We say that G is a graph on $[n]$, if $V(G) = [n]$, where $[n] = \{1, 2, \ldots, n\}$.

Let K be a field and $S = K[x_1, \ldots, x_n, y_1, \ldots, y_n]$ be the polynomial ring in $2n$ variables. For $1 \le i < j \le n$ we set $f_{ij} = x_i y_j - x_j y_i$. The binomials f_{ij} are the 2-minors of the matrix

$$\begin{pmatrix} x_1 & x_2 & \ldots & x_n \\ y_1 & y_2 & \ldots & y_n \end{pmatrix}.$$

© Springer International Publishing AG, part of Springer Nature 2018
J. Herzog et al., *Binomial Ideals*, Graduate Texts in Mathematics 279,
https://doi.org/10.1007/978-3-319-95349-6_7

Fig. 7.1 A labeled graph

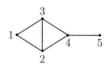

Definition 7.1 Let G be a graph on $[n]$. The *binomial edge ideal* $J_G \subset S$ of G is the ideal generated by the binomials $f_{ij} = x_i y_j - x_j y_i$ such that $i < j$ and $\{i, j\}$ is an edge of G.

Consider, for example, the graph G displayed in Figure 7.1. The binomial edge ideal of this graph is the ideal

$$J_G = (x_1 y_2 - x_2 y_1, x_1 y_3 - x_3 y_1, x_2 y_3 - x_3 y_2, x_2 y_4 - x_4 y_2, x_3 y_4 - x_4 y_3, x_4 y_5 - x_5 y_4).$$

7.1.1 Closed Graphs

We first study the question of when J_G has a quadratic Gröbner basis.

Theorem 7.2 *Let G be a graph on $[n]$, and let $<$ be the lexicographic order on $S = K[x_1, \ldots, x_n, y_1, \ldots, y_n]$ induced by $x_1 > x_2 > \cdots > x_n > y_1 > y_2 > \cdots > y_n$. Then the following conditions are equivalent:*

(i) *the generators f_{ij} of J_G form a quadratic Gröbner basis;*
(ii) *for all edges $\{i, j\}$ and $\{k, l\}$ with $i < j$ and $k < l$ one has $\{j, l\} \in E(G)$ if $i = k$, and $\{i, k\} \in E(G)$ if $j = l$.*

Proof (i) $\Rightarrow$ (ii): Suppose (b) is violated, say, $\{i, j\}$ and $\{i, k\}$ are edges with $i < j < k$, but $\{j, k\}$ is not an edge. Then $S(f_{ik}, f_{ij}) = y_i f_{jk}$ belongs to J_G, but none of the initial monomials of the quadratic generators of J_G divides $\mathrm{in}_<(y_i f_{jk})$.

(ii) $\Rightarrow$ (i): We apply Buchberger's criterion and show that all S-pairs $S(f_{ij}, f_{kl})$ reduce to 0. If $i \neq k$ and $j \neq l$, then $\mathrm{in}_<(f_{ij})$ and $\mathrm{in}_<(f_{kl})$ have no common factor. In this case, according to Lemma 1.27, $S(f_{ij}, f_{kl})$ reduces to zero. On the other hand, if $i = k$, we may assume that $l < j$. Then

$$S(f_{ij}, f_{il}) = y_i f_{lj}$$

is the standard expression of $S(f_{ij}, f_{il})$. Similarly, if $j = l$, we may assume that $i < k$. Then

$$S(f_{ij}, f_{kj}) = x_j f_{ik}$$

is the standard expression of $S(f_{ij}, f_{kj})$. In both cases the S-pair reduces to 0.   □

Fig. 7.2 The claw

Fig. 7.3 A net and a tent

Condition (ii) of Theorem 7.2 does not only depend on the isomorphism type of the graph, but also on the labeling of its vertices. For example, the graph G with edges $\{1, 2\}, \{2, 3\}$, and the graph G' with edges $\{1, 2\}, \{1, 3\}$ are isomorphic, but G satisfies condition (b), while G' does not.

Definition 7.3 A graph G satisfying the equivalent conditions of Theorem 7.2 is called closed with respect to the given labeling of the vertices, and G is called *closed* if it is closed with respect to a suitable labeling of its vertices.

The so-called *claw* shown in Figure 7.2 is the simplest example of a graph which is not closed. Indeed, suppose the claw is closed, and let $\{i, j\}, \{i, k\}$, and $\{i, l\}$ be the edges of the claw. Then $i \neq \min\{i, j, k, l\}$, since we assume that the claw is closed. If $j < i$, then $k > i$ and $l > i$, again since we assume the claw is closed. But then $\{k, j\}$ must be an edge of the claw, a contradiction.

A graph which does not contain any claw as an induced subgraph is called *claw-free*. Next follows a necessary condition for a graph to be closed.

Proposition 7.4 *If G is closed, then G is chordal and claw-free.*

Proof Suppose G is not chordal, then G contains a cycle C of length > 3 with no chord. Let i be the vertex of C with $i < j$ for all $j \in V(C) \setminus \{i\}$, and let $\{i, j\}$ and $\{i, k\}$ be the edges of C containing i. Then $i < j$ and $i < k$, but $\{j, k\} \notin E(G)$.

Since G is closed, any induced subgraph is closed as well. Since a claw is not closed, G must be claw-free.   □

The *path graph* on n vertices, denoted P_n, is the graph on $[n]$ with edges

$$\{1, 2\}, \{2, 3\}, \ldots, \{n - 1, n\}.$$

Any graph isomorphic to P_n is also called a path graph. The *length* of P_n is defined to be $n - 1$.

As a simple consequence of Proposition 7.4 we obtain

Corollary 7.5 *A bipartite graph is closed if and only if it is a path graph.*

Proof A bipartite graph has no odd cycles, see Lemma 5.1. Since a closed graph is chordal, and since a chordal graph has an odd cycle, unless it is a tree, a closed bipartite graph must be a tree. If the tree is not a path, then it is not claw-free. Thus a closed bipartite graph must be a path.

Conversely, if G is a path graph of length l, then G is closed for the labeling of the vertices such that $\{1, 2\}, \{2, 3\}, \ldots, \{l, l + 1\}$ are the edges of G.   □

The net and tent depicted in Figure 7.3 are chordal and claw-free graphs, but they are not closed. So Proposition 7.4 does not fully describe all closed graphs. Next we are aiming at giving a full classification of all closed graphs. In order to do this we have to introduce some terminology and concepts.

Let Δ be a simplicial complex. A facet F of Δ is called a *leaf*, if there exists a facet G of Δ with $G \neq F$ that $H \cap F \subset G \cap F$ for all facets H with $H \neq F$. The facet G is then called a *branch* of F. The simplicial complex Δ is called a *quasi-forest*, if the facets of Δ can be ordered $F_1, \ldots, F_m$ such that for each $i > 1$, the facet F_i is a leaf of $\langle F_1, \ldots, F_i \rangle$. Such an order of the facets of Δ is called a *leaf order*. A connected graph which is a quasi-forest is called a *quasi-tree*.

Let as before G be a graph on the vertex set $[n]$. A *clique* of G is a subset $F \subset [n]$ with the property that each 2-element subset of F is an edge of G. The set of all cliques forms a simplicial complex $\Delta(G)$, called the *clique complex* of G.

By a theorem of Dirac [54], G is chordal if and only if G has a *perfect elimination order* which means that its vertices can be labeled such that for every j, the set $F_j = \{i : i < j \text{ and } \{i, j\} \in E(G)\}$ is a clique of G. Equivalently, Dirac's theorem can be phrased as follows.

Theorem 7.6 *The graph G is chordal if and only if $\Delta(G)$ is a quasi-forest.*

With this preparation we obtain a first characterization of closed graphs.

Theorem 7.7 *Let G be a graph on $[n]$. The following conditions are equivalent:*

 (i) *G is closed;*
 (ii) *there exists a labeling of G such that all facets of $\Delta(G)$ are intervals $[a, b] \subset [n]$.*

Moreover, if the equivalent conditions hold and the facets $F_1, \ldots, F_r$ of $\Delta(G)$ are labeled such that $\min(F_1) < \min(F_2) < \cdots < \min(F_r)$, then $F_1, \ldots, F_r$ is a leaf order of $\Delta(G)$.

Proof For the proof of the theorem we may assume that G is connected.

(i) $\Rightarrow$ (ii): Let $F = \{j : \{j, n\} \in E(G)\} \cup \{n\}$, and let $k = \min\{j : j \in F\}$. Then $F = [k, n]$. Indeed, if $j \in F$ with $j < n$, then, by Problem 7.1, it follows that $\{j, j + 1\} \in E(G)$. Moreover, because G is closed and $\{j, n\} \in E(G)$, we see that also $\{j + 1, n\} \in E(G)$. Thus $j + 1 \in F$.

Next observe that F is a maximal clique of G, that is, a facet of $\Delta(G)$. First of all it is a clique, because if $i, j \in F$ with $i < j < n$, then, since $\{i, n\}$ and $\{j, n\}$ are edges of G, it follows that $\{i, j\}$ is an edge as well, since G is closed. Secondly, it is maximal, since $\{j, n\} \notin E(G)$, if $j \notin F$.

Let $H \neq F$ be a facet of $\Delta(G)$ with $H \cap F \neq \emptyset$, and let $\ell = \max\{j : j \in H \cap F\}$. We claim that $H \cap F = [k, \ell]$. There is nothing to prove if $k = \ell$. So now suppose that $k < \ell$ and let $k \leq t < \ell$ and $s \in H \setminus F$. Then $s, t < \ell$ and $\{s, \ell\}$ and $\{t, \ell\}$ are edges of G. Hence since G is closed it follows that $\{s, t\} \in E(G)$. This implies that $t \in H$, as desired.

It follows from the claim that the facet H for which $\max\{j : j \in H \cap F\}$ is maximal is a branch of F. In particular, F is a leaf. Let $H \cap F = [k, \ell]$, where H is a branch of F, and denote by G_ℓ the restriction of G to $[\ell]$. Since G_ℓ is again closed and since $\ell < n$, we may assume, by applying induction on the cardinality of the vertex set of G, that all facets of $\Delta(G_\ell)$ are intervals. Now let F' be any facet of $\Delta(G)$. If $F = F'$, then F is an interval, and if $F \neq F'$, then, as we have seen above, it follows that $F' \in \Delta(G_\ell)$. This yields the desired conclusion.

(ii) $\Rightarrow$ (i): Let $\{i, j\}$ and $\{k, \ell\}$ be edges of G with $i < j$ and $k < \ell$. If $i = k$, then $\{i, k\}$ and $\{i, \ell\}$ belong to the same maximal clique, that is, facet of $\Delta(G)$ which by assumption is an interval. Thus if $j \neq \ell$, then $\{j, \ell\} \in E(G)$. Similarly one shows that if $j = \ell$, but $i \neq k$, then $\{i, k\} \in E(G)$. Thus G is closed.

Finally it is obvious that the facets of $\Delta(G)$ ordered according to their minimal elements is a leaf order, because for this order F_{i-1} has maximal intersection with F_i for all i. $\qquad \square$

Definition 7.8 A graph G is called an *interval graph* if for all $v \in V(G)$ there exists an interval $I_v = [l_v, r_v]$ of the real line such that $I_v \cap I_w \neq \emptyset$ if and only if $\{v, w\} \in E(G)$. If, in addition, the intervals can be chosen such that there is no proper containment among them, then G is called a *proper interval graph* or simply a *PI graph*.

Let G be a graph. A set of intervals $\{I_v\}_{v \in V(G)}$ as in Definition 7.8 is called an *interval representation* of G.

Let G be a graph on the vertex set $[n]$. Then G satisfies the *proper interval ordering* with respect to the given labeling, if for all $i < j < k$ with $\{i, k\} \in E(G)$ it follows that $\{i, j\}, \{j, k\} \in E(G)$. We say G admits a *proper interval ordering* if G satisfies the proper interval ordering for a suitable relabeling of its vertices.

Theorem 7.9 *The following conditions are equivalent:*

(i) *G is a closed graph.*
(ii) *G is a proper interval graph.*

Proof We may assume that G is connected. For the proof we use the fact that G is a proper interval graph if and only if G admits a proper interval ordering, as shown in [141, Theorem 2.1]. Thus we need to show that G is closed if and only if G admits a proper interval ordering.

Let $[n]$ be the vertex set of G. Suppose first that G is closed. By Theorem 7.7 we may assume that the maximal cliques of G are (integral) intervals. Now let $i < j < k$ with $\{i, k\} \in E(G)$. Then the vertices i, k belong to a clique of G, say $[a, b]$. Then $j \in [a, b]$ and hence $\{i, j\}$ and $\{j, k\}$ are edges of G. This shows that the given labeling is a proper interval ordering.

Conversely, suppose G admits a proper interval ordering. We may assume that the given labeling has this property. Let $\{i, j\}$ and $\{i, k\}$ be two different edges of G with $i < j$ and $i < k$. We may assume that $j < k$. By the interval ordering property it follows that $\{j, k\} \in E(G)$. On the other hand, if $k, j < i$, the interval labeling property guarantees again that $\{j, k\} \in E(G)$. Thus G is closed.   □

A vertex $v \in V(G)$ is called a *simplicial vertex* of G if v belongs to exactly one maximal clique of $\Delta(G)$. The concept of simplicial vertices will be used in the proof of next theorem and in many more of the following results.

Theorem 7.10 *Let G be a graph. Then G is closed if and only if G is chordal, claw-free, net-free, and tent-free.*

Proof If G is closed, then G is chordal and any induced subgraph of G is closed as well. Hence, since the claw, the net, and the tent are not closed, none of them can be an induced subgraph of G.

We prove the converse by induction on the number of vertices of G. If G has two vertices, the statement is trivial. We now may assume that G is a connected chordal claw-free graph on the vertex set $[n]$, with $n \geq 3$, and that the converse is true for graphs with $n - 1$ vertices. Since G is chordal, we may choose a perfect elimination order on G. Then the vertex labeled with n is obviously a simplicial vertex.

Let G' be the restriction of G to the vertex set $[n - 1]$. Then G' is clearly chordal and claw-free and does neither contain a net or a tent as an induced subgraph. We claim that G' is also connected. Indeed, suppose G' has at least two connected components. Then, as G is connected, it follows that the vertex n must belong to at least two maximal cliques of G, a contradiction. Therefore, we may apply the inductive hypothesis to G' and conclude that G' is closed. By Theorem 7.7, it follows that we may relabel the vertices of G' with labels from 1 to $n - 1$ such that the facets of $\Delta(G')$ are $F_1 = [a_1, b_1], \ldots, F_r = [a_r, b_r]$ with $1 = a_1 < a_2 < \cdots < a_r < b_r = n - 1$. Of course, this new labeling may not be a perfect elimination order for G.

In order to prove that G is closed we will use the criterion given in Theorem 7.7. Let us first assume that G' itself is a clique. If the vertex n of G is adjacent to all the vertices of G', then G is a clique as well, thus it is closed. If not, then we may relabel the vertices of G' such that those which are adjacent to the vertex n of G have the largest labels among $1, \ldots, n - 1$. Then, we get $\Delta(G) = \langle [1, n - 1], [a, n] \rangle$ for some $1 < a \leq n - 1$. Thus G is a closed graph with two maximal cliques.

We now consider the case when G' has two maximal cliques, say, $\Delta(G') = \langle F_1, F_2 \rangle$ with $F_1 = [1, b]$, $F_2 = [a, n - 1]$ for some $1 < a \leq b < n - 1$.

Let $i_1, \ldots, i_\ell \in [n - 1]$ be the vertices of G' adjacent to n in G. Then $\{i_1, \ldots, i_\ell\}$ is a clique of G', and hence $\{i_1, \ldots, i_\ell\} \subset F_i$ for some i. We may assume that $\{i_1, \ldots, i_\ell\} \subset F_2$. Otherwise, we reduce to this case by relabeling the vertices of G' as follows: $i \mapsto n - i$ for $1 \leq i \leq n - 1$. If $\{i_1, \ldots, i_\ell\} = F_2$, then $\Delta(G) = \langle [1, b], [a, n] \rangle$, hence G is closed.

Now we assume that $\{i_1, \ldots, i_\ell\} \subsetneq F_2$. If all the vertices $i_1, \ldots, i_\ell$ are simplicial vertices of F_2 (in G'), then we may relabel all the simplicial vertices of F_2 such that

$\{i_1, \ldots, i_\ell\} = \{n - 1, n - 2, \ldots, n - \ell\}$. It follows that $\Delta(G) = \langle [1, b], [a, n - 1], [n - \ell, n] \rangle$, thus G is closed. We have to treat now the case when at least one of the vertices $i_1, \ldots, i_\ell$, let us say i_1, belongs to $F_1 \cap F_2$. If there is a simplicial vertex $k \in F_2$ which is not adjacent to n, then we get an induced claw graph in G with the edges $\{1, i_1\}, \{i_1, n\}, \{i_1, k\}$ which is impossible. Therefore, all the free vertices of F_2 are contained in the set $\{i_1, \ldots, i_\ell\}$. In this case we may permute the labels of the vertices in the intersection $F_1 \cap F_2$ such that the set $\{i_1, \ldots, i_\ell\}$ is an interval of the form $[c, n - 1]$ where $a \leq c \leq b$. Consequently, $\Delta(G) = \langle [1, b], [a, n - 1], [c, n] \rangle$, thus G is closed.

Finally, we discuss the case when $\Delta(G')$ has at least three facets, that is, the facets of $\Delta(G')$ are $F_1 = [a_1, b_1], \ldots, F_r = [a_r, b_r]$ with $1 = a_1 < a_2 < \cdots < a_r < b_r = n - 1$ and $r \geq 3$. Let, as before, $i_1, \ldots, i_\ell$ be the vertices adjacent to the vertex n. Since $F = \{i_1, \ldots, i_\ell\}$ is a clique in G', there exists a maximal clique of G' which contains F. We distinguish two cases.

Case 1. $\{i_1, \ldots, i_\ell\} \subset F_1$ or $\{i_1, \ldots, i_\ell\} \subset F_r$: If $\{i_1, \ldots, i_\ell\} \subset F_1$, then we may reduce to the case that $\{i_1, \ldots, i_\ell\} \subset F_r$ by reversing the labels of G', namely:$i \mapsto n - i$ for $1 \leq i \leq n - 1$.

If $\{i_1, \ldots, i_\ell\} = F_r$, then clearly G is closed since $\Delta(G) = \langle F_1, \ldots, F_{r-1}, F_r \cup \{n\} \rangle$. Next assume that $\{i_1, \ldots, i_\ell\} \subsetneq F_r$. We proceed as in the case when G' had two cliques. Indeed, if all the vertices $i_1, \ldots, i_\ell$ are free in G', then we may relabel the simplicial vertices of F_r such that $\{i_1, \ldots, i_\ell\} = \{n - \ell, n - \ell + 1, \ldots, n - 1\}$. With respect to this new labeling, $\Delta(G) = \langle F_1, \ldots, F_r, [n - \ell, n] \rangle)$, and thus G is closed. In contrast to the case when $\Delta(G')$ had two cliques, F_r may have non-empty intersection with several maximal cliques of G'. Let j be the smallest integer such that there exists an element in F, say i_1, such that $i_1 \in F_j \cap F_r$. We claim that in this case, the set $F_r \setminus F_j$ must be contained in $\{i_1, \ldots, i_\ell\}$. Indeed, let us assume that there exists $k \in F_r \setminus F_j$ such that k is not adjacent to the vertex n of G. Then $\{\min F_j, i_1\}, \{i_1, n\}, \{i_1, k\}$ is an induced claw of G, a contradiction.

Thus $F_r \setminus F_j \subset \{i_1, \ldots, i_\ell\}$. Then we may relabel (if necessary) the vertices of $F_j \cap F_r$ such that the set $\{i_1, \ldots, i_\ell\}$ is an interval of the form $[c, n - 1]$ where $a_r < c \leq b_j$. With respect to this new labeling, the maximal cliques of G are the intervals $F_1, \ldots, F_{r-1}$ and $F_r \cup \{n\}$, hence G is closed.

Case 2. $\{i_1, \ldots, i_\ell\} \subseteq F_i$ for some $2 \leq i \leq r - 1$.
If we have equality, namely $\{i_1, \ldots, i_\ell\} = F_i$ and $F_{i-1} \cap F_{i+1} = \emptyset$, then we may relabel the vertices of $F_i \cup \{n\}$ and of $F_{i+1}, \ldots, F_r$ such that $\Delta(G) = \langle F_1 = [a_1, b_1], \ldots, F_{i-1} = [a_{i-1}, b_{i-1}], F'_i = [a_i, b_i + 1], F'_{i+1} = [a_{i+1} + 1, b_{i+1} + 1], \ldots, F'_r = [a_r + 1, b_r + 1 = n] \rangle)$. The case that $F_{i-1} \cap F_{i+1} \neq \emptyset$ and $\{i_1, \ldots, i_\ell\} = F_i$ cannot occur. Indeed, let $j \in F_{i-1} \cap F_{i+1}$ and set $p = \min\{t : j \in F_t\}$, $q = \max\{t; j \in F_t\}$. Then G has the claw with edges $\{\min F_p, j\}, \{j, n\}, \{j, \max F_q\}$ as induced graph, which is impossible.

Let now $\{i_1, \ldots, i_\ell\} \subsetneq F_i$. We split the rest of the proof into two subcases.

Subcase 2 (a). The facet F_i of $\Delta(G')$ has a simplicial vertex. This implies, in particular, that $F_{i-1} \cap F_{i+1} = \emptyset$.

Let j be a simplicial vertex of F_i and assume $\{j, n\} \in E(G)$. If there exist some vertices $p \in F_i \cap F_{i-1}, q \in F_i \cap F_{i+1}$ which are not adjacent to n, we get an induced subgraph of G isomorphic to a net by choosing the triangle $\{j, p, q\}$ together with the edges $\{j, n\}, \{\min F_{i-1}, p\}, \{q, \max F_{i+1}.\}$ By our assumptions, this is impossible. Therefore, if F_i has a simplicial vertex adjacent to n, then we must have either $F_i \cap F_{i-1} \subset \{i_1, \ldots, i_\ell\}$ or $F_i \cap F_{i+1} \subset \{i_1, \ldots, i_\ell\}$. Obviously these two situations are symmetric. Let us assume that $F_i \cap F_{i+1} \subset \{i_1, \ldots, i_\ell\}$ and that there exists $p \in F_i \cap F_{i-1}$ which is not adjacent to n. Then we get the induced claw of G with edges

$$\{p, \min(F_i \cap F_{i+1})\}, \{n, \min(F_i \cap F_{i+1})\}, \{\min(F_i \cap F_{i+1}), \max F_{i+1}\}.$$

Thus, we have shown that if F_i has a simplicial vertex which is adjacent to n, then $\{i_1, \ldots, i_\ell\}$ must contain $(F_i \cap F_{i-1}) \cup (F_i \cap F_{i+1})$. In addition, if there exists another simplicial vertex of F_i, say u, which is not adjacent to n, we get the induced claw in G with the edges $\{u, \min(F_i \cap F_{i+1})\}, \{n, \min(F_i \cap F_{i+1})\}, \{\min(F_i \cap F_{i+1}), \max F_{i+1}\}$. Therefore, all the simplicial vertices of F_i must be adjacent to n. Summarizing, we showed that $\{i_1, \ldots, i_\ell\} = F_i$, contradicting our hypothesis of Subcase 2 (a).

Let us now assume that no simplicial vertex of F_i is adjacent to n. Then there exists a vertex $u \in F_i \cap F_{i+1}$ or $u \in F_i \cap F_{i-1}$ which is adjacent to n. There is no loss of generality in assuming that $u \in F_i \cap F_{i+1}$. Let k be any simplicial vertex of F_i. Then we find the induced claw subgraph of G with edges $\{k, u\}, \{u, n\}, \{u, \max F_{i+1}\}$, contradiction.

Subcase 2 (b). $F_{i-1} \cap F_{i+1} \neq \emptyset$. We will show that also this subcase cannot occur. If there exists a vertex $j \in F_{i-1} \cap F_{i+1}$ which is adjacent to n, then G has an induced claw with the edges $\{\min F_{i-1}, j\}, \{j, n\}, \{j, \max F_{i+1}\}$, contradiction. Consequently, n cannot be adjacent to any vertex of $F_{i-1} \cap F_{i+1}$.

Let now $j \in (F_i \cap F_{i-1}) \setminus F_{i+1}$ adjacent to n. If there is no vertex adjacent to n among the vertices of $F_i \cap F_{i+1}$, then we get the induced claw of G with the edges

$$\{\min F_{i-1}, j\}, \{j, n\}, \{j, \max F_i\}.$$

This implies that all the vertices in the set $(F_i \cap F_{i+1}) \setminus F_{i-1}$ must be adjacent to n. But, in this case, we reach a contradiction in the following way. Let $t \in (F_i \cap F_{i+1}) \setminus F_{i-1}$. The induced subgraph of G with the triangles

$$\{\min F_{i-1}, j, \max F_{i-1}\}, \{j, \max F_{i-1}, t\}, \{\max F_{i-1}, t, \max F_{i+1}\}, \text{ and } \{n, j, t\}$$

is isomorphic to H_2, contradiction to the hypothesis on G.

We end this subcase and the whole proof by observing that the situation when we choose $j \in (F_i \cap F_{i+1}) \setminus F_{i-1}$ adjacent to n is symmetric to the above one. $\quad\square$

7.1.2   The Computation of the Gröbner Basis

We now describe the reduced Gröbner basis of the binomial edge ideal of an arbitrary graph. For this we need to introduce the following concept: let G be a simple graph on $[n]$, and let i and j be two vertices of G with $i < j$. A path $i = i_0, i_1, \ldots, i_r = j$ from i to j is called *admissible*, if

(i) $i_k \neq i_\ell$ for $k \neq \ell$;
(ii)   for each $k = 1, \ldots, r - 1$ one has either $i_k < i$ or $i_k > j$;
(iii)   for any proper subset $\{j_1, \ldots, j_s\}$ of $\{i_1, \ldots, i_{r-1}\}$, the sequence $i, j_1, \ldots, j_s, j$ is not a path.

Given an admissible path

$$\pi : i = i_0, i_1, \ldots, i_r = j$$

from i to j, where $i < j$, we associate the monomial

$$u_\pi = \Big(\prod_{i_k > j} x_{i_k} \Big) \Big(\prod_{i_\ell < i} y_{i_\ell} \Big).$$

Theorem 7.11 *Let G be a graph on $[n]$. Let $<$ be the monomial order introduced in Theorem 7.2. Then the set of binomials*

$$\mathscr{G} = \bigcup_{i < j} \{ u_\pi f_{ij} \ : \ \pi \text{ is an admissible path from } i \text{ to } j \}$$

is the reduced Gröbner basis of J_G with respect to $<$.

Proof We organize this proof as follows: In the first step, we prove that $\mathscr{G} \subset J_G$. Then, since $\mathscr{G}$ is a system of generators, in the second step, we show that $\mathscr{G}$ is a Gröbner basis of J_G by using Buchberger's criterion. Finally, in the third step, it is proved that $\mathscr{G}$ is reduced.

First Step.   We show that for each admissible path π from i to j, where $i < j$, the binomial $u_\pi f_{ij}$ belongs J_G. Let $\pi : i = i_0, i_1, \ldots, i_{r-1}, i_r = j$ be an admissible path in G. We proceed with induction on r. Clearly the assertion is true if $r = 1$. Let $r > 1$ and $A = \{i_k : i_k < i\}$ and $B = \{i_\ell : i_\ell > j\}$. One has either $A \neq \emptyset$ or $B \neq \emptyset$. If $A \neq \emptyset$, then we set $i_{k_0} = \max A$. If $B \neq \emptyset$, then we set $i_{\ell_0} = \min B$. Suppose $A \neq \emptyset$. It then follows that each of the paths $\pi_1 : i_{k_0}, i_{k_0 - 1}, \ldots, i_1, i_0 = i$ and $\pi_2 : i_{k_0}, i_{k_0 + 1}, \ldots, i_{r-1}, i_r = j$ in G is admissible. Now, the induction hypothesis guarantees that each of $u_{\pi_1} f_{i_{k_0}, i}$ and $u_{\pi_2} f_{i_{k_0}, j}$ belongs to J_G. A routine computation says that the S-polynomial $S(u_{\pi_1} f_{i_{k_0}, i}, u_{\pi_2} f_{i_{k_0}, j})$ is equal to $u_\pi f_{ij}$. Hence $u_\pi f_{ij} \in J_G$, as desired. When $B \neq \emptyset$, the same argument applies as in the case $A \neq \emptyset$.

Second Step. It will be proven that the set of those binomials $u_\pi f_{ij}$, where π is an admissible path from i to j, forms a Gröbner basis of J_G. In order to show this we apply Buchberger's criterion, that is, we show that all S-pairs $S(u_\pi f_{ij}, u_\sigma f_{k\ell})$, where $i < j$ and $k < \ell$, reduce to zero. For this we will consider different cases.

Let $i < j$ and $k < \ell$. Suppose that the initial monomials $\mathrm{in}_<(f_{ij})$ and $\mathrm{in}_<(f_{k\ell})$ are relatively prime. Then the following four cases arise:

(i) $i < j = k < \ell$;
(ii) $i < j < k < \ell$;
(iii) $i < k < j < \ell$;
(iv) $i < k < \ell < j$.

Let π be an admissible path from i to j and σ an admissible path from k to ℓ. Since $\mathrm{in}_<(f_{ij})$ and $\mathrm{in}_<(f_{k\ell})$ are relatively prime, if the following conditions

(*) neither x_k nor y_ℓ appears in u_π;
(**) neither x_i nor y_j appears in u_σ

are satisfied, then $S((u_\pi/w)f_{ij}, (u_\sigma/w)f_{k\ell})$, where $w = \gcd(u_\pi, u_\sigma)$, reduces to zero. Hence $S(u_\pi f_{ij}, u_\sigma f_{k\ell})$ reduces to zero. Now, in each of the cases (i), (iii), and (iv), the above conditions (*) and (**) are satisfied. Thus only the case (ii) must be discussed. Let $i < j < k < \ell$.

If k belongs to π and if j does not belong to σ, then

$$S(u_\pi f_{ij}, u_\sigma f_{k\ell}) = S((u_\pi/x_k)x_k(x_i y_j - x_j y_i), u_\sigma(x_k y_\ell - x_\ell y_k))$$
$$= w(x_i y_j y_k x_\ell - y_i x_j x_k y_\ell),$$

where $w = \mathrm{lcm}(u_\pi/x_k, u_\sigma)$. Since u_σ divides w, it follows that $w \cdot y_i x_j x_k y_\ell$ can be divided by the initial monomial of $u_\sigma f_{k\ell}$. We then divide $w(x_i y_j y_k x_\ell - y_i x_j x_k y_\ell)$ by $u_\sigma f_{k\ell}$. Its reminder is

$$w(x_i y_j y_k x_\ell - y_i x_j y_k x_\ell) = w \cdot y_k x_\ell(x_i y_j - y_i x_j) = w \cdot y_k x_\ell f_{ij}.$$

Lemma 7.12(a) guarantees that $(u_\pi/x_k)y_k f_{ij}$ reduces to zero with respect to $\mathcal{G}$. Hence $w \cdot y_k x_\ell f_{ij}$ reduces to zero with respect to $\mathcal{G}$, as desired.

If k belongs to π and if j belongs to σ, then

$$S(u_\pi f_{ij}, u_\sigma f_{k\ell}) = S((u_\pi/x_k)x_k(x_i y_j - x_j y_i), (u_\sigma/y_j)y_j(x_k y_\ell - x_\ell y_k))$$
$$= w(x_i y_j y_k x_\ell - y_i x_j x_k y_\ell),$$

where $w = \mathrm{lcm}(u_\pi/x_k, u_\sigma/y_j)$. Since, by using Lemma 7.12(a) again, the monomial

$$(u_\pi/x_k)y_k f_{ij} = (u_\pi/x_k)y_k(x_i y_j - x_j y_i)$$

reduces zero with respect to $\mathcal{G}$, it follows that $(u_\pi/x_k)y_kx_iy_j$ and $(u_\pi/x_k)y_kx_jy_i$ can possess a common reminder with respect to $\mathcal{G}$. Thus in order to show that $w(x_iy_jy_kx_\ell - y_ix_jx_ky_\ell)$ reduces to zero, it suffices to prove that

$$w(y_ix_jy_kx_\ell - y_ix_jx_ky_\ell) = -w \cdot y_ix_j f_{k\ell}$$

reduces to zero. Lemma 7.12(b) guarantees that $(u_\sigma/y_j)x_j f_{k\ell}$ reduces to zero with respect to $\mathcal{G}$. Hence $w \cdot y_ix_j f_{k\ell}$ reduced to zero with respect to $\mathcal{G}$, as required.

It remains to consider the cases that either $i = k$ and $j \neq \ell$ or $i \neq k$ and $j = \ell$. Suppose we are in the first case. (The second case can be proved similarly.) We must show that $S(u_\pi f_{ij}, u_\sigma f_{i\ell})$ reduces to zero. We may assume that $j < \ell$, and must find a standard expression for $S(u_\pi f_{ij}, u_\sigma f_{i\ell})$ whose remainder is equal to zero.

Let $\pi : i = i_0, i_1, \ldots, i_r = j$ and $\sigma : i = i'_0, i'_1, \ldots, i'_s = \ell$. Then there exist indices a and b such that

$$i_a = i'_b \quad \text{and} \quad \{i_{a+1}, \ldots, i_r\} \cap \{i'_{b+1}, \ldots, i'_s\} = \emptyset.$$

Consider the path

$$\tau : j = i_r, i_{r-1}, \ldots, i_{a+1}, i_a = i'_b, i'_{b+1}, \ldots, i'_{s-1}, i'_s = \ell$$

from j to ℓ. To simplify the notation we write this path as

$$\tau : j = j_0, j_1, \ldots, j_t = \ell.$$

Let

$$j_{t(1)} = \min\{ j_c \; : \; j_c > j, \; c = 1, \ldots, t \},$$

and

$$j_{t(2)} = \min\{ j_c \; : \; j_c > j, \; c = t(1) + 1, \ldots, t \}.$$

Continuing these procedures yield the integers

$$0 = t(0) < t(1) < \cdots < t(q - 1) < t(q) = t.$$

It then follows that

$$j = j_{t(0)} < j_{t(1)} < \cdots < j_{t(q)-1} < j_{t(q)} = \ell$$

and, for each $1 \leq c \leq t$, the path

$$\tau_c : j_{t(c-1)}, j_{t(c-1)+1}, \ldots, j_{t(c)-1}, j_{t(c)}$$

is admissible.

It will be shown that

$$S(u_\pi f_{ij}, u_\sigma f_{i\ell}) = \sum_{c=1}^{q} v_{\tau_c} u_{\tau_c} f_{j_{t(c-1)} j_{t(c)}}$$

is a standard expression of $S(u_\pi f_{ij}, u_\sigma f_{i\ell})$ whose remainder is equal to 0. Here v_{τ_c} is the monomial defined as follows: let $w = y_i \operatorname{lcm}(u_\pi, u_\sigma)$. Thus $S(u_\pi f_{ij}, u_\sigma f_{i\ell}) = -w f_{j\ell}$. Then

(i) if $c = 1$, we set

$$v_{\tau_1} = \frac{x_\ell w}{u_{\tau_1} x_{j_{t(1)}}};$$

(ii) if $1 < c < q$, we set

$$v_{\tau_c} = \frac{x_j x_\ell w}{u_{\tau_c} x_{j_{t(c-1)}} x_{j_{t(c)}}};$$

(iii) if $c = q$, we set

$$v_{\tau_q} = \frac{x_j w}{u_{\tau_q} x_{j_{t(q-1)}}}.$$

Thus we have to show that

$$w f_{j\ell} = \frac{w x_\ell}{x_{j_{t(1)}}} f_{j j_{t(1)}} + \sum_{c=2}^{q-1} \frac{w x_j x_\ell}{x_{j_{t(c-1)}} x_{j_{t(c)}}} f_{j_{t(c-1)} j_{t(c)}} + \frac{w x_j}{x_{j_{t(q-1)}}} f_{j_{t(q-1)} \ell}$$

is a standard expression of $w f_{j\ell}$ with remainder 0. In other words, we must prove that

$$(\sharp) \qquad w(x_j y_\ell - x_\ell y_j) = \frac{w x_\ell}{x_{j_{t(1)}}} (x_j y_{j_{t(1)}} - x_{j_{t(1)}} y_j)$$

$$+ \sum_{c=2}^{q-1} \frac{w x_j x_\ell}{x_{j_{t(c-1)}} x_{j_{t(c)}}} (x_{j_{t(c-1)}} y_{j_{t(c)}} - x_{j_{t(c)}} y_{j_{t(c-1)}})$$

$$+ \frac{w x_j}{x_{j_{t(q-1)}}} (x_{j_{t(q-1)}} y_\ell - x_\ell y_{j_{t(q-1)}})$$

is a standard expression of $w(x_j y_\ell - x_\ell y_j)$ with remainder 0.
Since

$$w x_j y_\ell = \frac{w x_j}{x_{j_{t(q-1)}}} x_{j_{t(q-1)}} y_\ell > \frac{w x_j x_\ell}{x_{j_{t(q-2)}} x_{j_{t(q-1)}}} x_{j_{t(q-2)}} y_{j_{t(q-1)}}$$

$$> \cdots > \frac{w x_j x_\ell}{x_{j_{t(1)}} x_{j_{t(2)}}} x_{j_{t(1)}} y_{j_{t(2)}} > \frac{w x_\ell}{x_{j_{t(1)}}} x_j y_{j_{t(1)}},$$

it follows that, if the equality ($\sharp$) holds, then ($\sharp$) turns out to be a standard expression of $w(x_j y_\ell - x_\ell y_j)$ with remainder 0. If we rewrite ($\sharp$) as

$$w(x_j y_\ell - x_\ell y_j) = w(x_j x_\ell \frac{y_{j_{t(1)}}}{x_{j_{t(1)}}} - x_\ell y_j)$$

$$+ w x_j x_\ell \sum_{c=2}^{q-1} (\frac{y_{j_{t(c)}}}{x_{j_{t(c)}}} - \frac{y_{j_{t(c-1)}}}{x_{j_{t(c-1)}}})$$

$$+ w(x_j y_\ell - x_j x_\ell \frac{y_{j_{t(q-1)}}}{x_{j_{t(q-1)}}}),$$

then clearly the equality holds.

Third Step. Finally, we show that the Gröbner basis $\mathcal{G}$ is reduced. Let $u_\pi f_{ij}$ and $u_\sigma f_{k\ell}$, where $i < j$ and $k < \ell$, belong to $\mathcal{G}$ with $u_\pi f_{ij} \neq u_\sigma f_{k\ell}$. Let $\pi : i = i_0, i_1, \ldots, i_r = j$ and $\sigma : k = k_0, k_1, \ldots, k_s = \ell$. Suppose that $u_\pi x_i y_j$ divides either $u_\sigma x_k y_\ell$ or $u_\sigma x_\ell y_k$. Then $\{i_0, i_1, \ldots, i_r\}$ is a proper subset of $\{k_0, k_1, \ldots, k_s\}$.

Let $i = k$ and $j = \ell$. Then $\{i_1, \ldots, i_{r-1}\}$ is a proper subset of $\{k_0, k_1, \ldots, k_s\}$ and $k, i_1, \ldots, i_{r-1}, \ell$ is an admissible path. This contradicts the fact that σ is an admissible path.

Let $i = k$ and $j \neq \ell$. Then y_j divide u_σ. Hence $j < k$. This contradicts $i < j$.

Let $\{i, j\} \cap \{k, \ell\} = \emptyset$. Then $x_i y_j$ divide u_σ. Hence $i > \ell$ and $j < k$. This contradicts $i < j$. □

Lemma 7.12 *Let $i < j$ and π an admissible path from i to j.*

(a) *Let $k \in [n]$ belong to π with $j < k$. Then $(u_\pi/x_k) y_k f_{ij}$ reduces to zero with respect to $\mathcal{G}$.*

(b) *Let $k \in [n]$ belong to π with $k < i$. Then $(u_\pi/y_k) x_k f_{ij}$ reduces to zero with respect to $\mathcal{G}$.*

Proof A proof of (a) is given. The claim (b) can be proved similarly.

Let no vertex ξ with $j < \xi < k$ appear in π. Let π' be the subpath of π from i to k and π'' be the subpath of π from j to k. Then each of π' and π'' is admissible. Since $u_\pi/x_k = u_{\pi'} u_{\pi''}$, it follows that $(u_\pi/x_k) y_k (x_i y_j - x_j y_i)$ coincides with

$$u_{\pi''} u_{\pi'} (x_i y_k - x_k y_i) y_j - u_{\pi'} u_{\pi''} (x_j y_k - x_k y_j) y_i.$$

Hence

$$(u_\pi/x_k) y_k f_{ij} = y_j u_{\pi''} u_{\pi'} f_{ik} - y_i u_{\pi'} u_{\pi''} f_{jk},$$

as desired.

Let π possess a vertex ξ with $j < \xi < k$. One can find $j < \xi < k$ of π for which the subpath π' of π from ξ to k possesses no ξ' with $\xi < \xi' < k$. Then one can divide $(u_\pi/x_k)y_k f_{ij}$ by $u_{\pi'}f_{\xi k}$ and its reminder is $(u_\pi/x_\xi)y_\xi f_{ij}$. Continuing these procedures, it turns out that a reminder of $(u_\pi/x_k)y_k f_{ij}$ is of the form $(u_\pi/x_{k'})y_{k'}f_{ij}$, where $i < j < k'$ and where no ξ with $j < \xi < k'$ appears in π. Hence the argument in the previous paragraph says that $(u_\pi/x_k)y_k f_{ij}$ reduces to zero, as required.   □

Problems

7.1

(a) Let G be a finite simple graph on $[n]$. Show that G is closed with respect to the given labeling, if and only if for any two integers $1 \le i < j \le n$ the shortest walk $\{i_1, i_2\}, \{i_2, i_3\}, \ldots, \{i_{k-1}, i_k\}$ between i and j has the property that $i = i_1 < i_2 < \cdots < i_k = j$.
(b) Assume in addition to (a) that G is connected, and deduce by using (a), that for each $i < n$ one has that $\{i, i+1\} \in E(G)$.

7.2 Let G be a graph on $[n]$. Show that G is closed with respect to any labeling of the vertices if and only if G is a complete graph.

7.3 Let G be a path graph. By using Gröbner bases, show that J_G is generated by a regular sequence.

7.4 Let G be the 5-cycle labeled counterclockwise. Compute the Gröbner basis of J_G with respect to the lexicographic order induced by $x_1 > \cdots > x_5 > y_1 > \cdots > y_5$.

7.5 Let G be a graph on $[n]$, and let $\mathscr{G}$ be the reduced Gröbner basis of J_G with respect to the monomial order introduced in Theorem 7.2.

(a) Show that the maximal degree of an element of $\mathscr{G}$ is $\le n$.
(b) Show that there exists a suitable labeling of the vertices of G such that the maximal degree of an element of $\mathscr{G}$ is equal to n if and only G is a tree.

7.2   Primary Decomposition of Binomial Edge Ideals and Cohen-Macaulayness

Throughout this section, G will denote a finite simple graph on the vertex set $[n]$, unless otherwise stated.

Our objective to determine the primary decomposition of a binomial edge ideal is substantially simplified by the following remarkable consequence of Theorem 7.11.

Corollary 7.13 J_G is a radical ideal.

Proof By Theorem 7.11 we know that for a suitable monomial order, $\text{in}_<(J_G)$ is a squarefree monomial ideal. This implies that $\text{in}_<(J_G)$ is a radical ideal. Suppose now that $f^k \in J_G$ for some k. Then $\text{in}_<(f)^k = \text{in}_<(f^k) \in \text{in}_<(J_G)$, and hence $\text{in}_<(f) \in \text{in}_<(J_G)$. Thus there exists $g \in J_G$ with $\text{in}_<(g) = \text{in}_<(f)$, and hence $a \in K$ such that $\text{in}_<(f - ag) < \text{in}_<(f)$. Since $(f - ag)^k = f^k - gh$ for some h in S, it follows that $(f - ag)^k \in J_G$, and since $\text{in}_<(f - ag) < \text{in}_<(f)$ we may apply an induction argument to conclude that $f - ag \in J_G$. But then also $f \in J_G$. □

7.2.1   Primary Decomposition

For a radical ideal an irredundant primary decomposition is uniquely determined; it is just the intersection of all minimal prime ideals of the ideal. Therefore our next goal is to determine the minimal prime ideals of a binomial edge ideal.

For each subset $W \subset [n]$ we define a prime ideal $P_W(G)$. Let $T = [n] \setminus W$, and let $G_1, \ldots, G_{c(W)}$ be the connected components of G_T. Here G_T is the induced subgraph of G whose edges are exactly those edges $\{i, j\}$ of G for which $i, j \in T$. For each G_i we denote by $\tilde{G}_i$ the complete graph on the vertex set $V(G_i)$. We set

$$P_W(G) = (\bigcup_{i \in W} \{x_i, y_i\}, J_{\tilde{G}_1}, \ldots, J_{\tilde{G}_{c(W)}}).$$

In particular, $J_G = P_\emptyset(G)$, if G is complete.

Lemma 7.14 *The ideal $P_W(G)$ is a prime ideal.*

Proof We first reduce the polynomial ring S modulo the variables appearing in $P_W(G)$, to obtain the polynomial ring S' and a prime ideal $P \subset S'$ such that $S/P_W(G) \cong S'/P$. Furthermore, P is of the form $(P_1 + \cdots + P_{c(W)})S'$ with $P_i = P_\emptyset(\tilde{G}_i) \subset S_i$, where the $\tilde{G}_i$'s are complete graphs in disjoint sets of vertices, and where the S_i are polynomial rings over K in the corresponding variables. We prove by induction on i, that $S'/(P_1 + \cdots + P_i)S'$ is a domain. For $i = 1$, this follows from Problem 7.7. Let $i > 1$ and set $B = T/(P_1 + \cdots + P_{i-1})T$, where T is the polynomial ring over K in the variables of the polynomial rings $S_1, \ldots, S_{i-1}$. Then

$$B[x_j, y_j : j \in V(G_i)]/P_i B[x_j, y_j : j \in V(G_i)]$$
$$\cong T[x_j, y_j : j \in V(G_i)]/(P_1 + \cdots + P_i)T[x_j, y_j : j \in V(G_i)].$$

From Problem 7.7 it follows that $B[x_j, y_j : j \in V(G_i)]/P_i B[x_j, y_j : j \in V(G_i)]$, and hence also

$$A = T[x_j, y_j : j \in V(G_i)]/(P_1 + \cdots + P_i)T[x_j, y_j : j \in V(G_i)]$$

is a domain. This yields the desired conclusion, since $S'/(P_1 + \cdots + P_i)S'$ is just a polynomial extension of A.   □

Lemma 7.15 *We have* height $P_W(G) = |W| + (n - c(W))$.

Proof The height of $P_W(G)$ can be computed as follows: let $n_j = |V(G_j)| <$. Then

$$\text{height } P_W(G) = \text{height}(\bigcup_{i \in W}\{x_i, y_i\}) + \sum_{j=1}^{c(W)} \text{height } J_{\tilde{G}_j} = 2|W| + \sum_{j=1}^{c(W)}(n_j - 1)$$

$$= |W| + (|W| + \sum_{j=1}^{c(W)} n_j) - c(W) = |W| + (n - c(W)),$$

as required.   □

It is a general fact that all associated prime ideals of a binomial ideal in $K[x_1, \ldots, x_n]$ with K algebraically closed are binomial ideals in the sense that its generators are of the form $u - \lambda v$ with u and v monomials and $\lambda \in K$ with K the base field, see [58, Theorem 5.1]. In our particular case we have

Theorem 7.16 *Let G be a graph on the vertex set $[n]$. Then $J_G = \bigcap_{W \subset [n]} P_W(G)$.*

Proof It is obvious that each of the prime ideals $P_W(G)$ contains J_G. We will show by induction on n that each minimal prime ideal containing J_G is of the form $P_W(G)$ for some $W \subset [n]$. Since by Corollary 7.13, J_G is a radical ideal, and since a radical ideal is the intersection of its minimal prime ideals, the assertion of the theorem will follow.

Let P be a minimal prime ideal of J_G. We first show that $x_i \in P$ if and only $y_i \in P$. For this part of the proof we may assume that G is connected. Indeed, if $G_1, \ldots, G_r$ are the connected components of G, then each minimal prime ideal P of J_G is of the form $P_1 + \cdots + P_r$ where each P_i is a minimal prime ideal of J_{G_i}, see Problem 7.8. Thus if each P_i has the expected form, then so does P. Let $T = \{x_i : i \in [n], x_i \in P, y_i \notin P\}$. We will show that $T = \emptyset$. This will then imply that if $x_i \in P$, then $y_i \in P$. By symmetry it also follows that $y_i \in P$ implies $x_i \in P$, so that the final conclusion will be that $x_i \in P$ if and only $y_i \in P$.

We first observe that $T \neq \{x_1, \ldots, x_n\}$. Because otherwise we would have $J_G \subset J_{\tilde{G}} \subsetneq (x_1, \ldots, x_n) \subset P$, and P would not be a minimal prime ideal of J_G.

Suppose that $T \neq \emptyset$. Since $T \neq \{x_1, \ldots, x_n\}$, and since G is connected there exists $\{i, j\} \in E(G)$ such that $x_i \in T$ but $x_j \notin T$. Since $x_i y_j - x_j y_i \in J_G \subset P$, and since $x_i \in P$ it follows that $x_j y_i \in P$. Hence, since P is a prime ideal, we have $x_j \in P$ or $y_i \in P$. By the definition of T the second case cannot happen, and so $x_j \in P$. Since $x_j \notin T$, it follows that $y_j \in P$.

Let G' be the induced subgraph of G with vertex set $[n] \setminus \{j\}$. Then

$$(J_{G'}, x_j, y_j) = (J_G, x_j, y_j) \subset P.$$

Thus $\bar{P} = P/(x_j, y_j)$ is a minimal prime ideal of $J_{G'}$ with $x_i \in \bar{P}$ but $y_i \notin \bar{P}$ for all $x_i \in T \subset \bar{P}$. By induction hypothesis, $\bar{P}$ is of the form $P_W(G')$ for some subset $W \subset [n] \setminus \{j\}$. This contradicts the fact that $T \neq \emptyset$.

Now let G be again an arbitrary simple graph. By what we have shown it follows that there exists a subset $W \subset [n]$ such that $P = (\bigcup_{i \in W}\{x_i, y_i\}, \bar{P})$ where $\bar{P}$ is a prime ideal containing no variables. Let G' be the graph $G_{[n]\setminus W}$. Then reduction modulo the ideal $(\bigcup_{i \in W}\{x_i, y_i\})$ shows that $\bar{P}$ is a binomial prime ideal $J_{G'}$ which contains no variables. Let $G_1, \ldots, G_c$ be the connected components of G'. We will show that $\bar{P} = (J_{\tilde{G}_1}, \ldots, J_{\tilde{G}_c})$. This then implies that $P = (\bigcup_{i \in W}\{x_i, y_i\}, J_{\tilde{G}_1}, \ldots, J_{\tilde{G}_c})$, as desired.

To simplify notation we may as well assume that P itself contains no variables and have to show that $P = (J_{\tilde{G}_1}, \ldots, J_{\tilde{G}_c})$, where $G_1, \ldots, G_c$ are the connected components of G. In order to prove this we claim that if i, j with $i < j$ is an edge of $\tilde{G}_k$ for some k, then $f_{ij} \in P$. From this it will then follow that $(J_{\tilde{G}_1}, \ldots, J_{\tilde{G}_c}) \subset P$. Since $(J_{\tilde{G}_1}, \ldots, J_{\tilde{G}_c})$ is a prime ideal containing J_G, and P is a minimal prime ideal containing J_G, we conclude that $P = (J_{\tilde{G}_1}, \ldots, J_{\tilde{G}_c})$.

Let $i = i_0, i_1, \ldots, i_r = j$ be a path in G_k from i to j. We proceed by induction on r to show that $f_{ij} \in P$. The assertion is trivial for $r = 1$. Suppose now that $r > 1$. Our induction hypothesis says that $f_{i_1 j} \in P$. On the other hand, one has $x_{i_1} f_{ij} = x_j f_{i i_1} + x_i f_{i_1 j}$. Thus $x_{i_1} f_{ij} \in P$. Since P is a prime ideal and since $x_{i_1} \notin P$, we see that $f_{ij} \in P$. $\qquad\square$

Lemma 7.15 and Theorem 7.16 yield the following

Corollary 7.17 *Let G be a graph on $[n]$. Then*

$$\dim S/J_G = \max\{(n - |W|) + c(W) : W \subset [n]\}.$$

In particular, $\dim S/J_G \geq n + c$, where c is the number of connected components of G.

In general, the inequality given in Corollary 7.17 is strict. For example, if G is a claw, then $\dim S/J_G = 6$. On the other hand, we have

Corollary 7.18 *Let G be a graph on $[n]$ with c connected components. If S/J_G is Cohen–Macaulay, then $\dim S/J_G = n + c$.*

Proof Since $P_\emptyset(G)$ does not contain any monomials, it follows that $P_W(G) \not\subset P_\emptyset(G)$ for any nonempty subset $W \subset [n]$. Thus Theorem 7.16 implies that $P_\emptyset(G)$ is a minimal prime ideal of J_G. Since $\dim S/P_\emptyset(G) = n + c$ and since S/J_G is equidimensional, the assertion follows. $\qquad\square$

Now the question arises which of the prime ideals $P_W(G)$ are minimal prime ideals of J_G. The following result is the first important step to detect them.

Proposition 7.19 *Let G be a graph on $[n]$, and let W and T be subsets of $[n]$. Let $G_1, \ldots, G_s$ be the connected components of $G_{[n]\setminus W}$, and $H_1, \ldots, H_t$ the connected components of $G_{[n]\setminus T}$. Then the following conditions are equivalent:*

(i) $P_T(G) \subset P_W(G)$;

(ii) $T \subset W$ and for all $i = 1, \ldots, t$ one has $V(H_i) \setminus W \subset V(G_j)$ for some j.

Proof For a subset $U \subset [n]$ we let L_U be the ideal generated by the variables $\{x_i, y_i : i \in U\}$. With this notation introduced we have $P_W(G) = (L_W, J_{\tilde{G}_1}, \ldots, J_{\tilde{G}_s})$ and $P_T(G) = (L_T, J_{\tilde{H}_1}, \ldots, J_{\tilde{H}_t})$. Hence it follows that $P_T(G) \subset P_W(G)$, if and only if $T \subset W$ and $(L_W, J_{\tilde{H}_1}, \ldots, J_{\tilde{H}_t}) \subset (L_W, J_{\tilde{G}_1}, \ldots, J_{\tilde{G}_s})$.

Observe that $(L_W, J_{\tilde{H}_1}, \ldots, J_{\tilde{H}_t}) = (L_W, J_{\tilde{H}'_1}, \ldots, J_{\tilde{H}'_t})$ where $H'_i = (H_i)_{[n]\setminus W}$. It follows that $P_T(G) \subset P_W(G)$ if and only if $(L_W, J_{\tilde{H}'_1}, \ldots, J_{\tilde{H}'_t}) \subset (L_W, J_{\tilde{G}_1}, \ldots, J_{\tilde{G}_s})$ which is the case if and only if $(J_{\tilde{H}'_1}, \ldots, J_{\tilde{H}'_t}) \subset (J_{\tilde{G}_1}, \ldots, J_{\tilde{G}_s})$, because the generators of the ideals $(J_{\tilde{H}'_1}, \ldots, J_{\tilde{H}'_t})$ and $(J_{\tilde{G}_1}, \ldots, J_{\tilde{G}_s})$ have no variables in common with the x_i and y_i for $i \in W$.

Since $V(H'_i) = V(H_i) \setminus W$, the equivalence of (a) and (b) will follow once we have shown the following claim: let $A_1, \ldots, A_s$ and $B_1, \ldots, B_t$ be pairwise disjoint subsets of $[n]$. Then

$$(J_{\tilde{A}_1}, \ldots, J_{\tilde{A}_s}) \subset (J_{\tilde{B}_1}, \ldots, J_{\tilde{B}_t}),$$

if and only if for each $i = 1, \ldots, s$ there exists a j such that $A_i \subset B_j$.

It is obvious that if the conditions on the A_i and B_j are satisfied, then we have the desired inclusion of the corresponding ideals.

Conversely, suppose that $(J_{\tilde{A}_1}, \ldots, J_{\tilde{A}_s}) \subset (J_{\tilde{B}_1}, \ldots, J_{\tilde{B}_t})$. Without loss of generality we may assume that $\bigcup_{j=1}^t B_j = [n]$. Consider the surjective K-algebra homomorphism

$$\epsilon : S \to K[\{x_i, x_i z_1\}_{i \in B_1}, \ldots, \{x_i, x_i z_t\}_{i \in B_t}] \subset K[x_1, \ldots, x_n, z_1, \ldots, z_t]$$

with $\epsilon(x_i) = x_i$ for all i and $\epsilon(y_i) = x_i z_j$ for $i \in B_j$ and $j = 1, \ldots, t$. Then

$$\mathrm{Ker}(\epsilon) = (J_{\tilde{B}_1}, \ldots, J_{\tilde{B}_t}).$$

Now fix one of the sets A_i and let $k \in A_i$. Then $k \in B_j$ for some k. We claim that $A_i \subset B_j$. Indeed, let $\ell \in A_i$ with $\ell \neq k$ and suppose that $\ell \in B_r$ with $r \neq j$. Since $x_k y_\ell - x_\ell y_k \in J_{\tilde{A}_i} \subset (J_{\tilde{B}_1}, \ldots, J_{\tilde{B}_t})$, it follows that $x_k y_\ell - x_\ell y_k \in \mathrm{Ker}(\epsilon)$, so that $0 = \epsilon(x_k y_\ell - x_\ell y_k) = x_k x_\ell z_j - x_k x_\ell z_r$, a contradiction. $\square$

A vertex of G is called a *cut point* of G, if G has less connected components than $G_{[n]\setminus\{i\}}$. With this concept introduced, the final result regarding the minimal prime ideals can be formulated. For this purpose we may restrict ourselves to the case that G is connected, see Problem 7.8.

Theorem 7.20 *Let G be a connected graph on the vertex set $[n]$, and $W \subset [n]$. Then $P_W(G)$ is a minimal prime ideal of J_G if and only if $W = \emptyset$, or $W \neq \emptyset$ and* *for each $i \in W$ one has $c(W \setminus \{i\}) < c(W)$. In other words, this is the case, if and only if each $i \in W$ is a cut point of the graph $G_{([n]\setminus W)\cup\{i\}}$.*

Proof Assume that $P_W(G)$ is a minimal prime ideal of J_G and fix $i \in W$. Let $G_1, \ldots, G_r$ be the connected components of $G_{[n]\setminus W}$. We distinguish several cases.

Suppose that there is no edge $\{i, j\}$ of G such that $j \in G_k$ for some k. Set $T = W \setminus \{i\}$. Then the connected components of $G_{[n]\setminus T}$ are $G_1, \ldots, G_r, \{i\}$. Thus $c(T) = c(W) + 1$. However this case cannot happen, since Proposition 7.19 would imply that $P_T(G) \subset P_W(G)$.

Next suppose that there exists exactly one G_k, say G_1, for which there exists $j \in G_1$ such that $\{i, j\}$ is an edge of G. Then the connected components of $G_{[n]\setminus T}$ are $G'_1, G_2, \ldots, G_r$ where $V(G'_1) = V(G_1)\cup\{i\}$. Thus $c(T) = c(W)$. Again, this case cannot happen since Proposition 7.19 would imply that $P_T(G) \subset P_W(G)$.

It remains to consider the case that there are at least two components, say $G_1, \ldots, G_k, k \geq 2$, and $j_\ell \in G_\ell$ for $\ell = 1, \ldots, k$ such that $\{i, j_\ell\}$ is an edge of G. Then the connected components of $G_{[n]\setminus T}$ are $G'_1, G_{k+1}, \ldots, G_r$, where $V(G'_1) = \bigcup_{\ell=1}^k V(G_\ell)\cup\{i\}$. Hence in this case $c(T) < c(W)$.

Conversely, suppose that $c(W \setminus \{i\}) < c(W)$ for all $i \in W$. We want to show that $P_W(G)$ is a minimal rime ideal of J_G. Suppose this is not the case. Then there exists a proper subset $T \subset W$ with $P_T(G) \subset P_W(G)$. We choose $i \in W \setminus T$. By assumption, we have $c(W \setminus \{i\}) < c(W)$. The discussion of the three cases above shows that we may assume that $G'_1, G_{k+1}, \ldots, G_r$ are the components of $G_{[n]\setminus\{i\}}$ where $V(G'_1) = \bigcup_{\ell=1}^k V(G_\ell)\cup\{i\}$ and where $k \geq 2$. It follows that $G_{[n]\setminus T}$ has one connected component H which contains G'_1. Then $V(H) \setminus W$ contains the subsets $V(G_1)$ and $V(G_2)$. Hence $V(H) \setminus W$ is not contained in any $V(G_i)$. According to Proposition 7.19, this contradicts the assumption that $P_T(G) \subset P_W(G)$. □

The following example demonstrates Theorem 7.20.

Example 7.21 Let G be the path with n vertices. Then, for the monomial order used in Theorem 7.2, the initial terms $x_1 y_2, x_2 y_3, \ldots, x_{n-1} y_n$ of the generators form a regular sequence, and hence $\mathrm{in}_<(J_G)$ is generated by these monomials, see Corollary 1.30. In particular, $S/\mathrm{in}_<(J_G)$ is Cohen–Macaulay. This implies that S/J_G itself is Cohen–Macaulay, see Theorem 2.19. It follows from Corollary 7.18 that $\dim S/P = n + 1$ for all minimal prime ideals of J_G. Let W be any subset of $[n]$. Then Theorem 7.16 and Corollary 7.17 imply that the minimal prime ideals of J_G are exactly those prime ideals $P_W(G)$ for which $c(W) = |W| + 1$. Let $W \subset [n]$. Then there exists integers $0 \leq a_1 - 1 < b_1 < a_2 - 1 < b_2 < a_3 - 1 < b_3 < \cdots < a_r - 1 < b_r \leq n$ such that

$$W = \bigcup_{i=1}^r [a_i, b_i] \quad \text{where for each } i, \quad [a_i, b_i] = \{j \in \mathbb{Z}: a_i \leq j \leq b_i\}.$$

We see that $|W| = \sum_{i=1}^{r}(b_i - a_i + 1) = \sum_{i=1}^{r}(b_i - a_i) + r$, and that

$$c(W) = \begin{cases} r - 1, & \text{if } a_1 = 1 \text{ and } b_r = n, \\ r, & \text{if } a_1 \neq 1 \text{ and } b_r = n, \text{ or } a_1 = 1 \text{ and } b_r \neq n, \\ r + 1, & \text{if } a_1 \neq 1 \text{ and } b_r \neq n. \end{cases}$$

Thus $c(W) = |W| + 1$ if and only if $a_1 \neq 1$, $b_r \neq n$ and $a_i = b_i$ for all i. In other words, the minimal prime ideals of G are those $P_W(G)$ for which W is a subset of $[n]$ of the form $\{a_1, a_2, \ldots, a_r\}$ with $1 < a_1, a_r < n$ and $a_i < a_{i+1} - 1$ for all i.

Let $T \subset V(G)$. If each $i \in T$ is a cut point of the graph $G_{([n]\setminus T)\cup\{i\}}$, then we say that T has the *cut point property* for G. We denote by $\mathscr{C}(G)$ the set of all $T \subset V(G)$ such that T has the cut point property for G. By Theorem 7.20, $P_T(G)$ is a minimal prime ideal of J_G if and only if $T \in \mathscr{C}(G)$.

For later applications we need the following result.

Proposition 7.22 *Let G be a graph on $[n]$ and $v \in V(G)$. The following conditions are equivalent:*

(i) *There exists $T \in \mathscr{C}(G)$ such that $v \in T$;*
(ii) *v is not a simplicial vertex of $\Delta(G)$.*

Proof (i) $\Rightarrow$ (ii): Let us assume that v is a simplicial vertex and let F be the unique facet of $\Delta(G)$ such that $v \in F$. If $T \supseteq F \setminus \{v\}$, then $c(T \setminus \{v\}) > c(T)$ since, by removing v from T, we get a new connected component in $G_{[n]\setminus(T\setminus\{v\})}$, namely a trivial component which contains only the vertex v. On the other hand, if there exists $u \in (F \setminus \{v\}) \setminus T$, then u belongs to some connected component of $G_{[n]\setminus T}$. In this case, we get $c(T \setminus \{v\}) = c(T)$ since if we remove v from T, as u, v are adjacent, then v belongs to the same connected component as u in $G_{[n]\setminus(T\setminus\{v\})}$.

Therefore, in any case, we get $c(T \setminus \{v\}) \geq c(T)$, which is in contradiction to (i).

(ii) $\Rightarrow$ (i): Let us assume that $v \notin T$ for every $T \in \mathscr{C}(G)$. This implies that the indeterminates x_v and y_v do not belong to any minimal prime ideal of J_G. Consequently, x_v and y_v are regular on S/J_G. Since v is not a simplicial vertex of $\Delta(G)$, it follows that v belongs to at least two different maximal cliques of G. In particular, we may find two vertices u, w of G such that $\{u, v\}, \{v, w\} \in E(G)$ and $\{u, w\} \notin E(G)$.

It follows that $x_w(x_u y_v - x_v y_u) - x_u(x_w y_v - x_v y_w) = x_v(x_u y_w - x_w y_u) \in J_G$. As x_v is regular on S/J_G, we get $x_u y_w - x_w y_u \in J_G$ which is impossible since $\{u, w\} \notin E(G)$. □

7.2.2 Cohen–Macaulay Binomial Edge Ideals

In general it is hard to identify Cohen–Macaulay binomial edge ideals. A full classification of such ideals seems to be impossible. However for closed graphs a complete answer can be given. We first show

Proposition 7.23 *Let G be a connected graph on [n] which is closed with respect to the given labeling. Suppose further that G satisfies the condition that whenever* $\{i, j + 1\}$ *with* $i < j$ *and* $\{j, k + 1\}$ *with* $j < k$ *are edges of G, then* $\{i, k + 1\}$ *is an edge of G. Then* S/J_G *is Cohen–Macaulay.*

Proof We will show that $S/\operatorname{in}_<(J_G)$ is Cohen–Macaulay. This will then imply that S/J_G is Cohen–Macaulay as well.

Since the graph is closed, it follows from Theorem 7.2 that $\operatorname{in}_<(J_G)$ is generated by the monomials $x_i y_j$ with $\{i, j\} \in E(G)$ and $i < j$. Applying the automorphism $\varphi \colon S \to S$ which maps each x_i to x_i, and y_j to y_{j-1} for $j > 1$ and y_1 to y_n, $\operatorname{in}_<(J_G)$ is mapped to the ideal generated by all monomials $x_i y_j$ with $\{i, j+1\} \in E(G)$. This ideal has all its generators in $S' = K[x_1, \ldots, x_{n-1}, y_1, \ldots, y_{n-1}]$. Let $I \subset S'$ be the ideal generated by these monomials. Then $S/\operatorname{in}_<(J_G)$ is Cohen–Macaulay if and only if S'/I is Cohen–Macaulay. Note that I is the edge ideal of the bipartite graph Γ on the vertex set $\{x_1, \ldots, x_{n-1}, y_1, \ldots, y_{n-1}\}$, and with $\{x_i, y_j\} \in E(\Gamma)$ if and only if $\{i, j + 1\} \in E(G)$. Now we use the result from [92] that Cohen–Macaulay bipartite graphs are characterized as follows: suppose the edges of the bipartite graph can be labeled such that

(i) $\{x_i, y_i\}$ are edges for $i = 1, \ldots, n$;
(ii) if $\{x_i, y_j\}$ is an edge, then $i \le j$;
(iii) if $\{x_i, y_j\}$ and $\{x_j, y_k\}$ are edges, then $\{x_i, y_k\}$ is an edge.

Then the corresponding edge ideal is Cohen–Macaulay.

We are going to verify these conditions for our edge ideal. Condition (ii) is trivially satisfied, and condition (iii) is a consequence of our assumption that whenever $\{i, j + 1\}$ with $i < j$ and $\{j, k + 1\}$ with $j < k$ are edges of G, then $\{i, k + 1\}$ is an edge of G.

For condition (i) we have to show that $\{i, i+1\} \in E(G)$ for all i. But this follows from Problem 7.1. □

Examples 1

(a) Any complete graph satisfies the conditions of Proposition 7.23, so that S/J_G is Cohen–Macaulay. But of course this is known before because in this case J_G is the ideal of 2-minors of a generic $2 \times n$-matrix.
(b) The graph G with edges $\{1, 2\}$, $\{1, 3\}$, $\{2, 3\}$, $\{2, 4\}$, and $\{3, 4\}$ does not satisfy the conditions of Proposition 7.23. However, G is closed. But $\operatorname{in}_<(J_G)$ and J_G are not Cohen–Macaulay.
(c) A graph G need not be closed for S/J_G being Cohen–Macaulay. The tent displayed in Figure 7.3 is such an example.

Now we come the classification of closed graphs whose binomial edge ideal is Cohen–Macaulay.

Theorem 7.24 *Let G be a connected graph on [n] which is closed with respect to the given labeling. Then the following conditions are equivalent:*

(i) J_G *is unmixed;*
(ii) J_G *is Cohen-Macaulay;*

(iii) $\operatorname{in}_<(J_G)$ is Cohen-Macaulay;

(iv) G satisfies the condition that whenever $\{i, j + 1\}$ with $i < j$ and $\{j, k + 1\}$ with $j < k$ are edges of G, then $\{i, k + 1\}$ is an edge of G;

(v) there exist integers $1 = a_1 < a_2 < \cdots < a_r < a_{r+1} = n$ and a leaf order of the facets $F_1, \ldots, F_r$ of $\Delta(G)$ such that $F_i = [a_i, a_{i+1}]$ for all $i = 1, \ldots, r$.

Proof We begin by proving (i) $\Rightarrow$ (iv). By Theorem 7.7, $\Delta(G)$ has facets $F_1, \ldots, F_r$ where each facet is an interval. We may order the intervals $F_i = [a_i, b_i]$ such that $1 = a_1 < a_2 < \cdots < a_r \le b_r = n$. Since G is connected it follows that $a_{i+1} \le b_i$ for all i. Let $W = [a_r, b_{r-1}]$. Then $P_W(G)$ is a minimal prime ideal of J_G since W has the cut point property. Moreover, $c(W) = 2$, and so by Lemma 7.15, height $P_W(G) = n + (b_{r-1} - a_r + 1) - 2 = n + (b_{r-1} - a_r) - 1$. On the other hand, height $P_\emptyset(G) = n - 1$, since G is connected. Thus our assumption implies that $n + (b_{r-1} - a_r) - 1 = n - 1$ which implies that $b_{r-1} = a_r$. Let G' be the graph whose clique complex $\Delta(G')$ has the facets $F_1, \ldots, F_{r-1}$. Let $P_W(G')$ be a minimal prime ideal of G'. Then Proposition 7.22 implies that $b_{r-1} \notin W$. Therefore, $c_{G'}(W) = c_G(W)$, and hence $P_W(G)$ is a minimal prime ideal of J_G of same height as $P_W(G')$. Thus we conclude that $J_{G'}$ is unmixed as well. Induction on r concludes the proof.

In the sequence of implications (v) $\Rightarrow$ (iv) $\Rightarrow$ (iii) $\Rightarrow$ (ii) $\Rightarrow$ (i), the second follows from the proof of Proposition 7.23, and the third and the fourth are well known for any ideal.

It remains to prove (v) $\Rightarrow$ (iv). Let $i < j < k$ be three vertices of G such that $\{i, j + 1\}$ and $\{j, k + 1\}$ are edges of G. Then i and $j + 1$ belong to the same facet of $\Delta(G)$, let us say to F_ℓ. Then $k + 1$ must belong to F_ℓ as well since it is adjacent to j. Therefore, the condition from (iv) follows. $\square$

Closed graphs with Cohen-Macaulay binomial edge ideal have the following nice property.

Proposition 7.25 *Let G be a closed graph with Cohen–Macaulay binomial edge ideal, and let $<$ be the monomial order introduced in Theorem 7.2. Then $\beta_{ij}(J_G) = \beta_{ij}(\operatorname{in}_<(J_G))$ for all i and j.*

Proof We first assume that G is a connected. For a graded S-module W we denote by $B_W(s, t) = \sum_{i,j} \beta_{ij}(W)s^i t^j$ the Betti polynomial of W.

Since $\operatorname{in}_<(J_G)$ is Cohen–Macaulay, it follows from Theorem 7.24 that $[n] = \bigcup_{k=1}^r [a_k, a_{k+1}]$ with $1 = a_1 < a_2 < \ldots < a_r < a_{r+1} = n$ and such that each $G_k = G_{[a_k, a_{k+1}]}$ is a complete graph. It follows that $\operatorname{in}_<(J_G)$ is minimally generated by the set of monomials $\bigcup_{k=1}^r M_k$ where $M_k = \{x_i y_j : a_k \le i < j \le a_{k+1}\}$ for all k. Since for all $i \ne j$, the set of monomials of M_i and M_j are monomials in disjoint sets of variables, it follows that $\operatorname{Tor}_k(S/(M_i), S/(M_j)) = 0$ for all $i \ne j$ and all $k > 0$. From this we conclude that

$$B_{S/\operatorname{in}(J_G)}(s, t) = \prod_{i=1}^r B_{S/(M_i)}(s, t).$$

Since $\mathrm{Tor}_k(S/(M_i), S/(M_j)) = 0$ for all $k > 0$, and since $\mathrm{in}_<(J_{G_i}) = (M_i)$ for all i, we see that $\mathrm{Tor}_k(S/J_{G_i}, S/J_{G_j}) = 0$ for all $k > 0$ as well, see [24, Proposition 3.3]. Thus we have

$$B_{S/J_G}(s, t) = \prod_{i=1}^{r} B_{S/J_{G_i}}(s, t).$$

Hence it remains to be shown that if G is a complete graph, then $\beta_{ij}(J_G) = \beta_{ij}(\mathrm{in}(J_G))$ for all i and j. By Problem 2.15 we know that $\mathrm{in}_<(J_G)$ has a linear free S-resolution. Since for any graded ideal $\beta_{ij}(I) \leq \beta_{ij}(\mathrm{in}_<(I))$ (see Theorem 2.19), we conclude that J_G has a linear resolution. By Problem 2.11, the Betti numbers of an ideal with linear resolution are determined by the Hilbert function of the ideal. Now since, $\mathrm{Hilb}_{S/J_G}(t) = \mathrm{Hilb}_{S/\mathrm{in}_<(I)}(t)$ (see Proposition 2.6), the desired conclusion follows.

Finally assume that G is not connected, and let $G_1, \ldots, G_r$ be the connected components. Then $\mathrm{in}(J_i)$ and $\mathrm{in}(G_j)$ are monomials in distinct sets of variables. Hence we may use arguments similar as before to reduce the proof of the theorem to the case that G is connected. □

Proposition 7.25 yields

Corollary 7.26 *Let G be a closed graph with Cohen–Macaulay binomial edge ideal, and assume that $F_1, \ldots, F_r$ are the facets of $\Delta(G)$ with $k_i = |F_i|$ for $i = 1, \ldots, r$. Then the Cohen–Macaulay type of S/J_G is equal to $\prod_{i=1}^{r}(k_i - 1)$. In particular, S/J_G is Gorenstein if and only if G is a path graph.*

Proof Due to the proof of Proposition 7.25 it suffices to show that if G is a complete graph on $[n]$ (with $n \geq 2$), then the Cohen–Macaulay type of S/J_G is equal to $n - 1$. In this particular case, J_G is the ideal of 2-minors of a $2 \times n$-matrix whose resolution is given by the Eagon–Northcott complex. The type of S/J_G is the last Betti number in the resolution, which is $n - 1$. □

Problems

7.6 Let G be a graph and $v \in V(G)$. Show that $W = \{v\}$ has the cut point property for G if and only if there exists $u, w \in V(G)$ with $u, w \neq v$ such that v is in every path of G which connects u and w.

7.7 Let K_n be the complete graph on $[n]$, and let B be a domain. Show that $B[x_1, \ldots, x_n, y_1, \ldots, y_n]/P_\emptyset(K_n)B[x_1, \ldots, x_n, y_1, \ldots, y_n]$ is a domain.

7.8 Let G be a graph on $[n]$ and $G_1, \ldots, G_r$ be its connected components. Let $T \subset [n]$, and set $T_i = T \cap V(G_i)$ for $i = 1, \ldots, r$. Show

(i) $P_T(G) = \sum_{i=1}^{r} P_{T_i}(G_i)S$.

(ii) $P_T(G)$ is a minimal prime ideal of J_G if and only if each $P_{T_i}(G_i)$ is a minimal prime ideal of J_{G_i}.

7.9 Show that J_G is a prime ideal if and only if its connected components of G are complete graphs.

7.10 Let C_n be the n-cycle. Compute the minimal prime ideals of J_{C_n} and count them.

7.11 What are the minimal prime ideals of J_G when G is a complete bipartite graph?

7.12 Determine the number of minimal prime ideals of a path graph.

7.13 Give an example of a graph G for which J_G is unmixed, but S/J_G is not Cohen–Macaulay.

7.14 A graph G is called a *block graph*, if it is chordal and any two distinct maximal cliques intersect in at most one vertex. Show that if G is a block graph on $[n]$ with c connected components, depth $S/J_G = n + c$.

7.15 Let G be a block graph. Use Problem 7.14 to show that the following statements are equivalent:

(i) J_G is unmixed.
(ii) J_G is Cohen–Macaulay.
(iii) Each vertex of G belongs to at most two cliques.

7.3   On the Regularity of Binomial Edge Ideals

In general it is quite difficult to describe the resolution of a binomial ideal. The graded Betti numbers give the numerical data of the resolution and determine the regularity and projective dimension of the ideal. In this section we give lower and upper bounds of the regularity of a binomial edge ideal. But first we address the question of when a binomial edge ideal has linear relations or has a linear resolution.

7.3.1   Binomial Edge Ideals with Linear Resolution

As in the previous section we let G be a finite graph on the vertex set $[n]$, K a field, $J_G \subset S = K[x_1, \ldots, x_n, y_1, \ldots, y_n]$ the binomial edge ideal of G and $<$ the lexicographic order induced by $x_1 > x_2 > \cdots > x_n > y_1 > y_2 > \cdots > y_n$.

In what follows it is useful to note that J_G is naturally $\mathbb{Z}^n$-graded by setting $\deg x_i = \deg y_i = \epsilon_i$, where ϵ_i is the ith unit vector of $\mathbb{Z}^n$.

Theorem 7.27 *Let G be a graph. Then the following conditions are equivalent:*

(i) J_G *has a linear resolution;*
(ii) J_G *has linear relations;*
(iii) $\mathrm{in}_<(J_G)$ *is generated in degree 2 and has linear quotients;*
(iv) $\mathrm{in}_<(J_G)$ *has a linear resolution;*
(v) *G is a complete graph.*

Proof We notice that (iii) implies (iv), by Proposition 2.11, that (iv) implies (i), by Theorem 2.19, and of course (i) implies (ii).

(ii) $\Rightarrow$ (v): Let F be the graded free S-module with basis elements e_{ij} for $\{i, j\} \in E(G)$, and let $\psi: F \to J_G$ be the epimorphism with $\psi(e_{ij}) = f_{ij}$ for all $\{i, j\} \in E(G)$. We set $\deg e_{ij} = \epsilon_i + \epsilon_j$ for all $\{i, j\} \in E(G)$. Then ψ is a $\mathbb{Z}^n$-graded epimorphism, and hence $Z_1 = \mathrm{Ker}\, \psi$ is a $\mathbb{Z}^n$-graded S-module.

Assume that G is not complete. Then G contains a path over three vertices as an induced subgraph. Let $\{i, j, k\}$ be the vertices of this induced subgraph of G with edges $\{i, j\}$ and $\{j, k\}$. We may assume that $i < j < k$. We show that the degree 4 element $r = f_{ij}e_{jk} - f_{jk}e_{ij}$ of Z_1 cannot be reduced by elements of degree 3. Then we have $\beta_{1,4}(J_G) > 0$, and hence J_G does not only have linear relations which contradicts our assumption.

Indeed, the relation r has multidegree $\varepsilon_i + 2\varepsilon_j + \varepsilon_k$. If it is not a minimal relation, it must be reduced by generating relations of degree 3 involving basis elements e_{st} with $s \neq t$ and $s, t \in \{i, j, k\}$. Since the path with edges $\{i, j\}$ and $\{j, k\}$ is an induced subgraph of G, $\{i, k\}$ is not an edge of G. Thus the degree 3 relation must be a relation involving only e_{ij} and e_{jk}. But there is no such relation of degree 3, because f_{ij} and f_{jk} form a regular sequence.

(v) $\Rightarrow$ (iii): It follows from Theorem 7.10 that G is a closed graph. Hence $\mathrm{in}_<(J_G)$ is generated by the monomials x_iy_j with $1 \leq i < j \leq n$. We order the generators in lexicographical order induced by $x_1 > x_2 > \cdots > x_n > y_1 > y_2 > \cdots > y_n$. So, we have

$$x_1y_2 > x_1y_3 > \cdots > x_1y_n > x_2y_3 > x_2y_4 > \cdots > x_2y_n > \cdots > x_{n-1}y_n.$$

We let $u_1, \ldots, u_{\binom{n}{2}}$ be the generators of $\mathrm{in}_<(J_G)$ as listed above, that is, $u_1 > \cdots > u_{\binom{n}{2}}$, and claim that for each i, the ideal $(u_1, \ldots, u_{i-1}) : u_i$ is generated by a set of variables. This then implies that $\mathrm{in}_<(J_G)$ has linear quotients.

Note that the set of monomials $\{u_j / \gcd(u_j, u_i): 1 \leq j \leq i - 1\}$ is the set of monomial generators of $(u_1 \ldots, u_{i-1}) : u_i$. For each $1 \leq l \leq n - 2$, the ideal $(x_1y_2, \ldots, x_1y_n, x_2y_3 \ldots, x_2y_n, \ldots, x_ly_{l+1}, \ldots, x_ly_n) : x_{l+1}y_{l+2}$ is generated by the set $\{x_1, \ldots, x_l\}$, while for each $1 \leq l \leq n - 2$ and $l \leq t \leq n - 1$, the ideal

$$(x_1y_2, \ldots, x_1y_n, x_2y_3 \ldots, x_2y_n, \ldots, x_ly_{l+1}, \ldots, x_ly_t) : x_ly_{t+1}$$

is generated by the set $\{x_1, \ldots, x_{l-1}, y_{l+1}, \ldots, y_t\}$. This completes the proof of the implication (v) $\Rightarrow$ (iii) and proof of the theorem. $\square$

7.3.2 A Lower Bound for the Regularity

Our next goal is to present a lower and an upper bound for the regularity of a binomial edge ideal. Let $G_1, \ldots, G_r$ be the connected components of G. It is obvious that if S_i is the polynomial ring in the indeterminates indexed by the vertex set of G_i, then $S/J_G \cong \bigotimes_{i=1}^r (S_i/J_{G_i})$. Therefore, $\mathrm{reg}(S/J_G) = \sum_{i=1}^r \mathrm{reg}(S_i/J_{G_i})$. This equality shows that it is enough to consider connected graphs. The reader may then easily derive the bounds for the regularity of binomial edge ideals for arbitrary graphs.

Let G be a graph. An *induced path* of G is defined to be an induced subgraph of G which is isomorphic to a path graph.

Theorem 7.28 *Let G be a connected graph on $[n]$, and let ℓ be the length of the longest induced path of G. Then we have:*

(a) $\mathrm{reg}(J_G) \geq \ell + 1$.
(b) *If G is closed, then $\mathrm{reg}(J_G) = \mathrm{reg}(\mathrm{in}_<(J_G)) = \ell + 1$.*

Proof (a) We will use that

$$\beta_{ij}(J_G) \geq \beta_{ij}(J_{G_W})$$

for any subset $W \subset [n]$, where G_W is the induced subgraph on W. This inequality is an immediate consequence of the subsequent Lemma 7.30.

Now let P be an induced path of G of length ℓ. The binomial edge ideal of any path is generated by a regular sequence of binomials of degree 2 (see Problem 7.3). Thus, $\mathrm{reg}(J_P) = \ell + 1$. Since $\mathrm{reg}(J_G) \geq \mathrm{reg}(J_P)$, the desired result follows.

The proof of (b) of needs more preparations and will be postponed.   □

Remark 7.29 In the proof of part (a) we did not use that G is connected. Actually, if G has the connected components $G_1, \ldots, G_r$ with longest induced paths of length $\ell_1, \ldots, \ell_r$, respectively, then (a) can be improved to obtain: $\mathrm{reg}(J_G) \geq \ell_1 + \cdots + \ell_r + 1$. For the proof one uses the result of part (a) and arguments as in the proof of Proposition 7.25.

Let $\mathbf{a} = (a_1, \ldots, a_n) \in \mathbb{Z}^n$, we define the *support* of $\mathbf{a}$ as the set

$$\mathrm{supp}(\mathbf{a}) = \{i : a_i \neq 0\}.$$

Lemma 7.30 *Let $W \subset [n]$. Then for all $\mathbf{a} \in \mathbb{Z}^n$ with $\mathrm{supp}(\mathbf{a}) \subset W$ one has*

$$\beta_{i,\mathbf{a}}(J_G) = \beta_{i,\mathbf{a}}(J_{G_W}).$$

Proof Let

$$\mathbb{F} : \cdots \to \bigoplus_{\mathbf{a} \in \mathbb{Z}^n} S(-\mathbf{a})^{\beta_{i\mathbf{a}}} \to \cdots \to \bigoplus_{\mathbf{a} \in \mathbb{Z}^n} S(-\mathbf{a})^{\beta_{1\mathbf{a}}} \to S \to 0$$

be the $\mathbb{Z}^n$-graded minimal free resolution of S/J_G, and consider the subcomplex

$$\mathbb{F}': \quad \cdots \rightarrow \bigoplus_{\substack{\mathbf{a} \in \mathbb{Z}^n \\ \mathrm{supp}(\mathbf{a}) \subset W}} S(-\mathbf{a})^{\beta_{i\mathbf{a}}} \rightarrow \cdots \rightarrow \bigoplus_{\substack{\mathbf{a} \in \mathbb{Z}^n \\ \mathrm{supp}(\mathbf{a}) \subset W}} S(-\mathbf{a})^{\beta_{1\mathbf{a}}} \rightarrow S \rightarrow 0.$$

We claim that $\mathbb{F}'$ is a minimal $\mathbb{Z}^n$-graded free resolution of S/J_{G_W}. It is clear that J_{G_W} is the image of $\bigoplus_{\substack{\mathbf{a} \in \mathbb{Z}^n \\ \mathrm{supp}(\mathbf{a}) \subset W}} S(-\mathbf{a})^{\beta_{1\mathbf{a}}} \rightarrow S$. Next we show that $\mathbb{F}'$ is acyclic. To prove this, it suffices to show that the $\mathbb{Z}^n$-graded component $\mathbb{F}'_{\mathbf{a}}$ is acyclic for any $\mathbf{a} \in \mathbb{Z}^n$ with $\mathrm{supp}(\mathbf{a}) \subset W$. Indeed, let $\mathbf{a} \in \mathbb{Z}^n$ with $\mathrm{supp}(\mathbf{a}) \subset W$. Since, for any $\mathbf{b} \in \mathbb{Z}^n$, $S(-\mathbf{b})_{\mathbf{a}}$ is nonzero if and only if all components of $\mathbf{a} - \mathbf{b}$ are nonnegative, it follows that $\mathbb{F}_{\mathbf{a}} = F'_{\mathbf{a}}$. This implies that $\mathbb{F}'_{\mathbf{a}}$ is acyclic. Finally, $\mathbb{F}'$ is minimal since $\mathbb{F}$ is minimal. □

We now turn to the proof of Theorem 7.28(b). For this purpose we have to better understand the initial ideal of J_G when G is closed. We may assume that G is closed with respect to the given labeling of the vertices. In that case we have that

$$\mathrm{in}_<(J_G) = (x_i y_j : \{i, j\} \in E(G))$$

is the edge ideal of a bipartite graph on the vertex set $\{x_1, \ldots, x_n\} \cup \{y_1, \ldots, y_n\}$. We denote this bipartite graph by $\mathrm{in}_<(G)$.

A finite simple graph H is called *weakly chordal* if every induced cycle of H and of the complementary graph $\overline{H}$ has length at most 4.

Lemma 7.31 *Let G be a connected closed graph on $[n]$. Then the bipartite graph $\mathrm{in}_<(G)$ is weakly chordal.*

Proof We set $H = \mathrm{in}_<(G)$. That $\overline{H}$ has no induced cycle of length ≥ 5 is easy to see. Indeed, this is due to the fact that $\overline{H}$ consists of two complete graphs, say K_n^x on the vertex set $\{x_1, \ldots, x_n\}$ and K_n^y on the vertex set $\{y_1, \ldots, y_n\}$, together with the edges $\{x_i y_j : i \geq j\} \cup \{x_i y_j : i < j, \{i, j\} \notin E(G)\}$. Hence, if C is an induced cycle of $\overline{H}$ of length ≥ 5, then C contains at least three vertices either from K_n^x, or from K_n^y. Thus it cannot be an induced cycle of $\overline{H}$.

It remains to be shown that H has no induced cycle of length ≥ 5. Assume that this is not the case. Then there exist an integer $k \geq 3$ and an induced cycle C of H with vertices $x_{i_1}, y_{j_1}, \ldots, x_{i_k}, y_{j_k}$ (labeled clockwise). Then $\{i_\ell, j_\ell\}$ and $\{i_{\ell+1}, j_\ell\}$ are edges of G for $1 \leq \ell \leq k$, where we made the convention that $i_{k+1} = i_1$. Furthermore, since G is closed and since $i_{\ell+1} < j_\ell$, $j_{\ell+1}$ and $j_\ell > i_\ell, i_{\ell+1}$ we also have that $\{i_\ell, i_{\ell+1}\}$ and $\{j_\ell, j_{\ell+1}\}$ are edges of G for $\ell = 1, \ldots, k$.

We may assume that $i_1 < i_2$. Suppose there exists ℓ such that $i_\ell < j_{\ell+1} < j_\ell$. Since $\{i_\ell, j_\ell\}$ and $\{j_\ell, j_{\ell+1}\}$ are edges of G and since G is closed, it follows that $\{i_\ell, j_{\ell+1}\} \in E(G)$ which implies that $\{x_{i_\ell}, y_{j_{\ell+1}}\} \in E(H)$. This is a contradiction, since C is an induced subgraph of H. Similarly, $i_{\ell+1} < i_\ell < j_\ell$ is impossible. Therefore, for all ℓ, we must have either $i_\ell < i_{\ell+1} < j_\ell < j_{\ell+1}$ or $i_{\ell+1} < j_{\ell+1} \leq i_\ell < j_\ell$. As $i_1 < i_2$, we may choose t to be the largest index such that $i_t < i_{t+1}$.

Thus, we get $i_t < i_{t+1} < j_t < j_{t+1}$ and $i_{t+2} < j_{t+2} \leq i_{t+1} < j_{t+1}$, which implies that $i_{t+2} < j_{t+2} \leq i_{t+1} < j_t < j_{t+1}$. Since $\{i_{t+2}, j_{t+1}\}$ and $\{j_t, j_{t+1}\}$ are edges of G and G is closed, we obtain $\{i_{t+2}, j_t\} \in E(G)$ which leads to $\{x_{i_{t+2}}, y_{j_t}\} \in E(H)$, again contradicting the assumption that C is an induced cycle. $\square$

Let Γ be an arbitrary simple graph. An *induced matching* of Γ is an induced subgraph of Γ which consists of pairwise disjoint edges. The *induced matching number* of Γ, denoted indmatch(Γ), is the number of edges in a largest induced matching of Γ.

The following result is crucial for the proof of Theorem 7.28(b).

Proposition 7.32 *Let G be a connected graph on $[n]$ which is closed with respect to the given labeling. Then*

$$\text{indmatch(in}_<(G)) = \ell,$$

where ℓ is the length of the longest induced path of G.

Proof We set $H = \text{in}_<(G)$. First we show that indmatch(H) $\geq \ell$. This follows easily since it is obvious that if $i_0, \ldots, i_\ell$ is an induced path in G of length ℓ, then the edges

$$\{x_{i_0}, y_{i_1}\}, \{x_{i_1}, y_{i_2}\}, \ldots, \{x_{i_{\ell-1}}, y_{i_\ell}\}$$

form an induced subgraph of H.

We show now that indmatch(H) $\leq \ell$. Let indmatch(H) $= m$. Then H has m pairwise disjoint edges $\{x_{i_1}, y_{j_1}\}, \ldots, \{x_{i_m}, y_{j_m}\}$ that form an induced subgraph of H. To show the desired inequality we construct a path of length m in G.

As G is closed, we may assume, as we have seen in Theorem 7.7, that all the facets of the clique complex of G are intervals. We denote by $\mathscr{D}$ the set of induced matchings of H of the form $\{x_{i'_1}, y_{j_1}\}, \ldots, \{x_{i'_m}, y_{j_m}\}$, where we fix $y_{j_1}, \ldots, y_{j_m}$, and define a partial order on $\mathscr{D}$ by setting

$$\{x_{i'_1}, y_{j_1}\}, \ldots, \{x_{i'_m}, y_{j_m}\} \leq \{x_{i''_1}, y_{j_1}\}, \ldots, \{x_{i''_m}, y_{j_m}\},$$

if and only if $i'_k \leq i''_k$ for $k = 1, \ldots, m$.

Since $\mathscr{D}$ is a non-empty finite set, we may choose a minimal element in $\mathscr{D}$ which we may call again $\{x_{i_1}, y_{j_1}\}, \ldots, \{x_{i_m}, y_{j_m}\}$. After a reordering of the edges of this induced matching we may further assume that $i_1 < i_2 < \cdots < i_m$. Then $i_s \geq s$ for all s, and hence by construction it follows that

$$\{x_{i_1}, y_{j_1}\}, \ldots, \{x_{i_m}, y_{j_m}\}$$

is an induced matching, satisfying the following condition:

(∗)   for all $1 \leq s \leq m$, if $t < s$ and $\{t, j_s\} \in E(G)$,   then

$$\{x_{i_1}, y_{j_1}\}, \ldots, \{x_t, y_{j_s}\}, \{x_{i_{s+1}}, y_{j_{s+1}}\}, \ldots, \{x_{i_m}, y_{j_m}\}$$

is not an induced matching of H.

Note that we also have $j_t \le i_{t+1}$ for all $1 \le t \le m - 1$. Indeed, if there exists t such that $j_t > i_{t+1}$, then it follows that $i_t < i_{t+1} < j_t$. We obtain $\{i_{t+1}, j_t\} \in E(G)$ and $\{x_{i_{t+1}}, y_{j_t}\} \in E(H)$, a contradiction to our hypothesis.

Next we show that, under condition $(*)$ for the induced subgraph $\{x_{i_1}, y_{j_1}\}$, $\ldots, \{x_{i_m}, y_{j_m}\}$ of H, we have:

(i) i_s and i_{s+1} belong to the same clique of G for all $1 \le s \le m - 1$,
(ii) i_s, i_{s+1}, i_{s+2} do not belong to the same clique for any $1 \le s \le m - 2$.

Let us assume that we have already shown (i) and (ii). Then $L : i_1, i_2, \ldots, i_m, j_m$ is an induced path of G. Indeed, by (i), L is a path in G. Next, it is clear that we cannot have an edge $\{i_s, i_q\} \in E(G)$ with $q - s \ge 2$ by (ii). In addition, $\{i_s, j_m\} \notin E(G)$ for any $1 \le s \le m - 1$, since otherwise it follows that $\{x_{i_s}, y_{j_m}\} \in E(H)$ because $i_s < i_m < j_m$. This is a contradiction. Therefore, L is an induced path of G.

Let us first prove (ii). Suppose that there are three consecutive vertices i_s, i_{s+1}, i_{s+2} in L which belong to the same clique of G. Hence $\{i_s, i_{s+2}\} \in E(G)$. As $i_s < j_s \le i_{s+1} < j_{s+1} \le i_{s+2} < j_{s+2}$, we also have $\{i_s, j_{s+1}\} \in E(G)$, which is impossible.

Finally, we show (i). Let us assume that there exists s such that i_s and i_{s+1} do not belong to the same clique of G, in other words, $\{i_s, i_{s+1}\} \notin E(G)$. In particular, we have $i_s < j_s < i_{s+1}$. We need to consider the following two cases.

Case (a). $\{j_s, i_{s+1}\} \in E(G)$. We claim that

$$\{x_{i_1}, y_{j_1}\}, \ldots, \{x_{i_s}, y_{j_s}\}, \{x_{j_s}, y_{i_{s+1}}\}, \{x_{i_{s+1}}, y_{j_{s+1}}\}, \ldots, \{x_{i_m}, y_{j_m}\}$$

is an induced subgraph of H with pairwise disjoint edges. This will lead to a contradiction since indmatch$(H) = m$. To prove our claim, we note that $\{x_{j_s}, y_{j_{s+1}}\} \notin E(H)$ by $(*)$ and $\{x_{i_s}, y_{i_{s+1}}\} \notin E(H)$ since $\{i_s, i_{s+1}\} \notin E(G)$. Moreover, if $\{x_{i_q}, y_{i_{s+1}}\} \in E(H)$ for some $q < s$, then, as we have $i_q < i_s < j_s < i_{s+1}$, we get $\{i_q, j_s\} \in E(G)$, thus $\{x_{i_q}, y_{j_s}\} \in E(H)$, a contradiction. Similarly, if $\{x_{j_s}, y_{j_q}\} \in E(H)$ for some $q \ge s + 2$, as $j_s < i_{s+1} < i_q < j_q$, we get $\{i_{s+1}, j_q\} \in E(G)$, that is, $\{x_{i_{s+1}}, y_{j_q}\} \in E(H)$, again a contradiction.

Case (b). $\{j_s, i_{s+1}\} \notin E(G)$. Let then $j = \min\{t : \{t, i_{s+1}\} \in E(G)\}$. Since G is closed, we must have $j > j_s > i_s$. Let us consider the following disjoint edges of H :

$$\{x_{i_1}, y_{j_1}\}, \ldots, \{x_{i_s}, y_{j_s}\}, \{x_j, y_{i_{s+1}}\}, \{x_{i_{s+1}}, y_{j_{s+1}}\}, \ldots, \{x_{i_m}, y_{j_m}\}.$$

These edges determine an induced subgraph of H, which leads again to a contradiction to the fact that indmatch$(H) = m$. Indeed, since $j < i_{s+1}$, it follows that $\{j, j_{s+1}\} \notin E(G)$. As in the previous case, we get $\{x_{i_q}, y_{i_{s+1}}\} \notin$

$E(H)$ for $q < s$. Let us assume that $\{x_j, y_{j_q}\} \in E(H)$ for some $q \geq s+2$. Then $\{j, j_q\} \in E(G)$ and since $j < i_{s+1} < j_{s+1} < j_q$, we get $\{i_{s+1}, j_q\} \in E(G)$ or, equivalently, $\{x_{i_{s+1}}, y_{j_q}\} \in E(H)$, impossible. $\square$

Proof (of Theorem 7.28(b)) We use the fact, shown by Woodroofe [218, Theorem 14], that $\operatorname{reg}(I(\Gamma)) = \operatorname{indmatch}(\Gamma)+1$ for any weakly chordal graph Γ. Thus, together with part (a) of Theorem 7.28, Lemma 7.31 and Proposition 7.32 we get

$$\ell + 1 \leq \operatorname{reg}(J_G) \leq \operatorname{reg}(\operatorname{in}_<(J_G)) = \operatorname{reg}(I(H)) = \operatorname{indmatch}(H) + 1 = \ell + 1.$$

where $H = \operatorname{in}_<(G)$. This concludes the proof. $\square$

As a consequence of Theorem 7.28 we obtain the following upper bound for the regularity of J_G when G is a closed graph.

Corollary 7.33 *Let G be a closed graph, and let m be the number of maximal cliques of G. Then $\operatorname{reg}(J_G) \leq m + 1$.*

Proof Two different edges of an induced path of G cannot belong to the same clique. It follows that $\ell \leq m$, where ℓ is the length of the longest induced path of G. Thus Theorem 7.28(b) yields the desired conclusion. $\square$

Example 7.34 In general the inequality given in Corollary 7.33 may be strict. For example let G be the graph whose cliques are given by the intervals $[1, 3]$, $[2, 4]$, $[3, 5]$ and $[4, 6]$. Then $m = 4$. A longest induced path of G is $1, 2, 4, 5$. Its length is $\ell = 3$.

Corollary 7.35 *Let G be a closed graph on $[n]$. Then $\operatorname{reg}(J_G) \leq n$, and equality holds if and only if G is a path graph.*

Proof As before we denote by ℓ the length of the longest induced subgraph of G. By Theorem 7.28, $\operatorname{reg}(J_G) = \ell + 1$, and obviously $\ell + 1 \leq n$. If G is a path graph, then $\ell + 1 = n$. On the other hand, if $\operatorname{reg}(J_G) = n$ and if, as before, m denotes the number of maximal cliques of G, then Corollary 7.33 implies that $n = \operatorname{reg}(J_G) \leq m+1 \leq n$, so that $m + 1 = n$. This is only possible if G is a path graph. $\square$

7.3.3 An Upper Bound for the Regularity

Surprisingly, Corollary 7.35 holds true without the assumption that G is closed. Indeed, one has

Theorem 7.36 *Let G be a graph on $[n]$. Then*

(a) $\operatorname{reg}(\operatorname{in}_<(J_G)) \leq n$. *In particular, $\operatorname{reg}(J_G) \leq n$.*
(b) $\operatorname{reg}(J_G) = n$ *if and only if G is a path.*

The proof of the theorem requires some preparations.

We consider the $\mathbb{Z}^{2n}$-grading of $S = K[x_1, \ldots, x_n, y_1, \ldots, y_n]$ defined by $\deg x_i = \mathbf{e}_i$ and $\deg y_i = \mathbf{e}_{i+n}$. Binomial edge ideals are of course not $\mathbb{Z}^{2n}$-graded, but monomial ideals in S are $\mathbb{Z}^{2n}$-graded. To simplify the notation, we often identify the multidegree $(\mathbf{a}, \mathbf{b}) = (a_1, \ldots, a_n, b_1, \ldots, b_n) \in \mathbb{Z}^{2n}$ with the monomial

$$\mathbf{x}^{\mathbf{a}}\mathbf{y}^{\mathbf{b}} = x_1^{a_1} \cdots x_n^{a_n} y_1^{b_1} \cdots y_n^{b_n}.$$

For a $\mathbb{Z}^{2n}$-graded S-module M, we write

$$\beta_{i,\mathbf{x}^{\mathbf{a}}\mathbf{y}^{\mathbf{b}}}(M) = \beta_{i,(\mathbf{a},\mathbf{b})}(M),$$

and we set

$$P_M(t) = \sum_{k=0}^{2n} \sum_{(\mathbf{a},\mathbf{b})\in\mathbb{Z}^{2n}} \beta_{k,(\mathbf{a},\mathbf{b})}(M)\mathbf{x}^{\mathbf{a}}\mathbf{y}^{\mathbf{b}}t^k$$

for the $\mathbb{Z}^{2n}$-graded Poincaré series of M.

In what follows we shall need to follow general technical result.

Lemma 7.37 *Let $u_1, \ldots, u_g$ be monomials in S and $I = (u_1, \ldots, u_g)$. Then*

$$P_{S/I}(t) \le 1 + \sum_{u_j \notin (u_1,\ldots,u_{j-1})} P_{S/(u_1,\ldots,u_{j-1}):u_j}(t)u_j t,$$

where the inequality is understood to be coefficientwise.

Proof The assertion follows from the short exact sequences

$$0 \to S/\big((u_1,\ldots,u_{j-1}):u_j\big) \xrightarrow{u_j} S/(u_1,\ldots,u_{j-1}) \to S/(u_1,\ldots,u_j) \to 0$$

for $j = 2, 3, \ldots, g$, by applying mapping cones.   □

We call a path $\pi : s = i_0, i_1, \ldots, i_r = t$ of G *weakly admissible* (w-admissible, for short), if $s < t$ and, for $k = 1, 2, \ldots, r - 1$, one has either $i_k < s$ or $i_k > t$. The vertices s and t are called the *ends* of π and the vertices $i_1, \ldots, i_{r-1}$ are called the *inner vertices* of π.

For an w-admissible path $\pi : s = i_0, i_1, \ldots, i_r = t$, we define the monomial

$$v_\pi = \left(\prod_{v_k < s} y_{v_k}\right)\left(\prod_{v_k > t} x_{v_k}\right) x_s y_t.$$

Let $\mathscr{P}(G)$ be the set of all w-admissible paths of G, and let $<$ be the lexicographic order induced by $x_1 > \cdots > x_n > y_1 > \cdots > y_n$. As a consequence of Theorem 7.11 we have

Lemma 7.38 $\mathrm{in}_<(J_G) = (v_\pi : \pi \in \mathscr{P}(G))$.

The set of generators of $\mathrm{in}_<(J_G)$ given here may be not minimal because for w-admissible paths the binomials $u_\pi f_{ij}$ need not to form a reduced Gröbner basis.

The following result is crucial for the proof of part (a) of Theorem 7.36.

Lemma 7.39 *Let* $\pi : s = i_0, \ldots, i_r = t$ *be a w-admissible path and let* $1 \leq k \leq r - 1$. *Then following holds:*

(a) *If* $i_k < s$, *then there is an* $\ell > k$ *such that the path* $\pi' : i_k, i_{k+1}, \ldots, i_\ell$ *is a w-admissible path of G and* $v_{\pi'}$ *divides* $x_{i_k} v_\pi$.
(b) *If* $i_k > t$, *then there is an* $\ell < k$ *such that* $\pi' : i_\ell, i_{\ell+1}, \ldots, i_k$ *is a w-admissible path of G and* $v_{\pi'}$ *divides* $y_{v_k} v_\pi$.

Proof

(a) Let $\ell > k$ be the smallest integer satisfying $i_k < i_\ell \leq t$. Then the path $\pi' : i_k, i_{k+1}, \ldots, i_\ell$ satisfies the desired condition.
(b) is proved similarly.   □

We call a path π', satisfying for π condition (a) or (b) in Lemma 7.39, a *wedge* of π at i_k.

Let $g = |\mathscr{P}(G)|$, we now fix an ordering

$$\pi_1, \pi_2, \ldots, \pi_g$$

of the admissible paths of G, such that if the length of π_i is smaller than that of π_j then $i < j$. To simplify the notation, we write

$$v_k = v_{\pi_k}$$

for $k = 1, 2, \ldots, g$. Then $\mathrm{in}_<(J_G) = (v_1, \ldots, v_g)$. By the choice of the ordering, if π_i is a wedge of π_j then $i < j$. This fact immediately implies the following property.

Lemma 7.40 *Let* $1 < j \leq g$ *and let* s *and* t *be the ends of* π_j *with* $s < t$. *For any inner vertex* k *of* π_j, *one has* $x_k \in (v_1, \ldots, v_{j-1}) : v_j$ *if* $k < s$ *and* $y_k \in (v_1, \ldots, v_{j-1}) : v_j$ *if* $k > t$.

For a monomial $w \in S$, let

$$\mathrm{mult}(w) = \{k \in [n]: x_k y_k \text{ divides } w\}.$$

Note that, for a squarefree monomial $w \in S$, one has $\deg w \leq n + |\mathrm{mult}(w)|$.

The following proposition together with Theorem 2.19 yields the proof of Theorem 7.36(a).

Proposition 7.41 *For any monomial $w \in S$ and an integer $p > 0$, one has*

$$\beta_{p,w} (S/\operatorname{in}_<(J_G)) = 0 \quad if \quad |\operatorname{mult}(w)| \geq p.$$

In particular, $\operatorname{reg}(\operatorname{in}_<(J_G)) \leq n$.

Proof The second statement follows from the first statement together with the fact that the multigraded Betti numbers of a squarefree monomial ideal are concentrated in squarefree degrees.

In order to prove the first statement we first introduce the following definition. Set $M = \{v_1, v_2, \ldots, v_g\}$. We say that a subset $F = \{v_{i_1}, v_{i_2}, \ldots, v_{i_k}\} \subset M$ with $i_1 < \cdots < i_k$ is a *Lyubeznik subset* of M (of size k) if, for $j = 1, 2, \ldots, k$, any monomial v_ℓ with $\ell < i_j$ does not divide $\operatorname{lcm}(v_{i_j}, v_{i_{j+1}}, \ldots, v_{i_k})$.

The theorem will be a consequence of the following two claims.

Claim 1 Let $F = \{v_{i_1}, \ldots, v_{i_k}\}$ be a Lyubeznik subset of M. Then we have:

(i) $\operatorname{mult}(\operatorname{lcm}(F))$ contains no inner vertices of π_{i_1}.
(ii) If $\operatorname{mult}(\operatorname{lcm}(F))$ contains no inner vertices of π_{i_j} for $j = 2, 3, \ldots, k$, then

$$|\operatorname{mult}(\operatorname{lcm}(F))| \leq k - 1.$$

Claim 2 Let $F = \{v_{i_1}, \ldots, v_{i_k}\}$ be a Lyubeznik subset of M and w a monomial of S. Let $p > 0$ be an integer. Suppose that

(i) $\beta_{p,w}(S/((v_1, \ldots, v_{i_1-1}) : v_{i_1} \cdots v_{i_k})) \neq 0$, and that
(ii) $\operatorname{mult}(w \cdot \operatorname{lcm}(F))$ contains no inner vertices of π_{i_δ} for $\delta = 2, 3, \ldots, k$.

Then there is a Lyubeznik subset $\widetilde{F} = \{v_{j_1}, \ldots, v_{j_\ell}\}$ of M and a monomial $\widetilde{w}$ such that

(i') $\beta_{p-1,\widetilde{w}}(S/((v_1, \ldots, v_{j_1-1}) : v_{j_1} \cdots v_{j_\ell})) \neq 0$,
(ii') $\operatorname{mult}(\widetilde{w} \cdot \operatorname{lcm}(\widetilde{F}))$ contains no inner vertices of π_{j_δ} for $\delta = 2, 3, \ldots, \ell$, and
(iii') $|\operatorname{mult}(\widetilde{w} \cdot \operatorname{lcm}(\widetilde{F}))| - |\widetilde{F}| = |\operatorname{mult}(w \cdot \operatorname{lcm}(F))| - |F| - 1$.

We first show that these claims yield the desired result. Let $u \in S$ be a monomial such that $\beta_{p,u}(S/\operatorname{in}_<(J_G)) \neq 0$ with $p > 0$. We show that there is a Lyubeznik subset F such that

$$|\operatorname{mult}(u)| = |\operatorname{mult}(\operatorname{lcm}(F))| - |F| + p, \tag{7.1}$$

and that F satisfies the assumption of Claim 1(ii).

Note that this proves the desired statement by Claim 1(ii).

Recall $\operatorname{in}_<(J_G) = (v_1, \ldots, v_g)$. By Lemma 7.37, there is a Lyubeznik subset $\{v_j\}$ of size 1 such that $\beta_{p-1,u/v_j}(S/((v_1, \ldots, v_{j-1}) : v_j)) \neq 0$. If $p = 1$, then $u = v_j$, and the set $\{v_j\}$ has the desired property (7.1). Suppose $p > 1$. Then the pair of the Lyubeznik set $\{v_j\}$ and a monomial u/v_j satisfies the assumption (i) and (ii) of Claim 2. Thus, by applying Claim 2 repeatedly, one obtains a Lyubeznik subset $F = \{v_{i_1}, \ldots, v_{i_k}\}$ and a monomial w such that

$$\beta_{0,w}(S/((v_1, \ldots, v_{i_1-1}) : v_{i_1} \cdots v_{i_k})) \neq 0, \text{ and}$$

$$|\operatorname{mult}(w \cdot \operatorname{lcm}(F))| - |F| = |\operatorname{mult}(u)| - p.$$

The first condition says that $w = x^0 y^0$, where $\mathbf{0} = (0, \ldots, 0)$, and the second condition proves that F satisfies the (7.1).

It remains to prove Claims 1 and 2.

Proof of Claim 1: (i) Suppose to the contrary that there is an inner vertex k of π_{i_1} which belongs to $\operatorname{mult}(\operatorname{lcm}(F))$. Let π_j be a wedge of π_{i_1} at k. Then $j < i_1$ and v_j divides $\operatorname{lcm}(v_{i_1}, \ldots, v_{i_k})$ by Lemma 7.39. This contradicts the definition of Lyubeznik sets.

(ii) Let $s_1, t_1, s_2, t_2, \ldots, s_k, t_k$ be the ends of $\pi_{i_1}, \ldots, \pi_{i_k}$, where $s_j < t_j$ for all j. By (i) and the assumption, $\operatorname{mult}(\operatorname{lcm}(F))$ contains no inner vertices of π_{i_j} for all j. Hence

$$|\operatorname{mult}(\operatorname{lcm}(F))| \leq |\operatorname{mult}(x_{s_1} y_{t_1} x_{s_2} y_{t_2} \cdots x_{s_k} y_{t_k})| \leq k - 1,$$

where the last inequality follows from the fact that $s_1 < t_1, \ldots, s_k < t_k$.

Proof of Claim 2: We consider two cases.

Case 1: Suppose that $\operatorname{mult}(w \cdot \operatorname{lcm}(F))$ contains an inner vertex k of π_{i_1}. We may suppose that x_k divides v_{i_1} (the case that y_k divides v_{i_1} is similar). Since by Claim 1(i), y_k does not divide $\operatorname{lcm}(F)$, it follows that y_k divides w. Then, as $y_k \in (v_1, \ldots, v_{i_1-1}) : v_{i_1} \cdots v_{i_k}$, Lemma 7.40, implies that $\beta_{p,w}(S/((v_1, \ldots, v_{i_1-1}) : v_{i_1} \cdots v_{i_k})) \neq 0$ if and only if $\beta_{p-1,w/y_k}(S/((v_1, \ldots, v_{i_1-1}) : v_{i_1} \cdots v_{i_k})) \neq 0$. Then the pair of the set $\widetilde{F} = F$ and the monomial $\widetilde{w} = w/y_k$ satisfies (i'), (ii'), and (iii'), as desired.

Case 2: Suppose that $\operatorname{mult}(w \cdot \operatorname{lcm}(F))$ contains no inner vertices of π_{i_1}. For $j = 1, 2, \ldots, i_1 - 1$, let

$$\overline{v}_j = \frac{v_j}{\gcd(v_j, v_{i_1} \cdots v_{i_k})}.$$

Then

$$(\overline{v}_1, \ldots, \overline{v}_{i_1-1}) = (v_1, \ldots, v_{i_1-1}) : v_{i_1} \cdots v_{i_k}.$$

By Lemma 7.37 and (i), there is an $1 \leq i_0 < i_1$ such that $\overline{v}_{i_0} \notin (\overline{v}_1, \ldots, \overline{v}_{i_0-1})$ and

$$\beta_{p-1,w/\overline{v}_{i_0}}\left(S/((\overline{v}_1, \ldots, \overline{v}_{i_0-1}) : \overline{v}_{i_0})\right) \neq 0. \tag{7.2}$$

Let $\widetilde{w} = w/\overline{v}_{i_0}$ and $\widetilde{F} = \{v_{i_0}, v_{i_1} \dots, v_{i_k}\}$. Since, for $\ell < i_0$, $\overline{v}_\ell$ divides $\overline{v}_{i_0}$ if and only if v_ℓ divides $\mathrm{lcm}(v_{i_0}, v_{i_1}, \dots, v_{i_k})$, it follows that $\widetilde{F}$ is a Lyubeznik subset. Also, since

$$(\overline{v}_1, \dots, \overline{v}_{i_0-1}) : \overline{v}_{i_0} = (v_1, \dots, v_{i_0-1}) : v_{i_0} v_{i_1} \cdots v_{i_k},$$

(7.2) and the fact $w \cdot \mathrm{lcm}(F) = \widetilde{w} \cdot \mathrm{lcm}(\widetilde{F})$ say that the pair $\widetilde{F}$ and $\widetilde{w}$ satisfies (i'), (ii'), and (iii'), as desired. $\square$

We now turn to the proof of part (b) of Theorem 7.36. For that purpose we introduce some terminology and notation regarding graphs. Let G be a graph on $[n]$ and v a vertex of G. The set $N_G(v) = \{w : \{v, w\} \in E(G)\}$ is called the *neighborhood* of v. The *degree* of v, denoted $\deg_G v$, is the cardinality of $N(v)$.

Let $\{e_1, \dots, e_t\}$ be a set of edges of G. By $G \setminus \{e_1, \dots, e_t\}$, we mean the graph on the same vertex set as G in which the edges $e_1, \dots, e_t$ are omitted. Here, for an edge e of G, we simply write $G \setminus e$, instead of $G \setminus \{e\}$.

Let v, w be two distinct vertices of G, and assume that $e = \{v, w\}$ is not an edge of G. Then we denote by $G \cup e$ the graph on the same vertex set as G and with edge set $E(G \cup e) = E(G) \cup \{e\}$. Moreover, we let G_e be the graph on $[n]$ with edge set

$$E(G_e) = E(G) \cup E(G_1) \cup E(G_2),$$

where G_1 is the complete graph on $N_G(v)$ and G_2 is the complete graph on $N_G(w)$.

For an edge $e = \{i, j\}$ of G we denote the binomial $f_{ij} = x_i y_j - x_j y_i$ also be f_e. Inductive arguments will be used to prove Theorem 7.36(b). This requires the following technical results. The first of these results, Proposition 7.42, follows by standard arguments by considering the exact sequence

$$0 \longrightarrow S/(J_{H\setminus e} : f_e)(-2) \overset{f_e}{\longrightarrow} S/J_{H\setminus e} \longrightarrow S/J_H \to 0.$$

Proposition 7.42 *Let H be a graph and e be an edge of H. Then we have*

(a) $\mathrm{reg}(J_H) \le \max\{\mathrm{reg}(J_{H\setminus e}), \mathrm{reg}(J_{H\setminus e} : f_e) + 1\}$;
(b) $\mathrm{reg}(J_{H\setminus e}) \le \max\{\mathrm{reg}(J_H), \mathrm{reg}(J_{H\setminus e} : f_e) + 2\}$;
(c) $\mathrm{reg}(J_{H\setminus e} : f_e) + 2 \le \max\{\mathrm{reg}(J_{H\setminus e}), \mathrm{reg}(J_H) + 1\}$.

A reference for the proof of the next result is given in the notes at the end of this chapter.

Theorem 7.43 *Let G be a graph and $e = \{i, j\}$ be an edge of G. Then*

$$J_{G\setminus e} : f_e = J_{(G\setminus e)_e} + I_{G,e},$$

where $I_{G,e} = (g_{\pi,t} : \pi : i, i_1, \dots, i_s, j$ is a path between i, j in G and $0 \le t \le s)$ with $g_{\pi,0} = x_{i_1} \cdots x_{i_s}$ and $g_{\pi,t} = y_{i_1} \cdots y_{i_t} x_{i_{t+1}} \cdots x_{i_s}$ for $1 \le t \le s$.

Another tool for the proof of Theorem 7.36(b) is given by

Lemma 7.44 *Let G be a graph on $[n]$, v a simplicial vertex of G with $\deg_G(v) \geq 2$, and e an edge incident with v. Then $\mathrm{reg}(J_{G\setminus e} : f_e) \leq n - 2$.*

Proof Let $v_1, \ldots, v_t$ be all the neighbors of the simplicial vertex v, and $e_1, \ldots, e_t$ be the edges joining v to $v_1, \ldots, v_t$, respectively, where $t \geq 2$. Without loss of generality, assume that $e = e_t$. Note that for each $i = 1, \ldots, t - 1$, v, v_i, v_t is a path between v and v_t in G, so that for all $i = 1, \ldots, t - 1$, x_i and y_i are in the minimal monomial set of generators of the ideal $I_{G,e}$, as defined in Theorem 7.43. Also, all other paths between v and v_t in $G \setminus e$ contain v_i for some $i = 1, \ldots, t - 1$. Thus, all the monomials corresponding to these paths are divisible by either x_i or y_i for some $i = 1, \ldots, t - 1$. Hence, we have $I_{G,e} = (x_i, y_i : 1 \leq i \leq t - 1)$. So that $J_{G\setminus e} : f_e = J_{(G\setminus e)_e} + (x_i, y_i : 1 \leq i \leq t - 1)$. The binomial generators of $J_{(G\setminus e)_e}$ corresponding to the edges containing vertices $v_1, \ldots, v_{t-1}$, are contained in $I_{G,e}$. Let $H = (G \setminus e)_e$. Then, we have $J_{G\setminus e} : f_e = J_{H_{[n]\setminus\{v,v_1,\ldots,v_{t-1}\}}} + (x_i, y_i : 1 \leq i \leq t - 1)$, since v is an isolated vertex of $H_{[n]\setminus\{v_1,\ldots,v_{t-1}\}}$. Thus, $\mathrm{reg}(J_{G\setminus e} : f_e) = \mathrm{reg}(J_{H_{[n]\setminus\{v,v_1,\ldots,v_{t-1}\}}})$. But, $\mathrm{reg}(J_{H_{[n]\setminus\{v,v_1,\ldots,v_{t-1}\}}}) \leq n - 2$, by Theorem 7.36(a), since $t \geq 2$. Therefore, $\mathrm{reg}(J_{G\setminus e} : f_e) \leq n - 2$, as desired. $\square$

For any graph G we introduce the numerical invariant $\alpha_G = \min\{\alpha_G(v) : v \in V(G)\}$, where $\alpha_G(v)$ is defined to be $\binom{\deg_G(v)}{2} - |E(G_{N(v)})|$. Note that $\alpha_G = 0$ is equivalent to saying that G has a simplicial vertex. For example, let G be the graph which is shown in Figure 7.4. There we have $\alpha_G(1) = \alpha_G(5) = 0$, since the vertices 1 and 5 are both simplicial vertices. On the other hand, $\alpha_G(3) = \alpha_G(4) = 1$, and $\alpha_G(2) = 2$. Hence, $\alpha_G = 0$.

Proof (of Theorem 7.36(b)) By Corollary 7.35, $\mathrm{reg}(J_G) = n$, if G is a path. Thus we may now assume G is not a path and have to prove that $\mathrm{reg}\, J_G \leq n - 1$.

We first prove this when G contains a simplicial vertex, or equivalently when $\alpha_G = 0$. For that purpose we use induction on the number of vertices of G. If $n = 2$, then G consists of just two isolated vertices, and hence clearly $J_G = (0)$, and we are done. Now let $n > 1$, and assume that for any graph H over m vertices with $m < n$, which is not a path, and has a simplicial vertex, we have $\mathrm{reg}(J_H) \leq m - 1$. We distinguish two cases: either G has a vertex of degree 1 or G has no such vertex. Then, in the first case, by using our induction hypothesis, we show that the desired bound holds. Next, in the second case, roughly speaking, by removing certain edges, we reduce our problem to a graph with a vertex of degree 1, and hence we then conclude the proof of Theorem 7.36(b) in the case that G has a simplicial vertex by using the first case.

Fig. 7.4 A graph with $\alpha_G = 0$

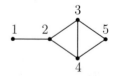

Case(i):   Suppose that G has a simplicial vertex v with $\deg(v) = 1$. Then v has
only one neighbor, say w. Let $e = \{v, w\}$ be the edge joining v and w. We have
$\operatorname{reg}(J_{G\backslash e}) = \operatorname{reg}(J_{(G\backslash e)_{[n]\backslash v}})$, since v is an isolated vertex of $G \backslash e$. Thus, by
Theorem 7.36(a), $\operatorname{reg}(J_{G\backslash e}) \leq n - 1$. On the other hand, we have $\operatorname{reg}(J_{G\backslash e} :$
$f_e) = \operatorname{reg}(J_{(G\backslash e)_e})$, by Theorem 7.43. Note that v is also an isolated vertex of
$(G \backslash e)_e$, so that we can disregard it in computing the regularity, and hence we
have $\operatorname{reg}(J_{G\backslash e} : f_e) = \operatorname{reg}(J_{((G\backslash e)_e)_{[n]\backslash v}})$. Thus, $\operatorname{reg}(J_{G\backslash e} : f_e) \leq n - 2$, by the
induction hypothesis, since $((G \backslash e)_e)_{[n]\backslash v}$ is a graph on $n - 1$ vertices and since
it has w as a simplicial vertex, and since $((G \backslash e)_e)_{[n]\backslash v}$, as well as $G \backslash e$, is not
a path. Hence, $\operatorname{reg}(J_{G\backslash e} : f_e) + 1 \leq n - 1$. Thus, by Lemma 7.42(a), we get
$\operatorname{reg}(J_G) \leq n - 1$.

Case(ii):   Suppose that all the simplicial vertices of G have degree greater than
one. Let v be a simplicial vertex of G and $v_1, \ldots, v_t$ be all the neighbors of v,
and $e_1, \ldots, e_t$ be the edges joining v to $v_1, \ldots, v_t$, respectively, where $t \geq 2$. By
Lemma 7.42(a) and Lemma 7.44, we have $\operatorname{reg}(J_G) \leq \max\{\operatorname{reg}(J_{G\backslash e_1}), n - 1\}$. If
$t > 2$, then by applying the same argument to the graph $G \backslash e_1$, we get $\operatorname{reg}(J_G) \leq$
$\max\{\operatorname{reg}(J_{G\backslash\{e_1, e_2\}}), n - 1\}$. Since $\deg_{G\backslash\{e_1,\ldots,e_l\}}(v) \geq 2$ for $l = 1, \ldots, t - 2$, we
can repeat this process to obtain $\operatorname{reg}(J_G) \leq \max\{\operatorname{reg}(J_{G\backslash\{e_1,\ldots,e_{t-1}\}}), n - 1\}$. Note
that $G \backslash \{e_1, \ldots, e_{t-1}\}$ is a graph on n vertices in which $\deg(v) = 1$. Thus, by
case (i), we have $\operatorname{reg}(J_{G\backslash\{e_1,\ldots,e_{t-1}\}})) \leq n - 1$. Thus, $\operatorname{reg}(J_G) \leq n - 1$.

In order to complete the proof of the theorem we now have to deal with the case
that G has no simplicial vertex. Assume that there exists a graph G on $[n]$ which
does not have any simplicial vertex (in particular, G is not a path) and for which
$\operatorname{reg}(J_G) \geq n$. We may assume that G has the least number of vertices, n, among the
graphs for which the desired inequality does not hold. Moreover, we assume that
α_G is the minimum among the graphs on n vertices with this property. Since G does
not contain any simplicial vertex, we have $\alpha_G \geq 1$, and hence there exists a vertex
v of G which has two neighbors, say v_1 and v_2, which are not adjacent in G, and
$\alpha_G = \alpha_G(v)$. Let $e = \{v_1, v_2\}$. By Lemma 7.42(b),

$$\operatorname{reg}(J_G) \leq \max\{\operatorname{reg}(J_{G\cup e}), \operatorname{reg}(J_G : f_e) + 2\}. \tag{7.3}$$

Moreover, $\alpha_{G\cup e}(v) = \alpha_G(v) - 1$, and hence $\alpha_{G\cup e} \leq \alpha_G - 1$. Since $G \cup e$ has n
vertices, we have $\operatorname{reg}(J_{G\cup e}) \leq n - 1$, by our choice of G. Note that $G \cup e$, as well
as G, is not a path.

Now, we show that $\operatorname{reg}(J_G : f_e) + 2 \leq n - 1$. By Theorem 7.43, we have
$J_G : f_e = J_{G_e} + I_{G\cup e}$. Since v_1, v, v_2 is a path between v_1 and v_2 in G, we
have $I_{G\cup e} = (x_v, y_v) + I_{(G\backslash v)\cup e}$, and hence $J_G : f_e = J_{G_e} + I_{G\cup e} = J_{(G\backslash v)_e} +$
$I_{(G\backslash v)\cup e} + (x_v, y_v)$. Thus, $\operatorname{reg}(J_G : f_e) = \operatorname{reg}(J_{(G\backslash v)_e} + I_{(G\backslash v)\cup e})$. By Theorem 7.43,
$\operatorname{reg}(J_{G\backslash v} : f_e) = \operatorname{reg}(J_{(G\backslash v)_e} + I_{(G\backslash v)\cup e})$, so that $\operatorname{reg}(J_G : f_e) = \operatorname{reg}(J_{G\backslash v} : f_e)$. On
the other hand, we have $\operatorname{reg}(J_{G\backslash v} : f_e) + 2 \leq \max\{\operatorname{reg}(J_{G\backslash v}), \operatorname{reg}(J_{(G\backslash v)\cup e}) + 1\}$, by
Lemma 7.42(c), and $\operatorname{reg}(J_{G\backslash v}) \leq n - 1$, by Theorem 7.36(a). Therefore, it remains
to be shown that $\operatorname{reg}(J_{(G\backslash v)\cup e}) \leq n - 2$.

We first claim that $(G \setminus v) \cup e$ is not a path. To prove the claim, suppose on the contrary that $(G \setminus v) \cup e$ is a path over $n - 1$ vertices. Then, $G \setminus v$ is the disjoint union of two path graphs π_t and π_s on two different sets of vertices, where $t + s = n - 1$. Note that e joins a vertex of minimum degree of π_t and a vertex of minimum degree of π_s, in $(G \setminus v) \cup e$. Moreover, v is adjacent to these two vertices in G. So, if $s \leq 2$ or $t \leq 2$, then G has a simplicial vertex which is a contradiction, by our choice of G. Suppose now that $t \geq 3$ and $s \geq 3$. Then v is adjacent to both of the degree 1 vertices of π_t and π_s in G, since otherwise G has a vertex of degree 1, and hence a simplicial vertex which contradicts the choice of G. Now suppose that $\{u, w\}$ is an edge of π_t with $\deg_{\pi_1} w = 1$. If u is adjacent to v, then w is a simplicial vertex of G which is a contradiction, because of our choice of G. So, suppose that u is not adjacent to v. Then u has just two neighbors in G which are not adjacent to each other, and hence $\alpha_G(u) = 1$. On the other hand, $\alpha_G(v) \geq 6$, because v is adjacent to at least four vertices, namely the vertices of degree 1 of π_t and π_s, and none of these vertices are adjacent to each other in G. So, we get a contradiction, since by the definition of α_G, we have $\alpha_G = \alpha_G(v) \leq \alpha_G(u)$. Therefore, $(G \setminus v) \cup e$ is not a path and the claim follows.

Thus, by the choice of G, we have $\operatorname{reg}(J_{(G \setminus v) \cup e}) \leq n - 2$, since $(G \setminus v) \cup e$ has $n - 1$ vertices. This yields the desired conclusion. $\square$

Problems

7.16 In Theorem 7.28 it was shown that if G is closed and connected, then $\operatorname{reg}(J_G) = \ell + 1$, where ℓ be the length of the longest induced path of G. Show by an example that the converse is not true.

7.17 Given integers $2 \leq m \leq n$, show there exists a graph on $[n]$ such that $\operatorname{reg}(J_G) = m$.

7.18 Let C_n be a cycle of length n. Compute $\operatorname{reg}(J_G)$.

7.19 Characterize those trees G on the vertex set $[n]$ for which $\operatorname{reg}(J_G) = n - 1$.

7.20 Let G be closed graph. Show that $\beta_{i,2i}(S/J_G) = \beta_{i,2i}(S/\operatorname{in}_<(J_G))$, where $<$ denotes the lexicographic order on $S = K[x_1, \ldots, x_n, y_1, \ldots, y_n]$ induced by $x_1 > \cdots > x_n > y_1 > \cdots > y_n$.

7.4   Koszul Binomial Edge Ideals

In this section we study the Koszul property of the K-algebras defined by binomial edge ideals. Let G be a finite simple graph on the vertex set $[n]$, K a field and $J_G \subset S = K[x_1, \ldots, x_n, , y_1, \ldots, y_n]$ the binomial edge ideal of G. We call G *Koszul*, if for some base field K, the standard graded K-algebra S/J_G is Koszul.

7.4.1 Koszul Graphs

For the study of Koszul graphs it is enough to consider connected graphs. Indeed, one has

Proposition 7.45 *Let G be a graph with connected components $G_1, \ldots, G_r$. Then G is Koszul if and only if G_i is Koszul for $1 \le i \le r$.*

Proof The proof follows from Problem 2.29 because $S/J_G \cong \bigotimes_{i=1}^{r} S_i/J_{G_i}$, where $S_i = K[\{x_j, y_j : j \in V(G_i)\}]$ for $1 \le i \le r$. □

The following result shows that Koszulness is inherited by induced subgraphs.

Proposition 7.46 *Let G be a Koszul graph, and let H be an induced subgraph of G. Then H is Koszul.*

Proof We may assume that $V(H) = [k]$. Let $S = K[x_1, \ldots, x_n, y_1, \ldots, y_n]$ and $T = K[x_1, \ldots, x_k, y_1, \ldots, y_k]$. Then T/J_H is an algebra retract of S/J_G. Indeed, let $L = (x_{k+1}, \ldots, x_n, y_{k+1}, \ldots, y_n)$. Then the composition $T/J_H \to S/J_G \to S/(J_G, L) \cong T/J_H$ of the natural K-algebra homomorphisms is an isomorphism. It follows therefore from Theorem 2.31 that H is again Koszul. □

As an application of Proposition 7.46 we have

Theorem 7.47 *Let G be a Koszul graph. Then G is chordal and claw free.*

Proof Suppose that G is not claw free. Then there exists an induced subgraph H of G which is isomorphic to a claw. We may assume that $V(H) = \{1, 2, 3, 4\}$, and let $R = K[x_1, \ldots, x_4, y_1, \ldots, y_4]$. A computation with Singular [49] shows that $\beta_{3,5}^{R/J_H}(K) \ne 0$. Thus H is not Koszul. By Proposition 7.46, this contradicts our assumption that G is Koszul.

Suppose that G is not chordal. Then there exists a cycle C of length ≥ 4 which has no chord. Then C is an induced subgraph and hence should be Koszul. We may assume that $V(C) = \{1, 2, \ldots, m\}$ with edges $\{i, i+1\}$ for $i = 1, \ldots, m-1$ and edge $\{1, m\}$ and set $T = K[x_1, \ldots, x_m, y_1, \ldots, y_m]$. We claim that $\beta_{2,m}^{T}(T/J_C) \ne 0$. For $m > 4$, this will imply that C is not Koszul, see Theorem 2.32(b). That a 4-cycle is not Koszul can again be directly checked with Singular [49]. Again, by Proposition 7.46, this contradicts the assumption that G is Koszul.

In order to prove the claim, we let $F = \bigoplus_{i=1}^{m} T e_i$ and consider the free presentation

$$\epsilon: F \to J_C \longrightarrow 0, \quad e_i \mapsto f_{i,i+1} \text{ for } i = 1, \ldots, m$$

For simplicity, here and in the following, we read $m + 1$ as 1.

Obviously, $g = \sum_{i=1}^{m} (\prod_{j=1}^{m} x_j)/(x_i x_{i+1}) e_i \in \operatorname{Ker} \epsilon$. We will show that g is a minimal generator of $\operatorname{Ker} \epsilon$. Indeed, let $g' = \sum_{i=1}^{m} g_i e_i \in \operatorname{Ker} \epsilon$ be an arbitrary relation, and suppose that some $g_j = 0$. Since the $f_{i,i+1}$ for $i \ne j$ form a regular sequence, it then follows that all the other g_i belong to J_C. However, since the coefficients of g do not belong to J_C, we conclude that g cannot be written as a linear combination of relations for which one of its coefficients is zero.

Now assume that all $g_i \neq 0$. Let ϵ_i denote the ith canonical unit vector of $\mathbb{Z}^n$. Since J_C is a $\mathbb{Z}^n$-graded ideal with $\deg_{\mathbb{Z}^n} x_i = \deg_{\mathbb{Z}^n} y_i = \epsilon_i$, we may assume that $g' = \sum_{i=1} g_i e_i$ is a homogeneous relation where $\deg_{\mathbb{Z}^n} e_i = \deg f_{i,i+1} = \epsilon_i + \epsilon_{i+1}$ and g_i is homogeneous satisfying $\deg_{\mathbb{Z}^n} g' = \deg_{\mathbb{Z}^n} g_i + \epsilon_i + \epsilon_{i+1}$ for all i. This is only possible if $\deg_{\mathbb{Z}^n} g' \geq \sum_{i=1}^m \epsilon_i$, coefficientwise. In particular it follows that $\deg g' \geq m$, where $\deg g'$ denotes the total degree of g'. Thus g cannot be a linear combination of relations of lower (total) degree and hence is a minimal generator of $\operatorname{Ker} \epsilon$. Since $\deg g = m$, we conclude that $\beta_{2,m}^T(T/J_C) \neq 0$. $\qquad\square$

Corollary 7.48 *Let G be a forest (i.e. a graph without cycles). Then G is Koszul if and only if each component of G is a path graph.*

Proof By Proposition 7.45 we may assume that G is connected and have to show that G is Koszul if and only if G is a path graph. If G is a path graph, then J_G is a closed graph, and hence Koszul. On the other hand, if G is not a path, then it contains an induced claw, and hence is not Koszul.

Let G be a graph. A vertex v of G is called a *simplicial vertex* of G, if v belongs to exactly one maximal clique of G.

Proposition 7.49 *Let G_1 be a graph with simplicial vertex v, G_2 a graph with simplicial vertex v', and assume that $V(G_1) \cap V(G_2) = \emptyset$. Then the linear forms $\ell_x = x_v - x_{v'}$ and $\ell_y = y_v - y_{v'}$ form a regular sequence on $S'/J_{G'}$ where G' is the graph whose connected components are G_1 and G_2 and where S' is the polynomial ring in which $J_{G'}$ is defined.*

Proof We first show that l_y is regular on $S'/J_{G'}$. Since

$$J_{G'} : l_y = \bigcap_{T \in \mathscr{C}(G')} (P_T : l_y),$$

it is sufficient to verify that

$$P_T : l_y = P_T \quad \text{for all} \quad T \in \mathscr{C}(G').$$

We actually show that $l_y \notin P_T$. Then this implies that $P_T : l_y = P_T$, because P_T is a prime ideal. We have

$$P_T = (\bigcup_{i \in T} \{x_i, y_i\}, J_{\tilde{G}'_1}, \ldots, J_{\tilde{G}'_{c(T)}}).$$

By Proposition 7.22 it follows that

$$l_y \notin (\bigcup_{i \in T} \{x_i, y_i\}),$$

since v and v' are simplicial vertices. Since l_y is a linear form, it cannot be obtained by a linear combination of the quadratic generators of P_T, hence $l_y \notin P_T$, as desired.

Next we claim that $(J_{G'}, l_y) : l_x = (J_{G'}, l_y)$. We may assume that $V(G_1) = \{1, 2, \ldots, n\}$, $V(G_2) = \{n+1, \ldots, m+n\}$, $v = n$ and $v' = n+1$. For the proof of the claim we describe the Gröbner basis of $J_{G'} + (l_y)$. We fix the lexicographic order induced by

$$x_1 > x_2 > \cdots > x_{n+m} > y_1 > y_2 \cdots > y_{n+m}. \tag{7.4}$$

Given an admissible path

$$\pi : i = i_0, i_1, \ldots, i_r = j$$

from i to j with $i < j$ we associate the monomial

$$u_\pi = (\prod_{i_k > j} x_{i_k})(\prod_{i_\ell < i} y_{i_\ell}).$$

Then, as shown in Theorem 7.11,

$$\mathscr{G}' = \{u_\pi f_{ij} : \pi \text{ is an admissible path from } i \text{ to } j\}. \tag{7.5}$$

is a Gröbner bases of $J_{G'}$.

We claim that

$$\mathscr{G} = \{l_y\} \cup \{u_\pi f_{ij} : \pi \text{ is an admissible path from } i \text{ to } j \neq n\} \cup$$
$$\cup \{u_\pi(x_i y_{n+1} - x_n y_i) : \pi \text{ is an admissible path from } i \text{ to } j = n\}. \tag{7.6}$$

is a Gröbner basis of $J_{G'} + (l_y)$.

Let $\mathscr{G}_0 = \mathscr{G}' \cup \{l_y\}$. By (7.5) and Buchberger's criterion all the S-pairs of polynomials in $\mathscr{G}'$ reduce to 0. Hence we only have to consider the S-pairs

$$S(l_y, u_\pi f_{ij})$$

for all $u_\pi f_{ij} \in \mathscr{G}'$. If y_n does not divide $\mathrm{in}(u_\pi f_{ij}) = u_\pi x_i y_j$, the S-pair reduces to 0. If y_n divides $u_\pi x_i y_j$, then π is an admissible path of G_1, and since n is the maximal in the labeling of $V(G_1)$, by the definition of an admissible path, $j = n$. Therefore,

$$S(l_y, u_\pi f_{in}) = -u_\pi(x_i y_{n+1} - x_n y_i)$$

with $\text{in}(-u_\pi(x_i y_{n+1} - x_n y_i)) = -u_\pi x_i y_{n+1}$. We want to show that

$$\mathcal{G}_1 = \{l_y\} \cup \{u_\pi f_{ij} : \pi \text{ is an admissible path from } i \text{ to } j\} \cup$$
$$\cup \{u_\pi(x_i y_{n+1} - x_n y_i) : \pi \text{ is an admissible path from } i \text{ to } j = n\}. \tag{7.7}$$

is a Gröbner basis of $J_{G'} + (l_y)$. Since $S(l_y, u_\pi f_{ij})$ reduce to 0 by the binomials described in the third set of (7.7) and since $S(u_\pi f_{ij}, u_\sigma f_{kl})$ reduce to 0 by the binomials described in the second set of (7.7), it remains to investigate the S-pairs of the form

(1) $S(u_\pi(x_i y_{n+1} - x_n y_i), u_\sigma(x_j y_{n+1} - x_n y_j))$ and
(2) $S(u_\pi(x_i y_{n+1} - x_n y_i), u_\sigma f_{kl})$.

Case (1):   If $i = j$, then the S-polynomial itself is 0. If $i \neq j$, then

$$S(u_\pi(x_i y_{n+1} - x_n y_i), u_\sigma(x_j y_{n+1} - x_n y_j)) = S(u_\pi f_{in}, u_\sigma f_{jn}),$$

and the assertion follows since $\mathcal{G}_1 \supset \mathcal{G}'$.
Case (2):   If $\{k, l\} \cap \{i, n+1\} = \emptyset$ or $i = l$, then $\text{in}(x_i y_{n+1} - x_n y_i)$ and $\text{in}(f_{kl})$ form a regular sequence. Hence the corresponding S-pair reduces to 0. If $n+1 \in \{k, l\}$, then σ is an admissible path in G_2 and $\text{in}(f_{kl}) = x_{n+1} y_l$. Therefore in this case the initial monomials form a regular sequence, too.

It remains to consider the case $i = k$. We observe that there exists a monomial w such that

$$S(u_\pi(x_i y_{n+1} - x_n y_i), u_\sigma f_{il}) = w(x_l y_{n+1} - x_n y_l), \tag{7.8}$$

and

$$S(u_\pi f_{in}, u_\sigma f_{il}) = w f_{ln} \tag{7.9}$$

Since (7.9) reduces to 0 in $\mathcal{G}_1$, there exists $f \in \mathcal{G}_1$ such that $\text{in}(f)$ divides $w x_l y_n$.
If y_n divides $\text{in}(f)$, then $f = u_\tau f_{jn}$, and this implies that $f' = u_\tau(x_j y_{n+1} - x_n y_j) \in \mathcal{G}_1$. Therefore the remainder of $w f_{ln}$ with respect to f is equal to the remainder of $w(x_l y_{n+1} - x_n y_l)$ with respect to f' and reduce to 0.
If y_n does not divide $\text{in}(f)$, then $\text{in}(f)$ divides $w x_l$ and hence divides the initial term of (7.8). That is, the remainder of $w f_{ln}$ with respect to f is

$$w' f_{l'n} \tag{7.10}$$

for some monomial w', and at the same time the remainder of $w(x_l y_{n+1} - x_n y_l)$ with respect of f is

$$w'(x_{l'} y_{n+1} - x_n y_{l'}). \tag{7.11}$$

Proceeding as before, since $w' f_{l'n}$ is not zero and reduces to 0, we can apply the same reduction step to $w' f_{l'n}$ and $w'(x_{l'} y_{n+1} - x_n y_{l'})$ following the arguments applied to the binomials in the second terms of Equations (7.8) and (7.9). Thanks to Buchberger's algorithm, since the expression (7.10) reduces to 0 in a finite number of steps, also the expression (7.11) reduces to 0 by the same number of steps.

Hence $\mathcal{G}_1$ is a Gröbner basis and we can remove the reducible polynomials $u_\pi f_{ij}$ with $j = n$ since their initial terms are divisible by $\text{in}(l_y) = y_n$. The claim follows.

Therefore

$$\text{in}(J_{G'} + (l_y)) = (y_n, u_\pi x_i y_j, u_\pi x_{i'} y_{n+1}) \text{ with } i < j \neq n, i' < n. \qquad (7.12)$$

Suppose that $f \in (J_{G'} + l_y) : l_x$, that is, $f(x_n - x_{n+1}) \in (J_{G'} + (l_y))$. Then $\text{in}(f(x_n - x_{n+1})) = \text{in}(f)x_n \in \text{in}(J_{G'} + (l_y))$. We observe that x_n does not divide any monomial in the minimal set of generators of $\text{in}(J_{G'} + (l_y))$. In fact, $i \neq n$ and $i' \neq n$ by (7.12). Let π be an admissible path such that there exists k with $1 \leq k < r$ and $i_k = n$. Since n is a simplicial vertex in a clique $F \in \Delta(G')$, π contains at least 2 vertices $u, w \in F$ with $n \notin \{u, w\}$. But since $\{u, w\} \in E(G)$, π is not admissible. Hence $\text{in}(f) \in \text{in}(J_{G'} + (l_y))$. Thus we have shown that $\text{in}(J_{G'} + (l_y) : l_x) \subset \text{in}(J_{G'} + (l_y))$. Since the other inclusion is trivially true, we get $\text{in}(J_{G'} + (l_y)) = \text{in}(J_{G'} + (l_y) : l_x)$, and since $J_{G'} + (l_y) \subset J_{G'} + (l_y) : l_x$ we finally deduce from this that $J_{G'} + (l_y) = J_{G'} + (l_y) : l_x$, as desired. $\square$

Let G_1 and G_2 be two graphs with $V(G_1) \cap V(G_2) = \{v\}$, and v is a simplicial vertex of G_1 and G_2. Let $G = G_1 \cup G_2$ with $V(G) = V(G_1) \cup V(G_2)$ and $E(G) = E(G_1) \cup E(G_2)$. We say that G is obtained by *gluing* G_1 and G_2 along the vertex v.

The following result allows us to construct families of Koszul graphs.

Theorem 7.50 *Let G be a graph obtained by gluing the graphs G_1 and G_2 along a vertex. Then G is Koszul if and only if G_1 and G_2 are Koszul.*

Proof Let $V(G) = [n]$ and assume that G_1 and G_2 are glued along the vertex $v \in [n]$. Let v' be a vertex which does not belong to $V(G)$ and let G'_2 be the graph with $V(G'_2) = (V(G_2)\setminus\{v\}) \cup \{v'\}$ whose edge set is $E(G'_2) = E(G_2\setminus\{v\}) \cup \{\{i, v'\} : \{i, v\} \in E(G_2)\}$. We set $S = K[x_1, \ldots, x_n, y_1, \ldots, y_n]$ and $S' = S[x_{v'}, y_{v'}]$. Let $\ell_x = x_v - x_{v'}$ and $\ell_y = y_v - y_{v'}$. By Proposition 7.49, ℓ_x, ℓ_y is a regular sequence on $S'/J_{G'}$, where G' is the graph whose connected components are G_1 and G'_2. Moreover, we obviously have

$$S'/(J_{G'}, \ell_x, \ell_y) \cong S/J_G.$$

Hence, Corollary 2.22 implies that G is Koszul if and only if G' is Koszul. Next, by Proposition 7.45, we see that G' is Koszul if and only its connected components, namely G_1 and G'_2, are Koszul. Finally, we observe that G'_2 is Koszul if and only if G_2 is so. $\square$

The following corollary presents a class of chordal and claw-free graphs which are Koszul.

Corollary 7.51 *Let G be a chordal and claw-free graph with the property that $\Delta(G)$ admits a leaf order $F_1, \ldots, F_r$ such that for all $i > 1$, the facet F_i intersects any of its branches in only one vertex. Then G is Koszul.*

Proof We proceed by induction on r. If $r = 1$, there is nothing to prove since any clique is Koszul. Let $r > 1$ and assume that the graph G' with $\Delta(G') = \langle F_1, \ldots, F_{r-1} \rangle$ is Koszul. We may assume that F_{r-1} is a branch of F_r and let $\{v\} = F_r \cap F_{r-1}$. The desired statement follows by applying Theorem 7.50 for G' and the clique F_r, once we show that v is a simplicial vertex of G'.

Let us assume that v is not free in G' and choose a maximal clique F_j with $j \leq r - 2$ such that $v \in F_j$. We may find three vertices $a, b, c \in V(G)$ such that $a \in F_r \setminus (F_{r-1} \cup F_j)$, $b \in F_{r-1} \setminus (F_r \cup F_j)$, and $c \in F_j \setminus (F_r \cup F_{r-1})$. If $\{a, b\} \in E(G)$, then there exists a maximal clique F_k with $k \leq r - 1$ such that $a, b \in F_k$. This implies that $a \in F_k \cap F_r \subset \{v\}$, contradiction. Therefore, $\{a, b\}$ is not an edge of G. Similarly, one proves that $\{a, c\} \notin E(G)$. Let us now assume that $\{b, c\} \in E(G)$. The clique on the vertices v, b, c is contained in some maximal clique F_k. We have $k \leq r - 2$ since $F_k \neq F_{r-1}$. Then it follows that $|F_k \cap F_{r-1}| \geq 2$ which is a contradiction to our hypothesis on G. Consequently, we have proved that $\{a, b\}, \{b, c\}, \{a, c\} \notin E(G)$. Hence, G contains a claw as an induced subgraph, contradiction. Therefore, v is a simplicial vertex of G'.   □

The net displayed in Figure 7.3 satisfies the conditions of Corollary 7.51 but is not closed, while the tent displayed in the same figure happens to be chordal but not Koszul. That the tent is not Koszul can be seen as follows: we label the tent as shown in Figure 7.5.

First observe that the graph G' restricted to the vertex set [4] is Koszul by Corollary 7.51, and that $B = K[x_1, \ldots, x_4, y_1, \ldots, y_4]/J_{G'}$ is an algebra retract of $A = K[x_1, \ldots, x_6, y_1, \ldots, y_6]/J_G$ with retraction map $A \rightarrow A/(x_5, x_6, y_5, y_6) \cong B$. Thus if A would be Koszul, the ideal (x_5, x_6, y_5, y_6) would have to have an A-linear resolution, see Theorem 2.31. It can be verified with Singular [49] that this is not the case.

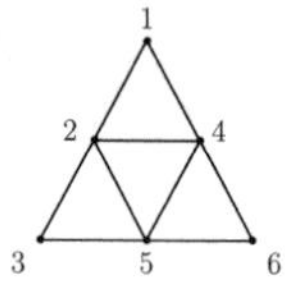

Fig. 7.5 The tent is not Koszul

7.4.2 Koszul Flags and Koszul Filtrations for Closed Graphs

In this subsection it will be shown that S/J_G has a Koszul filtration for any closed graph. We first characterize closed graphs by the property that the variables x_i form a Koszul flag of S/J_G for a suitable order of them. Here any chain of ideals $(0) = I_0 \subset I_1 \subset \ldots \subset I_m$ of S/J_G generated by linear forms is called a *Koszul flag* of S/J_G, if for all j, I_{j+1}/I_j is cyclic and the annihilator of I_{j+1}/I_j is generated by linear forms.

Consider the ideal I which is generated by the binomial $x_1 x_3 - x_2 x_3$. Then $I : x_3 = (I, x_1 - x_2) = (x_1 - x_2)$. Thus, in general, one cannot expect that the ideals $(I, x_{i+1}, \ldots, x_n) : x_i$ are generated by a subset of the variables modulo I, even when I is a binomial ideal. Therefore some additional assumptions on the Gröbner basis of I are required to have monomial colon ideals.

Theorem 7.52 *Let $I \subset R = K[x_1, \ldots, x_n]$ be an ideal generated by quadratic binomials, and let $<$ be the reverse lexicographic order induced by $x_1 > x_2 > \cdots > x_n$. Let $f_1, \ldots, f_m$ be the degree 2 binomials of the reduced Gröbner basis of I with respect to $<$. Let $f_i = u_i - v_i$ for $i = 1, \ldots, m$, and assume that $\gcd(u_i, v_i) = 1$ for all i. Then, for all i, we have:*

(a) $[(I, x_{i+1}, \ldots, x_n) : x_i]_1 = [(\operatorname{in}_<(I), x_{i+1}, \ldots, x_n) : x_i]_1$;

(b) *Suppose I has a quadratic Gröbner basis with respect to $<$. Then*

$$(I, x_{i+1}, \ldots, x_n) : x_i = (I, x_{i+1}, \ldots, x_n, \{x_j : \ j \le i, \ x_j x_i \in \operatorname{in}_<(I)\}),$$

and

$$(\operatorname{in}_<(I), x_{i+1}, \ldots, x_n) : x_i = (\operatorname{in}_<(I), x_{i+1}, \ldots, x_n, \{x_j : \ j \le i, \ x_j x_i \in \operatorname{in}_<(I)\}).$$

Proof

(a) Let $\ell = \sum_{k=1}^n a_k x_k$ be a linear form. First suppose that $\ell x_i \in (I, x_{i+1}, \ldots, x_n)$. We may assume that $a_k = 0$ for $k > i$. Let $x_j = \operatorname{in}_<(\ell)$. Then $j \le i$ and $x_j x_i \in \operatorname{in}_<(I, x_{i+1}, \ldots, x_n) = (\operatorname{in}_<(I), x_{i+1}, \ldots, x_n)$. Therefore, there exists f_k with $\operatorname{in}_<(f_k) = x_j x_i$. Thus, if $f_k = x_j x_i - x_r x_s$, then $s \ge i$. However, since $\gcd(u_k, v_k) = 1$, we see that $s > i$. This implies that $x_j x_i \in (I, x_{i+1}, \ldots, x_n)$ and, consequently, $(\ell - a_j x_j) x_i \in (I, x_{i+1}, \ldots, x_n)$. Since $x_j \in (\operatorname{in}_<(I), x_{i+1}, \ldots, x_n) : x_i$, induction on $\operatorname{in}_<(\ell)$ shows that $\ell \in (\operatorname{in}_<(I), x_{i+1}, \ldots, x_n) : x_i$.

Conversely, suppose $\ell \in (\operatorname{in}_<(I), x_{i+1}, \ldots, x_n) : x_i$. Since $(\operatorname{in}_<(I), x_{i+1}, \ldots, x_n)$ is a monomial ideal, we may assume that ℓ is a monomial and $\ell \notin (\operatorname{in}_<(I), x_{i+1}, \ldots, x_n)$, say, $\ell = x_j$. Then $j \le i$ and $x_j x_i \in (\operatorname{in}_<(I), x_{i+1}, \ldots, x_n)$. As before, there exists $f_k = x_j x_i - x_r x_s$ with $r \le s$ and $s > i$. It follows that $x_j x_i \in (I, x_{i+1}, \ldots, x_n)$. and hence $x_j \in (I, x_{i+1}, \ldots, x_n) : x_i$.

(b) Suppose that $\mathscr{G} = \{f_1, \ldots, f_m\}$ is the reduced Gröbner basis of I with respect to $<$. Let $J_i = (I, x_{i+1}, \ldots, x_n) : x_i$ and

$$J_i' = (I, x_{i+1}, \ldots, x_n, \{x_j : \ j \leq i, \ x_j x_i \in \text{in}_<(I)\}).$$

One has $J_i' \subset J_i$. To see why this is true, suppose that $x_j x_i \in \text{in}_<(I)$ with $j \leq i$. Then there is $f_k = x_j x_i - x_p x_q \in \mathscr{G}$ with $\text{in}_<(f_k) = x_j x_i$. Since $j \leq i$, it follows that either $p > i$ or $q > i$. Hence $x_p x_q \in (I, x_{i+1}, \ldots, x_n)$. Thus $x_j x_i \in (I, x_{i+1}, \ldots, x_n)$ and $x_j \in J_i$, as required.

Now, let $\mathscr{A}$ denote the set of homogeneous polynomials $f \in S$ of degree ≥ 1 which belong to J_i with the property that none of the monomials appearing in f belongs to J_i'. Suppose that $\mathscr{A} \neq \emptyset$. Among the polynomials belonging to $\mathscr{A}$, we choose $f \in \mathscr{A}$ such that $\text{in}_<(f) \leq \text{in}_<(g)$ for all $g \in \mathscr{A}$. Let $u = \text{in}_<(f)$. Since $x_i f \in (I, x_{i+1}, \ldots, x_n)$, one has $x_i u \in (\text{in}_<(I), x_{i+1}, \ldots, x_n)$. Since $u \notin J_i'$, it follows that $x_i u \in \text{in}_<(I)$. Thus there is $f_\ell = x_p x_q - x_r x_s$ with $\text{in}_<(f) = x_p x_q$ such that $x_p x_q$ divides $x_i u$. If, say, $p = i$, then x_q divides u. Thus $q \leq i$. Since $x_i x_q \in \text{in}_<(I)$, one has $x_q \in J_i'$. This contradicts our assumption that $u \notin J_i'$. Thus $p \neq i, q \neq i$ and $x_p x_q$ divides u. Let $w = (u/x_p x_q)x_r x_s$ and $f' = f - a(u - w)$, where $a \neq 0$ is the coefficient of u in f. Since $u - w \in I$, one has $f' \in J_i$. Since $u \notin J_i'$, one has $w \notin J_i'$. Thus $f' \in \mathscr{A}$ and $\text{in}_<(f') < \text{in}_<(f)$. This contradicts the choice of $f \in \mathscr{A}$. Hence $\mathscr{A} = \emptyset$ and $J_i = J_i'$, as desired.

The proof of the corresponding statement for $\text{in}_<(I)$ is obvious.   □

Before continuing we introduce some notation. For $k \in [n]$, we let

$$N^<(k) = \{j : j < k, \ \{j, k\} \in E(G)\} \text{ and } N^>(k) = \{j : j > k, \ \{k, j\} \in E(G)\}.$$

For the following proofs it will be useful to note that, provided that they are nonempty, each of these sets are intervals if the graph G is closed with respect to its labeling. Indeed, let us take $i \in N^<(k)$. In particular, we have $\{i, k\} \in E(G)$. Then, as all the maximal cliques of G are intervals (see Theorem 7.7), it follows that for any $i \leq j < k$, $\{j, k\} \in E(G)$, thus $j \in N^<(k)$. A similar argument works for $N^>(k)$.

Theorem 7.53 *Let G be a connected graph on the vertex set $[n]$. The following conditions are equivalent:*

(i) *G is closed with respect to the given labeling;*
(ii) *the sequence $x_n, x_{n-1}, \ldots, x_1$ has linear quotients modulo J_G, and hence establishes a Kosul flag.*

Proof (i) $\Rightarrow$ (ii): Let G be closed with respect to the given labeling. It follows that the generators of J_G form the reduced Gröbner basis of J_G with respect to the reverse lexicographic order induced by $y_1 > \cdots > y_n > x_1 > \cdots > x_n$. Let $i \leq n$.

The generators of $\text{in}_<(J_G)$ which are divisible by x_i are exactly $x_i y_j$ where $i < j$ and $\{i, j\} \in E(G)$. Hence, by using Theorem 7.52 (b), we get

$$(\bar{x}_n, \bar{x}_{n-1}, \ldots, \bar{x}_{i+1}) : \bar{x}_i = (\bar{x}_n, \bar{x}_{n-1}, \ldots, \bar{x}_{i+1}, \{\bar{y}_j : \ j \in N^>(i)\}). \quad (7.13)$$

Here $\bar{f}$ denotes the residue class for a polynomial $f \in S$ modulo J_G.

(ii) $\Rightarrow$ (i): We may suppose that $\bar{x}_n, \bar{x}_{n-1}, \ldots, \bar{x}_1$ has linear quotients and show that G is closed with respect to the given labeling. In fact, assume that G is not closed. Then there exist $\{i, j\}, \{i, k\} \in E(G)$ with $i < j < k$ or $i > j > k$ and such that $\{j, k\} \notin E(G)$.

Let us first consider the case that $i < j < k$. Since

$$\bar{x}_j \bar{y}_i \bar{y}_k = \bar{x}_i \bar{y}_j \bar{y}_k = \bar{x}_k \bar{y}_i \bar{y}_j,$$

we see that $\bar{y}_i \bar{y}_k \in (\bar{x}_n, \ldots, \bar{x}_{j+1}) : \bar{x}_j$.

We claim that $\bar{y}_i \bar{y}_k$ is a minimal generator of $(\bar{x}_n, \ldots, \bar{x}_{j+1}) : \bar{x}_j$, contradicting the assumption that $\bar{x}_n, \bar{x}_{n-1}, \ldots, \bar{x}_1$ has linear quotients. Indeed, suppose that $\bar{y}_i \bar{y}_k$ is not a minimal generator of $(\bar{x}_n, \ldots, \bar{x}_{j+1}) : \bar{x}_j$, then there exist linear forms ℓ_1 and ℓ_2 in S such that $\bar{\ell}_1 \bar{\ell}_2 = \bar{y}_i \bar{y}_k$ and at least one of the forms $\bar{\ell}_1, \bar{\ell}_2$ belongs to $(\bar{x}_n, \ldots, \bar{x}_{j+1}) : \bar{x}_j$.

Now we observe that J_G is $\mathbb{Z}^n$-graded with $\deg x_i = \deg y_i = \epsilon_i$ for all i, where ϵ_i is the ith canonical unit vector of $\mathbb{Z}^n$. It follows that the $\bar{\ell}_i$ are multi-homogeneous as well with $\deg \bar{\ell}_1 \bar{\ell}_2 = \epsilon_i + \epsilon_k$, say $\deg \bar{\ell}_1 = \epsilon_i$ and $\deg \bar{\ell}_2 = \epsilon_k$. Thus $\ell_1 = ax_i + by_i$ and $\ell_2 = cx_k + dy_k$ with $a, b, c, d \in K$. Let us first assume that $\bar{\ell}_1 \in (\bar{x}_n, \ldots, \bar{x}_{j+1}) : \bar{x}_j$. We get

$$ax_i x_j + bx_j y_i \in (J_G, x_n, \ldots, x_{j+1})$$

which implies that

$$\text{in}_<(ax_i x_j + bx_j y_i) \in \text{in}_<(J_G, x_n, \ldots, x_{j+1}) = ((\text{in}_< J_G), x_n, \ldots, x_{j+1}).$$

Here $<$ denotes the reverse lexicographic order induced by $y_1 > \cdots > y_n > x_1 > \cdots > x_n$. It follows that $x_i x_j \in \text{in}_<(J_G)$ or $x_j y_i \in \text{in}_<(J_G)$, which is impossible since the generators of degree 2 of $\text{in}_<(J_G)$ are of the form $x_k y_\ell$ with $\{k, \ell\} \in E(G)$ and $k < \ell$.

Let us now consider the case that $\bar{\ell}_2 \in (\bar{x}_n, \ldots, \bar{x}_{j+1}) : \bar{x}_j$. We get $cx_k x_j + dx_j y_k \in (J_G, x_n, \ldots, x_{j+1})$. If $d \neq 0$, we obtain $x_j y_k \in (J_G, x_n, \ldots, x_{j+1})$ and, therefore, $x_j y_k \in (\text{in}_<(J_G), x_n, \ldots, x_{j+1})$ which implies that $x_j y_k \in \text{in}_<(J_G)$, a contradiction since $\{j, k\} \notin E(G)$ by assumption. Therefore, we must have $\ell_2 = cx_k$ for some $c \in K \setminus \{0\}$. The equation $\bar{\ell}_1 \bar{\ell}_2 = \bar{y}_i \bar{y}_k$ implies that $cx_k(ax_i + by_i) - y_i y_k \in J_G$. It follows that one of the monomials $x_i x_k, x_k y_i, y_i y_k$ belongs to $\text{in}_<(J_G)$, contradiction.

Finally, we consider the case that $i > j > k$. Then $x_i f_{jk} \in J_G$, and so $\bar{f}_{jk} \in (\bar{x}_n, \ldots, \bar{x}_{i+1}) : \bar{x}_i$. By similar arguments as above, we show that $\bar{f}_{jk}$ is a minimal generator of $(\bar{x}_n, \ldots, \bar{x}_{i+1}) : \bar{x}_i$. Suppose that there exist linear forms $\ell_1 = ax_j + by_j$ and $\ell_2 = cx_k + dy_k$ such that $g = f_{jk} - \ell_1 \ell_2 \in J_G$. Since no monomial in the support of g belongs to $\mathrm{in}_<(G)$ (with the monomial order as in the previous paragraph), it follows that $g \notin J_G$, a contradiction. Hence, we see that $(\bar{x}_n, \ldots, \bar{x}_{i+1}) : \bar{x}_i$ is not generated by linear forms.   □

Lemma 7.54 *Let* $0 \le k \le n - 1$, $N^>(k) = \{k+1, \ldots, \ell\}$ *for some* $\ell \ge k+1$, *and* $N^<(k+1) = \{i, i+1, \ldots, k\}$ *for some* $i \le k$. *Then:*

(a) $(J_G, x_n, \ldots, x_{k+1}, y_{k+2}, \ldots, y_\ell) : y_{k+1} = (J_G, x_n, \ldots, x_{k+1}, x_k, \ldots, x_i, y_{k+2}, \ldots, y_\ell);$

(b) *for* $k + 2 \le s \le \ell$, y_s *is regular on* $(J_G, x_n, \ldots, x_i, y_{s+1}, \ldots, y_\ell)$.

Proof

(a) Let $r \in N^<(k+1)$. Then

$$x_r y_{k+1} = (x_r y_{k+1} - x_{k+1} y_r) + x_{k+1} y_r \in (J_G, x_n, \ldots, x_{k+1}).$$

This shows the inclusion $\supseteq$.

For the other inclusion, let $f \in S$ such that $f y_{k+1} \in (J_G, x_n, \ldots, x_{k+1}, y_{k+2}, \ldots, y_\ell)$. If H is the restriction of G to the set $[k]$, then $(J_G, x_n, \ldots, x_{k+1}, y_{k+2}, \ldots, y_\ell) = (J_H, x_n, \ldots, x_{k+1}, y_{k+2}, \ldots, y_\ell, \{x_r y_j : r \le k < j, \{r, j\} \in E(G)\})$. Let us observe that, if $\{r, j\} \in E(G)$ with $r \le k < j$, then, as G is closed, we have $\{k, j\} \in E(G)$, thus $j \in \{k+1, \ldots, \ell\}$. Therefore, we get

$$(J_G, x_n, \ldots, x_{k+1}, y_{k+2}, \ldots, y_\ell) = (J_H, x_n, \ldots, x_{k+1},$$

$$y_{k+2}, \ldots, y_\ell, x_i y_{k+1}, \ldots, x_k y_{k+1}).$$

By inspecting the S–polynomials of the generators in the right side of the above equality of ideals, it follows that

$$\mathrm{in}_<(J_G, x_n, \ldots, x_{k+1}, y_{k+2}, \ldots, y_\ell) =$$

$$(\mathrm{in}_<(J_H), x_n, \ldots, x_{k+1}, y_{k+2}, \ldots, y_\ell, x_i y_{k+1}, \ldots, x_k y_{k+1}).$$

Here $<$ denotes the lexicographic order on $S = K[x_1, \ldots, x_n, y_1, \ldots, y_n]$ induced by the natural order of the variables.

It follows that

$$\mathrm{in}_<(f) y_{k+1} \in (\mathrm{in}_<(J_H), x_n, \ldots, x_{k+1}, y_{k+2}, \ldots, y_\ell, x_i y_{k+1}, \ldots, x_k y_{k+1}),$$

which implies that $\mathrm{in}_<(f) \in (\mathrm{in}_<(J_H), x_n, \ldots, x_{k+1}, x_k, \ldots, x_i, y_{k+2}, \ldots, y_\ell)$. Hence, either $\mathrm{in}_<(f) \in (x_n, \ldots, x_{k+1}, x_k, \ldots, x_i, y_{k+2}, \ldots, y_\ell)$ or $\mathrm{in}_<(f) \in \mathrm{in}_<(J_H)$. In both cases we may proceed by induction on $\mathrm{in}_<(f)$. In the first case, let a be the coefficient of $\mathrm{in}_<(f)$ in f. Then $g = f - a\,\mathrm{in}_<(f)$ has

$\mathrm{in}_<(g) < \mathrm{in}_<(f)$ and $g y_{k+1} \in (J_G, x_n, \ldots, x_{k+1}, y_{k+2}, \ldots, y_\ell)$. In the second case, let $h \in J_H$ and $c \in K \setminus \{0\}$ such that $\mathrm{in}_<(h - cf) < \mathrm{in}_<(f)$. Thus, if $g = h - cf$, it follows that

$$g y_{k+1} \in (J_G, x_n, \ldots, x_{k+1}, y_{k+2}, \ldots, y_\ell)$$

as well.

(b). Let $k + 2 \le s \le \ell$. It is enough to show that y_s is regular on the initial ideal of $(J_G, x_n, \ldots, x_i, y_{s+1}, \ldots, y_\ell)$. Let H be the restriction of G to the set $[i]$. Then

$$\mathrm{in}_<(J_G, x_n, \ldots, x_i, y_{s+1}, \ldots, y_\ell) =$$

$$\mathrm{in}_<(J_H, x_n, \ldots, x_i, y_{s+1}, \ldots, y_\ell, \{x_r y_j : r < i < j, \{r, j\} \in E(G)\}) =$$

$$(\mathrm{in}_<(J_H), x_n, \ldots, x_i, y_{s+1}, \ldots, y_\ell, \{x_r y_j : r < i < j, \{r, j\} \in E(G)\}).$$

The last equality from above may be easily checked by observing that the S–polynomials $S(f_{r\ell}, x_r y_j)$ reduce to 0 for any $r < \ell \le i < j$ with $\{r, \ell\} \in E(H)$.

We claim that y_s does not divide any of the generators of

$$(\mathrm{in}_<(J_H), x_n, \ldots, x_i, y_{s+1}, \ldots, y_\ell, \{x_r y_j : r < i < j, \{r, j\} \in E(G)\}).$$

Obviously, y_s does not divide any of the generators of $\mathrm{in}_<(J_H)$. Next, if $\{r, s\} \in E(G)$ for some $r < i < k+1 < s$, then, as G is closed, we get $\{r, k+1\} \in E(G)$, contradiction to the fact that $i = \min N^<(k + 1)$. This shows that none of the generators $x_r y_j$ is divisible by y_s.   □

Theorem 7.55 *Let G be a closed graph. Then $R = S/J_G$ has a Koszul filtration.*

Proof Let G be closed with respect to its labeling. We set $\bar{f}$ for $f \mod(J_G) \in R = S/J_G$. For $k \in [n - 1]$, let $N^>(k) = \{k + 1, \ldots, \ell_k\}$ and $N^<(k + 1) = \{i_k, i_k + 1, \ldots, k\}$.

Let us consider the following families of ideals:

$$\mathcal{F}_1 = \bigcup_{k=1}^{n-1} \{(\bar{x}_n, \ldots, \bar{x}_1, \bar{y}_n, \ldots, \bar{y}_k), (\bar{x}_n, \ldots, \bar{x}_k)\},$$

$$\mathcal{F}_2 = \bigcup_{k=1}^{n-1} \{(\bar{x}_n, \ldots, \bar{x}_{k+1}, \bar{y}_{k+1}, \ldots, \bar{y}_{\ell_k}), (\bar{x}_n, \ldots, \bar{x}_{k+1}, \bar{y}_{k+2}, \ldots, \bar{y}_{\ell_k})\},$$

and

$$\mathcal{F}_3 = \bigcup_{k=1}^{n-1} \{(\bar{x}_n, \ldots, \bar{x}_{i_k}, \bar{y}_s, \ldots, \bar{y}_{\ell_k}) : k + 2 \le s \le \ell_k\}.$$

Fig. 7.6 The net has a
Koszul filtration

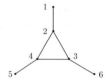

We claim that the family $\mathscr{F} = \mathscr{F}_1 \cup \mathscr{F}_2 \cup \mathscr{F}_3 \cup \{(0)\}$ is a Koszul filtration of R. We have to check that, for every $I \in \mathscr{F}$, there exists $J \in \mathscr{F}$ such that I/J is cyclic and $J : I \in \mathscr{F}$.

Let us consider $I = (\bar{x}_n, \ldots, \bar{x}_1, \bar{y}_n, \ldots, \bar{y}_k) \in \mathscr{F}_1$. Then, for $J = (\bar{x}_n, \ldots, \bar{x}_1, \bar{y}_n, \ldots, \bar{y}_{k+1}) \in \mathscr{F}_1$, we have $J : I = J$ since $\bar{y}_k$ is obviously regular on R/J.

For $I = (\bar{x}_n, \ldots, \bar{x}_k) \in \mathscr{F}_1$ with $1 \leq k \leq n - 1$, we take $J = (\bar{x}_n, \ldots, \bar{x}_{k+1}) \in \mathscr{F}_1$. Then, by (7.13), we get $J : I = (\bar{x}_n, \ldots, \bar{x}_{k+1}, \bar{y}_{k+1}, \ldots, \bar{y}_{\ell_k}) \in \mathscr{F}_2$. In addition, for $I = (\bar{x}_n)$, we have $(0) : I = (0)$ since $\bar{x}_n$ is regular on R.

Let us now choose $I \in \mathscr{F}_2$, $I = (\bar{x}_n, \ldots, \bar{x}_{k+1}, \bar{y}_{k+1}, \ldots, \bar{y}_{\ell_k})$ for some $1 \leq k \leq n - 1$. Then, $J = (\bar{x}_n, \ldots, \bar{x}_{k+1}, \bar{y}_{k+2}, \ldots, \bar{y}_{\ell_k}) \in \mathscr{F}_2$ and, by Lemma 7.54 (a), we have $J : I = (\bar{x}_n, \ldots, \bar{x}_{i_k}, \bar{y}_{k+2}, \ldots, \bar{y}_{\ell_k}) \in \mathscr{F}_3$.

Finally, if $I \in \mathscr{F}_3$, $I = (\bar{x}_n, \ldots, \bar{x}_{i_k}, \bar{y}_s, \ldots, \bar{y}_{\ell_k})$ for some $k + 2 \leq s \leq \ell_k$, we take $J = (\bar{x}_n, \ldots, \bar{x}_{i_k}, \bar{y}_{s+1}, \ldots, \bar{y}_{\ell_k}) \in \mathscr{F}_3$. By Lemma 7.54 (b), we get $J : I = J$ since $\bar{y}_s$ is regular on R/J. $\qquad \square$

The following example shows that the converse of Theorem 7.55 is not true. In other words, there exist Koszul graphs G which are not closed such that $R = S/J_G$ has a Koszul filtration.

Example 7.56 Let G be the net labeled as in Figure 7.6.

As we have seen in Section 7.1, the graph G is not closed. On the other hand, $K[x_1, \ldots, x_6, y_1, \ldots, y_6]/J_G$ possesses the following Koszul filtration:

(0),	(y_6),	(y_6, x_6),
(y_6, y_3),	(y_6, x_6, x_5),	(y_6, x_6, y_5, x_5),
(y_6, x_6, x_5, x_4),	$(y_6, y_4, x_6, x_5, x_4)$,	$(y_6, x_6, x_5, x_4, x_3)$,
$(y_6, x_6, x_5, x_4, x_3, x_2)$,	$(y_6, y_4, x_6, x_5, x_4, x_3)$,	$(y_6, y_4, y_3, x_6, x_5, x_4, x_3)$,
$(y_6, x_6, x_5, \ldots, x_1)$,	$(y_6, y_2, x_6, x_5, \ldots, x_2)$,	$(y_6, y_4, x_6, x_5, \ldots, x_2)$,
$(y_6, y_5, x_6, x_5, \ldots, x_1)$,	$(y_6, y_5, y_4, x_6, x_5, \ldots, x_1)$,	$(y_6, y_5, y_4, y_3 x_6, x_5, \ldots, x_1)$,
$(y_6, y_5, \ldots, y_2, x_6, x_5, \ldots, x_1)$,	$(y_6, y_5, \ldots, y_1, x_6, x_5, \ldots, x_1)$.	

Problems

7.21 Give an example of a chordal and claw free graph which is not Koszul.

7.22 Determine a Koszul filtration for a path graph.

7.23 Let G be a complete graph. Determine a Koszul filtration of S/J_G.

7.24 Let G be a closed graph. Show that the ideal $(x_1, \ldots, x_n)/J_G$ admits a linear S/J_G-resolution. Is this resolution finite?

7.25 Compute the S/J_{P_2}-resolution of the ideal $(x_1, x_2)/J_{P_2}$.

7.5   Permanental Edge Ideals and Lovász–Saks–Schrijver Ideals

In this section we study classes of ideals which are attached to a finite simple graph G on the vertex set $[n]$, and which are closely related to binomial edge ideals.

7.5.1   The Lovász–Saks–Schrijver Ideal L_G

Let K be a field and $S = K[x_1, \ldots, x_n, y_1, \ldots, y_n]$ be the polynomial ring over K in $2n$ variables.

Definition 7.57 The *permanental edge ideal* Π_G of G is the ideal generated by the polynomials $x_i y_j + x_j y_i$ with $\{i, j\} \in E(G)$, while the *Lovász–Saks–Schrijver ideal* is the ideal L_G generated by the polynomials $x_i x_j + y_i y_j$ with $\{i, j\} \in E(G)$.

More generally, if $X = (x_{ij})$ is an $n \times n$-matrix, then the *permanent* of X is the polynomial

$$p_X = \sum_\pi \prod_{i=1}^n x_{i\pi(i)},$$

where the sum is taken over all permutations π of $[n]$. Thus the permanental edge ideal Π_G is generated by the permanents of those 2×2-submatrices of the $2 \times n$-matrix $X = \begin{pmatrix} x_1 & \cdots & x_n \\ y_1 & \cdots & y_n \end{pmatrix}$ which correspond to the edges of G.

The Lovász–Saks–Schrijver ideals belong to a more general class of ideals which are related to orthogonal representations of graphs as introduced by Lovász [142] in 1979. Let $d \geq 1$ be an integer, and as in previous sections we denote by $\overline{G}$ the complementary graph of G with edge set $E(\overline{G}) = \binom{[n]}{2} \setminus E(G)$. An orthogonal representation of G in $\mathbb{R}^d$ is a map φ from $[n]$ to $\mathbb{R}^d$ such that for any edge $\{i, j\} \in E(\overline{G})$ in the complementary graph, the vectors $\varphi(i)$ and $\varphi(j)$ are orthogonal with respect to the standard scalar product in $\mathbb{R}^d$. Formulated differently, if we identify the image of the vertex i with the i-th row $(u_{i1}, \ldots, u_{id})$ of an $(n \times d)$-matrix $U = (u_{ij})_{(i,j) \in [n] \times [d]} \in \mathbb{R}^{n \times d}$, then the set of all orthogonal representations of the graph G is the vanishing set in $\mathbb{R}^{n \times d}$ of the ideal $L_{\overline{G}} \subset \mathbb{R}[x_{ij} : i = 1, \ldots, n, \ j = 1, \ldots, d]$, where $L_{\overline{G}}$ is generated by the homogeneous polynomials

$$\sum_{k=1}^{d} x_{ik} x_{jk} \tag{7.14}$$

In this context, the Lovász-Saks-Schrijver ideal L_G of G is just the ideal of orthogonal representations of $\overline{G}$ when $d = 2$.

Lovász, Saks, and Schrijver considered general-position orthogonal representations, that is, orthogonal representations in which any d representing vectors are linearly independent. In [143, Theorem 1.1] they proved the remarkable fact that G has such a representation in $\mathbb{R}^d$ if and only if G is $(n-d)$-connected in which case $L_{\overline{G}}$ is a prime ideal.

In the following remark we exhibit the relationship between binomial edge ideals, permanental edge ideals, and Lovász–Saks–Schrijver ideals.

Remark 7.58

(a) Assume that $\sqrt{-1} \in K$ and $\mathrm{char}(K) \neq 2$. We consider the following linear transformation φ with $\varphi(x_i) = x_i - y_i$ and $\varphi(y_i) = \sqrt{-1}(x_i + y_i)$ for all i. Then for every $i \neq j$, the binomial $x_i x_j + y_i y_j$ maps to $-2(x_i y_j + x_j y_i)$. Thus L_G is mapped to Π_G under this transformation.

(b) If $\sqrt{-1} \in K$ and G is a bipartite graph, then L_G may be identified with the binomial edge ideal J_G of G. Indeed, suppose $V(G) = V_1 \cup V_2$ is the bipartition of G with $|V_1| = m$ and $|V_2| = n$. We apply the automorphism of $K[x_1, \ldots, x_n, y_1, \ldots, y_n]$ to L_G defined by $x_i \mapsto x_i$ and $y_i \mapsto \sqrt{-1} y_i$ to obtain the binomial edge ideal J_G attached to the matrix

$$\begin{bmatrix} z_1 & \cdots & z_n \\ w_1 & \cdots & w_n \end{bmatrix},$$

where $z_i = x_i$ for $i = 1, \ldots, m$, $z_i = \sqrt{-1} y_i$ for $i = m+1, \ldots, n$, $w_i = \sqrt{-1} y_i$ for $i = 1, \ldots, m$, and $w_i = x_i$ for $i = m+1, \ldots, n$.

Like for binomial edge ideals it can be shown that Π_G is a radical ideal, provided $\mathrm{char}(K) \neq 2$. Indeed, in [125] it is shown that parity binomial edge ideals are radical provided the characteristic is not two. If $\sqrt{-1} \in K$, then the linear transformation $x_i \mapsto x_i$ for $i = 1, \ldots, n$ and $y_i \mapsto \sqrt{-1} y_i$ for $i = 1, \ldots, n$ maps permanental edge ideals to parity binomial edge ideals, and hence in this case permanental edge ideals are radical. The case that $\sqrt{-1} \notin K$ is treated similarly as in the proof of part (a) of the next theorem.

As a first consequence we obtain

Theorem 7.59 *Let G be a graph on $[n]$.*

(a) *If $\mathrm{char}(K) \neq 2$, then L_G is a radical ideal.*
(b) *If $\mathrm{char}(K) = 2$, then L_G is a radical ideal if and only if G is bipartite.*

Proof

(a) Assume first that $\sqrt{-1} \in K$. Then L_G and Π_G arise from each other by a linear change of coordinates, as we have seen in Remark 7.58. Hence L_G is a radical ideal if and only Π_G is.

Now suppose that $\sqrt{-1} \notin K$. We choose a field extension L/K with $\sqrt{-1} \in L$. Then $L_G \otimes_K L \subset L[x_1, \ldots, x_n, y_1, \ldots, y_n]$ is generated by the same binomials as L_G, and hence by the first part of the proof it follows that $L_G \otimes_K L$ is a radical ideal. Suppose L_G is not radical. Then there exists $f \notin L_G$ such that $f^k \in L_G$ for some k. It follows that $f^k \in L_G \otimes_K L$. It remains to show that $f \notin L_G \otimes_K L$. Suppose this is not the case. Let $\mu_f : S/L_G \to S/L_G$ be the S/L_G-module homomorphism induced by multiplication with f. Then $\operatorname{Im}(\mu_f \otimes_K L) = 0$, because $f \in L_G \otimes_K L$. Since L is a flat K-module, it follows that $\operatorname{Im}(\mu_f \otimes_K L) = (\operatorname{Im} \mu_f) \otimes_K L$, and hence $(\operatorname{Im} \mu_f) \otimes_K L = 0$. Since L is even faithfully flat over K, we conclude that $\operatorname{Im} \mu_f = 0$. This implies that $f = 0$, a contradiction.

(b) Since $\operatorname{char}(K) = 2$, we have that $\sqrt{-1} \in K$. Hence if G is bipartite, Remark 7.58 implies that L_G arises by a linear transformation from the binomial edge ideal J_G which is known to be radical by Corollary 7.13. Thus L_G is radical as well in this case.

It remains to consider the case that G is not bipartite. We want to show that L_G is not a radical ideal. According to the subsequent Lemma 7.60 it is enough to prove that $L_G S_Y$ is not a radical ideal. Here S_Y denotes localization of S with respect to the multiplicative set Y consisting of the powers of $y_1 y_2 \cdots y_n$. In S_Y all monomials in the y_i are units. Via the change of variables $x_i \mapsto z_i = \frac{x_i}{y_i}$ for $i = 1, \ldots, n$ we identify S_Y with $K[z_1, \ldots, z_n, y_1^{\pm 1}, \ldots, y_n^{\pm 1}]$. Thus the ideal $L_G S_Y$ is generated by the elements $z_i z_j + 1$ for $\{i, j\} \in E(G)$. We further transform $z_i \mapsto w_i := 1 + z_i$ for $i = 1, \ldots, n$. Then $L_G S_Y$ is generated by the elements $w_i + w_j + w_i w_j$ for $\{i, j\} \in E(G)$ in $S_Y = K[w_1, \ldots, w_n, y_1^{\pm 1}, \ldots, y_n^{\pm 1}]$. Since G is non-bipartite, there exists a subgraph of G which is an odd cycle, say C_m. We may assume that $V(C_m) = [m]$. Note that

$$\sum_{i=1}^{m-1}(w_i + w_{i+1} + w_i w_{i+1}) + (w_1 + w_m + w_1 w_m) = \sum_{i=1}^{m-1} w_i w_{i+1} + w_1 w_m, \quad (7.15)$$

since $\operatorname{char}(K) = 2$, and each w_i appears twice in the sum on the left-hand side of the equation. It follows that $\sum_{i=1}^{m-1} w_i w_{i+1} + w_1 w_m \in L_G S_Y$. We also have $w_{2i-1} w_{2i} + w_{2i} w_{2i+1} \in L_G S_Y$ for all $i = 1, \ldots, (m-1)/2$, because

$$w_{2i-1} w_{2i} + w_{2i} w_{2i+1} = w_{2i+1}(w_{2i-1} + w_{2i} + w_{2i} w_{2i-1}) \quad (7.16)$$
$$+ w_{2i-1}(w_{2i} + w_{2i+1} + w_{2i} w_{2i+1}).$$

From (7.15) and (7.16) we deduce that $w_1 w_m \in L_G S_Y$. By symmetry we also have $w_i w_{i+1} \in L_G S_Y$ for $i = 1, \ldots, m - 1$. This implies that $w_1 + w_m \in L_G S_Y$ and $w_i + w_{i+1} \in L_G S_Y$ for $i = 1, \ldots, m - 1$. Hence $w_{i+1}^2 \in L_G S_Y$ for $i = 1, \ldots, m - 1$, because $w_{i+1}^2 = w_{i+1}(w_i + w_{i+1}) + w_i w_{i+1}$. Similarly, $w_1^2 \in L_G S_Y$. In order to conclude the proof of the theorem, we show that $w_i \notin L_G S_Y$ for all $i = 1, \ldots, m$. Let F be the quotient field of $K[y_1^{\pm 1}, \ldots, y_n^{\pm 1}]$ and let $A = F[w_1, \ldots, w_n]/(w_{m+1}, \ldots, w_n) \cong F[w_1, \ldots, w_m]$. It is enough to show that $w_i \notin L_G A$ for all $i = 1, \ldots, m$. The above calculation has shown that $L_G A$ is a graded ideal generated by the linear forms $w_1 + w_m$ and $w_i + w_{i+1}$ for $i = 1, \ldots, m - 1$, and by the monomials $w_1 w_m$ and $w_i w_{i+1}$ for $i = 1, \ldots, m - 1$. Since $w_1 + w_m = \sum_{i=1}^{m-1}(w_i + w_{i+1})$ we see that $\dim_F (L_G A)_1 \le m - 1$, and hence not all w_i belong to $L_G A$. Say $w_1 \notin L_G A$. Since $w_i + w_{i+1} \in L_G A$ for $i = 1, \ldots, m$, it then follows that $w_i \notin L_G A$ for $i = 1, \ldots, m$.   □

In order to complete the proof of the preceding theorem we need

Lemma 7.60 *Let $T \subset S$ be a multiplicatively closed set, and let $I \subset S$ be an ideal such that $I S_T$ is not radical. Then I is not radical.*

Proof Since $I S_T$ is not radical, there exists $f/t \in S_T \setminus I S_T$ and an integer $k > 1$ such that $(f/t)^k \in I S_T$. It follows that $f^k/1 \in I S_T$. Therefore, there exist $g \in I$ and $t_0 \in T$ such that $f^k/1 = g/t_0$, and hence $(t_0 f)^k = t_0^{k-1} g \in I$. Assume that $t_0 f \in I$. Then $f/t = (t_0 f)/(t_0 t) \in I S_T$, a contradiction.   □

7.5.2 The Ideals I_{K_n} and $I_{K_{m,n-m}}$

It is our aim to understand the primary decomposition of the ideals L_G. As we have seen in Theorem 7.59, the ideal L_G is reduced when $\text{char}(K) \ne 2$. In particular, this is the case when $\sqrt{-1} \notin K$. Thus in this case L_G is the intersection of its minimal prime ideals.

Our first aim is to identify those minimal prime ideals of L_G which do not contain any variable. We denote by K_m the complete graph on $[m]$ and by $K_{m,n-m}$ complete bipartite graph on $[n]$ with vertex partition $[n] = \{1, \ldots, m\} \cup \{m + 1, \ldots, n\}$.

We define the ideals I_{K_n} and $I_{K_{m,n-m}}$ in $S = K[x_1, \ldots, x_n, y_1, \ldots, y_n]$ as follows:

We set $I_{K_1} = (0)$, $I_{K_2} = (x_1 x_2 + y_1 y_2)$ and for $n > 2$, we define I_{K_n} as the ideal generated by the binomials

$$
\begin{aligned}
f_{ij} &= x_i x_j + y_i y_j, & 1 \le i < j \le n, \\
g_{ij} &= x_i y_j - x_j y_i, & 1 \le i < j \le n, \\
h_i &= x_i^2 + y_i^2, & 1 \le i \le n.
\end{aligned}
\tag{7.17}
$$

For $1 \le m < n$ we define $I_{K_{m,n-m}}$ as the ideal generated by the binomials

$$f_{ij} = x_i x_j + y_i y_j, \quad 1 \le i \le m, \quad m+1 \le j \le n,$$

$$g_{ij} = x_i y_j - x_j y_i, \quad 1 \le i < j \le m \quad \text{or} \quad m+1 \le i < j \le n. \quad (7.18)$$

Throughout this and the following sections, when we refer to the standard generators of the ideals L_G, I_{K_n}, and $I_{K_{m,n-m}}$ we mean the generators introduced in Definition 7.57, and in (7.17) and (7.18), respectively.

Theorem 7.61 *Let G be a connected graph on [n], and let P be a minimal prime ideal of L_G which does not contain any variable. Suppose that $\sqrt{-1} \notin K$. Then $P = I_{K_{m,n-m}}$ for some m or $P = I_{K_n}$, depending on whether G is bipartite or G is non-bipartite.*

Proof Suppose first that G is bipartite with vertex bipartition $[n] = \{1, \ldots, m\} \cup \{m+1, \ldots, n\}$. We have $L_G \subset I_{K_{m,n-m}}$ and claim that $L_G S_y = I_{K_{m,n-m}} S_y$, where S_y denotes localization of S with respect to multiplicative set consisting of the powers of $y = y_1 y_2 \cdots y_n$. In the case that $G = K_{1,1}$ there is nothing to prove. Thus we may assume that G has at least three vertices. It suffices to show that $I_{K_{m,n-m}} S_y \subset L_G S_y$. The ideal $L_G S_y$ is generated by the elements $z_i z_j + 1$, where $\{i, j\} \in E(G)$ and where $z_i = x_i/y_i$ for $i = 1, \ldots, n$. We will show that $z_i - z_j \in L_G S_y$ for all $1 \le i < j \le m$ and for all $m+1 \le i < j \le n$. This together with the fact that $I_{K_{m,n-m}} S_y$ is generated by the polynomials

$$z_i z_j + 1, \quad 1 \le i \le m, \quad m+1 \le j \le n, \quad (7.19)$$

$$z_i - z_j, \, 1 \le i < j \le m \quad \text{or} \quad m+1 \le i < j \le n, \quad (7.20)$$

will then imply that indeed $I_{K_{m,n-m}} S_y \subset L_G S_y$. Let $1 \le i < j \le m$ (the case $m+1 \le i < j \le n$ can be treated similarly). Since G is connected, there exists a path $i = i_0, i_1, \ldots, i_{2s} = j$ in G. We have $z_i - z_j = \sum_{t=1}^{s}(z_{i_{2t-2}} - z_{i_{2t}})$. So it suffices to prove each of the summands $z_{i_{2t-2}} - z_{i_{2t}} \in L_G S_y$. Thus we may as well assume that $s = 1$. We have $z_i - z_j = z_{i_0} - z_{i_2} = z_{i_0}(z_{i_1} z_{i_2} + 1) - z_{i_2}(z_{i_0} z_{i_1} + 1)$ which is an element of $L_G S_y$. It proves the claim that $L_G S_y = I_{K_{m,n-m}} S_y$. It follows that $I_{K_{m,n-m}} S_y \subset P S_y$, and hence $I_{K_{m,n-m}} \subset P$, since P is a prime ideal and $y \notin P$. Finally, since P is a minimal prime ideal of L_G, and $I_{K_{m,n-m}}$ is a prime ideal as we shall in Theorem 7.65, we have $P = I_{K_{m,n-m}}$.

Next we consider the case that G is not bipartite. Similarly as in the bipartite case we have $L_G \subset I_{K_n}$ and claim that $L_G S_y = I_{K_n} S_y$. It suffices to show that $I_{K_n} S_y \subset L_G T_Y$. The ideal $L_G S_y$ is generated by the elements $z_i z_j + 1$, where $\{i, j\} \in E(G)$. We will show that $z_i - z_j \in L_G S_y$ for all $1 \le i < j \le n$. This together with the fact that $I_{K_m} S_y$ is generated by

$$z_i z_j + 1, z_i - z_j, \quad 1 \le i < j \le n, \quad (7.21)$$

$$z_i^2 + 1, \quad 1 \le i \le n \quad (7.22)$$

will then imply that indeed $I_{K_n} S_y \subset L_G S_y$. In fact, the polynomials (7.22) are linear combinations of the Equation (7.21), as can be seen from

$$z_i^2 + 1 = z_i(z_i - z_j) + (z_i z_j + 1)$$

for all $i < j$.

Let $1 \leq i < j \leq n$. Since G is non-bipartite, there exists an even walk (not necessarily a path) in G connecting i and j. As in the bipartite case, we deduce from this fact that $z_i - z_j \in L_G S_y$. As in the previous case it follows that $I_{K_n} \subset P$, and hence $P = I_{K_n}$ since by Theorem 7.66, I_{K_n} is a prime ideal. □

In order to complete the proof of Theorem 7.61 we need to show that the ideals I_{K_n} and $I_{K_{m,n-m}}$ are prime ideals in the case that $\sqrt{-1} \notin K$. The rest of the section is devoted to prove this. In a first step we show that the standard generators of these ideals form a Gröbner basis.

Lemma 7.62 *The standard generators of I_{K_n}, and the standard generators of $I_{K_{m,n-m}}$ form a Gröbner basis with respect to the lexicographic order induced by $x_1 > \cdots > x_n > y_1 > \cdots > y_n$.*

Proof The assertion of the lemma follows once we have shown that for either of the ideals all S-polynomials of the standard generators reduce to zero; see Theorem 1.29. If the initial monomials of a pair of binomials do not have a common factor, then this S-polynomial reduces to zero; see Corollary 1.30. Hence, in what follows, we only have to consider the case that the initial monomials have a common factor. In this case simple calculations show that such S-polynomials reduce to zero. We provide two examples and leave the remaining cases to the reader. First, for the standard generators h_i and f_{ij} of I_{K_n} we have $S(h_i, f_{ij}) = -y_i g_{ij}$, and second for the standard generators f_{ij} and f_{ik} of $I_{K_{m,n-m}}$ we have $S(f_{ij}, f_{ik}) = -y_i g_{jk}$ for $1 \leq i \leq m$ and $m + 1 \leq j < k \leq n$. □

Corollary 7.63 *Let $1 \leq m \leq n$. Then the variables $x_1, \ldots, x_n, y_1, \ldots, y_n$ are non zero-divisors modulo I_{K_n} and modulo $I_{K_{m,n-m}}$.*

Proof It follows from Lemma 7.62 that y_1 does not divide any of the monomial generators of $\mathrm{in}_<(I_{K_n})$, where $<$ is the lexicographic order induced by $x_1 > \cdots > x_n > y_1 > \cdots > y_n$. This implies that y_1 is a non zero-divisor modulo $\mathrm{in}_<(I_{K_n})$. Consequently, by Problem 2.24, y_1 is a non zero-divisor modulo I_{K_n}. By symmetry, all y_i are non zero-divisors modulo I_{K_n}. Furthermore, if we consider the initial ideal of I_{K_n} with respect to the lexicographic order induced by $y_1 > \cdots > y_n > x_1 > \cdots > x_n$, then as before it follows that x_1 is a non zero-divisor modulo $\mathrm{in}_<(I_{K_n})$, and hence modulo I_{K_n}. Again by symmetry it follows that all x_i are non zero-divisors modulo I_{K_n}.

We apply again Lemma 7.62 and deduce that y_1 and y_{m+1} do not divide any of the monomial generators of $\mathrm{in}_<(I_{K_{m,n-m}})$. This implies that y_1 and y_{m+1} are non zero-divisors modulo $\mathrm{in}_<(I_{K_{m,n-m}})$. Consequently, by Problem 2.24, y_1 and y_{m+1} are non zero-divisors modulo $I_{K_{m,n-m}}$. Again employing symmetry it follows that all y_i are non zero-divisors modulo $I_{K_{m,n-m}}$. The same arguments as used for the I_{K_n} now show that x_1 and x_{m+1}, and hence all x_i are non zero-divisors modulo $I_{K_{m,n-m}}$.   □

As another consequence of Lemma 7.62 we have

Corollary 7.64 $\text{height}(I_{K_n}) = n$ and $\text{height}(I_{K_{m,n-m}}) = n - 1$.

Proof Since $\text{height}(I) = \text{height}(\text{in}_<(I))$ for any graded ideal $I \subset S$ (see Theorem 2.19(b)), it suffices to show that $\text{height}(\text{in}_<(I_{K_n})) = n$ and $\text{height}(\text{in}_<(I_{K_{m,n-m}})) = n - 1$. By Lemma 7.62, $\text{in}_<(I_{K_n}) = J_1 + J_2$, where $J_1 = (x_1, \ldots, x_n)^2$ and $J_2 = (x_i y_j : 1 \le i < j \le n)$. Hence $(x_1, \ldots, x_n)$ is a minimal prime ideal of $\text{in}_<(I_{K_n})$, and any other monomial prime ideals $\text{in}_<(I_{K_n})$ has height $\ge n$. It follows that $\text{height}(\text{in}_<(I_{K_n})) = n$, as desired.

To compute the height of $I_{K_{m,n-m}}$ is more involved. By Lemma 7.62, we have

$$\text{in}_<(I_{K_{m,n-m}}) = (x_i x_j : 1 \le i \le m, m + 1 \le j \le n) \tag{7.23}$$
$$+ (x_i y_j : 1 \le i < j \le m \text{ or } m + 1 \le i < j \le n),$$

Thus $\text{in}_<(I_{K_{m,n-m}})$ may be viewed as the edge ideal of a bipartite graph H on the vertex set $V = V_1 \cup V_2$ with

$$V_1 = \{x_1, \ldots, x_m, y_{m+2}, \ldots, y_n\} \text{ and } V_2 = \{x_{m+1}, \ldots, x_n, y_2, \ldots, y_m\}.$$

We label the vertices of H such that $V_1 = \{v_1, \ldots, v_{n-1}\}$ and $V_2 = \{w_1, \ldots, w_{n-1}\}$, where $v_i = x_i$ for $1 \le i \le m$, $v_i = y_{n+m+1-i}$ for $m + 1 \le i \le n - 1$, $w_i = x_{m+i}$ for $1 \le i \le n - m$, and $w_i = y_{n-i+1}$ for $n - m + 1 \le i \le n - 1$. Figure 7.7 shows an example of such a graph for $m = 2$ and $n = 5$.

Recall that a Ferrers graph is a bipartite graph H' on $V = A \cup B$ with $A = \{a_1, \ldots, a_p\}$ and $B = \{b_1, \ldots, b_q\}$ such that $\{a_1, b_q\} \in E(H')$, $\{a_p, b_1\} \in E(H')$, and if $\{a_i, b_j\} \in E(H')$, then $\{a_t, b_l\} \in E(H')$ for all $1 \le t \le i$ and $1 \le l \le j$. Associated to a Ferrers graph H' is a sequence $\lambda = (\lambda_1, \ldots, \lambda_p)$ of nonnegative integers, where $\lambda_i = \deg_{H'} a_i$ which is the degree of the vertex a_i in H' for all $i = 1, \ldots, p$.

It can be seen that H with the labeling of the vertices as given above is a Ferrers graph.

We show that $S/\text{in}(I_{K_{m,n-m}})$ is Cohen–Macaulay. In particular it follows then that $\text{in}(I_{K_{m,n-m}})$ is an unmixed ideal of height $|V_1| = |V_2| = n - 1$, as desired.

Fig. 7.7 An example of a Ferrers graph

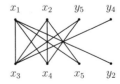

In order to prove that $S/\operatorname{in}(I_{K_{m,n-m}})$ is Cohen–Macaulay we refer to the algebraic theory of Ferrers graphs as developed by Corso and Nagel in [42]. According to [42, Corollary 2.7] we need to compute the sequence λ associated to H. By (7.23), $\deg_H v_i = \deg_H x_i = n - i$ for all $i = 1, \ldots, m$. Moreover, since by (7.23), $\deg_H y_j = j - 1 - m$, for all $j = m + 2, \ldots, n$, it follows that $\deg_H v_i = \deg_H y_{n+m+1-i} = n - i$ for all $i = m + 1, \ldots, n - 1$. Therefore, $\lambda = (n - 1, n - 2, \ldots, 2, 1)$ is the associated sequence to the Ferrers graph H, and hence by [42, Corollary 2.7], it follows that $S/\operatorname{in}_<(I_{K_{m,n-m}})$ is indeed Cohen–Macaulay. $\square$

Theorem 7.65 *The ideal $I_{K_{m,n-m}}$ is a prime ideal.*

Proof Because of Corollary 7.63 it suffices to show that $I_{K_{m,n-m}} S_y$ is a prime ideal in the ring S_y, where as before S_y denotes the localization with respect to $y = y_1 y_2 \cdots y_n$.

In order to see that $S_y/I_{K_{m,n-m}} S_y$ is a domain, we first consider the quotient R of S_y by the linear forms in (7.20) and denote by $\overline{I}$ the image of $I_{K_{m,n-m}} S_y$ in R. Notice that R is isomorphic to $K[z_1, z_{m+1}, y_1^{\pm 1}, \ldots, y_n^{\pm 1}]$, and that $S_y/I_{K_{m,n-m}} S_y \cong R/\overline{I}R$. Since the residue class map $S_y \to R$ identifies z_i with z_1 for $i = 1, \ldots, m$ and with z_{m+1} for $i = m + 1, \ldots, n$, we see that $\overline{I} = (z_1 z_{m+1} + 1)$. Since the polynomial $z_1 z_{m+1} + 1$ is irreducible, we conclude that $R/\overline{I}R$, and hence also $S_y/I_{K_{m,n-m}} S_y$ is a domain, as desired. $\square$

Theorem 7.66 *Let $n > 2$ be an integer.*

(a) *If $\sqrt{-1} \notin K$, then I_{K_n} is a prime ideal.*
(b) *If $\sqrt{-1} \in K$ and $\operatorname{char}(K) \neq 2$, then I_{K_n} is a radical ideal. More precisely,*

$$I_{K_n} = (x_1 + \sqrt{-1}y_1, \ldots, x_n + \sqrt{-1}y_n) \cap (x_1 - \sqrt{-1}y_1, \ldots, x_n - \sqrt{-1}y_n).$$

(c) *If $\operatorname{char}(K) = 2$, then I_{K_n} is a primary ideal with*

$$\sqrt{I_{K_n}} = (x_1 + y_1, \ldots, x_n + y_n).$$

Proof As in the proof of Theorem 7.65 we consider the image $I_{K_n} S_y$ of I_{K_n} in $S_y = K[z_1, \ldots, z_n, y_1^{\pm 1}, \ldots, y_n^{\pm 1}]$. It is generated by the polynomials (7.21) and (7.22).

Let R be the residue class ring of S_y modulo the linear forms given in (7.21). Then $R \cong K[z_1, y_1^{\pm 1}, \ldots, y_n^{\pm 1}]$ and $S_y/I_{K_n} S_y \cong R/(z_1^2 + 1)$.

(a) It follows that if $\sqrt{-1} \notin K$, then $S_y/I_{K_n} S_y$ is a domain, and hence I_{K_n} is a prime ideal in this case.
(b) Since $S_y/I_{K_n} S_y \cong R/((z_1 + \sqrt{-1})(z_1 - \sqrt{-1}))$ it follows that I_{K_n} is radical and has exactly two minimal prime ideals. The ideals $P_1 = (x_1 + \sqrt{-1}y_1, \ldots, x_n + \sqrt{-1}y_n)$ and $P_2 = (x_1 - \sqrt{-1}y_1, \ldots, x_n - \sqrt{-1}y_n)$ are prime ideals of height n containing I_{K_n}. By Corollary 7.64 we have $\operatorname{height}(K_n) = n$. It follows that $\{P_1, P_2\}$ is the set of minimal prime ideals of I_{K_n}.

(c) Since char(K) $= 2$, we have $x_i^2 + y_i^2 = (x_i + y_i)^2$ for all i. This shows that I_{K_n}
is not a prime ideal in this case. Furthermore, it follows that $x_i + y_i \in \sqrt{I_{K_n}}$
for all i. Since for all $i < j$, $g_{ij} = (x_i + y_i)x_j + (x_j + y_j)x_i$ and $f_{ij} =$
$(x_i + y_i)x_j + (x_j + y_j)y_i$, we see that $\sqrt{I_{K_n}} = (x_1 + y_1, \ldots, x_n + y_n)$, as
desired.   □

7.5.3   The Minimal Prime Ideals of L_G When $\sqrt{-1} \notin K$

By Theorem 7.59 the Lovász–Saks–Schrijver ideal L_G is reduced provided $\sqrt{-1} \notin$
K. Thus L_G is the intersection of its minimal prime ideals, and this intersection
represents its primary decomposition. It is the aim of this section to determine the
minimal prime ideals of L_G for the case that $\sqrt{-1} \notin K$.

Let H be a connected finite simple graph on the vertex set V. We define $\widetilde{H}$
as follows: if H is not bipartite, then $\widetilde{H}$ is the complete graph on V, and if H is
bipartite, then $\widetilde{H}$ is the complete bipartite graph on the given bipartition of H. Since
H is connected this bipartition is unique.

Let G be a finite graph on the vertex set $[n]$. For $T \subset [n]$ we set

$$Q_T(G) = (\{x_i, y_i\}_{i \in T}, I_{\widetilde{G}_1}, \ldots, I_{\widetilde{G}_{c(T)}}),$$

where $G_1, \ldots, G_{c(T)}$ are the connected components of $G_{[n] \setminus T}$. Note that if G_i is not
bipartite, then $I_{\widetilde{G}_i} = I_{K_{n_i}}$ for some n_i, and if G_i is bipartite, then $I_{\widetilde{G}_i} = I_{K_{m_i, n_i - m_i}}$
for some m_i and n_i.

It will turn out that the minimal prime ideals of L_G are all of the form $Q_T(G)$.
With the information given in Lemma 7.64 the height of these ideals can be easily
determined.

Proposition 7.67 *Let G be a graph on $[n]$ and let $T \subset [n]$. Then*

$$\text{height } Q_T(G) = |T| + n - b(T),$$

where $b(T)$ denotes the number of bipartite connected components of $G_{[n] \setminus T}$.

Proof We may assume that $G_1, \ldots, G_{b(T)}$ are the bipartite connected components
of G and $G_{b(T)+1}, \ldots, G_{c(T)}$ are the non-bipartite connected components of G. Let
$n_j = |V(G_j)|$ for all $j = 1, \ldots, c(T)$. Since the ideals $(x_i, y_i : i \in T)$ and $I_{\widetilde{G}_j}$ for
$j = 1, \ldots, c(T)$ are on pairwise disjoint sets of variables, it follows together with
Lemma 7.64 that

$$\text{height } Q_T(G) = \text{height}(x_i, y_i : i \in T) + \sum_{j=1}^{b(T)} \text{height}(I_{\widetilde{G}_j}) + \sum_{j=b(T)+1}^{c(T)} \text{height}(I_{\widetilde{G}_j})$$

$$= 2|T| + \sum_{j=1}^{b(T)} (n_j - 1) + \sum_{j=b(T)+1}^{c(T)} n_j$$

$$= |T| + (|T| + \sum_{j=1}^{b(T)} n_j + \sum_{j=b(T)+1}^{c(T)} n_j) - b(T)$$

$$= |T| + n - b(T),$$

as desired.   □

Next we have

Proposition 7.68 *Let G be a graph on $[n]$ and suppose that $\sqrt{-1} \notin K$. Then for all $T \subset [n]$, the ideal $Q_T(G)$ is a prime ideal and $L_G \subset Q_T(G)$.*

Proof That $L_G \subset Q_T(G)$ is an immediate consequence of Theorem 7.61.

Note that $Q_T(G)$ is of the form $I = \sum_i I_{K_{m_i,n_i-m_i}} + \sum_j I_{K_{t_j}} + U$, where U is generated by variables, and where $I_{K_{m_i,n_i-m_i}}$, $I_{K_{t_j}}$ and U are defined on pairwise disjoint sets of variables. Let $S' = S/U$. Then S' may be identified with a polynomial ring in the remaining variables and $I/U \subset S'$ identifies with $J = \sum_i I_{K_{m_i,n_i-m_i}} + \sum_j I_{K_{t_j}} \subset S'$. Thus it suffices to prove that J is a prime ideal. This will be a consequence of the following more general fact (∗) (similar to that of Lemma 7.14): for $j = 1, \ldots, m$, let I_j be an ideal in the polynomial ring

$$K[x_{11}, \ldots, x_{1n_1}, x_{21}, \ldots, x_{2n_2}, \ldots, x_{m1}, \ldots, x_{mn_m}]$$

satisfying the following properties:

  (i) the set of generators $\mathcal{G}_j$ of I_j is a subset of $K[x_{j1}, \ldots, x_{jn_j}]$;
  (ii) for all j the coefficients of the elements of $\mathcal{G}_j$ are $+1$ or -1;
  (iii) for any domain B with $\sqrt{-1} \notin B$ the ring $B[x_{j1}, \ldots, x_{jn_j}]/(\mathcal{G}_j)B[x_{j1}, \ldots, x_{jn_j}]$ is a domain and $\sqrt{-1} \notin B[x_{j1}, \ldots, x_{jn_j}]/(\mathcal{G}_j)B[x_{j1}, \ldots, x_{jn_j}]$.

Then $I_1 + \cdots + I_m$ is a prime ideal.

Before proving the (∗) let us use this fact to show that J is a prime ideal. In our particular case the ideals I_j are the ideals $I_{K_{m_i,n_i-m_i}}$ and $I_{K_{t_j}}$. Let $\mathcal{H}_i$ be the set of generators of $I_{K_{m_i,n_i-m_i}}$ as in (7.19) and (7.20) and $\mathcal{G}_j$ be the set of generators of $I_{K_{t_j}}$ as in (7.21) and (7.22). Clearly the conditions (i) and (ii) are satisfied. Let B be a domain with $\sqrt{-1} \notin B$. We first show that $B[x_{j1}, \ldots, x_{jn_j}]/(\mathcal{H}_j)B[x_{j1}, \ldots, x_{jn_j}]$ and $B[x_{j1}, \ldots, x_{jn_j}]/(\mathcal{G}_j)B[x_{j1}, \ldots, x_{jn_j}]$ are domains. As in the proofs of Theorem 7.65 and Theorem 7.66, where it was shown that $I_{K_{m,n-m}}$ and I_{K_n} are prime ideals, we need to show that $z_1 z_{m+1} + 1$ generates a prime ideal in $B[z_1, z_{m+1}, y_1^{\pm 1}, \ldots, y_n^{\pm 1}]$, and that $z_1^2 + 1$ generates a prime ideal in $B[z_1, y_1^{\pm 1}, \ldots, y_n^{\pm 1}]$. But this is obviously the case since $\sqrt{-1} \notin B$. Suppose $\sqrt{-1} \in B[x_{j1}, \ldots, x_{jn_j}]/(\mathcal{H}_j)B[x_{j1}, \ldots, x_{jn_j}]$. Then there exists

$f \in B[x_{j1}, \ldots, x_{jn_j}]$ such that $f^2 + 1 \in J_j$ where $J_j = (\mathcal{H}_j)B[x_{j1}, \ldots, x_{jn_j}]$. Since J_j is a graded ideal, all homogeneous components of $f^2 + 1$ belong to J_j. Therefore, if b is the constant term of f, then $b^2 + 1 \in J_j$, which is only possible if $b^2 + 1 = 0$. However since $\sqrt{-1} \notin B$, we obtain a contradiction.

Proof of (∗): We proceed by induction on m. The assertion is trivial for $m = 1$. Let $B = K[x_{11}, \ldots, x_{1n_1}, x_{21}, \ldots, x_{2n_2}, \ldots, x_{(m-1)1}, \ldots, x_{(m-1)n_{m-1}}]/(\mathcal{G}_1, \ldots, \mathcal{G}_{m-1})$. Then by our induction B is a domain and $\sqrt{-1} \notin B$. Moreover, we have $R/(I_1 + \cdots + I_m) \cong B[x_{m1}, \ldots, x_{jm_j}]/(\mathcal{G}_m)$, and hence (iii) implies that $I_1 + \cdots + I_m$ is a prime ideal. □

Theorem 7.69 *Suppose that $\sqrt{-1} \notin K$. Let G be a graph on $[n]$, and let P be a minimal prime ideal of L_G. Then there exists $T \subset [n]$ such that $P = Q_T(G)$.*

For the proof of this theorem we need

Lemma 7.70 *Let G be a connected graph on $[n]$, and let P be a minimal prime ideal of L_G containing a variable. Then there exists $k \in [n]$ such that $x_k, y_k \in P$.*

Proof If $G = K_2$, then L_G is a prime ideal, and hence $P = L_G$. Since L_G does not contain any variable, there is nothing to prove in this case. Now suppose that $G \neq K_2$ and that $x_i \in P$. Let us first assume that G is a bipartite graph on the vertex set $[n]$ with the bipartition $\{1, \ldots, m\} \cup \{m + 1, \ldots, n\}$. Suppose on the contrary that there exists no $k \in [n]$ such that $x_k, y_k \in P$. We claim that $(x_1, \ldots, x_m, y_{m+1}, \ldots, y_n) \subset P$. Given $j \in [m]$, there exists a path $i = i_0, i_1, \ldots, i_{2\ell} = j$. Here we used the fact that G is connected. We show by induction on ℓ that $x_j \in P$. Suppose that $\ell = 1$. Since $x_{i_0}x_{i_1} + y_{i_0}y_{i_1} \in P$ and $x_{i_0} \in P$ but $y_{i_0} \notin P$, it follows that $y_{i_1} \in P$. Similarly, since $x_{i_1}x_{i_2} + y_{i_1}y_{i_2} \in P$ and $y_{i_1} \in P$ but $x_{i_1} \notin P$, it follows that $x_{i_2} \in P$. Since $i_2, \ldots, i_{2\ell} = j$ is a path of length $2(\ell - 1)$ and $x_{i_2} \in P$, by induction hypothesis it follows that $x_j \in P$. By a similar argument for any $j \in \{m + 1, \ldots, n\}$, we have $y_j \in P$. Hence we have

$$L_G \subset I_{K_{m,n-m}} \subsetneq (x_1, \ldots, x_m, y_{m+1}, \ldots, y_n) \subset P,$$

which contradicts the assumption that P is a minimal prime ideal of L_G because $I_{K_{m,n-m}}$ is a prime ideal; see Theorem 7.65. Therefore, it follows that $x_k, y_k \in P$ for some k. Next assume that G is a non-bipartite graph. Since G is connected and non-bipartite, there exists $j \in [n]$ and an even path $i = i_0, i_1, \ldots, i_{2t} = j$, and an odd path $i = j_0, j_1, \ldots, j_{2s-1} = j$ in G connecting i and j. If there exists $\ell = i_0, \ldots, i_{2t}$ or $\ell = j_0, \ldots, j_{2s-1}$ with $x_\ell, y_\ell \in P$, then we are done. Otherwise, by an argument as in the bipartite case, we deduce from the generators $x_{i_r}x_{i_{r+1}} + y_{i_r}y_{i_{r+1}}$ for all $r = 0, \ldots, 2t - 1$, that $x_j \in P$. Similarly, we see that $y_j \in P$ by considering the generators attached to the odd path. □

Proof (of Theorem 7.69) We prove the theorem by induction on the number of vertices of G. If $|V(G)| = 2$, then $G = K_{1,1}$ and $L_G = Q_\emptyset(G)$. Now suppose that $|V(G)| > 2$, and let $G_1, \ldots, G_t$ be the connected components of G. Suppose first that $t > 1$. For $i = 1, \ldots, t$ let P_i be a minimal prime ideal of L_{G_i} which is contained in P. Then $\sum_{i=1}^{t} P_i \subset P$. Since $|V(G_i)| < |V(G)|$ for all i, our induction hypothesis implies that there exist subsets T_i such that $P_i = Q_{I_i}(G_i)$ for all i. Therefore, $\sum_{i=1}^{t} Q_{T_i}(G_i) \subset P$. Since $\sum_{i=1}^{t} Q_{T_i}(G_i) = Q_T(G)$ where $T = \bigcup_{i=1}^{t} T_i$, it follows that $Q_T(G) \subset P$. By Proposition 7.68, $Q_T(G)$ is a prime ideal, and hence $P = Q_T(G)$ because P is a minimal prime ideal of L_G. Next suppose that $t = 1$. If P does not contain any variable, then by Theorem 7.61 either $P = I_{K_n}$ or $P = I_{K_{m,n-m}}$ for suitable m. In either case, $P = Q_\emptyset(G)$. If P contains a variable, then by Lemma 7.70, there exists k such that $x_k, y_k \in P$. Let $\overline{P} = P/(x_k, y_k)$. Then $\overline{P}$ is a minimal prime ideal of $L_{G_{[n]\setminus\{k\}}}$. By induction hypothesis, there exists $\overline{T} \subset [n] \setminus \{k\}$ such that $\overline{P} = Q_{\overline{S}}(G_{[n]\setminus\{k\}})$. It follows that $P = Q_T(G)$ where $T = \overline{T} \cup \{k\}$. Since by Proposition 7.68 all $Q_T(G)$ are prime ideals and each $Q_T(G)$ contains L_G, the identity $L_G = \bigcap_{T \subset [n]} Q_T(G)$ follows from the first part of the theorem and the fact that $L_G = \sqrt{L_G}$, as noticed in Theorem 7.59.   □

As a consequence of Theorem 7.69 we obtain a primary decomposition of L_G which in general is highly redundant.

Corollary 7.71 *Let G be a graph on $[n]$ and suppose that $\sqrt{-1} \notin K$. Then*

$$L_G = \bigcap_{T \subset [n]} Q_T(G).$$

Combining Proposition 7.67 with Theorem 7.70 we obtain

Corollary 7.72 *Let G be a graph on $[n]$, and assume that $\sqrt{-1} \notin K$. Then*

$$\dim(S/L_G) = \max\{n - |T| + b(T) : \ T \subset [n]\}.$$

In particular, $\dim(S/L_G) \geq n + b$ where b is the number of bipartite connected components of G. Moreover, if L_G is unmixed, then $\dim(S/L_G) = n + b$.

Proof The equality follows from Proposition 7.67 and Theorem 7.70, and the equality implies the inequality $\dim(S/L_G) \geq n + b$. From Theorem 7.61 one deduces that $Q_\emptyset(G)$ is a minimal prime ideal of L_G. Hence if L_G is unmixed, then $\dim(S/L_G) = \dim(S/Q_\emptyset(G)) = n + b$.   □

Note that the lower bound given in Corollary 7.72 may be strict. For example, let G be the graph which is shown in Figure 7.8. Then $\dim(S/L_G) = 6$, while $n = 5$ and $b = 0$. On the other hand, $\dim(S/L_G) = n + b$ does not in general imply that L_G is unmixed. For instance, $\dim(S/L_{K_{2,2}}) = 5$ and in this case we have $n = 4$ and $b = 1$, but $L_{K_{2,2}}$ is not unmixed.

Fig. 7.8 The butterfly

In order to obtain an irredundant primary decomposition of L_G we have to identify those $T \subset [n]$ for which $Q_T(G)$ is minimal with respect to inclusion among the ideals $Q_{T'}(G)$ with $T' \subset [n]$.

The next result clarifies the inclusion relations between the ideals $Q_T(G)$ for $T \subset [n]$.

Proposition 7.73 *Let G be a graph on $[n]$, and let $T, T' \subset [n]$. Furthermore, let $H_1, \ldots, H_t$ and $G_1, \ldots, G_s$ be the connected components of $G_{[n] \setminus T}$ and $G_{[n] \setminus T'}$, respectively. Then $Q_T(G) \subset Q_{T'}(G)$ if and only if $T \subset T'$ and for all $i \in [t]$ with $|V(H_i)| > 1$ there exists $j \in [s]$ such that $V(H_i) \setminus T \subset V(G_j)$, and if H_i is bipartite (resp. non-bipartite), then G_j is also bipartite (resp. non-bipartite).*

Proof For every $A \subset [n]$, let $U_A = (x_i, y_i : i \in A)$. Then $Q_T(G) = (U_T, I_{\widetilde{H_1}}, \ldots, I_{\widetilde{H_t}})$ and $Q_{T'}(G) = (U_{T'}, I_{\widetilde{G_1}}, \ldots, I_{\widetilde{G_s}})$. One has $Q_T(G) \subset Q_{T'}(G)$ if and only if $T \subset T'$ and $(U_{T'}, I_{\widetilde{H_1}}, \ldots, I_{\widetilde{H_t}}) \subset (U_{T'}, I_{\widetilde{G_1}}, \ldots, I_{\widetilde{G_s}})$. For all $i = 1, \ldots, t$, let $I'_{\widetilde{H_i}}$ be the ideal generated by those generators of $I_{\widetilde{H_i}}$ which belong to $R = K[x_i, y_i : i \in [n] \setminus T']$. Then $(U_{T'}, I_{\widetilde{H_1}}, \ldots, I_{\widetilde{H_t}}) = (U_{T'}, I'_{\widetilde{H_1}}, \ldots, I'_{\widetilde{H_t}})$. It follows that $Q_T(G) \subset Q_{T'}(G)$ if and only if $T \subset T'$ and $(U_{T'}, I'_{\widetilde{H_1}}, \ldots, I'_{\widetilde{H_t}}) \subset (U_{T'}, I_{\widetilde{G_1}}, \ldots, I_{\widetilde{G_s}})$. The latter inclusion holds if and only if $(I'_{\widetilde{H_1}}, \ldots, I'_{\widetilde{H_t}}) \subset (I_{\widetilde{G_1}}, \ldots, I_{\widetilde{G_s}})$, since the generators of the ideals $(I'_{\widetilde{H_1}}, \ldots, I'_{\widetilde{H_t}})$ and $(I_{\widetilde{G_1}}, \ldots, I_{\widetilde{G_s}})$ belong to R. Now suppose $T \subset T'$. It is enough to show that the following conditions are equivalent:

(i) For all $i \in [t]$ with $|V(H_i)| > 1$, there exists $j \in [s]$ such that $V(H_i) \setminus T' \subset V(G_j)$, and if H_i is bipartite (resp. non-bipartite), then G_j is also bipartite (resp. non-bipartite).

(ii) $(I'_{\widetilde{H_1}}, \ldots, I'_{\widetilde{H_t}}) \subset (I_{\widetilde{G_1}}, \ldots, I_{\widetilde{G_s}})$.

The implication (i) $\Rightarrow$ (ii) is obvious. For the converse, let $i \in [t]$ with $|V(H_i)| > 1$, and let $k \in V(H_i) \setminus T'$. Then $k \in V(G_j)$ for some $j \in [s]$. We claim that $V(H_i) \setminus T' \subset V(G_j)$. If $V(H_i) \setminus T' = \{k\}$, there is nothing to prove. So we may assume that $|V(H_i) \setminus T'| \geq 2$. Suppose that there is an element $l \in V(H_i) \setminus T'$ such that $l \neq k$ and $l \notin V(G_j)$. Then there exists $r \in [s]$ with $r \neq j$ such that $l \in V(G_r)$. We may assume that $k < l$. First suppose that H_i is a bipartite graph on $A_1 \cup A_2$. Since $V(H_i) \setminus T'$ is nonempty, it follows that each connected component of $(H_i)_{[n] \setminus T'}$ is a connected component of $G_{[n] \setminus T'}$, and since H_i is bipartite, each of its components is bipartite as well. Hence

$$V(H_i) \setminus T' \subset \bigcup_{\substack{d=1 \\ G_d \text{ bipartite}}}^{s} V(G_d). \tag{7.24}$$

In particular, G_j and G_r are bipartite. If $k, l \in A_1$ or $k, l \in A_2$, then $g_{kl} = x_k y_l - x_l y_k \in I'_{\widetilde{H_i}}$. Hence by assumption (ii), $g_{kl} \in (I_{\widetilde{G_1}}, \ldots, I_{\widetilde{G_s}})$. Thus

$$g_{kl} = \sum_{t=1}^{p} r_t q_t, \tag{7.25}$$

where $r_t \in S$ and each q_t is a generator of $(I_{\widetilde{G}_1}, \ldots, I_{\widetilde{G}_s})$. Now, for $m \neq k, l$, we put all variables x_m and y_m equal to zero in the equality (7.25) and denote by $\overline{q}_t$ the image of q_t under this reduction. Then all $\overline{q}_t$ which are different from the binomials f_{kl}, g_{kl}, h_k and h_l listed in (7.17) and (7.18), are zero. Since k and l are contained in the different components G_j and G_r, respectively, it follows that $\overline{q}_t \neq f_{kl}, g_{kl}$. Also, since k and l belong to the bipartite components G_j and G_r, respectively, it follows that $\overline{q}_t \neq h_k, h_l$. Thus we see that after this reduction the right-hand side of (7.25) is zero while the left-hand side is nonzero, a contradiction. If $k \in A_1$ and $l \in A_2$, then $f_{kl} = x_k x_l + y_k y_l \in I'_{\widetilde{H}_i}$. Hence by assumption (ii), $f_{kl} \in (I_{\widetilde{G}_1}, \ldots, I_{\widetilde{G}_s})$. Then, similar to the previous case, we get a contradiction. Next, suppose that H_i is non-bipartite. So $g_{kl} = x_k y_l - x_l y_k \in I_{\widetilde{H}_i}$, and hence by assumption (ii), $g_{kl} \in (I_{\widetilde{G}_1}, \ldots, I_{\widetilde{G}_s})$. Thus $g_{kl} = \sum_{t=1}^{p} r_t q_t$, where $r_t \in S$ and each q_t is a generator of $(I_{\widetilde{G}_1}, \ldots, I_{\widetilde{G}_s})$. Now, as before, for $m \neq k, l$, we put all variables x_m and y_m equal to zero in this equality. After reduction it follows that g_{kl} can be written as $g_{kl} = r(x_k^2 + y_k^2) + r'(x_l^2 + y_l^2)$ for some polynomials $r, r' \in S$, which is a contradiction. Thus we see that indeed $V(H_i) \setminus T' \subset V(G_j)$. This proves the claim. By (7.24), it also follows that if H_i is bipartite, then G_j is bipartite.

Next we show that if H_i is non-bipartite, then G_j is also non-bipartite. Indeed, if H_i is non-bipartite, then $h_k = x_k^2 + y_k^2 \in I_{\widetilde{H}_i}$, and hence by the assumption (ii), $h_k \in (I_{\widetilde{G}_1}, \ldots, I_{\widetilde{G}_s})$. Thus

$$h_k = \sum_{t=1}^{p} r_t q_t, \tag{7.26}$$

where $r_t \in S$ and each q_t is a generator of $(I_{\widetilde{G}_1}, \ldots, I_{\widetilde{G}_s})$. If $h_k \neq q_t$ for all $t = 1, \ldots, p$, then by setting all variables x_m and y_m equal to zero for $m \neq k$, as before, we get $h_k = 0$, which is a contradiction. It follows that $h_k = q_t$ for some $t = 1, \ldots, p$. Therefore, h_k is a generator of $I_{\widetilde{G}_j}$, and hence G_j is a non-bipartite graph, too. $\square$

Now we are ready to determine the minimal prime ideals of L_G in the case that $\sqrt{-1} \notin K$.

Let G be a graph on $[n]$. In Section 7.2 we introduced a cut point of G as a vertex $i \in [n]$ with the property that G has less connected components than $G_{[n] \setminus \{i\}}$. In addition, we now call a vertex $i \in [n]$ a *bipartition point* of G if G has less bipartite connected components than $G_{[n] \setminus \{i\}}$. Let $\mathcal{M}(G)$ be the set of all sets $T \subset [n]$ such that each $i \in T$ is either a cut point or a bipartition point of the graph $G_{([n] \setminus T) \cup \{i\}}$. In particular, we have $\emptyset \in \mathcal{M}(G)$.

Theorem 7.74 *Let G be a graph on* $[n]$. *Suppose* $\sqrt{-1} \notin K$. *Then*

$$\{Q_T(G)\colon\ T \in \mathscr{M}(G)\}$$

is the set of minimal prime ideals of L_G.

Proof By Theorem 7.69 the minimal prime ideals of L_G are of the form $Q_T(G)$ with $T \subset [n]$. Assume first that $Q_T(G)$ is a minimal prime ideal of L_G. We want to show that $T \in \mathscr{M}(G)$. We may assume that $T \neq \emptyset$. Let $G_1, \ldots, G_r$ be the connected components of $G_{[n]\backslash T}$. Let $i \in T$ and $T' = T \setminus \{i\}$. Now we show that i is either a cut point or a bipartition point of the graph $G_{[n]\backslash T'}$. If i is not adjacent to any vertex of $G_1, \ldots, G_r$, then the connected components of $G_{[n]\backslash T'}$ are $G_1, \ldots, G_r$ together with the isolated vertex i. So Proposition 7.73 implies that $Q_{T'}(G) \subsetneq Q_T(G)$, a contradiction. Hence there exist some connected components of $G_{[n]\backslash T}$, say $G_1, \ldots, G_k$, which have at least one vertex adjacent to i. Then $G_1', G_{k+1}, \ldots, G_r$ are the connected components of $G_{[n]\backslash T'}$, where G_1' is the induced subgraph of $G_{[n]\backslash T'}$ on $(\bigcup_{j=1}^{k} V(G_j)) \cup \{i\}$. First suppose that $k = 1$. Then i is not a cut point of $G_{[n]\backslash T'}$, and if G_1' is bipartite, then G is also bipartite. Therefore, by Proposition 7.73, we have $Q_{T'}(G) \subsetneq Q_T(G)$, which is again a contradiction. Similarly, if G_1' and G_1 are both non-bipartite, we get a contradiction. If G_1' is non-bipartite and G_1 is bipartite, then i is a bipartition point of $G_{[n]\backslash T'}$. Next suppose that $k \geq 2$. Then clearly i is a cut point of $G_{[n]\backslash T'}$. Thus, indeed $T \in \mathscr{M}(G)$.

Conversely, suppose $T \in \mathscr{M}(G)$. Since $Q_\emptyset(G)$ does not contain any variable, it is not contained in any other $Q_{T'}(G)$. So $Q_\emptyset(G)$ is a minimal prime ideal of L_G. Now let $\emptyset \neq T \in \mathscr{M}(G)$ and let $G_1, \ldots, G_r$ be the connected components of $G_{[n]\backslash T}$. Suppose that $Q_T(G)$ is not a minimal prime ideal of L_G. Then by Theorem 7.69, there exists some $T' \subsetneq T$ such that $Q_{T'}(G) \subsetneq Q_T(G)$. Let $i \in T \setminus T'$. Then i is either a cut point or a bipartition point of $G_{([n]\backslash T)\cup\{i\}}$, since $T \in \mathscr{M}(G)$. If i is a cut point of $G_{([n]\backslash T)\cup\{i\}}$, then by a similar argument as in the first part of the proof, $G_1', G_{k+1}, \ldots, G_r$ are the connected components of $G_{([n]\backslash T)\cup\{i\}}$, where $k \geq 2$ and G_1' is the induced subgraph of $G_{([n]\backslash T)\cup\{i\}}$ on $(\bigcup_{j=1}^{k} V(G_j)) \cup \{i\}$. Thus $G_{[n]\backslash T'}$ has a connected component H which contains G_1' as an induced subgraph, and so $\bigcup_{j=1}^{k} V(G_j) \subset V(H) \setminus T$. Proposition 7.73, this contradicts the fact that $Q_{T'}(G) \subset Q_T(G)$. If i is a bipartition point but not a cut point of $G_{([n]\backslash T)\cup\{i\}}$, then by a similar argument as in the first part of the proof, $G_1', G_2, \ldots, G_r$ are the connected components of $G_{([n]\backslash T)\cup\{i\}}$, where G_1' is the induced subgraph of $G_{([n]\backslash S)\cup\{i\}}$ on $V(G_1) \cup \{i\}$ such that G_1' is non-bipartite and G_1 is bipartite. Thus $G_{[n]\backslash T'}$ has a connected component H which contains G_1' as an induced subgraph, and hence H is also a non-bipartite graph. Moreover, $V(G_1) \subset V(H) \setminus T$. Hence by Proposition 7.73, $V(G_1) = V(H) \setminus T$, since $Q_{T'}(G) \subset Q_T(G)$. Applying Proposition 7.73 once again we obtain a contradiction, since H is non-bipartite but G_1 is bipartite.   □

Corollary 7.75 *Let K be a field with* $\mathrm{char}(K) = 0$ *or* $\mathrm{char}(K) \not\equiv 1, 2 \bmod 4$. *Then the ideal L_G is prime if and only if G is a disjoint union of edges and isolated vertices.*

Proof Suppose first that G be a disjoint union of edges and isolated vertices. It suffices to prove that L_G is a prime ideal in the case that K is algebraically closed. The ideal L_G is the sum of ideals of the form $(x_i x_j + y_i y_j)$ for $i \neq j$ which are defined on pairwise disjoint sets of variables, and hence S/L_G is a tensor product of copies of $K[x_i, x_j, y_i, y_j]/(x_i x_j + y_i y_j)$ for $i \neq j$ and a polynomial ring. Since $x_i x_j + y_i y_j$ for $i \neq j$ is irreducible over any field, each factor is a domain. Now it follows from [147, Proposition 5.17] that S/L_G is a domain and hence that L_G is prime.

Conversely, suppose that L_G is a prime ideal. We may assume that K is a prime field. Then our hypothesis implies that $\sqrt{-1} \notin K$. Since by Theorem 7.74, $Q_\emptyset(G)$ is a minimal prime ideal of L_G, and since L_G is a prime ideal, it follows that $L_G = Q_\emptyset(G)$. By Theorem 7.74, this implies that $\mathscr{M}(G) = \{\emptyset\}$.

Let H be a connected component of G. The desired conclusion follows once we have shown that $H = K_2$ or a single vertex. First suppose that H is not a complete graph. Then there exists a minimal non-empty subset T of $V(H)$ with the property that $H_{V(H)\setminus T}$ is a disconnected graph. It follows that each element i of T is a cut point of $H_{([n]\setminus T)\cup\{i\}}$, and hence a cut point of $G_{([n]\setminus)\cup\{i\}}$. Therefore, by Theorem 7.74, $T \in \mathscr{M}(G)$, which contradicts the fact that $\mathscr{M}(G) = \{\emptyset\}$. Thus H is complete. Let $H = K_m$ where $V(H) = [m]$ and $m \geq 3$. Then $T' = [m] \setminus \{1, 2\} \in \mathscr{M}(G)$, since each element i of T' is a bipartition point of the graph $G_{([n]\setminus T')\cup\{i\}} = K_3$. Therefore, we get a contradiction, and hence $H = K_2$ or a single vertex, as desired. $\square$

As before, we denote by $b(T)$ the number of bipartite components of $G_{([n]\setminus T)}$. By using the correspondence between the set of minimal prime ideals of L_G and the set $\mathscr{M}(G)$ given in Theorem 7.74, one obtains the following criterion for unmixedness of the ideal L_G when $\sqrt{-1} \notin K$.

Corollary 7.76 *Let G be a graph with b bipartite connected components, and suppose that $\sqrt{-1} \notin K$. Then L_G is unmixed if and only if $b(T) = |T| + b$ for every $\emptyset \neq T \in \mathscr{M}(G)$.*

Proof The ideal L_G is unmixed if and only if all the minimal prime ideals of L_G have the same height. By Theorem 7.74, this is equivalent to say that for all $\emptyset \neq T \in \mathscr{M}(G)$, height$(Q_T(G)) = $ height$(Q_\emptyset(G))$. By Proposition 7.67, this is the case if and only if for every $\emptyset \neq T \in \mathscr{M}(G)$, we have $b(T) = |T| + b$. $\square$

Problems

7.26 Let $n \geq 3$ be an integer, and let K be field with $\sqrt{-1} \notin K$. We denote by C_n the cycle on $[n]$. Show that L_{C_n} is unmixed if and only if n is odd.

7.27 Let $n \geq 2$ be an integer, K_n the complete graph on n, and let K be a field with $\sqrt{-1} \notin K$. Then L_{K_n} is unmixed if and only if $n = 2$ or 3.

7.28 Show that $I_{K_{m,n-m}}$ is a Cohen-Macaulay ideal. Is I_{K_n} a Cohen–Macaulay ideal as well?

7.29 Determine the minimal prime ideals of L_{P_n} for the path graph P_n on $[n]$ when $\sqrt{-1} \notin K$.

7.30 Which are the minimal prime ideals of the complete bipartite graph on [7] with vertex decomposition $\{1, 2, 3\} \cup \{4, 5, 6, 7\}$ when $\sqrt{-1} \notin K$.

7.31 Suppose char$(K) = 2$. Find a nilpotent element in S/L_{C_n} when $n > 1$ is odd.

Notes

Binomial edge ideals were introduced by Herzog, Hibi, Hreinsdóttir, Kahle and Rauh [97], and independently by Ohtani [169]. In both of these papers it was shown that binomial edge ideals admit a squarefree initial ideal. Later, Badiane, Burke, and Sköldberg determined in [8] the universal Gröbner basis of a binomial edge ideal and showed it coincides with its Graver basis.

The closed graphs are by definition those whose binomial edge ideals have a quadratic Gröbner basis with respect to the lexicographic order. Crupi and Rinaldo [45] showed that a graph is closed if and only if there exists a monomial order for which the binomial edge ideal has a quadratic Gröbner basis.

In Theorem 7.7, a characterization of closed graphs in terms of the clique complex of the graph was given by in Ene, Herzog, and Hibi [61]. This characterization was used by Crupi and Rinaldo [46] to identify closed graphs as the so-called PI graphs, as defined by Hajós [87]. The proof is given in Theorem 7.9. Combining this theorem with Golumbic's [82, p. 195] characterization of PI graphs (proper interval graphs), one obtains Theorem 7.10. We present a self-contained proof by Ene, relying only on Theorem 7.7. Several other properties and characterizations of PI graphs can be found in [43, 74, 78, 88, 181, 182].

The minimal prime ideals of a binomial edge ideal are determined in [97]. This information is used in [61] to classify all closed graphs whose binomial edge ideal is Cohen–Macaulay. In [9], Banerjee and Núñez-Betancourt relate the projective dimension of S/J_G as well as the Cohen–Macaulay property of J_G to invariants that measure the connectivity of G. In general not so much is known about the free resolution of binomial edge ideals. However for some special cases the extremal Betti numbers, the regularity, and in some cases the graded Betti numbers are known, see [56, 219]. On the other hand, Kiani and Saeedi Madani [183] showed that the binomial edge ideal J_G has a linear resolution if and only if G is a complete graph, cf. Theorem 7.27. In [100] this result was generalized by describing the linear strand of any binomial edge ideal. In Theorem 7.28(a) a lower bound for the regularity of a binomial edge ideal is given. The proof is taken from the paper [144] by Matsuda and Murai, while part (b) is due to Ene and Zarojanu [70]. They showed that the lower bound is achieved by closed graphs. In the same paper Matsuda and Murai proved (see Theorem 7.36(a)) that the regularity of the binomial edge ideal of a graph G is bounded above by its number vertices, and they conjectured that this bound is achieved if and only if G is a path graph. This conjecture was proved by Kiani and Saeedi Madani [129], and is presented in Theorem 7.36(b). For their proof they use essentially Theorem 7.43 which is due to Mohammadi and Sharifan [148]. An alternative proof of Theorem 7.36(a) is given by Conca, De Negri, and Gorla in [37]. In [184] Kiani and Saeedi Madani conjectured that the regularity of a binomial edge ideal is also bounded by the number of maximal cliques of G, increased by one. At present this conjecture is widely open.

There are conjectures regarding the comparison of the graded Betti numbers of J_G and that of $\mathrm{in}_<(J_G)$. In [61] it is conjectured that the extremal Betti numbers of J_G and $\mathrm{in}_<(J_G)$ coincide, and that even all their graded Betti numbers coincide if G is closed. In Proposition 7.25, which is taken from [61], the expected equality of graded Betti numbers was proved for Cohen–Macaulay closed graphs. Moreover, for any closed graph it is known that the linear strands of J_G and $\mathrm{in}_<(J_G)$ have the same Betti numbers. This is a consequence of the result in [100], mentioned above, and a result in [128]. The equality of the extremal Betti numbers of J_G and $\mathrm{in}_<(J_G)$ is proved in [56] for very special cases. The proof is based on results of [186, 219]. Other results supporting these conjectures can be found in [33, 70].

A graph is called Koszul, if for some base field K, the standard graded K-algebra S/J_G is Koszul. The classification of Koszul graphs is still incomplete. Theorem 7.47, which asserts that a Koszul graph is chordal and claw free, is taken from [62]. Proposition 7.49 is due to Rauf and Rinaldo [174], while Theorem 7.50 which describes the glueing of Koszul graphs is again taken from [62]. Subsection 7.4.2 reflects the results of [63].

There are several generalizations of binomial edge ideals. One of these generalizations by Ene, Herzog, Hibi, and Qureshi deals with pairs of graphs [67] which is an extension of a construction of Rauh [175]. The paper [65] on determinantal facet ideals by Ene, Herzog, Hibi, and Mohammadi generalizes binomial edge ideals in a different direction: the binomials corresponding the edges of a graph are replaced by maximal minors of an $m \times n$-matrix corresponding to the facets of a pure simplicial complex.

Related to binomial edge ideals are the so-called permanental edge ideals and Lovász–Saks–Schrijver ideals. Orthogonal representations of graphs, introduced by Lovász, are maps from the vertex set of a graph to $\mathbb{R}^d$ where non-adjacent vertices are sent to orthogonal vectors. The Lovász–Saks–Schrijver ideals are the ideals expressing this condition. The first study of $L_{\overline{G}}$ and the geometry of the variety of orthogonal representations of G can be found in [143]. For that reason the ideal $L_{\overline{G}}$ of orthogonal graph representations of G is called the Lovász-Saks-Schrijver ideal of G. It is observed in [101] that under certain conditions these ideals and binomial edge ideals are related via linear transformations. The theory of permanental edge ideals and Lovász–Saks–Schrijver ideals has been independently developed by Herzog, Macchia, Saeedi Madani, and Welker [101] and also by Kahle and Sarmiento and Windisch [125] for $d = 2$. Permanental ideals have first been studied by Laubenbacher and Swanson [138]. For $d > 2$, Lovász–Saks–Schrijver ideals are investigated in [40].

Chapter 8
Ideals Generated by 2-Minors

Abstract In this chapter, we study ideals generated by 2-minors. Classical classes of ideals of this type are the ideals of 2-minors of an $m \times n$-matrix of indeterminates. The ideals considered here are generated by certain subsets of 2-minors of such a matrix. Any of these subsets is defined by a collection $\mathscr{C}$ of cells and include 2-sided ladders. Two types of such ideals are considered: those which are generated by the 2-minors corresponding to the cells in $\mathscr{C}$, called the adjacent minors, and those which are generated by all inner minors of $\mathscr{C}$. The Gröbner basis of such ideals will be studied, and it will be discussed when these ideals are prime ideals. Furthermore, algebraic properties, like normality or Cohen–Macaulayness, of the algebras defined by these ideals will be considered.

8.1 Configurations of Adjacent 2-Minors

Let $X = (x_{ij})_{\substack{i=1,\ldots,m \\ j=1,\ldots,n}}$ be a matrix of indeterminates, and let S be the polynomial ring over the field K in the variables x_{ij}. Let $\delta = [a_1, a_2 | b_1, b_2]$ be the 2-minor with rows a_1, a_2 and columns b_1, b_2. The elements $(a_i, b_j) \in \mathbb{Z}_{\geq 0}^2$ are called the *vertices* and the sets $\{(a_1, b_1), (a_1, b_2)\}, \{(a_1, b_1), (a_2, b_1)\}, \{(a_1, b_2), (a_2, b_2)\}$, and $\{(a_2, b_1), (a_2, b_2)\}$ the *edges* of the minor $[a_1, a_2 | b_1, b_2]$, see Figure 8.1. The set of vertices of δ will be denoted by $V(\delta)$.

The 2-minor $\delta = [a_1, a_2 | b_1, b_2]$ is called *adjacent* if $a_2 = a_1 + 1$ and $b_2 = b_1 + 1$. Let $\mathscr{C}$ be any set of adjacent 2-minors of X. We call such a set a *configuration* of adjacent 2-minors, and denote by $J_{\mathscr{C}}$ the ideal generated by the elements of $\mathscr{C}$.

Given $a = (i, j)$ and $b = (k, l)$ in $\mathbb{Z}_{\geq 0}^2$, we write $a \leq b$ if $i \leq k$ and $j \leq l$. The set $[a, b] = \{c \in \mathbb{Z}_{\geq 0}^2 : a \leq c \leq b\}$ is called an *interval*. An interval of the form $C = [a, a + (1, 1)]$ is called a *cell*. As can be seen from Figure 8.2, any configuration of adjacent 2-minors is defined by a collection of cells. This is the perspective that we will take when we study polyominoes in the next section.

© Springer International Publishing AG, part of Springer Nature 2018
J. Herzog et al., *Binomial Ideals*, Graduate Texts in Mathematics 279,
https://doi.org/10.1007/978-3-319-95349-6_8

Fig. 8.1 The edges of a
2-minor

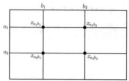

Fig. 8.2 A connected
configuration

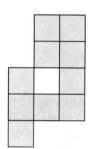

8.1.1 Prime Configurations of Adjacent 2-Minors

In this subsection, we classify the configurations whose ideals of adjacent 2-minors
are prime ideals.

Proposition 8.1 *Let $\mathscr{C}$ be a configuration of adjacent 2-minors. Then, the following
holds:*

(a) *$I_{\mathscr{C}}$ is a lattice basis ideal;*
(b) *$I_{\mathscr{C}}$ is a prime ideal if and only if all x_{ij} are nonzerodivisors modulo $I_{\mathscr{C}}$.*

Proof

(a) Let ϵ_{ij} be the element (ϵ_i, ϵ_j) of $\mathbb{Z}^m \times \mathbb{Z}^n$ where the ϵ_i, respectively, ϵ_j denote
   the canonical basis elements of $\mathbb{Z}^m$, respectively, of $\mathbb{Z}^n$. Then, for $i = 1, \ldots, m$
   and $j = 1, \ldots, n$, the elements ϵ_{ij} form a basis of the free $\mathbb{Z}$-module $\mathbb{Z}^m \times \mathbb{Z}^n$.
   For $i = 1, \ldots, m-1$ and $j = 1, \ldots, n-1$, we set $v_{ij} = \epsilon_{ij} + \epsilon_{i+1,j+1} - \epsilon_{i,j+1} -$
   $\epsilon_{i+1,j}$. Then, these elements form the basis $\mathscr{B}$ of a lattice L, and the lattice ideal
   I_L is just the ideal $I_2(X)$ of 2-minors of the matrix X, cf. Problem 3.13. It is
   a classical result that $I_2(X)$ is a prime ideal (see also Problem 7.9). Therefore,
   Theorem 3.17 implies that $(\mathbb{Z}^m \times \mathbb{Z}^n)/L$ is torsionfree. Now, let $\mathscr{B}'$ be the subset
   of $\mathscr{B}$ consisting of those v_{ij} for which $x_{ij}x_{i+1,j+1} - x_{i,j+1}x_{i+1,j}$ belongs to $\mathscr{C}$.
   Then, we see that $I_{\mathscr{C}}$ is equal to the lattice basis ideal $I_{\mathscr{B}'}$.
(b) Let $L' \subset L$ be the lattice with basis $\mathscr{B}'$. Then, L' is a direct summand of L.
   This implies that $(\mathbb{Z}^m \times \mathbb{Z}^n)/L'$ is torsionfree as well. Hence, by Theorem 3.17,
   $I_{L'}$ is a prime ideal. Now, we use Corollary 3.22 and deduce that $I_{L'} = I_{\mathscr{C}}$:
   $(\prod_{ij} x_{ij})^{\infty}$. Thus, if all x_{ij} are nonzerodivisors modulo $I_{\mathscr{C}}$, then $I_{\mathscr{C}} = I_{L'}$, and
   hence $I_{\mathscr{C}}$ is a prime ideal. The converse direction of statement (b) is obvious.

$\square$

Fig. 8.3 A chessboard
configuration

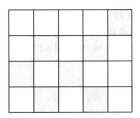

The set of vertices of $\mathscr{C}$, denoted $V(\mathscr{C})$, is the union of the vertices of its
adjacent 2-minors. Two distinct adjacent 2-minors $\delta, \gamma \in \mathscr{C}$ are called *connected*,
respectively, *weakly connected* if there exist $\delta_1, \ldots, \delta_r \in \mathscr{C}$ such that $\delta = \delta_1$,
$\gamma = \delta_r$, and such that for $i = 1, \ldots, r - 1$, δ_i and δ_{i+1} have a common edge,
respectively, a common vertex.

A maximal subset $\mathscr{D}$ of $\mathscr{C}$ with the property that any two minors of $\mathscr{D}$ are
connected is called a *connected component of $\mathscr{C}$*. A configuration $\mathscr{C}$ is called
connected, if $\mathscr{C}$ has only one connected component. A connected configuration of
adjacent 2-minors is displayed in Figure 8.2

To any configuration of adjacent 2-minors $\mathscr{C}$, we attach a graph $G_\mathscr{C}$ as follows:
the vertices of $G_\mathscr{C}$ are the connected components of $\mathscr{C}$. Let $\mathscr{A}$ and $\mathscr{B}$ be two
connected components of $\mathscr{C}$. Then, there is an edge between $\mathscr{A}$ and $\mathscr{B}$ if there exists
a minor $\delta \in \mathscr{A}$ and a minor $\gamma \in \mathscr{B}$ which have exactly one vertex in common. Note
that $G_\mathscr{C}$ may have multiple edges.

A set of adjacent 2-minors is called a *chessboard configuration*, if any two minors
of this set meet in at most one vertex. An example of a chessboard configuration is
given in Figure 8.3. An ideal $I \subset S$ is called a *chessboard ideal* if $I = I_\mathscr{C}$ where $\mathscr{C}$ is
a chessboard configuration. Note that the graph $G_\mathscr{C}$ of a chessboard configuration is
a simple bipartite graph. Indeed, in the case of a chessboard configuration the set of
vertices V of the graph $G_\mathscr{C}$ corresponds to the set of 2-minors of the configuration.
We define the vertex decomposition $V = V_1 \cup V_2$ of V by letting V_1 be the set of
2-minors located in the odd floors, and V_2 the set of 2-minors located in the even
floors.

Theorem 8.2 *Let $\mathscr{C}$ be a configuration of adjacent 2-minors. Then, the following
conditions are equivalent:*

(i) *$I_\mathscr{C}$ is a prime ideal.*
(ii) *$\mathscr{C}$ is a chessboard configuration and $G_\mathscr{C}$ has no cycle of length 4.*

For the proof of Theorem 8.2, we need the following two lemmata.

Lemma 8.3 *Let I be an ideal generated by adjacent 2-minors. For each of the
minors, we mark one of the monomials appearing in the minor as a potential initial
monomial. Then, there exists an ordering of the variables such that the marked
monomials are indeed the initial monomials with respect to the lexicographic order
induced by the given ordering of the variables.*

Proof In general, suppose that, in the set $[N] = \{1, 2, \ldots, N\}$, for each pair $(i, i+1)$ an ordering either $i < i + 1$ or $i > i + 1$ is given. We claim that there is a total order $<$ on $[N]$ which preserves the given ordering. Working by induction on N, we may assume that there is a total order $i_1 < \ldots < i_{N-1}$ on $[N - 1]$ which preserve the given ordering for the pairs $(1, 2), \ldots, (N - 2, N - 1)$. If $N - 1 < N$, then $i_1 < \ldots < i_{N-1} < N$ is a required total order $<$ on $[N]$. If $N - 1 > N$, then $N < i_1 < \ldots < i_{N-1}$ is a required total order $<$ on $[N]$.

The above fact guarantees the existence of an ordering of the variables such that the marked monomials are indeed the initial monomials with respect to the lexicographic order induced by the given ordering of the variables.   □

The following example demonstrates the construction of the monomial order given in the proof of Lemma 8.3.

Example 8.4 In Figure 8.4, each of the squares represents an adjacent 2-minor, and the diagonal in each of the squares indicates the marked monomial of the corresponding 2-minor. For a lexicographic order for which the marked monomials in Figure 8.4 are the initial monomials, the numbering of the variables in the top row must satisfy the following inequalities:

$$1 < 2 > 3 < 4 > 5 > 6.$$

By using the general strategy given in the proof of Lemma 8.3, we relabel the top row of the vertices by the numbers 1 up to 6, and proceed in the same way in the next rows. The final result can be seen in Figure 8.5

We call a vertex of a 2-minor in $\mathscr{C}$ *free*, if it does not belong to any other 2-minor of $\mathscr{C}$, and we call the 2-minor $\delta = ad - bc$ *free*, if either (i) a and d are free or (ii) b and c are free.

Lemma 8.5 *Let $\mathscr{C}$ be a chessboard configuration with $|\mathscr{C}| \geq 2$. Suppose $G_{\mathscr{C}}$ does not contain any cycle of length 4. Then, the $\mathscr{C}$ contains at least two free 2-minors.*

Fig. 8.4 Marked initial monomials

Fig. 8.5 Relabeling of the variables

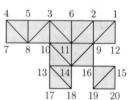

Fig. 8.6 A sequence of
adjacent 2-minors

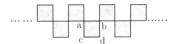

Proof We may assume there is at least one nonfree 2-minor in $\mathscr{C}$, say $\delta = ad - bc$. Since we do not have a cycle of length 4, there exists a sequence of 2-minors in $\mathscr{C}$ as indicated in Figure 8.6. Then, the leftmost and the rightmost 2-minor of this sequence is free.   □

Proof (Proof of Theorem 8.2) (i) $\Rightarrow$ (ii): Let $\delta, \gamma \in I_{\mathscr{C}}$ be two adjacent 2-minors which have an edge in common. Say, $\delta = ae - bd$ and $\gamma = bf - ce$. Then, $b(af - cd) \in I_{\mathscr{C}}$, but neither b nor $af - cd$ belongs to $I_{\mathscr{C}}$. Therefore, $\mathscr{C}$ must be a chessboard. Suppose $G_{\mathscr{C}}$ contains a cycle of length 4. Then, there exist in $I_{\mathscr{C}}$ adjacent two minors $\delta_1 = ae - bd$, $\delta_2 = ej - fi$, $\delta_3 = hl - ik$, and $\delta_4 = ch - dg$. Then, $h(bcjk - afgl) \in I_{\mathscr{C}}$, but neither h nor $bcjk - afgl$ belongs to $I_{\mathscr{C}}$.

(ii) $\Rightarrow$ (i): By virtue of Proposition 8.1, what we must prove is that all variables x_{ij} are nonzerodivisors of $S/I_{\mathscr{C}}$. Let $\mathscr{G}$ be the set of generating adjacent 2-minors of $I_{\mathscr{C}}$. Fix an arbitrary vertex x_{ij}. We claim that for each of the minors in $\mathscr{G}$ we may mark one of the monomials in the support as a potential initial monomial such that the variable x_{ij} appears in none of the potential initial monomials and that any two potential initial monomials are relatively prime.

We are going to prove this claim by induction on $|\mathscr{G}|$. If $|\mathscr{G}| = 1$, then the assertion is obvious. Now assume that $|\mathscr{G}| \geq 2$. Then, Lemma 8.5 says that there exist at least two free adjacent 2-minors in $\mathscr{G}$. Let $\delta = ad - bc$ be one of them and assume that a and d are free vertices of δ. We may assume that $x_{ij} \neq a$ and $x_{ij} \neq d$. Let $\mathscr{G}' = \mathscr{G} \setminus \{\delta\}$. By assumption of induction, for each of the minors of $\mathscr{G}'$ we may mark one of the monomials in the support as a potential initial monomial such that the variable x_{ij} appears in none of the potential initial monomials and that any two potential initial monomials are relatively prime. Then, these markings together with the marking ad are the desired markings of the elements of $\mathscr{G}$.

According to Lemma 8.3, there exists an ordering of the variables such that with respect to the lexicographic order induced by this ordering the potential initial monomials become the initial monomials. Since the initial monomials are relatively prime, it follows that $\mathscr{G}$ is a Gröbner basis of $I_{\mathscr{C}}$, and since x_{ij} does not divide any initial monomial of an element in $\mathscr{G}$ it follows that x_{ij} is a nonzerodivisor of $S/\operatorname{in}(I_{\mathscr{C}})$, where $\operatorname{in}(I_{\mathscr{C}})$ is the initial ideal of $I_{\mathscr{C}}$. But then, x_{ij} is a nonzerodivisor of $S/I_{\mathscr{C}}$ as well.   □

8.1.2 Configurations of Adjacent 2-Minors with Quadratic Gröbner Basis

The goal of this section is to identify all configurations $\mathscr{C}$ of adjacent 2-minors for which $I_{\mathscr{C}}$ has a quadratic Gröbner basis. To achieve this goal, several preliminary steps are required.

A configuration $\mathscr{C}$ of adjacent 2-minors is called a *path*, if there exists an ordering $\delta_1, \ldots, \delta_r$ of the elements of $\mathscr{C}$ such that for all i,

$$\delta_j \cap \delta_i \subset \delta_{i-1} \cap \delta_i \quad \text{for all} \quad j < i, \quad \text{and} \quad \delta_{i-1} \cap \delta_i \quad \text{is an edge of } \delta_i.$$

Such an ordering is called a *path ordering*. A vertex of δ_1 or of δ_r which does not belong to any other δ_j of the path is called an *end point* of the path.

A path $\mathscr{C}$ with path ordering $\delta_1, \ldots, \delta_r$ where $\delta_i = [a_i, a_i + 1 | b_i, b_i + 1]$ for $i = 1, \ldots, r$ is called *monotone*, if the sequences of integers $a_1, \ldots, a_r$ and $b_1, \ldots, b_r$ are monotone sequences. The monotone path $\mathscr{C}$ is called *decreasing* if the sequences $a_1, \ldots, a_r$ and $b_1, \ldots, b_r$ are both increasing or both decreasing, and the monotone path is called *increasing*, if one of the sequences is increasing and the other one is decreasing, see Figure 8.7.

If for $\mathscr{C}$ we have $a_1 = a_2 = \cdots = a_r$, or $b_1 = b_2 = \cdots = b_r$, then we call $\mathscr{C}$ a *line path*. Notice that a line path is both monotone increasing and monotone decreasing.

Let $\delta = ad - bc$ be an adjacent 2-minor with $a = x_{ij}$, $b = x_{ij+1}$, $c = x_{i+1j}$, and x_{i+1j+1}. Then, the monomial ad is called the *diagonal* of δ.

Lemma 8.6 *Let $\mathscr{C}$ be a monotone increasing (decreasing) path of 2-minors. Then, for any monomial order $<$ for which $I_{\mathscr{C}}$ has a quadratic Gröbner basis, the initial monomials of the generators are all diagonals (anti-diagonals).*

Proof Suppose first that $\mathscr{C}$ is a line path. If $I_{\mathscr{C}}$ has a quadratic Gröbner basis, then the initial monomials of the 2-minors of $\mathscr{C}$ are all diagonals or all anti-diagonals, because otherwise there would be two 2-minors δ_1 and δ_2 in $\mathscr{C}$ connected by an edge such that $\text{in}(\delta_1)$ is a diagonal and $\text{in}(\delta_2)$ is an anti-diagonal. The S-polynomial of δ_1 and δ_2 is a binomial of degree 3 which belongs to the reduced Gröbner basis of $I_{\mathscr{C}}$, a contradiction. If all initial monomials of the 2-minors in $\mathscr{C}$ are diagonals, we interpret $\mathscr{C}$ as a monotone increasing path, and if all initial monomials of the 2-minors in $\mathscr{C}$ are anti-diagonals, we interpret $\mathscr{C}$ as a monotone decreasing path.

Fig. 8.7 Monotone paths

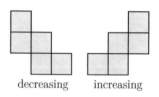

decreasing   increasing

Fig. 8.8 Sub-paths of a
monotone path

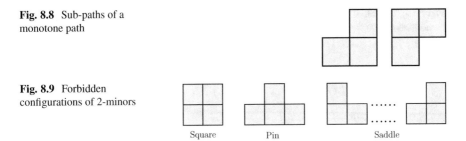

Fig. 8.9 Forbidden
configurations of 2-minors

Square   Pin   Saddle

Now, assume that $\mathscr{C}$ is not a line path. We may assume that $\mathscr{C}$ is monotone
increasing. (The argument for a monotone decreasing path is similar). Then, since $\mathscr{C}$
is not a line path it contains one of the following sub-paths displayed in Figure 8.8.

For both sub-paths, the initial monomials must be diagonals, otherwise $I_{\mathscr{C}}$ would
not have a quadratic Gröbner basis. Then, as in the case of line paths one sees that
all the other initial monomials of $\mathscr{C}$ must be diagonals. □

A configuration of adjacent 2-minors which is of the form shown in Figure 8.9,
or which is obtained by rotation from them, is called a *square*, a *pin*, and a *saddle*,
respectively.

Lemma 8.7 *Let $\mathscr{C}$ be a connected configuration of adjacent 2-minors. Then, $\mathscr{C}$ is
a monotone path if and only if $\mathscr{C}$ contains neither a square nor a pin nor a saddle.*

Proof Assume that $\mathscr{C} = \delta_1, \delta_2, \ldots, \delta_r$ with $\delta_i = [a_i, a_i + 1 | b_i, b_i + 1]$ for
$i = 1, \ldots, r$ is a monotone path. Without loss of generality, we may assume that
the both sequences $a_1, \ldots, a_r$ and $b_1, \ldots, b_r$ are monotone increasing. We will
show by induction on r that it contains no square, no pin, and no saddle. For
$r = 1$, the statement is obvious. Now, let us assume that the assertion is true
for $r - 1$. Since $\mathscr{C}' = \delta_1, \delta_2, \ldots, \delta_{r-1}$ is monotone increasing, it follows that
the coordinates of the minors δ_i for $i = 1, \ldots, r - 1$ sit inside the rectangle R
with corners $(a_1, b_1), (a_{r-1} + 1, b_1), (a_{r-1} + 1, b_{r-1} + 1), (a_1, b_{r-1} + 1)$, and
$\mathscr{C}'$ has no square, no pin, and no saddle. Since $\mathscr{C}$ is monotone increasing, $\delta_r =
[a_{r-1}, a_{r-1} + 1 | b_{r-1} + 1, b_{r-1} + 2]$ or $\delta_r = [a_{r-1} + 1, a_{r-1} + 2 | b_{r-1}, b_{r-1} + 1]$. It
follows that if $\mathscr{C}$ would contain a square, a pin, or a saddle, then the coordinates of
one of the minors $\delta_i, i = 1, \ldots, r - 1$ would not be inside the rectangle R.

Conversely, suppose that $\mathscr{C}$ contains no square, no pin, and no saddle. Then, $\mathscr{C}'$
contains no square, no pin, and no saddle as well. Thus, arguing by induction on
r, we may assume that $\mathscr{C}'$ is a monotone path. Without loss of generality, we may
even assume that $a_1 \leq a_2 \leq \cdots \leq a_{r-1}$ and $b_1 \leq b_2 \leq \cdots \leq b_{r-1}$. Now, let δ_r be
connected to δ_i (via an edge). If $i \in \{2, \ldots, r - 2\}$, then $\mathscr{C}$ contains a square, a pin,
or a saddle which involves δ_r, a contradiction. If $i = 1$ or $i = r - 1$, and $\mathscr{C}$ is not
monotone, then $\mathscr{C}$ contains a square or a saddle involving δ_r. □

Now, we are ready to classify all configurations of adjacent 2-minors for which
$I_{\mathscr{C}}$ admits a quadratic Gröbner basis,

Fig. 8.10 The initial
monomials of two adjacent
2-minors

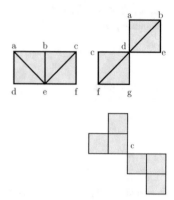

Fig. 8.11 Two connected
components with the
common corner c

Theorem 8.8 *Let $\mathscr{C}$ be a configuration of adjacent 2-minors. Then, the following conditions are equivalent:*

(i) *$I_{\mathscr{C}}$ has a quadratic Gröbner basis with respect to the lexicographic order induced by a suitable order of the variables.*

(ii) (α) *Each connected component of $\mathscr{C}$ is a monotone path.*

    (β) *If $\mathscr{A}$ and $\mathscr{B}$ are components of $\mathscr{C}$ which meet in a vertex which is not an end point of $\mathscr{A}$ or not an end point of $\mathscr{B}$, and if $\mathscr{A}$ is monotone increasing, then $\mathscr{B}$ must be monotone decreasing, and vice versa.*

(ii) *The initial ideal of $I_{\mathscr{C}}$ with respect to the lexicographic order induced by a suitable order of the variables is a complete intersection.*

Proof (i) $\Rightarrow$ (ii): (α) Suppose there is component $\mathscr{A}$ of $\mathscr{C}$ which is not a monotone path. Then, according to Lemma 8.7, $\mathscr{A}$ contains a square, a pin, or a saddle. In all three cases, no matter how we label the vertices of the component $\mathscr{A}$, it will contain, up to a rotation or reflection, two adjacent 2-minors with initial monomials as indicated in Figure 8.10.

In the first case, the S-polynomial of the two minors is $abf - bcd$ and in the second case it is $aef - bcg$. We claim that in both cases these binomials belong to the reduced Gröbner basis of $I_{\mathscr{C}}$, which contradicts our assumption (a).

Indeed, first observe that the adjacent 2-minors generating the ideal $I_{\mathscr{C}}$ is the unique minimal set of binomials generating $I_{\mathscr{C}}$. Therefore, the initial monomials of degree 2 are exactly the initial monomials of these binomials. Suppose now that $abf - bcd$ does not belong to the reduced Gröbner basis of $I_{\mathscr{C}}$. Then, one of the monomials ab, af, or bf must be the initial monomial of an adjacent 2-minor, which is impossible. In the same way, one argues in the second case.

(β) Assume that $\mathscr{A}$ and $\mathscr{B}$ have a vertex c in common. Then, c must be a corner of $\mathscr{A}$ and $\mathscr{B}$, that is, a vertex which belongs to exactly one 2-minor of $\mathscr{A}$ and exactly one 2-minor of $\mathscr{B}$, see Figure 8.11.

If for both components of the initial monomials are the diagonals (anti-diagonals), then the S-polynomial of the 2-minor in $\mathscr{A}$ with vertex c and the 2-minor of $\mathscr{B}$ with vertex c is a binomial of degree three whose initial monomial is

Fig. 8.12 Monotone increasing paths meeting in a vertex

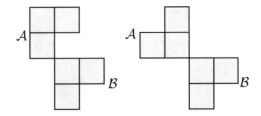

not divisible by any initial monomial of $\mathscr{C}$, unless c is an end point of both $\mathscr{A}$ and $\mathscr{B}$. Thus, the desired conclusion follows from Lemma 8.6.

(ii) $\Rightarrow$ (iii): The condition (b) implies that any pair of initial monomials of two distinct binomial generators of $I_{\mathscr{C}}$ are relatively prime. Hence, the initial ideal is a complete intersection.

(iii) $\Rightarrow$ (i): Since the initial monomial of the 2-minors generating $I_{\mathscr{C}}$ belong to any reduced Gröbner basis of $I_{\mathscr{C}}$, they must form a regular sequence. This implies that S-polynomials of any two generating 2-minors of I_C reduce to 0. Therefore, $I_{\mathscr{C}}$ has a quadratic Gröbner basis. $\quad\square$

Corollary 8.9 *Let $\mathscr{C}$ be a configuration of adjacent 2-minors satisfying the conditions of Theorem 8.8(ii). Then, $I_{\mathscr{C}}$ is a radical ideal generated by a regular sequence.*

Proof Let $\mathscr{C} = \delta_1, \ldots, \delta_r$. By Theorem 8.8, there exists a monomial order $<$ such that $\mathrm{in}_<(\delta_1), \ldots, \mathrm{in}_<(\delta_r)$ is a regular sequence. It follows that $\delta_1, \ldots, \delta_r$ is a regular sequence. Since the initial monomials are squarefree and form a Gröbner basis of $I_{\mathscr{C}}$, it follows that $I_{\mathscr{C}}$ is a radical ideal, cf. proof of Corollary 7.13. $\quad\square$

To demonstrate Theorem 8.8, we consider the following two examples displayed in Figure 8.12

In both examples, the component $\mathscr{A}$ and the component $\mathscr{B}$ are monotone increasing paths. In the first example, $\mathscr{A}$ and $\mathscr{B}$ meet in a vertex which is an end point of $\mathscr{A}$, therefore condition (ii)(β) of Theorem 8.8 is satisfied, and the ideal $I_{\mathscr{A} \cup \mathscr{B}}$ has a quadratic Gröbner basis. However, in the second example $\mathscr{A}$ and $\mathscr{B}$ meet in a vertex which is not an end point of $\mathscr{A}$ and not an end point of $\mathscr{B}$. Therefore, condition (ii)(β) of Theorem 8.8 is not satisfied, and the ideal $I_{\mathscr{A} \cup \mathscr{B}}$ does not have a quadratic Gröbner basis for the lexicographic order induced by any order of the variables.

8.1.3 Minimal Prime Ideals of Convex Configurations of Adjacent 2-Minors

Let $[a_1, a_2 | b_1, b_2]$ be a 2-minor. Each of the adjacent 2-minors $[a, a + 1 | b, b + 1]$ with $a_1 \leq a < a_2$ and $b_1 \leq b < b_2$ is called an adjacent 2-minor of $[a_1, a_2 | b_1, b_2]$.

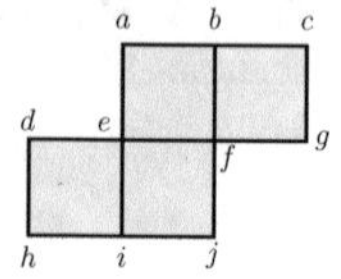

Fig. 8.13 A convex
configuration

Let $\mathscr{C}$ be a configuration of adjacent 2-minors, and let $\delta = [a_1, a_2|b_1, b_2]$ be a
2-minor whose vertices belong to $V(\mathscr{C})$. Then, δ is called an *inner minor* of $\mathscr{C}$, if
all adjacent 2-minors of δ belong to $\mathscr{C}$. The set of inner minors of $\mathscr{C}$ will be denoted
by $\text{In}(\mathscr{C})$ and the ideal they generate by $J_{\mathscr{C}}$.

A weakly connected configuration $\mathscr{C}$ of adjacent 2-minors is called *convex*, if
each minor $[a_1, a_2|b_1, b_2]$ whose vertices belong to $V(\mathscr{C})$ is an inner minor of $\mathscr{C}$. An
arbitrary configuration $\mathscr{C}$ of adjacent 2-minors is called convex, if each of its weakly
connected components is convex. For example, the configurations of adjacent 2-
minors displayed in Figure 8.12 are both convex, while the configuration shown in
Figure 8.2 is not convex.

We want to determine the minimal prime ideals of $I_{\mathscr{C}}$ when $\mathscr{C}$ is a convex
configuration of adjacent 2-minors. For this purpose, we have to introduce some
terminology: let $\mathscr{C} = \delta_1, \delta_2, \dots, \delta_r$ be an arbitrary configuration of adjacent 2-
minors. A subset W of the vertex set of $\mathscr{C}$ is called *admissible*, if for each δ_i either
$W \cap V(\delta_i) = \emptyset$ or $W \cap V(\delta_i)$ contains an edge of δ_i.

The admissible sets of the convex configuration of 2-minors displayed in
Figure 8.13 are the following:

$$\emptyset, \{c, g\}, \{d, h\}, \{a, e, i\}, \{b, f, j\}, \{a, b, c\}, \dots, \{a, b, c, d, e, f, g, h, i, j\}.$$

Let $W \subset V(\mathscr{C})$ be an admissible set. We define the ideal $P_W(\mathscr{C})$ as follows: let
$\mathscr{C}' = \{\delta \in \mathscr{C} : V(\delta) \cap W = \emptyset\}$. Then, $P_W(\mathscr{C})$ is defined to be generated by the
variables corresponding to W together with the inner 2-minors of $\mathscr{C}'$. Obviously,
$I_{\mathscr{C}} \subset P_W(\mathscr{C})$. Note that

$$P_W(\mathscr{C}) = (W, \text{In}(\mathscr{C}')) = (W, J_{\mathscr{C}'}) = (W, P_{\emptyset}(\mathscr{C}')).$$

For the configuration displayed in Figure 8.13, we have

$$P_{\emptyset}(\mathscr{C}) = (af - be, aj - bi, ej - fi, ag - ce, bg - cf, di - eh, dj - fh),$$

$$P_{\{d,h\}}(\mathscr{C}) = (d, h, af - be, aj - bi, ej - fi, ag - ce, bg - cf).$$

Lemma 8.10 *Let $\mathscr{C}$ be a convex configuration of adjacent 2-minors, and let $W \subset
V(\mathscr{C})$ be an admissible set of $\mathscr{C}$, and let $\mathscr{C}' = \{\delta \in \mathscr{C} : V(\delta) \cap W = \emptyset\}$. Then, $\mathscr{C}'$
is again a convex configuration of 2-adjacent minors. Moreover, for any admissible
set $W \subset V(\mathscr{C})$ the ideal $P_W(\mathscr{C})$ is a prime ideal.*

Proof Let $\mathscr{C}''$ be one of the weakly connected components of $\mathscr{C}'$. Let $[a_1, a_2|b_1, b_2]$ be a minor whose vertices belong to $V(\mathscr{C}'')$. We want to show that $[a_1, a_2|b_1, b_2]$ is an inner minor of $\mathscr{C}''$, in other words, that all adjacent 2-minors $\delta = [a, a + 1|b, b + 1]$ of $[a_1, a_2|b_1, b_2]$ belong to $\mathscr{C}''$. Suppose one of these adjacent 2-minors, say $\delta = [i, i + 1|j, j + 1]$, does not belong to $\mathscr{C}'$. Then, one of the edges of δ belongs to W, say $\{x_{i+1,j}, x_{i+1,j+1}\}$. If δ does not meet the vertices on the border lines connecting the corners x_{a_1,b_2} and x_{a_2,b_2}, and x_{a_2,b_1} and x_{a_2,b_2}, then $\delta' = [i + 1, i + 2|j + 1, j + 2]$ belongs to $[a_1, a_2|b_1, b_2]$, and hence it belongs to $\mathscr{C}$, since $\mathscr{C}$ is convex. Since $V(\delta') \cap W \neq \emptyset$ and W is an admissible set of $\mathscr{C}$, we see that either $x_{i+1,j+2} \in W$ or $x_{i+2,j+1} \in W$. Proceeding in this way, we see that W meets a border line of $[a_1, a_2|b_1, b_2]$. We may assume that $x_{j,b_2} \in W$ for some j with $a_1 + 1 < j < a_2 - 1$.

Now, if the adjacent 2-minor $[j, j + 1|b_2, b_2 + 1] \in \mathscr{C}$, then either $x_{j+1,b_2} \in W$ or $x_{j,b_2+1} \in W$. Proceeding in this way, we find a sequence of elements $x_{i_1,j_1}, \ldots, x_{i_r,j_r}$ which belongs to W with the property that (i) $(i_1, j_1) = (j, b_2)$, (ii) for all k with $1 \leq k < r$ we have $(i_{k+1}, j_{k+1}) = (i_k + 1, j_k)$ or $(i_{k+1}, j_{k+1}) = (i_k, j_k + 1)$, and (iii) the adjacent 2-minor $[i_r, i_r + 1|j_r, j_r + 1]$ does not belong to $\mathscr{C}$ (otherwise the sequence could be extended). Moreover, for $1 \leq k < r$ we have that $\delta_k = [i_k, i_k + 1|j_k, j_k + 1] \in \mathscr{C}$ and $\delta_k \cap W \neq \emptyset$ for all k. By construction, $\delta_{r-1} = [i_r - 1, i_r|j_r, j_r + 1]$ or $\delta_{r-1} = [i_r, i_r + 1|j_r - 1, j_r]$ belong to $\mathscr{C}$. We may assume that $\delta_{r-1} = [i_r - 1, i_r|j_r, j_r + 1]$. Then, it follows that all the adjacent 2-minors $\gamma_k = [k, k + 1|j_r, j_r + 1]$ for $k = i_r, \ldots, m - 1$ do not belong to $\mathscr{C}$. Indeed, if $\gamma_k \in \mathscr{C}$ for some k, then since $\delta_{r-1} = [i_r - 1, i_r|j_r, j_r + 1]$ belongs to $\mathscr{C}$ and since $\mathscr{C}$ is convex, it would follow that $[i_r, i_r + 1|j_r, j_r + 1]$ belongs to $\mathscr{C}$, a contradiction. Similarly, there exists $x_{k_1,l_1}, \ldots, x_{k_s,l_s}$ which belongs to W with the property that (i) $(k_1, l_1) = (j, b_2)$, (ii) for all t with $1 \leq t < s$ we have $(i_{t+1}, j_{t+1}) = (i_t - 1, j_t)$ or $(i_{t+1}, j_{t+1}) = (i_t, j_t - 1)$, and either the adjacent 2-minors $[i_s - 1, i_s|k - 1, k]$ do not belong to $\mathscr{C}$ for $k = 1, \ldots, j_s$, or the adjacent 2-minors $[k - 1, k|j_s - 1, j_s]$ do not belong to $\mathscr{C}$ for $k = 1, \ldots, i_s$.

Since the vertices x_{a_1,b_2} and x_{a_2,b_2} belong to the weakly connected component $\mathscr{C}''$ of $\mathscr{C}'$, there exists a chain $\sigma_1, \ldots, \sigma_v$ of adjacent 2-minors in $\mathscr{C}'$ with $V(\sigma_i) \cap V(\sigma_{i+1}) \neq \emptyset$ for all i and such that $x_{a_1,b_2} \in \sigma_1$ and $x_{a_2,b_2} \in \sigma_v$. It follows that $\{x_{i_1,j_1}, \ldots, x_{i_r,j_r}, x_{k_1,l_1}, \ldots, x_{k_s,l_s}\} \cap \sigma_i \neq \emptyset$ for some i. Therefore, $V(\sigma_i) \cap W \neq \emptyset$, a contradiction since $\sigma_i \in \mathscr{C}'$.

Now, since $\mathscr{C}'$ is a convex configuration, Corollary 8.23 implies that $P_\emptyset(\mathscr{C}')$ is a prime ideal. Therefore, $P_W(\mathscr{C})$ is a prime ideal as well. □

Theorem 8.11 *Let $\mathscr{C}$ be a convex configuration of adjacent 2-minors. Let P be a minimal prime ideal of $I_\mathscr{C}$. Then, there exists an admissible set $W \subset V(\mathscr{C})$ such that $P = P_W(\mathscr{C})$. In particular,*

$$\sqrt{I(\mathscr{C})} = \bigcap_W P_W(\mathscr{C}),$$

where the intersection is taken over all admissible sets $W \subset V(\mathscr{C})$.

Fig. 8.14 A non-convex
configuration

Proof Let P be any minimal prime ideal of $I(\mathscr{C})$, and let W be the set of variables among the generators of P. We claim that W is admissible. Indeed, suppose that $W \cap V(\delta) \neq \emptyset$ for some adjacent 2-minor of $\mathscr{C}$. Say, $\delta = ad - bc$ and $a \in W$. Then, $bc \in P$. Hence, since P is a prime ideal, it follows that $b \in P$ or $c \in P$. Thus, W contains the edge $\{a, c\}$ or the edge $\{a, b\}$ of δ.

Since $I(\mathscr{C}) \subset P$, it follows that $(W, I(\mathscr{C})) \subset P$. Observe that $(W, I(\mathscr{C})) = (W, I(\mathscr{C}'))$, where $W \cap V(\mathscr{C}') = \emptyset$ and $\mathscr{C}'$ is again a convex configuration; see Lemma 8.10. Modulo W we obtain a minimal prime ideal $\bar{P}$ of the ideal $I(\mathscr{C}')$ which contains no variables.

By Corollary 8.23, the ideal $P_\emptyset(\mathscr{C}')$ is a prime ideal containing $I(\mathscr{C}')$. Thus, the assertion of the theorem follows once we have shown that $P_\emptyset(\mathscr{C}') \subset \bar{P}$.

Since $P_\emptyset(\mathscr{C}')$ is generated by the union of the set of 2-minors of certain $r \times s$-matrices, it suffices to show that if P is a prime ideal having no variables among its generators and containing all adjacent 2-minors of the $r \times s$-matrix X, then it contains all 2-minors of X. In order to prove this, let $\delta = [a_1, a_2 | b_1, b_2]$ be an arbitrary 2-minor of X. We prove that $\delta \in P$ by induction on $(a_2 - a_1) + (b_2 - b_1)$. For $(a_2 - a_1) + (b_2 - b_1) = 2$, this is the case by assumption. Now, let $(a_2 - a_1) + (b_2 - b_1) > 2$. We may assume that $a_2 - a_1 > 1$. Let $\delta_1 = [a_1, a_2 - 1 | b_1, b_2]$ and $\delta_2 = [a_2 - 1, a_2 | b_1, b_2]$. Then, $x_{a_2-1, b_1} \delta = x_{a_2, b_1} \delta_1 + x_{a_1, b_1} \delta_2$. Therefore, by induction hypothesis $x_{a_2-1, b_1} \delta \in P$. Since P is a prime ideal, and $x_{a+k-1, 1} \notin P$ it follows that $\delta \in P$, as desired. $\qquad \square$

In general, it seems to be pretty hard to find the primary decomposition for ideals generated by adjacent 2-minors. For example, the primary decomposition (computed with the help of Singular [49]) of the ideal $I(\mathscr{C})$ of adjacent 2-minors, shown in Figure 8.14, is the following:

$$I(\mathscr{C}) = (ae - bd, ch - dg, ej - fi, hl - ik)$$
$$= (ik - hl, fi - ej, dg - ch, bd - ae, bcjk - afgl) \cap (d, e, h, i).$$

It turns out that $I(\mathscr{C})$ is a radical ideal. On the other hand, if we add the minor $di - eh$, we get a connected configuration $\mathscr{C}'$ of adjacent 2-minors. The ideal $I(\mathscr{C}')$ is not radical, because it contains a pin, see Proposition 8.14. Indeed, one has

$$\sqrt{I(\mathscr{C}')} = (ae - bd, ch - dg, ej - fi, hl - ik, di - eh, fghl - chjl,$$
$$bfhl - aejl, bchk - achl, bcfh - acej).$$

By applying the next theorem, we can determine the minimal prime ideals of $I(\mathscr{C}')$. We get

$$\sqrt{I(\mathscr{C}')} = (ae - bd, ch - dg, ej - fi, hl - ik, di - eh, fghl - chjl,$$
$$bfhl - aejl, bchk - achl, bcfh - acej)$$
$$= (-ik + hl, -fi + ej, -ek + dl, -fh + dj, -eh + di, -fg + cj,$$
$$-eg + ci, -dg + ch, -bk + al, -bh + ai, -bd + ae)$$
$$\cap\ (d, e, h, i) \cap (a, d, h, i, j) \cap (d, e, f, h, k) \cap (c, d, e, i, l) \cap (b, e, g, h, i)$$
$$\cap\ (a, d, h, k, ej - fi) \cap (c, d, e, f, hl - ik) \cap (b, e, i, l, ch - dg)$$
$$\cap\ (g, h, i, j, ae - bd).$$

The presentation of $\sqrt{I(\mathscr{C})}$ as an intersection of prime ideals as given in Theorem 8.11 is usually not irredundant. In order to obtain an irredundant intersection, we have to identify the minimal prime ideals of $I(\mathscr{C})$ among the prime ideals $P_W(\mathscr{C})$.

For any configuration $\mathscr{C}$, we denote by $\mathscr{G}(\mathscr{C})$ the set of adjacent 2-minors generating $P_\emptyset(\mathscr{C})$.

Theorem 8.12 *Let $\mathscr{C}$ be a convex configuration of adjacent 2-minors, and let $V, W \subset V(\mathscr{C})$ be admissible sets of $\mathscr{C}$, and let $P_V(\mathscr{C}) = (V, \mathscr{G}(\mathscr{C}'))$ and $P_W(\mathscr{C}) = (W, \mathscr{G}(\mathscr{C}''))$ where $\mathscr{C}' = \{\delta \in \mathscr{C} : V(\delta) \cap V = \emptyset\}$ and $\mathscr{C}'' = \{\delta \in \mathscr{C} : V(\delta) \cap W = \emptyset\}$. Then,*

(a) $P_V(\mathscr{C}) \subset P_W(\mathscr{C})$ *if and only if* $V \subset W$, *and for all elements*

$$\delta \in \mathscr{G}(\mathscr{C}') \setminus \mathscr{G}(\mathscr{C}'')$$

one has that $W \cap V(\delta)$ *contains an edge of* δ.

(b) $P_W(\mathscr{C}) = (W, \mathscr{G}(\mathscr{C}''))$ *is a minimal prime ideal of* $I(\mathscr{C})$ *if and only if for all admissible subsets* $V \subset W$ *with* $P_V(\mathscr{C}) = (V, \mathscr{G}(\mathscr{C}'))$ *there exists*

$$\delta \in \mathscr{G}(\mathscr{C}') \setminus \mathscr{G}(\mathscr{C}'')$$

such that the set $W \cap V(\delta)$ *does not contain an edge of* δ.

Proof

(a) Suppose that $P_V(\mathscr{C}) \subset P_W(\mathscr{C})$. The only variables in $P_W(\mathscr{C})$ are those belonging to W. This shows that $V \subset W$. The inclusion $P_V(\mathscr{C}) \subset P_W(\mathscr{C})$ implies that $\delta \in (W, \mathscr{G}(\mathscr{C}''))$ for all $\delta \in \mathscr{G}(\mathscr{C}')$. Suppose $W \cap V(\delta) = \emptyset$. Then, δ belongs to $P_\emptyset(\mathscr{C}'') = (\mathscr{G}(\mathscr{C}''))$. Let $f = u - v \in \mathscr{G}(\mathscr{C}'')$. Neither u nor v appears in another element of $\mathscr{G}(\mathscr{C}'')$. Therefore, any binomial of degree 2 in $P_\emptyset(\mathscr{C}'')$ belongs to $\mathscr{G}(\mathscr{C}'')$. In particular, $\delta \in \mathscr{G}(\mathscr{C}'')$, a contradiction. Therefore, $W \cap V(\delta) \neq \emptyset$.

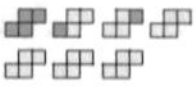

Fig. 8.15 The admissible sets shown in Figure 8.13

Suppose that $W \cap V(\delta)$ does not contain an edge of $\delta = ad - bc$. We may assume that $a \in W \cap V(\delta)$. Then, since $\delta \in P_W(\mathscr{C})$, it follows that $bc \in P_W(\mathscr{C})$. Since $P_W(\mathscr{C})$ is a prime ideal, we conclude that $b \in P_W(\mathscr{C})$ or $c \in P_W(\mathscr{C})$. Then, $b \in W$ or $c \in W$ and hence either the edge $\{a, b\}$ or the edge $\{a, c\}$ belongs to $W \cap V(\delta)$.

The "if" part of statement (a) is obvious.

(b) is a simple consequence of Theorem 8.11 and statement (a).   □

In Figure 8.15, we display all the minimal prime ideals of $I(\mathscr{P})$ for the path $\mathscr{P}$ shown in Figure 8.13. The fat dots mark the admissible sets and the dark shadowed areas, the regions where the inner 2-minors have to be taken.

8.1.4 Strongly Connected Configurations Which Are Radical

We call a connected configuration of adjacent 2-minors *strongly connected*, if the following condition is satisfied: for any two adjacent 2-minors $\delta_1, \delta_2 \in \mathscr{C}$ which have exactly one vertex in common, there exists $\delta \in \mathscr{C}$ which has a common edge with δ_1 and a common edge with δ_2.

This section is devoted to study strongly connected configuration of adjacent 2-minors $\mathscr{C}$ for which $I(\mathscr{C})$ is a radical ideal.

We call a configuration $\mathscr{C}$ of adjacent 2-minors a *cycle*, if for each $\delta \in \mathscr{C}$ there exist exactly two $\delta_1, \delta_2 \in \mathscr{C}$ such that δ and δ_1 have a common edge and δ and δ_2 have a common edge.

Lemma 8.13 *Let $\mathscr{C}$ be a strongly connected configuration which does not contain a pin. Then, $\mathscr{C}$ is a path or a cycle.*

Proof If $\mathscr{C}$ does not contain a pin, then for each adjacent 2-minor $\delta \in \mathscr{C}$ there exist at most two adjacent 2-minors in $\mathscr{C}$ which have a common edge with δ. Thus, if $\mathscr{C}$ is not a cycle but connected, there exists $\delta_1, \delta_2 \in \mathscr{C}$ such that δ_1 has a common edge only with δ_1. Now, in the configuration $\mathscr{C}' = \mathscr{C} \setminus \{\delta_1\}$ the element δ_2 has at most one edge in common with another element of $\mathscr{C}'$. If δ_2 has no edge in common with another element of $\mathscr{C}'$, then $\mathscr{C} = \{\delta_1, \delta_2\}$. Otherwise, continuing this argument, a simple induction argument yields the desired conclusion.   □

Proposition 8.14 *Let $\mathscr{C}$ be a strongly connected configuration of adjacent 2-minors. If $I(\mathscr{C})$ is a radical ideal, then $\mathscr{C}$ is a path or a cycle.*

Proof By Lemma 8.13, it is enough to prove that $\mathscr{C}$ does not contain a pin. Suppose $\mathscr{C}$ contains the pin $\mathscr{C}'$ as shown in Figure 8.16.

Fig. 8.16 A labeled pin

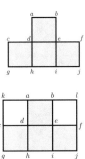

Fig. 8.17 A pin with
neighbors

Then, $q = acej - bcfh \notin I(\mathscr{C}')$ but $q^2 \in I(\mathscr{C}') \subset I(\mathscr{C})$. We consider two cases. In the first case, suppose that the adjacent 2-minors $kd - ac$ and $bf - le$ do not belong to $\mathscr{C}$, see Figure 8.17.

Then, $q \notin (I(\mathscr{C}), W)$ where W is the set of vertices which do not belong to $\mathscr{C}'$. It follows that $q \notin I(\mathscr{C})$. In the second case, we may assume that $ac - kd \in \mathscr{C}$. Let $\mathscr{C}''$ be the configuration with the adjacent 2-minors $kd - ac, ae - bd, ch - dg, di - eh$. Then, $r = kdi - aeg \notin I(\mathscr{C}'')$ but $r^2 \in I(\mathscr{C}'') \subset I(\mathscr{C})$. Then, $r \notin (I(\mathscr{C}), V)$ where V is the set of vertices in $\mathscr{C}$ which do not belong to $\mathscr{C}'$. It follows that $r \notin I(\mathscr{C})$. Thus, in both cases we see that $I(\mathscr{C})$ is not a radical ideal. $\square$

Problems

8.1 Let $\mathscr{C}$ be a path with more than one 2-minor. Find a zerodivisor modulo $I(\mathscr{C})$.

8.2 Show that there is no converse to the statement of Proposition 8.14. In other words, show that there is a cycle configuration $\mathscr{C}$ such that $I(\mathscr{C})$ is not a radical ideal.

8.3 Let $\mathscr{C}$ be the configuration of 2-minors whose vertices belong to the interval $[(1, 1), (3, 3)]$. Determine the minimal prime ideals of $I(\mathscr{C})$ and compute $\sqrt{I(\mathscr{C})}$.

8.2 Polyominoes

Polyominoes are, roughly speaking, plane figures obtained by joining squares of equal size edge to edge. To explain this more precisely, we first recall the concept of cells introduced in the previous section. We consider $(\mathbb{R}^2, \leq)$ as a partially ordered set with $(x, y) \leq (z, w)$ if and only if $x \leq z$ and $y \leq w$. Let $a, b \in \mathbb{Z}^2$. Then, the set $[a, b] = \{c \in \mathbb{Z}^2 : a \leq c \leq b\}$ is called an *interval*. In what follows, it is convenient also to define $[a, b]$ to be $[b, a]$ if $b \leq a$. Furthermore, we set $\overline{[a, b]} = \{x \in \mathbb{R}^2 : a \leq x \leq b\}$.

Fig. 8.18 Polyomino

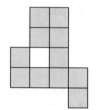

Let $a = (i, j), b = (k, l) \in \mathbb{Z}^2$ with $i < k$ and $j < l$. Then, the elements a and b are called *diagonal corners*, and the elements $c = (i, l)$ and $d = (k, j)$ are called *anti-diagonal corners* of $[a, b]$.

A *cell* is an interval of the form $[a, b]$, where $b = a + (1, 1)$. The cell $C = [a, a + (1, 1)]$ consists of the elements $a, a + (0, 1), a + (1, 0)$, and $a + (1, 1)$, which are called the *vertices* of C. We denote the set of vertices of C by $V(C)$. The intervals $[a, a + (1, 0)], [a + (1, 0), a + (1, 1)], [a + (0, 1), a + (1, 1)]$, and $[a, a + (0, 1)]$ are called the *edges* of C. Each edge consists of two elements, called the *corners of the edge*.

We now consider a finite collection of cells $\mathscr{P}$ in $\mathbb{Z}^2$. Let C and D be two cells of $\mathscr{P}$. Then, C and D are said to be *connected*, if there is a sequence of cells $\mathscr{C} = C_1, \ldots, C_m = D$ of $\mathscr{P}$ such that $C_i \cap C_{i+1}$ is an edge of C_i for $i = 1, \ldots, m - 1$. If, in addition, $C_i \neq C_j$ for all $i \neq j$, then $\mathscr{C}$ is called a *path* (connecting C and D). The collection of cells $\mathscr{P}$ is called a *polyomino* if any two cells of $\mathscr{P}$ are connected, see Figure 8.18. The set $V(\mathscr{P}) = \bigcup_{C \in \mathscr{P}} V(C)$ is called the set of vertices of $\mathscr{P}$.

Let $\mathscr{Q}$ be an arbitrary collection of cells. Then, each connected component of $\mathscr{Q}$ is a polyomino.

Let $\mathscr{P}$ be a polyomino, and let K be a field. We denote by S the polynomial ring over K with variables x_{ij} where $(i, j) \in V(\mathscr{P})$. A 2-minor $x_{ij}x_{kl} - x_{il}x_{kj} \in S$ is called an *inner minor* of $\mathscr{P}$ if all the cells $[(r, s), (r + 1, s + 1)]$ with $i \leq r \leq k - 1$ and $j \leq s \leq l - 1$ belong to $\mathscr{P}$. In that case, the interval $[(i, j), (k, l)]$ is called an *inner interval* of $\mathscr{P}$. The ideal $I_{\mathscr{P}} \subset S$ generated by all the inner minors of $\mathscr{P}$ is called the *polyomino ideal* of $\mathscr{P}$. We also set $K[\mathscr{P}] = S/I_{\mathscr{P}}$, and call it the *coordinate ring* of $\mathscr{P}$.

8.2.1 Balanced Polyominoes

Among the polyominoes, the balanced polyominoes admit coordinate rings with many nice properties. An interval $[a, b]$ with $a = (i, j)$ and $b = (k, l)$ is called a *horizontal edge interval* of $\mathscr{P}$ if $j = l$ and the sets $\{r, r + 1\}$ for $r = i, \ldots, k - 1$ are edges of cells of $\mathscr{P}$. Similarly, one defines vertical edge intervals of $\mathscr{P}$.

An integer value function $\alpha : V(\mathscr{P}) \to \mathbb{Z}$ is called *admissible*, if for all maximal horizontal or vertical edge intervals $\mathscr{I}$ of $\mathscr{P}$ one has

Fig. 8.19 An admissible labeling

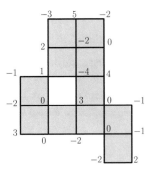

$$\sum_{a \in \mathscr{I}} \alpha(a) = 0.$$

In Figure 8.19, an admissible labeling of the polyomino is shown. Given an admissible labeling α, we define the binomial

$$f_\alpha = \prod_{\substack{a \in V(\mathscr{P}) \\ \alpha(a)>0}} x_a^{\alpha(a)} - \prod_{\substack{a \in V(\mathscr{P}) \\ \alpha(a)<0}} x_a^{-\alpha(a)}.$$

Let $J_{\mathscr{P}}$ be the ideal generated by the binomials f_α where α is an admissible labeling of $\mathscr{P}$. It is obvious that $I_{\mathscr{P}} \subset J_{\mathscr{P}}$. We call a polyomino *balanced* if for any admissible labeling α, the binomial $f_\alpha \in I_{\mathscr{P}}$. This is the case if and only if $I_{\mathscr{P}} = J_{\mathscr{P}}$.

Consider the free abelian group $G = \bigoplus_{(i,j) \in V(\mathscr{P})} \mathbb{Z}e_{(i,j)}$ with basis elements $e_{(i,j)}$. To any cell $C = [(i, j), (i + 1, j + 1)]$ of $\mathscr{P}$ we attach the element $b_C = e_{(i,j)} + e_{(i+1,j+1)} - e_{(i+1,j)} - e_{(i,j+1)}$ in G and let $\Lambda \subset G$ be the lattice spanned by these elements.

Lemma 8.15 *The elements b_C form a K-basis of Λ and hence $\mathrm{rank}_{\mathbb{Z}} \Lambda = |\mathscr{P}|$. Moreover, Λ is saturated. In other words, G/Λ is torsionfree.*

Proof We order the basis elements $e_{(i,j)}$ lexicographically. Then, the initial basis element of b_C is $e_{(i,j)}$. This shows that the elements b_C are linearly independent and hence form a $\mathbb{Z}$-basis of Λ. We may complete this basis of Λ by the elements $e_{(i,j)}$ for which (i, j) is not a lower-left corner of a cell of $\mathscr{P}$ to obtain a basis of G. This shows that G/Λ is free, and hence torsionfree. □

The lattice ideal I_Λ attached to the lattice Λ is the ideal generated by all binomials

$$f_v = \prod_{\substack{a \in V(\mathscr{P}) \\ v_a>0}} x_a^{v_a} - \prod_{\substack{a \in V(\mathscr{P}) \\ v_a<0}} x_a^{-v_a}$$

with $v = \sum_{a \in V(\mathscr{P})} v_a e_a \in \Lambda$.

Proposition 8.16 *Let $\mathscr{P}$ be a balanced polyomino. Then, $I_{\mathscr{P}} = I_\Lambda$.*

Proof The assertion follows once we have shown that for any $v \in \Lambda$ there exists an admissible labeling α of $\mathscr{P}$ such that $v_a = \alpha(a)$ for all $a \in V(\mathscr{P})$. Indeed, since the elements $b_C \in \Lambda$ form a $\mathbb{Z}$-basis of Λ, there exist integers $z_C \in \mathbb{Z}$ such that $v = \sum_C z_C b_C$. We set $\alpha = \sum_{C \in \mathscr{P}} z_C \alpha_C$ where for $C = [(i, j), (i + 1, j + 1)]$,

$$\alpha_C((k, l)) = \begin{cases} 1, & \text{if } (k, l) = (i, j) \text{ or } (k, l) = (i + 1, j + 1), \\ -1, & \text{if } (k, l) = (i + 1, j) \text{ or } (k, l) = (i, j + 1), \\ 0, & \text{otherwise.} \end{cases}$$

Then, $\alpha(a) = v_a$ for all $a \in V(\mathscr{P})$. Since each α_C is an admissible labeling of $\mathscr{P}$ and since any linear combination of admissible labelings is again an admissible labeling, the desired result follows.   □

Corollary 8.17 *If $\mathscr{P}$ is a balanced polyomino, then $I_{\mathscr{P}}$ is a prime ideal of height $|\mathscr{P}|$.*

Proof By Proposition 8.16, $I_{\mathscr{P}} = I_\Lambda$ and by Lemma 8.15, Λ is saturated. It follows that $I_{\mathscr{P}}$ is a prime ideal, see Theorem 3.17. Next, it follows from Proposition 3.1 (see also Problem 3.12) that height $I_{\mathscr{P}} = \operatorname{rank}_{\mathbb{Z}} \Lambda$. Hence, the desired conclusion regarding height $I_{\mathscr{P}}$ follows from Lemma 8.15.   □

Next, for any balanced polyomino $\mathscr{P}$, we will identify the primitive binomials in $I_{\mathscr{P}}$. This will allow us to show that the initial ideal of $I_{\mathscr{P}}$ is a squarefree monomial ideal for any monomial order.

The primitive binomials in $\mathscr{P}$ are determined by cycles. A sequence of vertices $\mathscr{C} = a_1, a_2, \ldots, a_m$ in $V(\mathscr{P})$ with $a_m = a_1$ and such that $a_i \neq a_j$ for all $1 \leq i < j \leq m - 1$ is called a *cycle* in $\mathscr{P}$ if the following conditions hold:

(i) $[a_i, a_{i+1}]$ is a horizontal or vertical edge interval of $\mathscr{P}$ for all $i = 1, \ldots, m - 1$;
(ii) for $i = 1, \ldots, m$, one has: if $[a_i, a_{i+1}]$ is a horizontal interval of $\mathscr{P}$, then $[a_{i+1}, a_{i+2}]$ is a vertical edge interval of $\mathscr{P}$ and vice versa. Here, $a_{m+1} = a_2$.

It follows immediately from the definition of a cycle that $m - 1$ is even. Given a cycle $\mathscr{C}$, we attach to $\mathscr{C}$ the binomial

$$f_{\mathscr{C}} = \prod_{i=1}^{(m-1)/2} x_{a_{2i-1}} - \prod_{i=1}^{(m-1)/2} x_{a_{2i}}$$

Theorem 8.18 *Let $\mathscr{P}$ be a balanced polyomino.*

(a) *Let $\mathscr{C}$ be a cycle in $\mathscr{P}$. Then, $f_{\mathscr{C}} \in I_{\mathscr{P}}$.*
(b) *Let $f \in I_{\mathscr{P}}$ be a primitive binomial. Then, there exists a cycle $\mathscr{C}$ in $\mathscr{P}$ such that each maximal interval of $\mathscr{P}$ contains at most two vertices of $\mathscr{C}$ and $f = \pm f_{\mathscr{C}}$.*

Proof

(a) Let $\mathscr{C} = a_1, a_2, \ldots, a_m$ be a cycle in $\mathscr{P}$. We define a labeling α of $\mathscr{P}$ by setting $\alpha(a) = 0$ if $a \notin \mathscr{C}$ and $\alpha(a_i) = (-1)^{i+1}$ for $i = 1, \ldots, m$, and claim that α is an admissible labeling of $\mathscr{P}$. To see this, we consider a maximal horizontal edge interval I of $\mathscr{P}$. If $I \cap \mathscr{C} = \emptyset$, then $\alpha(a) = 0$ for all $a \in I$. On the other hand, if $I \cap \mathscr{C} \neq \emptyset$, then there exist integers i such that $a_i, a_{i+1} \in I$ (where $a_{i+1} = a_1$ if $i = m - 1$), and no other vertex of I belongs to $\mathscr{C}$. It follows that $\sum_{a \in I} \alpha(a) = 0$. Similarly, we see that $\sum_{a \in I} \alpha(a) = 0$ for any vertical edge interval. It follows form the definition of α that $f_{\mathscr{C}} = f_\alpha$, and hence since $\mathscr{P}$ is balanced it follows that $f_{\mathscr{C}} \in I_{\mathscr{P}}$.

(b) Let $f \in I_{\mathscr{P}}$ be a primitive binomial. Since $\mathscr{P}$ is balanced and f is irreducible, there exists an admissible labeling α of $\mathscr{P}$ such that

$$ f = f_\alpha = \prod_{\substack{a \in V(\mathscr{P}) \\ \alpha(a) > 0}} x_a^{\alpha(a)} - \prod_{\substack{a \in V(\mathscr{P}) \\ \alpha(a) < 0}} x_a^{-\alpha(a)}. $$

Choose $a_1 \in V(\mathscr{P})$ such that $\alpha(a_1) > 0$. Let I_1 be the maximal horizontal edge interval with $a_1 \in I_1$. Since α is admissible, there exists some $a_2 \in I_1$ with $\alpha(a_2) < 0$. Let I_2 be the maximal vertical edge interval containing a_2. Then, similarly as before, there exists $a_3 \in I_2$ with $\alpha(a_3) > 0$. In the next step, we consider the maximal horizontal edge interval containing a_3 and proceed as before. Continuing in this way, we obtain a sequence $a_1, a_2, a_3, \ldots$, of vertices of $\mathscr{P}$ such that $\alpha(a_1), \alpha(a_2), \alpha(a_3), \ldots$ is a sequence with alternating signs. Since $V(\mathscr{P})$ is a finite set, there exists a number m such that $a_i \neq a_j$ for all $1 \leq i < j \leq m$ and $a_m = a_i$ for some $i < m$. It follows that $\alpha(a_m) = \alpha(a_i)$ which implies that $m - i$ is even. Then, the sequence $\mathscr{C} = a_i, a_{i+1}, \ldots a_m$ is a cycle in $\mathscr{P}$, and hence by (a), $f_{\mathscr{C}} \in I_{\mathscr{P}}$.

For any binomial $g = u - v$ we set $g^{(+)} = u$ and $g^{(-)} = v$. Now, if i is odd, then $f_{\mathscr{C}}^{(+)}$ divides $f^{(+)}$ and $f_{\mathscr{C}}^{(-)}$ divides $f^{(-)}$, while if i is even, then $f_{\mathscr{C}}^{(+)}$ divides $f^{(-)}$ and $f_{\mathscr{C}}^{(-)}$ divides $f^{(+)}$. Since f is primitive, we see that $f = \pm f_{\mathscr{C}}$, as desired. □

Corollary 8.19 *Let $\mathscr{P}$ be a balanced polyomino. Then, for any monomial order, the ideal $I_{\mathscr{P}}$ admits a squarefree initial ideal.*

Proof By Corollary 8.17, $I_{\mathscr{P}}$ is a prime ideal. This implies that $I_{\mathscr{P}}$ is a toric ideal, see Theorem 3.4. Now, we use the fact, shown in Theorem 3.13, that the reduced Gröbner basis of a toric ideal with respect to any monomial order consists of primitive binomials. Since by Theorem 8.18, the primitive binomials of $I_{\mathscr{P}}$ have squarefree initial terms for any monomial order, the desired conclusion follows. □

The preceding corollary has nice consequences.

Corollary 8.20 *Let $\mathscr{P}$ be a balanced polyomino. Then, $K[\mathscr{P}]$ is a normal Cohen–Macaulay domain of dimension $|V(\mathscr{P})| - |\mathscr{P}|$.*

Fig. 8.20 A simple
polyomino

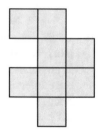

Proof A toric ring whose toric ideal admits a squarefree initial ideal is normal, see
Corollary 4.26. By a theorem of Hochster [27, Theorem 6.3.5], a normal toric ring is
Cohen–Macaulay. Thus, the first part of the assertion follows from Corollary 8.19.
The second part is a consequence of Corollary 8.17.   □

8.2.2 Simple Polyominoes

In this section, we introduce simple polyominoes. Roughly speaking these are the
polyominoes without holes. As a main result, we will show that the coordinate ring
of a simple polyomino is a domain.

A polyomino $\mathscr{P}$ is called *simple*, if for any two cells C and D with vertices in
$\mathbb{Z}^2$ which do not belong to $\mathscr{P}$, there exists a path $\mathscr{C} : C = C_1, C_2, \ldots, C_t = D$
with $C_i \notin \mathscr{P}$ for all $i = 1, \ldots, t$. For example, the polyomino which is shown in
Figure 8.18 is not simple, while Figure 8.20 shows a simple polyomino.

The purpose of this section is to prove the following:

Theorem 8.21 *A polyomino is simple if and only if it is balanced.*

Combining this result with Corollary 8.17 we obtain:

Corollary 8.22 *Let $\mathscr{P}$ be a simple polyomino. Then, $I_{\mathscr{P}}$ is a prime ideal.*

The following result is an important special case of this corollary: let C and D
be two cells with lower-left corners (i, j) and (k, l). Then, the *cell interval*, denoted
by $[C, D]$, is the set of cells

$$[C, D] = \{E : E \in \mathbb{Z}_{\geq 0}^2 \text{ with lower-left corner } (r, s), \text{ for } i \leq r \leq k, j \leq s \leq l\}$$

If (i, j) and (k, l) are in horizontal position, then the cell interval $[A, B]$ is called a
horizontal cell interval. Similarly, one defines a vertical cell interval.

A polyomino $\mathscr{P}$ is called *row convex*, if the horizontal cell interval $[C, D]$ is
contained in $\mathscr{P}$ for any two cells C and D of $\mathscr{P}$ whose lower-left corners are
in horizontal position. Similarly, one defines *column convex*. A collection of cells
$\mathscr{P}$ is called *convex*, if it is row and column convex. Figure 8.20 displays a simple
polyomino which is row convex but not column convex.

For the proof of the next result, it suffices to notice that convex polyominoes are simple.

Corollary 8.23 *Let $\mathscr{P}$ be a convex polyomino. Then, $I_{\mathscr{P}}$ is a prime ideal.*

The proof of Theorem 8.21 requires some preparation. Let $\mathscr{P}$ be a polyomino. There exists an interval $I = [a, b]$ with $V(\mathscr{P}) \subset I$. Let $\mathscr{H}$ be the collection of cells C with the property that $C \notin \mathscr{P}$. The connected components of $\mathscr{H}$ with the property that the vertices of all its cells belong to I are called the *holes* of $\mathscr{P}$. For example, the polyomino which is shown in Figure 8.18 has exactly one hole consisting of just one cell. Note that this definition does not depend on the particular choice of I. $\mathscr{P}$ is simple if and only if it is hole free. Each hole of $\mathscr{P}$ is a polyomino. In fact, one even has:

Lemma 8.24 *Each hole of a polyomino is a simple polyomino.*

Proof We use the notation just introduced. Let $\mathscr{P}'$ be a hole of the polyomino $\mathscr{P}$, and assume that $\mathscr{P}'$ is not simple. Let $\mathscr{P}''$ be a hole of $\mathscr{P}'$. Then, $\mathscr{P}''$ is again a polyomino, and $\mathscr{P}'$ as well as $\mathscr{P}''$ belong to I. Let C be a cell of $\mathscr{P}''$ which shares an edge with a cell $D \in \mathscr{P}'$. Suppose $C \notin \mathscr{P}$, then $C \in \mathscr{P}'$ because C is connected to D, and $\mathscr{P}'$ is a connected component of $\mathscr{H}$. This is a contradiction, and hence $C \in \mathscr{P}$. Let $\mathscr{Q}$ the connected component of $\mathscr{H}$ whose cells do not all belong to I. Let E be a cell of $\mathscr{P}$ which has an edge in common with a cell F of $\mathscr{Q}$ and which belongs to the same connected component of $\mathscr{P}$ as C. Furthermore, let G be a cell in $\mathscr{Q}$ which does not belong to I. Then, there is a path of cells $C, \ldots, E$ for which all cells belong to $\mathscr{P}$, and a path of cells $F, \ldots, G$ for which all cells belong to $\mathscr{Q}$. Composing these two paths, we obtain a path $C, \ldots, G$ for which no cell belongs to $\mathscr{P}'$. This contradicts the fact that $\mathscr{P}''$ is a hole of $\mathscr{P}'$. $\qquad\square$

The polyomino in Figure 8.18 has two cells intersecting at only one vertex which does not belong to any other cell. This cannot happen if the polyomino is simple. Indeed, we have:

Lemma 8.25 *Let $\mathscr{P}$ be a simple polyomino. Then, there does not exist any vertex v which belongs to exactly two cells C and C' of $\mathscr{P}$ such that $C \cap C' = \{v\}$.*

Proof Suppose on the contrary that there exists such a vertex v. According to Figure 8.21, the only cells of $\mathscr{P}$ which contain v could be the four cells C, C', D, and D'. By our assumption, we may assume that C and C' belong to $\mathscr{P}$ and D and D' do not belong to $\mathscr{P}$. Since $\mathscr{P}$ is a polyomino, there exists a path of cells of $\mathscr{P}$ connecting C and C'. Thus, either D or D' is contained in a hole of $\mathscr{P}$. It contradicts the fact that $\mathscr{P}$ is a simple polyomino. $\qquad\square$

Let $\mathscr{P}$ be a polyomino. We recall from Section 8.2.1 that an interval $[a, b]$ with $a = (i, j)$ and $b = (k, j)$ is called a *horizontal edge interval* of $\mathscr{P}$, if the intervals $[(t, j), (t + 1, j)]$ for $t = i, \ldots, k - 1$ are edges of cells of $\mathscr{P}$. Similarly, a *vertical edge interval* of $\mathscr{P}$ is defined to be an interval $[a, b]$ with $a = (i, j)$ and $b = (i, l)$ such that the intervals $[(i, t), (i, t + 1)]$ for $t = j, \ldots, l - 1$ are edges of cells of $\mathscr{P}$.

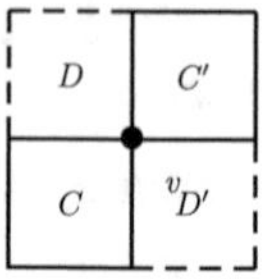

Fig. 8.21 Two cells C and
C' belong to $\mathscr{P}$

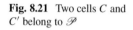

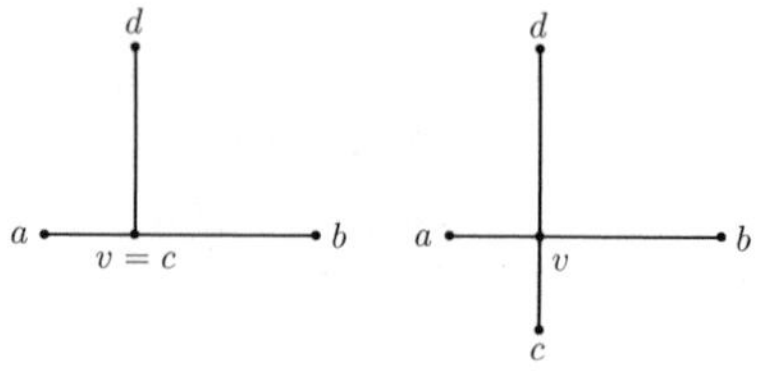

Fig. 8.22 The vertex v is not
a common endpoint of I and
I'

We call an edge of a cell C of $\mathscr{P}$ a *border edge*, if it is not an edge of any other cell, and define the *border* of $\mathscr{P}$ to be the union of all border edges of $\mathscr{P}$. A *horizontal border edge interval* of $\mathscr{P}$ is defined to be a horizontal edge interval of $\mathscr{P}$ whose edges are border edges. Similarly, we define a *vertical border edge interval* of $\mathscr{P}$.

Corollary 8.26 *Let $\mathscr{P}$ be a simple polyomino and let I and I' be two distinct maximal border edge intervals of $\mathscr{P}$ with $I \cap I' \neq \emptyset$. Then, their intersection is a common endpoint of I and I'. Furthermore, at most two maximal border edge intervals of $\mathscr{P}$ have a nontrivial intersection.*

Proof Let $I = [a, b]$ and $I' = [c, d]$. The edge intervals I and I' are not both horizontal or vertical edge intervals, since otherwise their maximality implies that they are disjoint. Suppose that I is a horizontal edge interval and I' is a vertical edge interval. So, obviously, they intersect in one vertex, say v. Suppose that v is not an endpoint of I or I'. If v is an endpoint of just one of them, then without loss of generality, we may assume that we are in the case which is shown on the left-hand side of Figure 8.22. Thus, since I and I' are maximal border edge intervals, it follows that among the four possible cells of $\mathbb{Z}^2$ which contain v, exactly one of them belongs to $\mathscr{P}$, which is a contradiction. If v is not an endpoint of any of I and I', then we are in the case which is displayed on the right-hand side of Figure 8.22. Among four possible cells of $\mathbb{Z}^2$ which contain v, only a pair of them, say C and C', with $C \cap C' = \{v\}$, belong to $\mathscr{P}$, since the edges of I and I' are all border edges. But, by Lemma 8.25, this is impossible, since $\mathscr{P}$ is simple. Thus, v has to be a common endpoint of I and I'.

Now, suppose more than two maximal border edge intervals have a nontrivial intersection. Then, this intersection is a common endpoint of these intervals. Thus, at least two of these intervals are either horizontal or vertical, contradicting the fact that they are all maximal. □

Next, we introduce some concepts and facts about rectilinear polygons which are used in the study of simple polyominoes. A *rectilinear polygon* is a polygon whose edges meet orthogonally. It is easily seen that the number of edges of

Fig. 8.23 A rectilinear
polygon

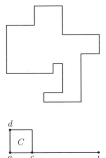

Fig. 8.24 The interval $[a, b]$
and a cell C

a rectilinear polygon is even. Note that rectilinear polygons are also known as *orthogonal polygons*. A rectilinear polygon is shown in Figure 8.23.

A rectilinear polygon is called *simple* if it does not self-intersect. The rectilinear polygon in Figure 8.23 is a simple rectilinear polygon.

Let R be a simple rectilinear polygon. The bounded area whose border is R is called the *interior* of R. By the *open interior* of R, we mean the interior of R without its boundary.

A simple rectilinear polygon has two types of corners: the corners in which the smaller angle (90 degrees) is interior to the polygon are called *convex corners*, and the corners in which the larger angle (270 degrees) is interior to the polygon are called *concave corners*.

Let $E_1, \ldots, E_m$ be the border edges of $\mathcal{P}$. Then, we set $B(\mathcal{P}) = \bigcup_{i=1}^{m} \overline{E}_i$. Observe that the border of $\mathcal{P}$ as defined before is the set of lattice points which belong to $B(\mathcal{P})$.

Lemma 8.27 *Let $\mathcal{P}$ be a simple polyomino. Then, $B(\mathcal{P})$ is a simple rectilinear polygon.*

Proof First, we show that for each maximal horizontal (resp., vertical) border edge interval $I = [a, b]$ of $\mathcal{P}$, there exists a unique maximal vertical (resp., horizontal) border edge interval I' such that a is an endpoint of it. By Corollary 8.26, the vertex a is then the endpoint of precisely I and I'. Without loss of generality, let $I = [a, b]$ be a horizontal maximal border edge interval of $\mathcal{P}$. Let C be the only cell of $\mathcal{P}$ for which a is a vertex, and which has a border edge contained in I. First, we assume that a is a diagonal corner of C which implies that C is upside of I, see Figure 8.24. The argument of the other case in which a is an anti-diagonal corner of C, and hence C is downside of I, is similar.

Referring to Figure 8.24, we distinguish two cases: either the unique cell D, different from C sharing the edge $[a, d]$ with C, belongs to $\mathcal{P}$ or not.

Let us first assume that $D \notin \mathcal{P}$. Then, $[a, d]$ is a border edge of $\mathcal{P}$, and hence it is contained in a maximal vertical border edge interval I' of $\mathcal{P}$ such that by Corollary 8.26, a is an endpoint of I'. Hence, I' is the unique maximal vertical border edge interval of $\mathcal{P}$ for which a is an endpoint.

Fig. 8.25 The intervals $[a, c]$
and $[f, a]$ are two border
edges

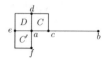

Next, assume that $D \in \mathscr{P}$. Then, the cell C' belongs to $\mathscr{P}$ (see Figure 8.25), because $[a, b]$ is a maximal horizontal border edge interval, so that $[e, a]$ cannot be a border edge. The edge $[f, a]$ is a border edge, since otherwise there is a cell containing both of the edges $[f, a]$ and $[a, c]$, contradicting the fact that $[a, c]$ is a border edge. Therefore, there exists the unique maximal vertical border edge interval I' which contains $[f, a]$ such that a is an endpoint of I'.

The same argument can be applied for b to show that b is also just the endpoint of I and of a unique maximal vertical border edge interval I' of $\mathscr{P}$.

Now, let I_1 be a maximal horizontal border edge interval of $\mathscr{P}$. By what we have shown before, there exists a unique sequence of maximal border edge intervals $I_1, I_2, \ldots$ of $\mathscr{P}$ with $I_i = [a_i, a_{i+1}]$ such that they are alternatively horizontal and vertical. Since $V(\mathscr{P})$ is finite, there exists a smallest integer r such that for some $i < r - 1$, $I_i \cap I_r \neq \emptyset$. Since I_i and I_r are distinct maximal border edge intervals of $\mathscr{P}$, they intersect in one of their endpoints, by Corollary 8.26. Thus, $I_i \cap I_r = \{a_i\}$, since $r \neq i$ and by Corollary 8.26, a_{i+1} cannot be a common vertex between three maximal border edge intervals I_i, I_{i+1}, and I_r. It follows that $i = 1$, since otherwise a_i also belong to I_{i-1} which is a contradiction, by Corollary 8.26.

Our discussion shows that $R = \bigcup_{j=1}^{r} I_j$ is a simple rectilinear polygon. Suppose that $R \neq B(\mathscr{P})$. Then, there exists a maximal border edge interval I_1' which is different from the intervals I_j. As we did for I_1, we may start with I_1' to construct a sequence of border edge intervals I_j' to obtain a simple rectilinear polygon R' whose edges are formed by some maximal border edge intervals of $\mathscr{P}$. We claim that $R \cap R' = \emptyset$. Suppose this is not the case, then $I_j \cap I_k' \neq \emptyset$ for some j and k, and hence by Corollary 8.26 these two intervals meet at a common endpoint. Thus, it follows that I_k' also has a common intersection with one of the neighbor intervals I_t of I_j, contradicting the fact that no three maximal border edge intervals intersect nontrivially, see Corollary 8.26. Hence, $R \cap R' = \emptyset$, as we claimed.

All the cells of the interior of R must belong to $\mathscr{P}$, because otherwise $\mathscr{P}$ is not simple. It follows that R' does not belong to the interior of R, and vice versa. Thus, the interior cells of R and R' form two disjoint sets of cells of $\mathscr{P}$. Since $\mathscr{P}$ is a polyomino, there exists a path of cells connecting the interior cells of R with those of R'. The edges where this path meets R and R' cannot be border edges, a contradiction. Thus, we conclude that $R = B(\mathscr{P})$. □

For the proof of the main theorem of this section (Theorem 8.21), special admissible labelings of polyominoes are required.

An *inner interval* I of a polyomino $\mathscr{P}$ is an interval with the property that all cells inside I belong to $\mathscr{P}$.

Let I be an inner interval of a polyomino $\mathscr{P}$. Then, we introduce the admissible labeling $\alpha_I : V(\mathscr{P}) \to \mathbb{Z}$ of $\mathscr{P}$ as follows:

Fig. 8.26 A border labeling

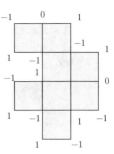

$$\alpha_I(a) = \begin{cases} -1, & \text{if } a \text{ is a diagonal corner of } I, \\ 1, & \text{if } a \text{ is an anti-diagonal corner of } I, \\ 0, & \text{otherwise.} \end{cases}$$

Next, we introduce a special labeling of a simple polyomino $\mathscr{P}$, called a *border labeling*. By Lemma 8.27, $B(\mathscr{P})$ is a rectilinear polygon. While walking counterclockwise around $B(\mathscr{P})$, we label the corners alternatively by $+1$ and -1 and label all the other vertices of $\mathscr{P}$ by 0. Since $B(\mathscr{P})$ has even number of vertices, this labeling is always possible for $\mathscr{P}$. Also, it is obvious that every simple polyomino has exactly two border labelings. Figure 8.26 shows a border labeling of the polyomino which was displayed in Figure 8.20.

Lemma 8.28 *A border labeling of a simple polyomino is admissible.*

Proof Let $\mathscr{P}$ be a simple polyomino, and let α be a border labeling of $\mathscr{P}$. Let I be a maximal horizontal edge interval of $\mathscr{P}$. We show that $\sum_{a \in I} \alpha(a) = 0$. Let $I_1, \ldots, I_t$ be all maximal horizontal border edge intervals of $\mathscr{P}$ which are contained in I. Note that the intervals I_j are pairwise disjoint. Then, $\sum_{a \in I} \alpha(a) = \sum_{\substack{a \in I_i \\ 1 \le i \le t}} \alpha(a)$, since the only elements of I for which $\alpha(a) \ne 0$ are the corners of the rectilinear polygon $B(\mathscr{P})$, and since the endpoints of $I_1, \ldots, I_t$ are corners of $B(\mathscr{P})$. But, $\sum_{\substack{a \in I_i \\ 1 \le i \le t}} \alpha(a) = 0$, since by definition of a border labeling, we have $\sum_{a \in I_i} \alpha(a) = 0$, for each $i = 1, \ldots, t$. Similarly, for a maximal vertical edge interval I of $\mathscr{P}$, we have $\sum_{a \in I} \alpha(a) = 0$. Hence, α is admissible. $\qquad\square$

Now, let $\mathscr{P}$ be a polyomino contained in the rectangular polyomino $\mathscr{I}$ with $V(\mathscr{I}) = [(1, 1), (m, n)]$ for some positive integers m and n. Let I be an inner interval of $\mathscr{P}$, and set $\mathbf{u}_I = (u_I^{(i,j)})_{\substack{1 \le i \le m \\ 1 \le j \le n}} \in \mathbb{Z}^{m \times n}$ where

$$u_I^{(i,j)} = \begin{cases} -1, & \text{if } (i, j) \text{ is a diagonal corner of } I, \\ 1, & \text{if } (i, j) \text{ is an anti-diagonal corner of } I, \\ 0, & \text{otherwise.} \end{cases}$$

Note that if I is just a cell C of $\mathscr{P}$, then with the notation of Lemma 8.15, $\mathbf{u}_I = b_C$. There, it is also shown that the elements b_C with $C \in \mathscr{I}$ are linearly independent over $\mathbb{Z}$.

We set

$$\mathscr{M}(\mathscr{P}) = \{\mathbf{u} : \mathbf{u} = \pm\mathbf{u}_I \text{ for some inner interval } I \text{ of } \mathscr{P}\}.$$

In the next proposition, which provides a new characterization of balanced polyominoes, we refer to the connectedness of vectors via $\mathscr{M}(\mathscr{P})$. We refer the reader to Corollary 3.10 and the definitions preceding it.

Proposition 8.29 *Let $\mathscr{P}$ be a polyomino. Then, the following conditions are equivalent:*

   (i) *$\mathscr{P}$ is balanced;*
  (ii) *For each admissible labeling α of $\mathscr{P}$, α^+ and α^- are connected via $\mathscr{M}(\mathscr{P})$;*
 (iii) *For each admissible labeling α of $\mathscr{P}$, there exist $\mathbf{u}_1, \ldots, \mathbf{u}_t \in \mathscr{M}(\mathscr{P})$ such that $\alpha^- + \mathbf{u}_1 + \cdots + \mathbf{u}_i \in \mathbb{Z}_{\geq 0}^n$ for all $i = 1, \ldots, t$, and $\alpha^+ = \alpha^- + \mathbf{u}_1 + \cdots + \mathbf{u}_t$.*

Proof The conditions (ii) and (iii) are equivalent by the definition of $G_{\mathscr{M}(\mathscr{P})}$. We show that (i) and (ii) are equivalent. Let α be an admissible labeling of $\mathscr{P}$. Then, $f_\alpha = \mathbf{x}^{\alpha^+} - \mathbf{x}^{\alpha^-} \in I(\mathscr{M}(\mathscr{P}))$ if and only if α^+ and α^- are connected via $\mathscr{M}(\mathscr{P})$. But, note that $I_\mathscr{P} = I(\mathscr{M}(\mathscr{P}))$. So, $f_\alpha \in I_\mathscr{P}$ if and only if α^+ and α^- are connected via $\mathscr{M}(\mathscr{P})$. Thus, the assertion follows, since $\mathscr{P}$ is balanced if and only if $I_\mathscr{P} = J_\mathscr{P}$.   □

Now, we are ready for the proof of the main result of this section.

Proof (of Theorem 8.21) Let $\mathscr{P}$ be a polyomino. First, suppose $\mathscr{P}$ is simple. We have to show that for any admissible labeling α of $\mathscr{P}$ we have that $f_\alpha \in I_\mathscr{P}$, and we show this by induction on deg f_α. Suppose deg $f_\alpha = 2$. Then, $\alpha = \pm\alpha_I$ for some inner interval I, because $\mathscr{P}$ is simple. Thus, by definition $f_\alpha \in I_\mathscr{P}$.

Now, suppose that deg $f_\alpha > 2$. We choose $a_0 \in V(\mathscr{P})$ with $\alpha(a_0) > 0$. Since α is admissible, there exists a horizontal edge interval $[a_0, a_1]$ of $\mathscr{P}$ with $\alpha(a_1) < 0$. By using again that α is admissible, there exists a vertical edge interval $[a_1, a_2]$ of $\mathscr{P}$ with $\alpha(a_2) > 0$. Proceeding in this way, we obtain a sequence of edge intervals of $\mathscr{P}$,

$$[a_0, a_1], [a_1, a_2], [a_2, a_3], \ldots$$

which are alternatively horizontal and vertical and such that $\text{sign}(\alpha(a_i)) = (-1)^i$ for all i.

Since $V(\mathscr{P})$ is a finite set, there exists a smallest integer r such that $[a_r, a_{r+1}]$ intersects $[a_j, a_{j+1}]$ for some $j < r - 1$. We may assume that $j = 0$. If $[a_r, a_{r+1}]$ is a vertical interval, then $[a_r, a_{r+1}]$ and $[a_0, a_1]$ intersect in precisely one vertex, which we call a. If $[a_r, a_{r+1}]$ is horizontal, then we let $a = a_1$. In this way, we obtain a simple rectilinear polygon R whose edges are edge intervals of $\mathscr{P}$ with corner sequence $a, a_1, a_2, \ldots, a_{r-1}, a$ if $[a_r, a_{r+1}]$ is vertical and corner sequence

Fig. 8.27 A good corner and
its rectangle

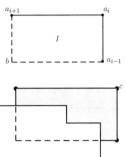

Fig. 8.28 R intersects the
rectangle

$a, a_2, a_3, \ldots, a_{r-1}, a$ if $[a_r, a_{r+1}]$ is horizontal. Moreover, we have $\mathrm{sign}(\alpha(a_i)) = (-1)^i$ for all i. The cells in the interior of R all belong to $\mathcal{P}$ because $\mathcal{P}$ is simple. We may assume that the orientation of R given by the order of the corner sequence is counterclockwise. Then, with respect to this orientation the interior of R meets R on the left-hand side, see Figure 8.23.

We call a convex corner c of R *good* if the rectangle which is spanned by c and its neighbor corners is in the interior of R. We claim that R has at least four good corners. We will prove the claim later and first discuss its consequences. Since R has at least four good corners, there is at least one good corner c such that c and its neighbor corners are all different from a. Let I be the rectangle in the interior of R spanned by c and its neighbor corners. Without loss of generality, we may assume that this corner looks like the one displayed in Figure 8.27 with $c = a_i$.

Since all cells in the interior of I belong to the interior of R and since all those cells belong to $\mathcal{P}$, it follows that $f_{\alpha_I} \in I_{\mathcal{P}}$. Without loss of generality, we may assume that $\alpha(a_i) < 0$, and hence $\alpha(a_{i-1}), \alpha(a_{i+1}) > 0$. Then, the homogeneous binomial $g = f_\alpha - (\mathbf{x}^{\alpha^+}/x_{a_{i-1}}x_{a_{i+1}})f_{\alpha_I}$ has the same degree as f_α and belongs to $J_{\mathcal{P}}$, since f_α and f_{α_I} belong to $J_{\mathcal{P}}$. Furthermore, $g = x_{a_i}h$, where $h = x_b(\mathbf{x}^{\alpha^+}/x_{a_{i-1}}x_{a_{i+1}}) - \mathbf{x}^{\alpha^-}/x_{a_i}$. It follows that $h \in J_{\mathcal{P}}$, since $x_{a_i} \notin J_{\mathcal{P}}$ and since $J_{\mathcal{P}}$ is a prime ideal. Since $J_{\mathcal{P}}$ is generated by the binomials f_β with β an admissible labeling of $\mathcal{P}$, there exist $f_{\beta_l} \in J_{\mathcal{P}}$ such that $h = \sum_{l=1}^{s} r_l f_{\beta_l}$, where $\deg f_{\beta_l} \leq \deg h$ and $r_l \in S$ for all l. Since $\deg h < \deg f_\alpha$, we also have $\deg f_{\beta_l} < \deg f_\alpha$ for all l. Thus, our induction hypothesis implies that $f_{\beta_l} \in I_{\mathcal{P}}$ for all l. It follows that $h \in I_{\mathcal{P}}$, and hence $f_\alpha \in I_{\mathcal{P}}$, since $f_{\alpha_I} \in I_{\mathcal{P}}$.

In order to complete the proof that $\mathcal{P}$ is balanced, it remains to prove that indeed any rectilinear polygon R has at least four good convex corners. We prove this by defining an injective map γ which assigns to each convex corner of R which is not good a concave corner of R. Since, as is well known and easily seen, for any simple rectilinear polygon the number of convex corners is four more than the number of concave corners, it will follow that there are at least four good corners.

The map γ is defined as follows: let c be a convex corner of R which is not good. Then, the polygon R crosses the open interior of the rectangle which is spanned by c and the neighbor corners of c. The gray area in Figure 8.28 belongs to the interior of R.

Fig. 8.29 L_{t_0} defines $\gamma(c)$

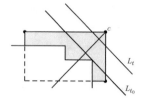

Now, we let L be the angle bisector of the 90 degrees angle centered in c. Next, we consider the set $\mathscr{L}_c$ of all lines perpendicular to L. The unique line in $\mathscr{L}_c$, which intersects L in the point p and such that the distance from c to p is t, will be denoted by L_t. There is a smallest number t_0 such that L_{t_0} has a nontrivial intersection with R in the open interior of the rectangle. This intersection with L_{t_0} consists of at least one and at most finitely many concave corners of R, see Figure 8.29.

We define γ to assign to c one of these concave corners. The map γ is injective. Indeed, if d is another convex corner of R with $\gamma(d) = \gamma(c)$, then the line in $\mathscr{L}_d$ which hits $\gamma(c)$ must be identical with L_{t_0}, and this implies that d lies in the intersection of the rectangle with the linear half space defined by L_{t_0} containing c. But, in this area there is no other corner of R which is not good. Hence, $d = c$.

Conversely, suppose now that $\mathscr{P}$ is balanced and assume that $\mathscr{P}$ is not simple. Let $\mathscr{P}'$ be a hole of $\mathscr{P}$. Then, by Lemma 8.24, $\mathscr{P}'$ is a simple polyomino. Let α be a border labeling of $\mathscr{P}'$. We consider the labeling β of $\mathscr{P}$ which for each $a \in V(\mathscr{P})$ is defined as follows:

$$\beta(a) = \begin{cases} \alpha(a) & \text{if } a \in V(\mathscr{P}'), \\ 0 & \text{if } a \notin V(\mathscr{P}'). \end{cases}$$

Then, β is an admissible labeling of $\mathscr{P}$, by a similar argument as in the proof of Lemma 8.28. Indeed, let I be a maximal horizontal (vertical) edge interval of $\mathscr{P}$ and let $\mathscr{S}$ be the set of all horizontal (vertical) border edge intervals of $\mathscr{P}'$ such that $I_j \cap I \neq \emptyset$. If $\mathscr{S} = \emptyset$, then $\beta(a) = 0$ for all $a \in I$. If $\mathscr{S} \neq \emptyset$ and $I_j \in \mathscr{S}$, then $I_j \subset I$. Since the intervals I_j are disjoint, we have $\sum_{a \in I} \beta(a) = \sum_{\substack{a \in I_j \\ I_j \in \mathscr{S}}} \beta(a) = \sum_{\substack{a \in I_j \\ I_j \in \mathscr{S}}} \alpha(a)$. Hence, $\sum_{a \in I} \beta(a) = 0$, because by definition of α, we have $\sum_{a \in I_j} \alpha(a) = 0$ for all $I_j \in \mathscr{S}$.

Note that we may consider α and β as vectors in $\mathbb{Z}^{m \times n}$ where m and n are positive integers with $V(\mathscr{P}) \subset [(1, 1), (m, n)]$. Since $\mathscr{P}$ is a balanced polyomino, it follows that there exist $\mathbf{u}_1, \ldots, \mathbf{u}_t \in \mathcal{M}(\mathscr{P})$ such that $\beta^+ = \beta^- + \mathbf{u}_1 + \cdots + \mathbf{u}_t$, by Proposition 8.29. On the other hand, since $\mathscr{P}'$ is a simple polyomino, it follows from the first part of the proof that $\mathscr{P}'$ is also balanced. Thus, by Proposition 8.29 there exist $\mathbf{u}'_1, \ldots, \mathbf{u}'_l \in \mathcal{M}(\mathscr{P}')$ such that $\alpha^+ = \alpha^- + \mathbf{u}'_1 + \cdots + \mathbf{u}'_l$, since α is admissible by Lemma 8.28. Note that by the construction of the labeling β, it is clear that $\beta^+ = \alpha^+$ and $\beta^- = \alpha^-$ are vectors in $\mathbb{Z}^{m \times n}$. So, we have $\mathbf{u}_1 + \cdots + \mathbf{u}_t = \mathbf{u}'_1 + \cdots + \mathbf{u}'_l$. For each $i = 1, \ldots, t$, we have $\mathbf{u}_i = \pm \mathbf{u}'_{l_i}$, and for each $j = 1, \ldots, l$, we have $\mathbf{u}'_j = \pm\mathbf{u}_{I'_j}$, where I_i and I'_j are inner intervals of $\mathscr{P}$ and $\mathscr{P}'$, respectively. So, it follows that for each i, j, $\mathbf{u}_i$, and $\mathbf{u}'_j$ are linear combination of the b_C's and $b_{C'}$'s, respectively, where C stands for cells of $\mathscr{P}$ and C' stands for cells of $\mathscr{P}'$. But, the b_C's and $b_{C'}$'s are linearly independent, so that $\mathbf{u}_1 + \cdots + \mathbf{u}_t = \mathbf{u}'_1 + \cdots + \mathbf{u}'_l = 0$, which is a contradiction, since obviously we have $\beta^+ \neq \beta^-$. Therefore, $\mathscr{P}$ is a simple polyomino. $\qquad\qquad\square$

8.2.3 A Toric Presentation of Simple Polyominoes

Let $\mathscr{P}$ be a polyomino. In this section, we will identify $K[\mathscr{P}]$ as the edge ring of a suitable graph if $\mathscr{P}$ is simple. This will then yield another proof of the fact, shown in the previous section, that $K[\mathscr{P}]$ is a domain.

Let $\{V_1, \ldots, V_m\}$ be the set of maximal vertical edge intervals and $\{H_1, \ldots, H_n\}$ be the set of maximal horizontal edge intervals of $\mathscr{P}$. We denote by $G(\mathscr{P})$, the associated bipartite graph of $\mathscr{P}$, whose vertex set is $\{v_1, \ldots, v_m\} \bigsqcup \{h_1, \ldots, h_n\}$ and whose edge set defined as follows:

$$E(G(\mathscr{P})) = \{\{v_i, h_j\} \mid V_i \cap H_j \in V(\mathscr{P})\}.$$

Example 8.30 Figure 8.30 shows a polyomino $\mathscr{P}$ with maximal vertical and maximal horizontal edge intervals labeled as $\{V_1, \ldots, V_4\}$ and $\{H_1, \ldots, H_4\}$, respectively, and Figure 8.31 shows the associated bipartite graph $G(\mathscr{P})$ of $\mathscr{P}$.

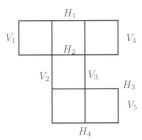

Fig. 8.30 Maximal intervals of $\mathscr{P}$

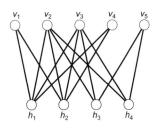

Fig. 8.31 The associate bipartite graph of $\mathscr{P}$

Given a cycle $\mathscr{C}_{\mathscr{P}} : a_1, a_2, \ldots, a_m$ of $\mathscr{P}$, we call $\{a_1, a_2, \ldots, a_m\}$ the vertex set of $\mathscr{C}_P$, and denote this set by $V(\mathscr{C}_P)$. As before, in Section 8.2.1, we attach to the cycle $\mathscr{C}_P$ the binomial

$$f_{\mathscr{C}_{\mathscr{P}}} = \prod_{i=1}^{(m-1)/2} x_{a_{2i-1}} - \prod_{i=1}^{(m-1)/2} x_{a_{2i}}$$

The cycle $\mathscr{C}_P$ is called *primitive*, if each maximal interval of $\mathscr{P}$ contains at most two vertices of $\mathscr{C}_P$.

Note that $\mathscr{C} : v_{i_1}, h_{j_1}, v_{i_2}, h_{j_2}, \ldots, v_{i_r}, h_{j_r}$ defines a cycle in $G(\mathscr{P})$, if and only if the sequence of vertices $\mathscr{C}_{\mathscr{P}} : V_{i_1} \cap H_{j_1}, V_{i_2} \cap H_{j_1}, V_{i_2} \cap H_{j_2}, \ldots, V_{i_r} \cap H_{j_r}, V_{i_1} \cap H_{j_r}$ is a primitive cycle in $\mathscr{P}$.

Let $K[G(\mathscr{P})] = K[v_p h_q \mid \{p, q\} \in E(G(\mathscr{P}))] \subset T = K[v_1, \ldots, v_m, h_1, \ldots, h_n]$ be the edge ring of $G(\mathscr{P})$, and let S be the polynomial ring over K with variables x_{ij} with $(i, j) \in V(\mathscr{P})$. We define a K-algebra homomorphism $\varphi : S \to T$ with image $K[G(\mathscr{P})]$ by $\varphi(x_{ij}) = v_p h_q$, where p and q are uniquely determined by the identity $\{(i, j)\} = V_p \cap H_q$. We denote by $L_{\mathscr{P}}$ the toric ideal of $K[G(\mathscr{P})]$. By Corollary 5.12, $L_{\mathscr{P}}$ is generated by the binomials associated with cycles in $G(\mathscr{P})$.

Theorem 8.31 *Let $\mathscr{P}$ be a simple polyomino. Then, $I_{\mathscr{P}} = L_{\mathscr{P}}$. In other words, $K[\mathscr{P}] \cong K[G(\mathscr{P})]$.*

For the proof of the theorem, we need a lemma. We recall from graph theory that a graph is called *weakly chordal* if every cycle of length greater than 4 has a chord. We say that a cycle $\mathscr{C}_{\mathscr{P}} : a_1, a_2, \ldots, a_m$ in $\mathscr{P}$ with $a_m = a_1$ has a *self-crossing*, if there exist indices i and j such that $a_i, a_{i+1} \in V_k$ and $a_j, a_{j+1} \in H_l$ and $a_i, a_{i+1}, a_j, a_{j+1}$ are all distinct and $V_k \cap H_l \neq \emptyset$. In this situation, if $\mathscr{C}$ is the associated cycle in $G(\mathscr{P})$, then $\{v_k, h_l\} \in E(G(\mathscr{P}))$ is a chord in $\mathscr{C}$.

Let $\mathscr{C}_{\mathscr{P}} : a_1, a_2, \ldots, a_r$ be a cycle in $\mathscr{P}$ which does not have any self-crossing. Then, we call the area bounded by the edge intervals $[a_i, a_{i+1}]$ and $[a_r, a_1]$ for $i \in \{1, r-1\}$, the *interior* of $\mathscr{C}_{\mathscr{P}}$. Moreover, we call a cell C is an *interior cell* of $\mathscr{C}_{\mathscr{P}}$ if C belongs to the interior of $\mathscr{C}_{\mathscr{P}}$.

Lemma 8.32 *Let $\mathscr{P}$ be a simple polyomino. Then, the graph $G(\mathscr{P})$ is weakly chordal.*

Proof Let $\mathscr{C}$ be a cycle of $G(\mathscr{P})$ of length $2n$ with $n \geq 3$ and $\mathscr{C}_{\mathscr{P}}$ be the associated primitive cycle in $\mathscr{P}$. We may assume that $\mathscr{C}_{\mathscr{P}}$ does not have any self-crossing. Otherwise, by following the definition of self-crossing, we know that $\mathscr{C}$ has a chord.

Let $\mathscr{C} : v_{i_1}, h_{j_1}, v_{i_2}, h_{j_2}, \ldots, v_{i_r}, h_{j_r}$ and $\mathscr{C}_{\mathscr{P}} : V_{i_1} \cap H_{j_1}, V_{i_2} \cap H_{j_1}, V_{i_2} \cap H_{j_2}, \ldots, V_{i_r} \cap H_{j_r}, V_{i_1} \cap H_{j_r}$. We may write $a_1 = V_{i_1} \cap H_{j_1}, a_2 = V_{i_2} \cap H_{j_1}, a_3 = V_{i_2} \cap H_{j_2}, \ldots, a_{2r-1} = V_{i_r} \cap H_{j_r}, a_{2r} = V_{i_1} \cap H_{j_r}$. Also, we may assume that a_1 and a_2 belong to the same maximal horizontal edge interval. Then, a_{2r} and a_1 belong to the same maximal vertical edge interval.

Fig. 8.32 A maximal inner interval

First, we show that every interior cell of $\mathcal{C}_{\mathcal{P}}$ belongs to $\mathcal{P}$. Suppose that we have an interior cell C of $\mathcal{C}_{\mathcal{P}}$ which does not belong to $\mathcal{P}$. Let $\mathcal{J}$ be any interval such that $\mathcal{P} \subset \mathcal{J}$. Then, by using the definition of simple polyomino, we obtain a path of cells $C = C_1, C_2, \ldots, C_t$ with $C_i \notin P, i = 1, \ldots t$ and C_t is a boundary cell in $\mathcal{J}$. It shows that $V(C_1) \cup V(C_2) \cup \ldots \cup V(C_t)$ intersects at least one of $[a_i, a_{i+1}]$ for $i \in \{1, \ldots, r-1\}$ or $[a_r, a_1]$, which is not possible because $\mathcal{C}_{\mathcal{P}}$ is a cycle in $\mathcal{P}$. Hence, $C \in \mathcal{P}$. It shows that an interval in the interior of $\mathcal{C}_{\mathcal{P}}$ is an inner interval of $\mathcal{P}$.

Let $\mathcal{I}$ be the maximal inner interval of $\mathcal{C}_{\mathcal{P}}$ to which a_1 and a_2 belong and let b, c the corner vertices of $\mathcal{I}$. We may assume that a_1 and c are the diagonal corners and a_2 and b are the anti-diagonal corners of $\mathcal{I}$. If $b, c \in V(\mathcal{C}_{\mathcal{P}})$, then primitivity of $\mathcal{C}$ implies that $\mathcal{C}$ is a cycle of length 4. We may assume that $b \notin V(\mathcal{C}_{\mathcal{P}})$. Let H' be the maximal horizontal edge interval which contains b and c. The maximality of $\mathcal{I}$ implies that $H' \cap V(\mathcal{C}_{\mathcal{P}}) \neq \emptyset$. For example, see Figure 8.32. Therefore, $\{v_{i_1}, h'\}$ is a chord in $\mathcal{C}$. $\qquad\square$

Proof (of Theorem 8.31) First, we show that $I_{\mathcal{P}} \subset L_{\mathcal{P}}$. Let $f = x_{ij}x_{kl} - x_{il}x_{kj} \in I_{\mathcal{P}}$. Then, there exist maximal vertical edge intervals V_p and V_q and maximal horizontal edge intervals H_m and H_n of $\mathcal{P}$ such that $(i, j), (i, l) \in V_p$, $(k, j), (k, l) \in V_q$, and $(i, j), (k, j) \in H_m$, $(i, l), (k, l) \in H_n$. It follows that $\varphi(x_{ij}x_{kl}) = v_p h_m h_n v_q = \varphi(x_{il}x_{kj})$. This shows $f \in L_P$.

Next, we show that $L_{\mathcal{P}} \subset I_{\mathcal{P}}$. By Corollary 5.15, the toric ideal of a weakly chordal bipartite graph is minimally generated by quadratic binomials associated with cycles of length 4. Thus, it suffices to show that $f_{\mathcal{C}} \in I_{\mathcal{P}}$ where $\mathcal{C}$ is a cycle of length 4 in $G(\mathcal{P})$.

Let $\mathcal{I}$ be an interval such that $\mathcal{P} \subset \mathcal{I}$. Let $\mathcal{C} : h_1, v_1, h_2, v_2$. Then, $\mathcal{C}_{\mathcal{P}} : a_{11} = H_1 \cap V_1, a_{21} = H_2 \cap V_1, a_{22} = H_2 \cap V_2$, and $a_{12} = H_1 \cap V_2$ is the associated cycle in $\mathcal{P}$ which also determines an interval of $\mathcal{I}$. Let a_{11} and a_{22} be the diagonal corners of this interval. We need to show that $[a_{11}, a_{22}]$ is an inner interval in $\mathcal{P}$. Assume that $[a_{11}, a_{22}]$ is not an inner interval of $\mathcal{P}$, that is, there exists a cell $C \in [a_{11}, a_{22}]$ which does not belong to $\mathcal{P}$. Using the fact that $\mathcal{P}$ is a simple polyomino, we obtain a path of cells $C = C_1, C_2, \ldots, C_r$ with $C_i \notin \mathcal{P}, i = 1, \ldots, r$ and C_r is not a cell of $\mathcal{I}$. Then, $V(C_1) \cup \ldots \cup V(C_r)$ intersects at least one of the maximal intervals H_1, H_2, V_1, V_2, say H_1, which contradicts the fact that H_1 is an interval in $\mathcal{P}$. Hence, $[a_{11}, a_{22}]$ is an inner interval of $\mathcal{P}$ and $f_{\mathcal{C}} \in I_{\mathcal{P}}$, as desired. $\qquad\square$

Problems

8.4 Give a direct proof of the fact (avoiding Theorem 8.21) that a row or column convex polyomino is balanced.

8.5 Show that convex polyominoes are simple.

8.6 Show that a simple polyomino has no holes.

8.7 Show that each hole of a polyomino is a polyomino.

Notes

Ideals generated by the t-minors of an $m \times n$-matrix of a matrix $X = (x_{ij})$ of indeterminates is a classical subject of studies, cf. [29, 117]. Some of these results have been extended to ideals generated by the t-minors of one- and two-sided ladders [36]. Ideals generated by sets of 2-minors are binomial ideals. In Chapter 7, any set of 2-minors of a $2 \times n$ matrix is considered. The study of general sets of 2-minors of an $m \times$-matrix with $m, n \geq 3$ requires some additional assumptions on this set. Among the first papers dealing with such class of ideals are those of Hoşten and Sullivant [120] and of Qureshi [172]. Motivated by applications in algebraic statistics, ideals generated by adjacent 2-minors have been introduced by Hoşten and Sullivant, while Qureshi was the first to consider polyomino ideals. The results on ideals generated by adjacent 2-minors are presented in Sections 8.1, 8.1.2 and 8.1.3 and are taken from [96].

One of the central problems in the algebraic theory of polyominoes is the classification of prime polyominoes which are those polyominoes whose polyomino ideal is a prime ideal. A first result in this direction was presented in Qureshi's paper [172], where it is shown that convex polyominoes are prime. In [103], balanced polyominoes were introduced and it was shown that balanced polyominoes are prime and that their residue class ring is a normal Cohen–Macaulay domain, see Corollary 8.17 and Corollary 8.20. Later, Herzog and Saeedi Madani showed in [102] that a polyomino is balanced if and only if it is prime. This result is presented in Section 8.2.2, see Theorem 8.21. As a consequence, one obtains that simple polyominoes are prime, see Corollary 8.22. An independent proof of this fact is given by Qureshi, Shibuta, and Shikama in [173]. Their proof is presented in Subsection 8.2.3. It is still an open problem to classify all prime polyominoes. A few other classes of polyominoes, other than simple polyominoes, are known to be prime [113]. Since the residue class ring of a polyomino ideal of a simple polyomino is a normal domain, it would be of interest to know the class group of such rings. From [172], this is only known only for the so-called stack polyominoes in which case also the Gorenstein polyominoes among them are determined.

Chapter 9
Statistics

Abstract Diaconis–Sturmfels (Ann. Statist. 26:363–397, 1998) introduced a Markov chain Monte Carlo method based on the algebraic theory of toric ideals. This approach turned out to be one of the origins of Algebraic Statistics which has become a very active and interesting research area. In this chapter, we give a survey on this concept and on related topics. In Section 9.1, we introduce the basic facts on contingency tables. In particular, "the p-value" is explained by an example of a 2-way contingency table. In Section 9.2, Markov bases are introduced. The Markov chain Monte Carlo method of Diaconis–Sturmfels is discussed, emphasizing the fact that any Markov basis in their theory corresponds to a set of binomial generators of a toric ideal. In addition, in Section 9.3, we discuss the method of a sequential importance sampling, and its relationship with the normality of toric rings. In Section 9.4, we study the toric rings and ideals of hierarchical models. In particular, the toric rings of no m-way interaction models are related to the notion of rth Lawrence liftings. Finally, in Section 9.5, as generalizations of Segre products and Veronese subrings, the so-called Segre–Veronese configurations are considered. This kind of configurations are applied to a test for the independence in group-wise selections.

9.1 Basic Concepts of Statistics (2-Way Case)

We start with an example of a 2-way contingency table appearing in [106].

A class has 26 students and all students take subjects "Geometry" and "Probability." The score of each subject is one of 5, 4, 3, 2, 1. The following table T_0 classifies the students according to their scores on Geometry and Probability.

© Springer International Publishing AG, part of Springer Nature 2018
J. Herzog et al., *Binomial Ideals*, Graduate Texts in Mathematics 279,
https://doi.org/10.1007/978-3-319-95349-6_9

Geom. \ Prob.	5	4	3	2	1	Total
5	2	1	1	0	0	4
4	8	3	3	0	0	14
3	0	2	1	1	1	5
2	0	0	0	1	1	2
1	0	0	0	0	1	1
Total	10	6	5	2	3	26

For example, there are 8 students whose score on Geometry is 4 and whose score on Probability is 5. The sequence 10, 6, 5, 2, 3 is the number of students whose score on Probability is 5, 4, 3, 2, 1, respectively. This sequence and the sequence 4, 14, 5, 2, 1 are called the *marginals* of the contingency table. The table T_0 is a 2-way contingency table of size 5×5 whose sample set consists of 26 students. The entries of the contingency table are called *cells*. There are two factors (scores on Geometry and on Probability) and each factor has 5 categories (5, 4, 3, 2, 1). Here, we call the contingency table "2-way" since there are 2 factors considered.

We would like to know whether the scores on the two subjects are correlated. First, we suppose the *null hypothesis* H_0 which says that the scores on the two subjects are independent. If H_0 is true, then we can compute the expected value of each cell by knowing the marginals of the contingency table. For example, the expected value of the (2, 1) cell (i.e., the number of the students whose score of Geometry is 4 and whose score of Probability is 5) is

$$26 \cdot \frac{14}{26} \cdot \frac{10}{26} = 5.38 \cdots .$$

Let e_{ij} denote the expected value of the (i, j) cell. Then, the table $T_e = (e_{ij})_{1 \le i, j \le 5}$ of the expected values is

Geom. \ Prob.	5	4	3	2	1	Total
5	1.54	0.92	0.77	0.31	0.46	4
4	5.38	3.23	2.69	1.08	1.62	14
3	1.92	1.15	0.96	0.38	0.58	5
2	0.77	0.46	0.38	0.15	0.08	2
1	0.38	0.23	0.19	0.08	0.12	1
Total	10	6	5	2	3	26

One of the common methods, the so-called χ^2 test, compares T_0 with T_e by the χ^2 statistics which measures the difference of T_0 and T_e. For the table $T_0 = (t_{ij})$, the χ^2 *statistics* is given by the formula:

$$\chi^2(T_0) = \sum_{i=1}^{5} \sum_{j=1}^{5} \frac{(t_{ij} - e_{ij})^2}{e_{ij}} = 25.338.$$

Let $\mathscr{F}_{T_0}$ denote the set of tables with the same marginals as T_0:

$$\mathscr{F}_{T_0} = \left\{ T = (t_{ij}) \; : \; 0 \le t_{ij} \in \mathbb{Z}, \begin{array}{|ccccc|c} t_{11} & t_{12} & t_{13} & t_{14} & t_{15} & 4 \\ t_{21} & t_{22} & t_{23} & t_{24} & t_{25} & 14 \\ t_{31} & t_{32} & t_{33} & t_{34} & t_{35} & 5 \\ t_{41} & t_{42} & t_{43} & t_{44} & t_{45} & 2 \\ t_{51} & t_{52} & t_{53} & t_{54} & t_{55} & 1 \\ \hline 10 & 6 & 5 & 2 & 3 & 26 \end{array} \right\}.$$

Note that if $T = (t_{ij})$ belongs to $\mathscr{F}_{T_0}$, then $t_{15}, t_{25}, t_{35}, t_{45}, t_{55}, t_{51}, t_{52}, t_{53}$, and t_{54} are determined by t_{ij} ($1 \le i \le 4, 1 \le j \le 4$). We say that the *degrees of freedom* is $(5-1)(5-1) = 16$. If the hypothesis H_0 is true, then it is known that the χ^2 statistic has an asymptotic χ^2 distribution with degrees of freedom $v = 16$. More precisely, consider the function

$$f(x) = \begin{cases} \frac{x^{v/2-1} e^{-x/2}}{2^{v/2} \Gamma(v/2)} & \text{if } x \ge 0, \\ 0 & \text{otherwise} \end{cases}$$

with $v = 16$, see Figure 9.1. Then, $\int_a^b f(x)dx$ approximates the probability that $T \in \mathscr{F}_{T_0}$ satisfies $a \le \chi^2(T) \le b$. In particular, we have $\int_0^\infty f(x)dx = 1$. In Figure 9.2, the shadowed area represents the upper 5% of the distribution, that is, $\int_{26.30}^\infty f(x)dx = 0.05$. Thus, for $\xi \ge 26.30$, $\int_\xi^\infty f(x)dx \le 0.05$. So, any T with $\chi^2(T) \ge 26.30$ is considered to be rare, because T appears with probability ≤ 0.05. Since $\chi^2(T_0) = 25.338$ is less than 26.30, our conclusion is "we cannot reject H_0."

If $\chi^2(T_0) \ge 26.30$, then we conclude that H_0 is rejected and hence the scores on the two subjects are correlated.

However, there is a problem with this method. There might not be a good fit with the asymptotic distribution if, for example, one of the following conditions is satisfied:

(i) The cardinality of the sample set is small;
(ii) The contingency table is *sparse*, that is, it has many zero cells.

The cardinality of the sample set is 26 and this is small. Moreover, the contingency table T_0 has many cells which are zero. Hence, χ^2 test may be not good for the contingency table T_0.

For this reason, we may use Fisher's exact test. We now assume that H_0 is true and that $\mathscr{F}_{T_0}$ follows the *multiple hypergeometric distribution*, that is, for each $T = (t_{ij}) \in \mathscr{F}_{T_0}$, the probability of the occurrence of T is

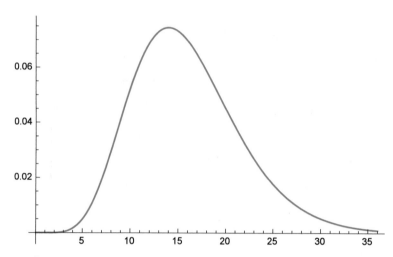

Fig. 9.1 χ^2 distribution

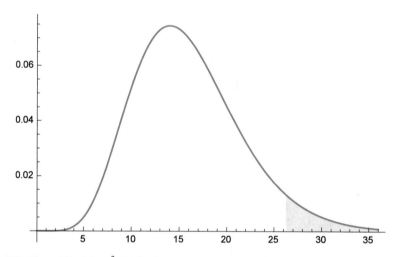

Fig. 9.2 Upper 5% of the χ^2 distribution

$$h(T) = \frac{4!14!5!2!1!10!6!5!2!3!}{26! \prod_{i,j} t_{ij}!},$$

where the numbers appearing in the numerator correspond to the marginals of T_0. Note that $h(T)$ equals to the following probability:

(a) There are 26 balls in a large box;
(b) Among 26 balls, there are 4 balls labeled "1," 14 balls labeled "2," 5 balls labeled "3," 2 balls labeled "4," and 1 ball labeled "5" (each ball has exactly one label);

(c) There are 5 small boxes labeled from "1" to "5";
(d) We pick the balls from the large box one by one and at random (without seeing the labels), and put them into the small boxes so that 10 balls are in the box labeled "1," 6 balls are in the box labeled "2," 5 balls are in the box labeled "3," 2 balls are in the box labeled "4," 3 balls are in the box labeled "5," respectively.
(e) The number $h(T)$ equals to the probability that the number of the balls labeled "i" in the box labeled "j" is t_{ij} for all $1 \leq i, j \leq 5$. In fact, the number of permutations of 26 balls (ignoring the small boxes) is

$$\alpha = \frac{26!}{4!14!5!2!1!}$$

and the number of permutations of 26 balls which satisfies condition "the number of the balls labeled "i" in the box labeled "j" is t_{ij} for all $1 \leq i, j \leq 5$" is

$$\beta = \frac{10!}{\prod_{i=1}^{5} t_{i1}!} \cdot \frac{6!}{\prod_{i=1}^{5} t_{i2}!} \cdot \frac{5!}{\prod_{i=1}^{5} t_{i3}!} \cdot \frac{2!}{\prod_{i=1}^{5} t_{i4}!} \cdot \frac{3!}{\prod_{i=1}^{5} t_{i5}!} = \frac{10!6!5!2!3!}{\prod_{i,j} t_{ij}!}.$$

Thus, the probability is $\beta/\alpha = h(T)$.

For Fisher's exact test, we compute the *p-value*:

$$p = \sum_{\substack{T \in \mathscr{F}_{T_0} \\ \chi^2(T) \geq \chi^2(T_0) = 25.338}} h(T).$$

The *p*-value is the probability that the χ^2 statistics of $T \in \mathscr{F}_0$ is greater than or equal to $\chi^2(T_0) = 25.338$. If $p < 0.05$, then it means that the probability is very small and $\chi^2(T_0)$ is very rare. Thus, we conclude that H_0 is rejected and hence the scores on the two subjects are correlated. In our particular case, $p = 0.0609007$ and hence our conclusion is "we cannot reject H_0" again. However, in order to compute *p*-value, we have to

- compute the χ^2 statistics for all tables in $\mathscr{F}_{T_0}$ which consists of 229174 tables,
- select all the tables T with $\chi^2(T) \geq 25.338$, and
- compute the sum of the value $h(T)$ of them.

It is impossible to compute the *p*-value if the cardinality of the set $\mathscr{F}_{T_0}$ is very big.
    If the χ^2 test cannot be applied to provide a reliable result, for example, if the contingency table is sparse, and if, in addition, we cannot compute the exact *p*-value, then we can use the *Markov chain Monte Carlo method* (called MCMC method for short). The MCMC method is a sampling from $\mathscr{F}_{T_0}$ by using "Markov bases."

Problems

9.1 A class has 70 students and all students take subjects "Set Theory" and "Group Theory." The score of each subject is one of 5, 4, 3, 2, 1. The following table T_0 classifies the students according to their scores on Set Theory and Group Theory.

S. \ G.	5	4	3	2	1	Total
5	4	2	0	0	0	6
4	1	6	2	1	0	10
3	0	1	11	6	1	19
2	0	2	4	12	3	21
1	0	0	0	3	11	14
Total	5	11	17	22	15	70

Suppose the null hypothesis H_0 which says that the scores on the two subjects are independent.

(a) For the table $T_0 = (t_{ij})$, compute the χ^2 statistics $\chi^2(T_0)$.
(b) Test H_0 by using χ^2 distribution with degrees of freedom 16.

9.2   Markov Bases for m-Way Contingency Tables

Let $\mathscr{S}$ be a finite set of N elements, which is called the *sample set*. For $k = 1, \ldots, m$, let $X_k = \{Y_1^{(k)}, \ldots, Y_{r_k}^{(k)}\}$ such that $\mathscr{S} = \bigcup_{j=1}^{r_k} Y_j^{(k)}$ and that $Y_p^{(k)} \cap Y_q^{(k)} = \emptyset$ for $1 \le p < q \le r_k$, $1 \le k \le m$. Each X_k is called a *factor*, and each $Y_j^{(k)}$ is called a *category*. An m-way contingency table of size $r_1 \times \cdots \times r_m$ is an m-dimensional array

$$T = (t_{i_1 \cdots i_m})_{1 \le i_1 \le r_1, \ldots, 1 \le i_m \le r_m}$$

such that $N = \sum_{i_1, \ldots, i_m} t_{i_1 \cdots i_m}$ and $\left| \bigcap_{k=1}^{m} Y_{i_k}^{(k)} \right| = t_{i_1 \cdots i_m}$ for $1 \le i_1 \le r_1, \ldots, 1 \le i_m \le r_m$.

For example, for the contingency table T_0 in the previous section, $\mathscr{S}$ consists of $N = 26$ students, there are $m = 2$ factors, the score on Geometry is the factor $X_1 = \{Y_1^{(1)}, \ldots, Y_5^{(1)}\}$, and the score on Probability is the factor $X_2 = \{Y_1^{(2)}, \ldots, Y_5^{(2)}\}$, where the category $Y_1^{(1)}$ consists of the students whose scores of Geometry is 5, the category $Y_1^{(2)}$ consists of the students whose scores of Probability is 5, and so on. Recall that, for the 2-way contingency table $T_0 = (x_{ij})_{1 \le i, j \le 5}$ of size 5×5 discussed in the previous section, we defined $\mathscr{F}_{T_0}$ to be

$$\mathscr{F}_{T_0} = \left\{ T = (t_{ij}) \in \mathbb{Z}_{\geq 0}^{5 \times 5} : \sum_{i=1}^{5} t_{ij} = \sum_{i=1}^{5} x_{ij} \ (1 \leq j \leq 5), \right.$$

$$\left. \sum_{j=1}^{5} t_{ij} = \sum_{j=1}^{5} x_{ij} \ (1 \leq i \leq 5) \right\}.$$

Let T_0 be the following 3-way contingency table of size $2 \times 2 \times 2$.

X_3	$Y_1^{(3)}$		$Y_2^{(3)}$	
$X_1 \backslash X_2$	$Y_1^{(2)}$	$Y_2^{(2)}$	$Y_1^{(2)}$	$Y_2^{(2)}$
$Y_1^{(1)}$	x_{111}	x_{121}	x_{112}	x_{122}
$Y_2^{(1)}$	x_{211}	x_{221}	x_{212}	x_{222}

For the definition of $\mathscr{F}_{T_0}$, there are many other natural choices. For example,

$$\mathscr{F}_{T_0} = \left\{ T=(t_{ijk}) \in \mathbb{Z}_{\geq 0}^{2 \times 2 \times 2} : \begin{array}{l} \sum_{j,k} t_{1jk} = \sum_{j,k} x_{1jk}, \ \ \sum_{i,k} t_{i1k} = \sum_{i,k} x_{i1k}, \\ \sum_{i,j} t_{ij1} = \sum_{i,j} x_{ij1}, \ \ \sum_{i,j,k} t_{ijk} = \sum_{i,j,k} x_{ijk} \end{array} \right\}$$

(9.1)

$$\mathscr{F}_{T_0} = \left\{ T=(t_{ijk}) \in \mathbb{Z}_{\geq 0}^{2 \times 2 \times 2} : \begin{array}{l} t_{ij1} + t_{ij2} = x_{ij1} + x_{ij2} \\ t_{1jk} + t_{2jk} = x_{1jk} + x_{2jk} \quad (1 \leq i, j, k \leq 2) \\ t_{i1k} + t_{i2k} = x_{i1k} + x_{i2k} \end{array} \right\}$$

(9.2)

are possible in this case. We identify $2 \times 2 \times 2$ array $T = (t_{ijk})$ with the vector

$$(t_{111}, t_{112}, t_{121}, t_{122}, t_{211}, t_{212}, t_{221}, t_{222})^t.$$

Then, the above $\mathscr{F}_{T_0}$ is of the form

$$\mathscr{F}_{T_0} = \left\{ T = (t_{ijk}) \in \mathbb{Z}_{\geq 0}^{2 \times 2 \times 2} : AT = AT_0 \right\}$$

for a suitable integer matrix A whose column vectors are indexed by ijk with $1 \leq i, j, k \leq 2$. In fact, in our example we have

$$A = \begin{pmatrix} 1 & 1 & 1 & 1 & 0 & 0 & 0 & 0 \\ 1 & 1 & 0 & 0 & 1 & 1 & 0 & 0 \\ 1 & 0 & 1 & 0 & 1 & 0 & 1 & 0 \\ 1 & 1 & 1 & 1 & 1 & 1 & 1 & 1 \end{pmatrix}$$

for (9.1), and

$$A = \begin{pmatrix} 1 & 1 & & & & & & & & \\ & & 1 & 1 & & & & & & \\ & & & & 1 & 1 & & & & \\ & & & & & & 1 & 1 & & \\ \hline 1 & & & & 1 & & & & & \\ & 1 & & & & & 1 & & & \\ & & 1 & & & & & 1 & & \\ & & & 1 & & & & & & 1 \\ \hline 1 & & 1 & & & & & & & \\ & 1 & & 1 & & & & & & \\ & & & & 1 & & 1 & & & \\ & & & & & 1 & & & 1 & \end{pmatrix}$$

for (9.2). Such a matrix A is called a *model matrix*. For an m-way contingency table, we can define a model matrix in the same way. In particular, the columns of a model matrix of an m-contingency table of size $r_1 \times \cdots \times r_m$ are indexed by the sequences $i_1 i_2 \cdots i_m$ with $1 \le i_1 \le r_1, \ldots, 1 \le i_m \le r_m$

Let A be a model matrix for m-way contingency tables of size $r_1 \times \cdots \times r_m$ and set

$$\mathrm{Ker}_{\mathbb{Z}}(A) = \{M \in \mathbb{Z}^{r_1 \times \cdots \times r_m} : AM = \mathbf{0}\}.$$

It is easy to see that if T and T' belong to $\mathscr{F}_{T_0}$ for a contingency table T_0, then we have $T - T' \in \mathrm{Ker}_{\mathbb{Z}}(A)$.

Definition 9.1 Let $\{M_1, \ldots, M_\ell\}$ be a finite subset of $\mathrm{Ker}_{\mathbb{Z}}(A)$. Then, $\{M_1, \ldots, M_\ell\}$ is called a *Markov basis* for A, if for any m-way contingency table T_0 of size $r_1 \times \cdots \times r_m$, and for any $T, T' \in \mathscr{F}_{T_0}$, there exist $M_{i_1}, \ldots, M_{i_\Lambda}$ such that

$$T' = T + \sum_{k=1}^{\Lambda} \varepsilon_k M_{i_k}, \text{ where } \varepsilon_k \in \{\pm 1\} \quad (1 \le k \le \Lambda),$$

$$T + \sum_{k=1}^{\lambda} \varepsilon_k M_{i_k} \in \mathscr{F}_{T_0} \quad (1 \le \lambda \le \Lambda).$$

We assume that, for each $T \in \mathscr{F}_{T_0}$, the probability of the occurrence of T is defined by some distribution $h(T)$. (For example, in the previous section, $h(T)$ is the multiple hypergeometric distribution.) Using the Markov basis $\{M_1, \ldots, M_\ell\}$, we have a sampling by the following algorithm:

Algorithm (Metropolis–Hastings)

1. Choose $T \in \mathscr{F}_{T_0}$ at random and set $T' = T$;
2. Repeat the following:

   2.1. Select M_i from $\{M_1, \ldots, M_\ell\}$ at random (with probability $1/\ell$);
   2.2. Select ε from $\{\pm 1\}$ at random (with probability $1/2$);
   2.3. If $T' + \varepsilon M_i$ is a nonnegative matrix, then set $T' = T' + \varepsilon M_i$ with probability

$$\min \left\{ \frac{h(T' + \varepsilon M_i)}{h(T')}, 1 \right\}.$$

Since $\{M_1, \ldots, M_\ell\}$ is a Markov basis, there exist no unreachable elements of $\mathscr{F}_{T_0}$ in the Metropolis–Hastings algorithm. By this algorithm, we have a sequence of tables

$$T^{(1)}, \ T^{(2)}, \ \ldots, T^{(s)} \in \mathscr{F}_{T_0},$$

which follows the distribution $h(T)$. Recall that the p-value of T_0 is

$$p = \sum_{\substack{T \in \mathscr{F}_{T_0} \\ \chi^2(T) \geq \chi^2(T_0)}} h(T).$$

Since $h(T)$ is the probability of the appearance of T in $\mathscr{F}_{T_0}$ following the distribution $h(T)$, the p-value is the probability of the appearance of the tables in $\mathscr{F}_{T_0}$ such that their χ^2 statistics is greater than or equal to that of T_0. From the sample $\{T^{(1)}, \ldots, T^{(s)}\}$, we estimate the p-value of T_0 by:

$$\frac{|\{k \in \{1, \ldots, s\} \ : \ \chi^2(T^{(k)}) \geq \chi^2(T_0)\}|}{s}, \tag{9.3}$$

which is the percentage of the tables in $\{T^{(1)}, \ldots, T^{(s)}\}$ such that their χ^2 statistics is greater than or equal to that of T_0. Since $\{T^{(1)}, \ldots, T^{(s)}\}$ follows the distribution $h(T)$, the value (9.3) approximates p-value.

Example 9.2 A Markov basis for the model in the previous section is well known. Let $\mathscr{M}_{5 \times 5}$ denote the set of all integer 5×5 matrices which satisfy that the sum of all entries of each rows and each columns is zero. Let $\{M_1, \ldots, M_{100}\}$ be the set of all 5×5 matrices of the form

$$
\begin{array}{cc}
 & j_1 \quad j_2 \\
\begin{array}{c} i_1 \\ \\ i_2 \end{array}
\left(
\begin{array}{cc}
+1 & -1 \\
 & \\
-1 & +1
\end{array}
\right) \in \mathscr{M}_{5 \times 5}.
\end{array}
$$

Then, $\{M_1, \ldots, M_{100}\}$ is a Markov basis. For example,

$$T = \begin{pmatrix} 1 & 2 & 1 & 0 & 0 \\ 7 & 3 & 3 & 1 & 0 \\ 1 & 1 & 1 & 1 & 1 \\ 1 & 0 & 0 & 0 & 1 \\ 0 & 0 & 0 & 0 & 1 \end{pmatrix}, \quad T' = \begin{pmatrix} 2 & 1 & 1 & 0 & 0 \\ 6 & 3 & 4 & 1 & 0 \\ 1 & 2 & 0 & 1 & 1 \\ 1 & 0 & 0 & 0 & 1 \\ 0 & 0 & 0 & 0 & 1 \end{pmatrix}$$

belong to $\mathscr{F}_{T_0}$ and

$$M = \begin{pmatrix} 1 & -1 & 0 & 0 & 0 \\ -1 & 1 & 0 & 0 & 0 \\ 0 & 0 & 0 & 0 & 0 \\ 0 & 0 & 0 & 0 & 0 \\ 0 & 0 & 0 & 0 & 0 \end{pmatrix}, \quad M' = \begin{pmatrix} 0 & 0 & 0 & 0 & 0 \\ 0 & 1 & -1 & 0 & 0 \\ 0 & -1 & 1 & 0 & 0 \\ 0 & 0 & 0 & 0 & 0 \\ 0 & 0 & 0 & 0 & 0 \end{pmatrix} \in \{M_1, \ldots, M_{100}\}$$

satisfy

$$T = \begin{pmatrix} 1 & 2 & 1 & 0 & 0 \\ 7 & 3 & 3 & 1 & 0 \\ 1 & 1 & 1 & 1 & 1 \\ 1 & 0 & 0 & 0 & 1 \\ 0 & 0 & 0 & 0 & 1 \end{pmatrix} \xrightarrow{+M} \begin{pmatrix} 2 & 1 & 1 & 0 & 0 \\ 6 & 4 & 3 & 1 & 0 \\ 1 & 1 & 1 & 1 & 1 \\ 1 & 0 & 0 & 0 & 1 \\ 0 & 0 & 0 & 0 & 1 \end{pmatrix} \in \mathscr{F}_{T_0} \xrightarrow{-M'} T' = \begin{pmatrix} 2 & 1 & 1 & 0 & 0 \\ 6 & 3 & 4 & 1 & 0 \\ 1 & 2 & 0 & 1 & 1 \\ 1 & 0 & 0 & 0 & 1 \\ 0 & 0 & 0 & 0 & 1 \end{pmatrix}.$$

In general, it is difficult to compute a Markov basis for a given model matrix A. Diaconis and Sturmfels [53] found the relationship between a Markov basis and the toric ideal of a model matrix.

Example 9.3 Let T_0 be a 2×3 contingency table. Consider the model matrix

$$A = \begin{pmatrix} 1 & 1 & 1 & 0 & 0 & 0 \\ 0 & 0 & 0 & 1 & 1 & 1 \\ 1 & 0 & 0 & 1 & 0 & 0 \\ 0 & 1 & 0 & 0 & 1 & 0 \\ 0 & 0 & 1 & 0 & 0 & 1 \end{pmatrix}.$$

Then,

$$\left\{ M_1 = \begin{pmatrix} 1 & -1 & 0 \\ -1 & 1 & 0 \end{pmatrix}, \quad M_2 = \begin{pmatrix} 0 & 1 & -1 \\ 0 & -1 & 1 \end{pmatrix}, \quad M_3 = \begin{pmatrix} 1 & 0 & -1 \\ -1 & 0 & 1 \end{pmatrix} \right\}$$

is a Markov basis. We identify these matrices with the integer column vectors

$$M_1 = (1, -1, 0, -1, 1, 0)^t, \quad M_2 = (0, 1, -1, 0, -1, 1)^t, \quad M_3 = (1, 0, -1, -1, 0, 1)^t.$$

The column vectors belong to $\mathrm{Ker}_{\mathbb{Z}}(A)$, and the toric ideal I_A of A is generated by the binomials

$$f_{M_1} = x_1 x_5 - x_2 x_4,$$
$$f_{M_2} = x_2 x_6 - x_3 x_5,$$
$$f_{M_3} = x_1 x_6 - x_4 x_6.$$

The correspondence in Example 9.3 holds in general.

Theorem 9.4 *Let A be a model matrix. Then, a finite subset $\mathscr{B} = \{M_1, \dots, M_\ell\}$ of $\mathrm{Ker}_{\mathbb{Z}}(A)$ is a Markov basis for A if and only if I_A is generated by $f_{M_1}, \dots, f_{M_\ell}$.*

Proof Note that $I_{\mathscr{B}} \subset I_A$. For any contingency tables T and T',

$$T, T' \in \mathscr{F}_{T_0} \text{ for some } T_0 \iff AT = AT' \iff \mathbf{x}^T - \mathbf{x}^{T'} \in I_A.$$

On the other hand, by Corollary 3.10, T and T' are connected via $\mathscr{B}$ if and only if $\mathbf{x}^T - \mathbf{x}^{T'}$ belongs to $I_{\mathscr{B}}$. Thus, $\mathscr{B}$ is a Markov basis for A if and only if $I_A = I_{\mathscr{B}}$, which is equivalent to say that I_A is generated by $f_{M_1}, \dots, f_{M_\ell}$.   □

Problems

9.2 Show that any proper subset of $\{M_1, M_2, M_3\}$ in Example 9.3 does not satisfy the condition in Definition 9.1. (Do not use Theorem 9.4.)

9.3 Consider the model of the 2-way contingency table $T_0 = (x_{ij})_{1 \le i, j \le 5}$ of size 5×5 discussed in the previous section.

(a) What is the model matrix A in this case?
(b) Let $\{M_1, \dots, M_{100}\}$ be the Markov basis defined in Example 9.2. Show that every proper subset of $\{M_1, \dots, M_{100}\}$ does not satisfy the condition in Definition 9.1.
(c) Estimate the p-value in this case by the MCMC method using the Markov basis $\{M_1, \dots, M_{100}\}$.

9.3 Sequential Importance Sampling and Normality of Toric Rings

Sequential importance sampling is another method to estimate the p-value. We go back to the first example of this chapter: let T_0 be the table
Then, we can choose $T = (t_{ij}) \in \mathscr{F}_{T_0}$ randomly as follows:

Geom. \ Prob.	5	4	3	2	1	Total
5	2	1	1	0	0	4
4	8	3	3	0	0	14
3	0	2	1	1	1	5
2	0	0	0	1	1	2
1	0	0	0	0	1	1
Total	10	6	5	2	3	26

Choose t_{11} from $\{0, 1, \ldots, \min\{4, 10\}\}$ randomly, say $t_{11} = 3$. Then, we consider

Geom. \ Prob.	5	4	3	2	1	Remainder
5	✓	-	-	-	-	1
4	-	-	-	-	-	14
3	-	-	-	-	-	5
2	-	-	-	-	-	2
1	-	-	-	-	-	1
Remainder	7	6	5	2	3	23

Geom. \ Prob.	5	4	3	2	1	Total
5	3					3
4						0
3						0
2						0
1						0
Total	3	0	0	0	0	3

Choose t_{21} from $\{0, 1, \ldots, \min\{14, 7\}\}$ randomly, say $t_{21} = 5$. Then, we consider

Geom. \ Prob.	5	4	3	2	1	Remainder
5	✓	-	-	-	-	1
4	✓	-	-	-	-	9
3	-	-	-	-	-	5
2	-	-	-	-	-	2
1	-	-	-	-	-	1
Remainder	2	6	5	2	3	18

Geom. \ Prob.	5	4	3	2	1	Total
5	3					3
4	5					5
3						0
2						0
1						0
Total	8	0	0	0	0	8

...

Finally, we get a table $T = (t_{ij})$, say,

Geom. \ Prob.	5	4	3	2	1	Remainder
5	✓	✓	✓	✓	✓	0
4	✓	✓	✓	✓	✓	0
3	✓	✓	✓	✓	✓	0
2	✓	✓	✓	✓	✓	0
1	✓	✓	✓	✓	✓	0
Remainder	0	0	0	0	0	0

Geom. \ Prob.	5	4	3	2	1	Total
5	3	0	1	0	0	4
4	5	3	3	2	1	14
3	1	2	0	0	2	5
2	1	0	1	0	0	2
1	0	1	0	0	0	1
Total	10	6	5	2	3	26

For any 2-way contingency tables, this method does not get stuck. However, there exists a 3-way contingency table for which this method gets stuck.

Consider, for example, the following $4 \times 4 \times 4$ contingency table T_0:

$$T_0 = (x_{ijk}) = \begin{array}{|cccc|cccc|cccc|cccc|} \hline 1\,0\,0\,0 & 0\,0\,0\,1 & 0\,1\,0\,0 & 0\,0\,0\,1 \\ 0\,0\,1\,0 & 0\,0\,0\,0 & 1\,0\,0\,0 & 0\,1\,0\,0 \\ 0\,1\,0\,0 & 0\,0\,1\,0 & 0\,0\,0\,1 & 0\,0\,0\,0 \\ 0\,0\,0\,0 & 1\,0\,0\,0 & 0\,0\,0\,1 & 0\,0\,1\,0 \\ \hline \end{array}$$

and let

$$\mathcal{F}_{T_0} = \left\{ T = (t_{ijk}) \in \mathbb{Z}_{\geq 0}^{4 \times 4 \times 4} : \begin{array}{l} \sum_{\ell=1}^{4} t_{ij\ell} = \sum_{\ell=1}^{4} x_{ij\ell} \\ \sum_{\ell=1}^{4} t_{\ell jk} = \sum_{\ell=1}^{4} x_{\ell jk} \\ \sum_{\ell=1}^{4} t_{i\ell k} = \sum_{\ell=1}^{4} x_{i\ell k} \end{array} (1 \leq i, j, k \leq 4) \right\}.$$

(9.4)

Then, $\mathcal{F}_{T_0}$ is the set of contingency tables $T = (t_{ijk})$ whose marginals are:

$$\begin{array}{|cc|}\hline & 1 \\ t_{1jk} & 1 \\ & 1 \\ & 0 \\ \hline 1\ \ 1\ \ 1\ 0 \end{array} \quad \begin{array}{|cc|}\hline & 1 \\ t_{2jk} & 0 \\ & 1 \\ & 1 \\ \hline 1\ \ 0\ \ 1\ 1 \end{array} \quad \begin{array}{|cc|}\hline & 1 \\ t_{3jk} & 1 \\ & 1 \\ & 1 \\ \hline 1\ \ 1\ \ 0\ 2 \end{array} \quad \begin{array}{|cc|}\hline & 1 \\ t_{4jk} & 1 \\ & 0 \\ & 1 \\ \hline 0\ \ 1\ \ 1\ 1 \end{array} \quad \begin{array}{|cccc|}\hline 1\ 1\ 0\ 2 \\ 1\ 1\ 1\ 0 \\ 0\ 1\ 1\ 1 \\ 1\ 0\ 1\ 1 \\ \hline \end{array}.$$

Here, the rightmost table is $(m_{jk})_{\substack{j=1,\ldots,4 \\ k=1,\ldots,4}}$ where $m_{jk} = \sum_{i=1}^{4} x_{ijk}$. Then, we try to choose $T = (t_{ijk})$ in $\mathcal{F}_{T_0}$ randomly. For example, choose t_{314} from $\{0, 1\}$ randomly, say $t_{314} = 1$. Then, we consider the incomplete contingency table with $t_{314} = 1$ fixed, the other entries undetermined, and the bold marked marginal changed accordingly.

$$\begin{array}{|cccc|c}\hline - - - - & 1 \\ - - - - & 1 \\ - - - - & 1 \\ - - - - & 0 \\ \hline 1\ 1\ 1\ 0 & 3 \end{array} \quad \begin{array}{|cccc|c}\hline - - - - & 1 \\ - - - - & 0 \\ - - - - & 1 \\ - - - - & 1 \\ \hline 1\ 0\ 1\ 1 & 3 \end{array} \quad \begin{array}{|cccc|c}\hline - - - \checkmark & 0 \\ - - - - & 1 \\ - - - - & 1 \\ - - - - & 1 \\ \hline 1\ 1\ 0\ 1 & 3 \end{array} \quad \begin{array}{|cccc|c}\hline - - - - & 1 \\ - - - - & 1 \\ - - - - & 0 \\ - - - - & 1 \\ \hline 0\ 1\ 1\ 1 & 3 \end{array} \quad \begin{array}{|cccc|}\hline 1\ 1\ 0\ 1 \\ 1\ 1\ 1\ 0 \\ 0\ 1\ 1\ 1 \\ 1\ 0\ 1\ 1 \\ \hline \end{array}$$

However, at this point, we can predict that we will be stuck. There is no $4 \times 4 \times 4$ table whose marginals are as above and $t_{314} = 1$. What is the difference between two examples above? The toric ring of the model matrix of the 2-way contingency table is *normal* since it is the edge ring of the complete bipartite graph (Theorem 5.20). On the other hand, the toric ring of the model matrix of the $4 \times 4 \times 4$ table is *not normal*. It will turn that the sequential importance sampling does not work for all cases when the toric ring of the model matrix is not normal, while it works in all cases when it is normal.

Let A_{444} be the model matrix which defines $\mathscr{F}_{T_0}$ as in (9.4). Since each entry of T_0 is a nonnegative integer, it is clear that $A_{444}T_0$ belongs to $\mathbb{Z}_{\geq 0}A_{444}$, where, as in Chapter 4, $\mathbb{Z}_{\geq 0}A_{444}$ denotes the linear combinations of the column vectors of A_{444} with nonnegative integer coefficients.

Let $\mathbf{v}$ be the vector corresponding to the (new) marginals:

$$
\begin{array}{|c|}\hline 1 \\ 1 \\ 1 \\ 0 \\ \hline 1\,1\,1\,0 \end{array}
\quad
\begin{array}{|c|}\hline 1 \\ 0 \\ 1 \\ 1 \\ \hline 1\,0\,1\,1 \end{array}
\quad
\begin{array}{|c|}\hline \mathbf{0} \\ 1 \\ 1 \\ 1 \\ \hline 1\,1\,0\,\mathbf{1} \end{array}
\quad
\begin{array}{|c|}\hline 1 \\ 1 \\ 0 \\ 1 \\ \hline 0\,1\,1\,1 \end{array}
\quad
\begin{array}{|c|}\hline 1\,1\,0\,\mathbf{1} \\ 1\,1\,1\,0 \\ 0\,1\,1\,1 \\ 1\,0\,1\,1 \end{array}
\tag{9.5}
$$

Then, $\mathbf{v}$ belongs to $\mathbb{Z}A_{444}$, since $\mathbf{v} = A_{444}T_0 - \mathbf{a}_{314}$. However, one can check that the vector $\mathbf{v}$ does not belong to $\mathbb{Z}_{\geq 0}A_{444}$, see Problem 9.4. This is equivalent to say that there exists no table T such that $A_{444}T = v$. Of course, v should at least belong to $\mathbb{Q}_{\geq 0}A_{444}$. Then, if $\mathbb{Q}_{\geq 0}A_{444} \cap \mathbb{Z}A_{444} = \mathbb{Z}_{\geq 0}A_{444}$, then $\mathbf{v}$ is a possible marginal vector. However, in general, one only has $\mathbb{Z}_{\geq 0}A_{444} \subsetneqq \mathbb{Q}_{\geq 0}A_{444} \cap \mathbb{Z}A_{444}$. Indeed, in our example $\mathbf{v} \in \mathbb{Q}_{\geq 0}A_{444} \cap \mathbb{Z}A_{444} \setminus \mathbb{Z}_{\geq 0}A_{444}$. In fact,

$$
\begin{array}{|cccc|c|}\hline 1/2 & 1/2 & 0 & 0 & 1 \\ 1/2 & 0 & 1/2 & 0 & 1 \\ 0 & 1/2 & 1/2 & 0 & 1 \\ 0 & 0 & 0 & 0 & 0 \\ \hline 1 & 1 & 1 & 0 \end{array}
\quad
\begin{array}{|cccc|c|}\hline 1/2 & 0 & 0 & 1/2 & 1 \\ 0 & 0 & 0 & 0 & 0 \\ 0 & 0 & 1/2 & 1/2 & 1 \\ 1/2 & 0 & 1/2 & 0 & 1 \\ \hline 1 & 0 & 1 & 1 \end{array}
\quad
\begin{array}{|cccc|c|}\hline 0 & 0 & 0 & 0 & 0 \\ 1/2 & 1/2 & 0 & 0 & 1 \\ 0 & 1/2 & 0 & 1/2 & 1 \\ 1/2 & 0 & 0 & 1/2 & 1 \\ \hline 1 & 1 & 0 & 1 \end{array}
\quad
\begin{array}{|cccc|c|}\hline 0 & 1/2 & 0 & 1/2 & 1 \\ 0 & 1/2 & 1/2 & 0 & 1 \\ 0 & 0 & 0 & 0 & 0 \\ 0 & 0 & 1/2 & 1/2 & 1 \\ \hline 0 & 1 & 1 & 1 \end{array}
\quad
\begin{array}{|c|}\hline 1\,1\,0\,1 \\ 1\,1\,1\,0 \\ 0\,1\,1\,1 \\ 1\,0\,1\,1 \end{array}
$$

Problems

9.4 Check that there exists no $4 \times 4 \times 4$ table T such that $A_{444}T = \mathbf{v}$, where A_{444} and $\mathbf{v}$ are defined as in (9.4) and in (9.5).

9.5 Verify whether the model matrix A_{444} above is very ample or not.

9.4 Toric Rings and Ideals of Hierarchical Models

Let $T = (t_{i_1 \cdots i_m})$ be any m-way contingency table of size $r_1 \times \cdots \times r_m$. With each subset $F = \{i_1, \ldots, i_s\}$ of $[m] = \{1, \ldots, m\}$ and each $(\ell_{i_1}, \ldots, \ell_{i_s}) \in [r_{i_1}] \times \cdots \times [r_{i_s}]$, we associate the number

$$t^{F}_{\ell_{i_1}\cdots\ell_{i_s}} = \sum_{(\ell_{j_1},\dots,\ell_{j_{m-s}})\in[r_{j_1}]\times\cdots\times[r_{j_{m-s}}]} t_{\ell_1\cdots\ell_m},$$

where $\{j_1, \dots, j_{m-s}\} = [m] \setminus F$.

This concept can be used to describe various models for T. To explain this, we consider our example (9.1), which for $T_0 = (x_{ijk}) \in \mathbb{Z}_{\geq 0}^{2\times2\times2}$ is given by

$$\mathscr{F}_{T_0} = \left\{ T = (t_{ijk}) \in \mathbb{Z}_{\geq 0}^{2\times2\times2} : \begin{array}{l} \sum_{j,k} t_{1jk} = \sum_{j,k} x_{1jk}, \quad \sum_{i,k} t_{i1k} = \sum_{i,k} x_{i1k}, \\ \sum_{i,j} t_{ij1} = \sum_{i,j} x_{ij1}, \quad \sum_{i,j,k} t_{ijk} = \sum_{i,j,k} x_{ijk} \end{array} \right\}.$$

By using the above notation, $\mathscr{F}_{T_0}$ can also be expressed by

$$\mathscr{F}_{T_0} = \left\{ T = (t_{ijk}) \in \mathbb{Z}_{\geq 0}^{2\times2\times2} : t_1^{\{1\}} = x_1^{\{1\}},\ t_1^{\{2\}} = x_1^{\{2\}},\ t_1^{\{3\}} = x_1^{\{3\}},\ t^{\emptyset} = x^{\emptyset} \right\},$$

or by

$$\mathscr{F}_{T_0} = \left\{ T = (t_{ijk}) \in \mathbb{Z}_{\geq 0}^{2\times2\times2} : \begin{array}{l} t_i^{\{1\}} = x_i^{\{1\}}, \quad t_j^{\{2\}} = x_j^{\{2\}}, \quad t_k^{\{3\}} = x_k^{\{3\}} \\ (1 \leq i, j, k \leq 2) \end{array} \right\}.$$

Similarly, the model for T_0 given in (9.2) can be expressed as

$$\mathscr{F}_{T_0} = \left\{ T = (t_{ijk}) \in \mathbb{Z}_{\geq 0}^{2\times2\times2} : \begin{array}{l} t_{ij}^{\{1,2\}} = x_{ij}^{\{1,2\}},\ t_{jk}^{\{2,3\}} = x_{jk}^{\{2,3\}},\ t_{ik}^{\{1,3\}} = x_{ik}^{\{1,3\}} \\ (1 \leq i, j, k \leq 2) \end{array} \right\}.$$

Thus, these models are characterized by the sets $D_1 = \{\{1\}, \{2\}, \{3\}\}$ and $D_2 = \{\{1, 2\}, \{2, 3\}, \{1, 3\}\}$, respectively. In fact, the model for T_0 given in (9.1) can be expressed as:

$$\mathscr{F}_{T_0} = \left\{ T = (t_{ijk}) \in \mathbb{Z}_{\geq 0}^{2\times2\times2} : t_i^F = x_i^F \text{ for all } F \in D_1 \text{ and } 1 \leq i \leq 2 \right\},$$

and the model for T_0 given in (9.2) can be expressed as:

$$\mathscr{F}_{T_0} = \left\{ T = (t_{ijk}) \in \mathbb{Z}_{\geq 0}^{2\times2\times2} : t_{ij}^F = x_{ij}^F \text{ for all } F \in D_2 \text{ and } 1 \leq i, j \leq 2 \right\}.$$

We now consider again the general case, and let F' be a subset of $F = \{i_1, \dots, i_s\} \subset [m]$. With a loss of generality, we may assume $F' = \{i_1, \dots, i_t\}$. Then,

$$t^{F'}_{\ell_{i_1}\cdots\ell_{i_t}} = \sum_{\ell_{i_{t+1}}\cdots\ell_{i_s}\in[r_{i_{t+1}}]\times\cdots\times[r_{i_s}]} t^{F}_{\ell_{i_1}\cdots\ell_{i_t}}.$$

Hence, for each $(\ell_{i_1}, \ldots, \ell_{i_t}) \in [r_{i_1}] \times \cdots \times [r_{i_t}]$, if

$$t^F_{\ell_{i_1} \cdots \ell_{i_s}} = x^F_{\ell_{i_1} \cdots \ell_{i_s}} \text{ for all } (\ell_{i_{t+1}}, \ldots, \ell_{i_s}) \in [r_{i_{t+1}}] \times \cdots \times [r_{i_s}],$$

then $t^{F'}_{\ell_{i_1} \cdots \ell_{i_t}} = x^{F'}_{\ell_{i_1} \cdots \ell_{i_t}}$.

For example,

$$t^{\{1\}}_1 = t^{\{1,2\}}_{1,1} + t^{\{1,2\}}_{1,2}$$

and hence,

$$t^{\{1,2\}}_{1,1} = x^{\{1,2\}}_{1,1} \text{ and } t^{\{1,2\}}_{1,2} = x^{\{1,2\}}_{1,2} \implies t^{\{1\}}_1 = x^{\{1\}}_1.$$

In conclusion, we see that the above two models for T_0 are determined by the simplicial complexes:

$$\Delta_1 = \{\emptyset, \{1\}, \{2\}, \{3\}\}, \quad \Delta_2 = \{\emptyset, \{1\}, \{2\}, \{3\}, \{1, 2\}, \{2, 3\}, \{1, 3\}\},$$

respectively. In fact, the model for T_0 given in (9.1) can be expressed as:

$$\mathscr{F}_{T_0} = \left\{ T = (t_{ijk}) \in \mathbb{Z}^{2 \times 2 \times 2}_{\geq 0} : \begin{array}{c} t^F_{\ell_{i_1} \cdots \ell_{i_s}} = x^F_{\ell_{i_1} \cdots \ell_{i_s}} \\ \text{for all } F = \{i_1, \ldots, i_s\} \in \Delta_1 \\ \text{and } 1 \leq \ell_{i_1}, \ldots, \ell_{i_s} \leq 2 \end{array} \right\},$$

and the model for T_0 given in (9.2) can be expressed as:

$$\mathscr{F}_{T_0} = \left\{ T = (t_{ijk}) \in \mathbb{Z}^{2 \times 2 \times 2}_{\geq 0} : \begin{array}{c} t^F_{\ell_{i_1} \cdots \ell_{i_s}} = x^F_{\ell_{i_1} \cdots \ell_{i_s}} \\ \text{for all } F = \{i_1, \ldots, i_s\} \in \Delta_2 \\ \text{and } 1 \leq \ell_{i_1}, \ldots, \ell_{i_s} \leq 2 \end{array} \right\}.$$

Definition 9.5 A model of an m-way contingency table T_0 of size $r_1 \times \cdots \times r_m$ is called a *hierarchical model*, if there exists a simplicial complex Δ on $[m]$ such that

$$\mathscr{F}_{T_0} = \left\{ T = (t_{i_1 \cdots i_m})_{\substack{1 \leq i_k \leq r_k \\ 1 \leq k \leq m}} : \begin{array}{c} t^F_{\ell_{i_1} \cdots \ell_{i_s}} = x^F_{\ell_{i_1} \cdots \ell_{i_s}} \\ \text{for all } F = \{i_1, \ldots, i_s\} \in \Delta \\ \text{and } (\ell_{i_1}, \ldots, \ell_{i_s}) \in [r_{i_1}] \times \cdots \times [r_{i_s}] \end{array} \right\}.$$

The model matrix of this hierarchical model given by Δ will be denoted by $A_{r_1 \cdots r_m}(\Delta)$.

In this section, we study the toric ring and ideal of hierarchical models. One reason for doing this is that any finite binomial system of generators of these toric ideals gives us a Markov basis for the model. Moreover, if we can show that for a hierarchical model matrix $A_{r_1 \cdots r_m}(\Delta)$ the associated toric ring is normal, then this implies that $\mathbb{Q}_{\geq 0} A_{r_1 \cdots r_m}(\Delta) \cap \mathbb{Z} A_{r_1 \cdots r_m}(\Delta) = \mathbb{Z}_{\geq 0} A_{r_1 \cdots r_m}(\Delta)$. This property then simplifies the sequential importance sampling process, as explained in the previous section. In the next two subsections, we consider special classes of hierarchical models.

9.4.1 Decomposable Graphical Models

For a simplicial complex Δ on $\{1, 2, \ldots, m\}$, let Facet(Δ) be the set of all facets of Δ. Given an m-way $r_1 \times \cdots \times r_m$ contingency table with the hierarchical model given by Δ, let

$$K[\mathbf{x}] = K[x_{i_1 \cdots i_m} : 1 \leq i_j \leq r_j \ (1 \leq j \leq m)],$$

$$K[\mathbf{t}] = K[t^F_{\ell_1 \cdots \ell_n} : F = \{k_1, \ldots, k_n\} \in \text{Facet}(\Delta), 1 \leq \ell_j \leq r_{k_j} \ (1 \leq j \leq n)]$$

be polynomial rings over a field K. The toric ideal of the model matrix $A_{r_1 \cdots r_m}(\Delta)$ is denoted by $I_{r_1 \cdots r_m}(\Delta)$. Then, the toric ideal $I_{r_1 \cdots r_m}(\Delta)$ is the kernel of homomorphism

$$\pi : K[\mathbf{x}] \longrightarrow K[\mathbf{t}]$$

defined by:

$$\pi(x_{i_1 \cdots i_m}) = \prod_{F = \{k_1, \ldots, k_n\} \in \text{Facet}(\Delta)} t^F_{i_{k_1} \cdots i_{k_n}}.$$

Example 9.6 We consider a 3-way $2 \times 2 \times 3$ contingency table with the hierarchical model given by the simplicial complex $\Delta = \{\emptyset, \{1\}, \{2\}, \{3\}, \{1, 2\}, \{2, 3\}\}$.
Then, Facet(Δ) = $\{\{1, 2\}, \{2, 3\}\}$, and

$$K[\mathbf{x}] = K[x_{111}, x_{112}, x_{113}, x_{121}, x_{122}, x_{123}, x_{211}, x_{212}, x_{213}, x_{221}, x_{222}, x_{223}],$$

$$K[\mathbf{t}] = K[t^{\{1,2\}}_{11}, t^{\{1,2\}}_{12}, t^{\{1,2\}}_{21}, t^{\{1,2\}}_{22}, t^{\{2,3\}}_{11}, t^{\{2,3\}}_{12}, t^{\{2,3\}}_{13}, t^{\{2,3\}}_{21}, t^{\{2,3\}}_{22}, t^{\{2,3\}}_{23}].$$

Furthermore, $\pi : K[\mathbf{x}] \longrightarrow K[\mathbf{t}]$ is defined by $\pi(x_{ijk}) = t^{\{1,2\}}_{ij} t^{\{2,3\}}_{jk}$.

Proposition 9.7 *Let Δ be a simplicial complex and let $r_1, \ldots, r_m$ and $s_1, \ldots, s_m$ be integers such that $s_i \leq r_i$ for all $1 \leq i \leq m$. Then, $A_{s_1 \cdots s_m}(\Delta)$ is a combinatorial pure subconfiguration of $A_{r_1 \cdots r_m}(\Delta)$.*

The proof is left as a problem to the reader (Problem 9.6).
Let Δ be a simplicial complex on $[m] = \{1, 2, \ldots, m\}$. Recall the definition of a leaf of Δ, a branch of a leaf, a leaf order, a quasi-forest, and a quasi-tree, given

in Chapter 7. A *simplicial vertex* of a leaf F is a vertex $j \in F$ such that $j \notin H$ for all facets H of Δ with $H \neq F$. It is clear that a vertex $j \in F$ is a simplicial vertex of F if and only if j belongs to $F \setminus G$, where G is a branch of F. Given a leaf order $F_1, F_2, \ldots, F_r$, let Δ_i be the subcomplex $\langle F_1, F_2, \ldots, F_i \rangle$ of Δ. A *separator* of Δ is a subset $W \subset [m]$ with the property that there are subsets U_a and U_b of $[m]$ satisfying the following conditions:

$$[m] = U_a \cup U_b, \quad W = U_a \cap U_b, \quad U_a \setminus W \neq \emptyset, \quad U_b \setminus W \neq \emptyset,$$

$$\{i, j\} \notin \Delta \text{ for all } i \in U_a \setminus W \text{ and } j \in U_b \setminus W.$$

Let Δ be a quasi-forest on $[m]$ with a leaf order $F_1, \ldots, F_r$. For each leaf F_q of the subcomplex Δ_q, fix a branch $F_{q'}$ of F_q, where $1 \leq q' < q$. Let T denote the finite graph on the vertex set $[r]$ with the edges $\{2', 2\}, \{3', 3\}, \ldots, \{r', r\}$. It then follows that T is connected. Since T has r vertices and $r - 1$ edges, T is a tree. The tree T is called a *relation tree* of Δ.

Example 9.8 Let $\Delta = \{\{1, 2, 3\}, \{3, 4, 5\}, \{2, 4, 6\}, \{2, 3, 4\}\}$. Then, Δ is a quasi-forest on [6] with a leaf order

$$F_1 = \{1, 2, 3\}, \quad F_2 = \{3, 4, 5\}, \quad F_3 = \{2, 4, 6\}, \quad F_4 = \{2, 3, 4\}.$$

With respect to this order, the edge set of the relation tree is:

$$\{\{1, 4\}, \{2, 4\}, \{3, 4\}\}.$$

On the other hand, Δ is a quasi-forest on [6] with a leaf order

$$F_1 = \{1, 2, 3\}, \quad F_2 = \{3, 4, 5\}, \quad F_3 = \{2, 3, 4\}, \quad F_4 = \{2, 4, 6\}.$$

With respect to this order, the edge set of the relation tree is:

$$\{\{1, 3\}, \{2, 3\}, \{3, 4\}\}.$$

Let $\{q', q\}$ with $q' < q$ be an edge of T. By deleting the edge $\{q', q\}$ from T, one obtains two trees $T_{q'}$ and T_q, where the vertex q' belongs to $T_{q'}$ and where the vertex q belongs to T_q.

Lemma 9.9 *Work with the same notation as above. Then, one has:*

(i) *If $j \in [r]$ is a vertex of T_q, then $j \geq q$;*

(ii) *The set $F_{q'} \cap F_q$ is a separator of Δ.*

Proof Let V_i denote the set of vertices of T_i for $i \in \{q, q'\}$.

(i) Let $j = \min\{i \in [r] : i \in V_q\}$. Suppose $j \neq q$. Since $j \in V_q$ and since $\{j', j\}$ is an edge of T with $\{j', j\} \neq \{q', q\}$, it follows that $\{j', j\}$ is an edge of T_q. In particular, j' belongs to V_q. However, since $j' < j$, we have $j' \notin V_q$, which is a contradiction. Hence, we have $j = q$.

(ii) Let $U_q = \bigcup_{j \in V_q} F_j$ and $U_{q'} = \bigcup_{j \in V_{q'}} F_j$. Then, we have $[m] = U_{q'} \cup U_q$. Let $W = F_{q'} \cap F_q$. We will show that $W = U_{q'} \cap U_q$.

First, we consider the case $q = r$. By (i), it follows that $V_r = \{r\}$ and $V_{r'} = \{1, 2, \ldots, r-1\}$. Thus, we have $U_r = F_r$ and $U_{r'} \cap U_r = \bigcup_{j=1}^{r-1}(F_j \cap F_r)$. Since F_r is a leaf of Δ, it follows that $\bigcup_{j=1}^{r-1}(F_j \cap F_r) = F_{r'} \cap F_r = W$.

It remains to show the case $q < r$. By induction on r, suppose that the assertion holds for Δ_{r-1}. Then, we may assume that $r \in V_q$ and $U_{q'} \cap (\bigcup_{r \neq j \in V_q} F_j) = W$. Since $r \in V_q$, we have $r' \in V_q$. Let $k \in V_{q'}$. Since $F_k \cap F_r \subset U_{r'} \cap F_r = F_{r'} \cap F_r$, it follows that $F_k \cap F_r \subset F_k \cap F_{r'}$. Hence, have $U_{q'} \cap U_q = U_{q'} \cap (\bigcup_{r \neq j \in V_q} F_j) = W$.

Note that $q \in U_q \setminus W$ and $q' \in U_{q'} \setminus W$. Thus, it is enough to show that $\{i, j\} \notin \Delta$ for all $i \in U_q \setminus W$ and $j \in U_{q'} \setminus W$. Suppose that $\{i, j\} \in \Delta$. Then, there exists a facet F_k with $\{i, j\} \subset F_k$. Let, say, $F_k \subset U_q$. Then, $j \in U_{q'} \cap U_q = W$, which is a contradiction. $\square$

Let Δ be a quasi-forest on $[m]$. Fix a leaf order $F_1, F_2, \ldots, F_r$ of Δ. By relabeling the vertices of Δ, if necessary, we may assume that the simplicial vertices of the leaf F_r of Δ are $m, m-1, \ldots, m_r$, and that, for each $1 \leq i < r$, the simplicial vertices of the leaf F_i of the quasi-forest Δ_i are $m_{i+1} - 1, m_{i+1} - 2, \ldots, m_i$. In particular, $F_1 = \{m_2 - 1, m_2 - 2, \ldots, 1\}$. Fix a relation tree T of Δ on the vertex set $[r]$ with the edges $\{2', 2\}, \{3', 3\}, \ldots, \{r', r\}$, where $q' < q$ for each $2 \leq q \leq r$. By Lemma 9.9, each edge $\{q', q\}$ of T yields the decomposition $\mathrm{Facet}(\Delta) = C_{q'} \cup C_q$, where $C_{q'} = \{F_j : j \text{ is a vertex of } T_{q'}\}$ and $C_q = \{F_j : j \text{ is a vertex of } T_q\}$. Note that $j \geq q$ if $F_j \in C_q$.

Given the separator $W = F_{q'} \cap F_q$ of Δ with $U_q = \bigcup_{j \in V_q} F_j$ and $U_{q'} = \bigcup_{j \in V_{q'}} F_j$, and given the integer vectors $\delta = (\delta_1, \ldots, \delta_m)$, $\rho = (\rho_1, \ldots, \rho_m) \in [r_1] \times \cdots \times [r_m]$ with the property that $\delta_j = \rho_j$ for all $j \in W$, we associate the quadratic binomial

$$f_\rho^\delta(C_{q'}, C_q) = x_\delta x_\rho - x_{\delta'} x_{\rho'}$$

belonging to $I_{r_1 \cdots r_m}(\Delta)$, where

$$\delta_i' = \begin{cases} \delta_i & \text{if } i \in U_{q'}, \\ \rho_i & \text{otherwise}, \end{cases} \qquad \rho_i' = \begin{cases} \rho_i & \text{if } i \in U_{q'}, \\ \delta_i & \text{otherwise}. \end{cases}$$

Example 9.7 (continued) Let, as before, $\Delta = \{\emptyset, \{1\}, \{2\}, \{3\}, \{1, 2\}, \{2, 3\}\}$. Then, $F_1 = \{1, 2\}$, $F_2 = \{2, 3\}$ is a leaf order and $W = F_1 \cap F_2 = \{2\}$ is a separator of Δ. For this W, $U_a = \{1, 2\}$, $U_b = \{2, 3\}$, $C_a = \{\{1, 2\}\}$, and $C_b = \{\{2, 3\}\}$. Hence,

$$f_\rho^\delta(C_a, C_b) = x_{113} x_{212} - x_{112} x_{213}$$

for $\delta = (1, 1, 3)$ and $\rho = (2, 1, 2)$.

Let $\mathscr{G}_{r_1 \cdots r_m}(\Delta, T)$ be the finite set of quadratic binomials defined by:

$$\mathscr{G}_{r_1 \cdots r_m}(\Delta, T) = \left\{ f_\rho^\delta(C_{q'}, C_q) \neq 0 : \begin{array}{c} \{q', q\} \text{ is an edge of } T \\ \delta, \rho \in [r_1] \times \cdots \times [r_m], \\ \delta_k = \rho_k \text{ for all } k \in F_{q'} \cap F_q \end{array} \right\}.$$

Let $<_{\text{lex}}$ be the lexicographic order on $K[\mathbf{x}]$ induced by the ordering of variables defined by:

$x_\delta < x_\rho \Longleftrightarrow$ the left-most nonzero component of $(\rho_1 - \delta_1, \ldots, \rho_m - \delta_m)$ is positive.

Theorem 9.10 gives an explicit description of Markov bases for the hierarchical model arising from a quasi-forest.

Theorem 9.10 *Let Δ be a quasi-forest and T a relation tree of Δ. Then, $\mathscr{G}_{r_1 \cdots r_m}(\Delta, T)$ is a Gröbner basis of the toric ideal $I_{r_1 \cdots r_m}(\Delta)$ with respect to $<_{\text{lex}}$.*

Proof Suppose that $\mathscr{G}_{r_1 \cdots r_m}(\Delta, T)$ is not a Gröbner basis. Let $\mathscr{B}$ denote the set of monomials w of $K[\mathbf{x}]$ that do not belong to $(\text{in}_{<_{\text{lex}}}(f) : f \in \mathscr{G}_{r_1 \cdots r_m}(\Delta, T))$. By Theorem 3.11, there exists a nonzero binomial $x_{\delta^{(1)}} \cdots x_{\delta^{(d)}} - x_{\rho^{(1)}} \cdots x_{\rho^{(d)}}$ in $I_{r_1 \cdots r_m}(\Delta)$ such that both $x_{\delta^{(1)}} \cdots x_{\delta^{(d)}}$ and $x_{\rho^{(1)}} \cdots x_{\rho^{(d)}}$ belong to $\mathscr{B}$. Assume that $x_{\delta^{(d)}} \leq_{\text{lex}} \cdots \leq_{\text{lex}} x_{\delta^{(1)}}$ and $x_{\rho^{(d)}} \leq_{\text{lex}} \cdots \leq_{\text{lex}} x_{\rho^{(1)}}$. Let $\delta^{(i)} = (\delta_1^{(i)}, \ldots, \delta_m^{(i)})$ and $\rho^{(i)} = (\rho_1^{(i)}, \ldots, \rho_m^{(i)})$ for $1 \leq i \leq d$.

Suppose that $\delta_j^{(i)} \neq \rho_j^{(i)}$ for some $1 \leq i \leq d$ and for some $1 \leq j \leq m$. Let $j^* = \min\{j \in [m] : \delta_j^{(i)} \neq \rho_j^{(i)} \text{ for some } i \in [d]\}$. Let $m_{q^*} \leq j^* < m_{q^*+1}$. Then, j^* is a simplicial vertex of the leaf F_{q^*} of the quasi-forest Δ_{q^*}. Let $i^* = \min\{i \in [d] : \delta_{j^*}^{(i)} \neq \rho_{j^*}^{(i)}\}$ and let $i_* = \max\{i \in [d] : \delta_j^{(i^*)} = \delta_j^{(i)} \text{ for all } 1 \leq j < m_{q^*}\}$. It is clear that $i^* \leq i_*$. Since $x_{\delta^{(d)}} \leq_{\text{lex}} \cdots \leq_{\text{lex}} x_{\delta^{(1)}}$, it follows that $\delta_j^{(i^*)} = \delta_j^{(i)} = \rho_j^{(i)}$ for all $i^* \leq i \leq i_*$ and for all $1 \leq j < m_{q^*}$.

Assume $\delta_{j^*}^{(i^*)} < \rho_{j^*}^{(i^*)}$. Let

$$M = \left\{ c \in [d] : \begin{array}{c} \text{there exists } i \text{ with } i^* \leq i \leq i_* \\ \text{such that } \rho_j^{(c)} = \delta_j^{(i)} \text{ for all } j^* \geq j \in F_{q^*} \end{array} \right\},$$

$$N = \left\{ c \in [d] : \begin{array}{c} \text{there exists } i \text{ with } i^* \leq i \leq i_* \\ \text{such that } \delta_j^{(c)} = \delta_j^{(i)} \text{ for all } j^* \geq j \in F_{q^*} \end{array} \right\}.$$

Since $\pi(x_{\delta^{(1)}} \cdots x_{\delta^{(d)}}) = \pi(x_{\rho^{(1)}} \cdots x_{\rho^{(d)}})$, we have $|M| = |N|$. Suppose that $c < i^*$ belongs to M. Then, $\delta_j^{(c)} = \rho_j^{(c)}$ for all $1 \leq j \leq j^*$. Hence, c belongs to N. Thus, we have

$$|\{c \in M : c < i^*\}| \leq |\{c \in N : c < i^*\}|. \tag{9.6}$$

On the other hand, it is trivial that $\{i^*, i^* + 1, \ldots, i_*\} \subset N$. Suppose $i^* \in M$, i.e., there exists $i^* \leq i \leq i_*$ such that $\rho_j^{(i^*)} = \delta_j^{(i)}$ for all $j^* \geq j \in F_{q^*}$. Then, $i \neq i^*$ and $\rho_j^{(i^*)} = \delta_j^{(i)}$ for all $1 \leq j \leq j^*$. Hence, we have $\delta_{i^*} <_{\mathrm{lex}} \delta_i$, which is a contradiction. Thus, $i^* \notin M$. Therefore,

$$|\{c \in M : i^* \leq c \leq i_*\}| < |\{c \in N : i^* \leq c \leq i_*\}|. \tag{9.7}$$

By Equations (9.6) and (9.7) together with $|M| = |N|$, it follows that there exists $c > i_*$ with $c \in M$.

Let $c^* > i_*$ belong to M. Then, there exists $i^* \leq i \leq i_*$ such that $\rho_j^{(c^*)} = \delta_j^{(i)}$ for all $j^* \geq j \in F_{q^*}$. If $1 \leq j < m_{q^*}$ and $j \in F_{q^*}$, then $\rho_j^{(c^*)} = \delta_j^{(i)} = \rho_j^{(i^*)}$. Suppose that $\rho_j^{(c^*)} = \rho_j^{(i^*)}$ for all $1 \leq j < m_{q^*}$ with $j \notin F_{q^*}$. Then, $\delta_j^{(c^*)} = \rho_j^{(c^*)} = \rho_j^{(i^*)} = \delta_j^{(i^*)}$ for all $1 \leq j < m_{q^*}$. This contradicts $c^* > i_*$. Hence, $\rho_j^{(c^*)} \neq \rho_j^{(i^*)}$ for some $1 \leq j < m_{q^*}$ with $j \notin F_{q^*}$. Since $\rho_{c^*} \leq_{\mathrm{lex}} \rho_{i^*}$, $\rho_j^{(c^*)} < \rho_j^{(i^*)}$ for some $1 \leq j < m_{q^*}$ with $j \notin F_{q^*}$. Suppose that $\rho_j^{(c^*)} \geq \rho_j^{(i^*)}$ for all $m_{q^*} \leq j \leq j^*$. Then, $\delta_j^{(i)} = \rho_j^{(c^*)} \geq \rho_j^{(i^*)} = \delta_j^{(i^*)}$ for all $m_{q^*} \leq j < j^*$. Hence, $\delta_j^{(i)} \geq \delta_j^{(i^*)}$ for all $1 \leq j < j^*$. Moreover, $\delta_{j^*}^{(i)} = \rho_{j^*}^{(c^*)} \geq \rho_{j^*}^{(i^*)} > \delta_{j^*}^{(i^*)}$. Thus, we have $\delta_{i^*} <_{\mathrm{lex}} \delta_i$, which is a contradiction. Therefore, $\rho_j^{(c^*)} < \rho_j^{(i^*)}$ for some $m_{q^*} \leq j \leq j^*$.

Thus,

$$f_{\rho_{c^*}}^{\rho_{i^*}}(C_{q^*}, C_{(q^*)'}) = x_{\rho_{i^*}} x_{\rho_{c^*}} - x_{(\rho_{i^*})'} x_{(\rho_{c^*})'}$$

is nonzero and a binomial belonging to $\mathscr{G}_{r_1 \cdots r_m}(\Delta, T)$. Since $x_{(\rho_{i^*})'} <_{\mathrm{lex}} x_{\rho_{i^*}}$ and $x_{(\rho_{c^*})'} <_{\mathrm{lex}} x_{\rho_{i^*}}$, the initial monomial of $f_{\rho_{c^*}}^{\rho_{i^*}}(C_{q^*}, C_{(q^*)'})$ with respect to $<_{\mathrm{lex}}$ is $x_{\rho_{i^*}} x_{\rho_{c^*}}$. This contradicts $x_{\rho^{(1)}} \cdots x_{\rho^{(d)}} \in \mathscr{B}$. □

Let G be a finite simple graph on the vertex set $[m]$ and $E(G)$ the set of edges of G. Let $\Delta(G)$ be the clique complex of G defined in Chapter 7. A model of an m-way contingency table characterized by $\Delta(G)$ is called a *graphical model* of G.

Example 9.11 Let C_4 be a cycle of length 4. Then, the toric ideal $I_{2222}(\Delta(C_4))$ is minimally generated by 8 quadratic binomials and 8 binomials of degree 4.

Theorem 9.12 *Let $\Delta(G)$ be a clique complex of a graph G and fix positive integers $r_1, \ldots, r_m \geq 2$. Then, the following conditions are equivalent:*

(i) *$\Delta(G)$ is quasi-forest;*
(ii) *G is chordal;*
(iii) *$I_{r_1 \cdots r_m}(\Delta(G))$ is generated by quadratic binomials;*
(iv) *$I_{r_1 \cdots r_m}(\Delta(G))$ possesses a quadratic Gröbner basis.*

Proof First, (i) $\Longleftrightarrow$ (ii) follows from Theorem 7.6. By Theorem 9.10, we have (i) $\Longrightarrow$ (iv). The implication (iv) $\Longrightarrow$ (iii) holds in general. Thus, it is enough to show (iii) $\Longrightarrow$ (ii). Suppose that G is not a chordal graph. Then, G has an induced cycle

C of length $\ell \geq 4$. We may assume that $C = (1, 2, \ldots, \ell)$. By Proposition 9.7, the toric ring

$$K[A_{\underbrace{2\ldots 2}_{\ell \text{ times}}1\ldots 1}(\Delta(G))] \tag{9.8}$$

is a combinatorial pure subring of the toric ring $K[A_{r_1\cdots r_m}(\Delta(G))]$. Since the toric ring in (9.8) is isomorphic to the toric ring $K[A_{2\ldots 2}(\Delta(C))]$, it is enough to show that the toric ideal $I_{2\ldots 2}(\Delta(C))$ is not generated by quadratic binomials. One can show that $K[A_{2222}(\Delta(C_4))]$ is a combinatorial pure subring of $K[A_{2\ldots 2}(\Delta(C))]$ (see Problem 9.7). As stated in Example 9.11, $I_{2222}(\Delta(C_4))$ is not generated by quadratic binomials. □

9.4.2 No m-Way Interaction Models and Higher Lawrence Liftings

First, we introduce the notion of rth Lawrence liftings which is a generalization of Lawrence liftings.

Definition 9.13 Given an integer matrix $A \in \mathbb{Z}^{d \times n}$, the rth Lawrence lifting of A is the configuration

$$\Lambda^{(r)}(A) = \underbrace{\begin{pmatrix} A & & & \\ & A & & \\ & & \ddots & \\ & & & A \\ I_n & I_n & \cdots & I_n \end{pmatrix}}_{r \text{ times}}$$

where I_n is the $n \times n$ identity matrix.

In particular, the toric ring of $\Lambda^{(2)}(A)$ is isomorphic to the toric ring of the Lawrence lifting $\Lambda(A)$ of A. Indeed, $\mathrm{Ker}_{\mathbb{Z}}(\Lambda^{(2)}(A)) = \mathrm{Ker}_{\mathbb{Z}}(\Lambda(A))$ holds in general.

The first result is a simple observation.

Proposition 9.14 Let A' be a subconfiguration of a configuration A. Then, $K[\Lambda^{(r')}(A')]$ is a combinatorial pure subring of $K[\Lambda^{(r)}(A)]$ for all $2 \leq r' \leq r$.

Proposition 9.15 Let $K[\Lambda^{(r)}(A)]$ be the rth Lawrence lifting of a configuration A. If $K[\Lambda^{(r)}(A)]$ is very ample, then A is unimodular.

Proof Suppose that $K[\Lambda^{(r)}(A)]$ is very ample. By Proposition 9.14, $K[\Lambda^{(2)}(A)] \cong K[\Lambda(A)]$ is a combinatorial pure subring of $K[\Lambda^{(r)}(A)]$. Hence, by Lemma 4.40, $K[\Lambda(A)]$ is very ample. Finally, by Theorem 4.42, A is unimodular. $\qquad\square$

For an integer matrix $A \in \mathbb{Z}^{d \times n}$, let $\mathrm{Ker}_{\mathbb{Z}}(A) = \{\mathbf{b} \in \mathbb{Z}^n : A\mathbf{b} = \mathbf{0}\}$. It then follows that we have:

$$\mathrm{Ker}_{\mathbb{Z}}(\Lambda^{(r)}(A)) = \left\{ \begin{pmatrix} \mathbf{b}^{(1)} \\ \vdots \\ \mathbf{b}^{(r)} \end{pmatrix} \in \mathbb{Z}^{rn} : \mathbf{b}^{(i)} \in \mathrm{Ker}_{\mathbb{Z}}(A) \ (1 \le i \le r), \ \sum_{i=1}^{r} \mathbf{b}^{(i)} = \mathbf{0} \right\}.$$

In what follows, for the rest of this section, we write the column vector

$$\mathbf{b} = \begin{pmatrix} \mathbf{b}^{(1)} \\ \vdots \\ \mathbf{b}^{(r)} \end{pmatrix}$$

as $\mathbf{b} = \{\mathbf{b}^{(1)}, \dots, \mathbf{b}^{(r)}\}$. The *type* of $\mathbf{b} = \{\mathbf{b}^{(1)}, \dots, \mathbf{b}^{(r)}\} \in \mathrm{Ker}_{\mathbb{Z}}(\Lambda^{(r)}(A))$ is defined by:

$$\mathrm{type}(\mathbf{b}) = |\{i \in [r] : \mathbf{b}^{(i)} \ne \mathbf{0}\}|.$$

Theorem 9.16 *For any configuration $A \in \mathbb{Z}^{d \times n}$, there exists a constant m such that, for any $r \ge 2$, the toric ideal of $\Lambda^{(r)}(A)$ is generated by binomials $f_{\mathbf{b}}$ with* $\mathrm{type}(\mathbf{b}) \le m$.

The minimum value $m(A)$ of such m is called the *Markov complexity* of A. We prove Theorem 9.16 by showing the stronger Theorem 9.17 below.

A sum $\mathbf{c} + \mathbf{d}$ of integer vectors $\mathbf{c} = (c_1, \dots, c_n)$ and $\mathbf{d} = (d_1, \dots, d_n)$ is called *conformal* if $|c_i + d_i| = |c_i| + |d_i|$ for all $1 \le i \le n$. For an integer matrix $A \in \mathbb{Z}^{d \times n}$, the Graver basis of A is the set of all vectors $\mathbf{b} \in \mathrm{Ker}_{\mathbb{Z}}(A)$ such that $\mathbf{b}$ has no conformal decomposition $\mathbf{b} = \mathbf{c} + \mathbf{d}$ with $\mathbf{0} \ne \mathbf{c}, \mathbf{d} \in \mathrm{Ker}_{\mathbb{Z}}(A)$. Note that, for a configuration A, $\{f_{\mathbf{b}_1}, \dots, f_{\mathbf{b}_k}\}$ is the Graver basis of I_A if and only if $\{\mathbf{b}_1, \dots, \mathbf{b}_k\}$ is the Graver basis of A.

Theorem 9.17 *Let $A \in \mathbb{Z}^{d \times n}$ be a configuration such that $\mathrm{Ker}_{\mathbb{Z}}(A) \ne \{\mathbf{0}\}$. Then, there exists a constant g such that, for any $r \ge 2$, the Graver basis of $\Lambda^{(r)}(A)$ consists of the vectors $\mathbf{b}$ with $\mathrm{type}(\mathbf{b}) \le g$.*

Let $g(A)$ be the minimum value of such g, and let $B = (\mathbf{b}_1, \dots, \mathbf{b}_k)$, where $\{\mathbf{b}_1, \dots, \mathbf{b}_k\}$ is the Graver basis of A. If $\mathrm{Ker}_{\mathbb{Z}}(B) = \{\mathbf{0}\}$, then $g(A) = 2$. If $\mathrm{Ker}_{\mathbb{Z}}(B) \ne \{\mathbf{0}\}$, then $g = \max\{|\mathbf{c}_1|, \dots, |\mathbf{c}_\ell|\}$, where $\{\mathbf{c}_1, \dots, \mathbf{c}_\ell\}$ is the Graver basis of B.

The number $g(A)$ is called the *Graver complexity* of A.

Lemma 9.18 *Suppose that a nonzero vector* $\mathbf{b} = (\mathbf{b}^{(1)}, \ldots, \mathbf{b}^{(r)})$ *belongs to the Graver basis of* $\Lambda^{(r)}(A)$ *and that* $\mathbf{b}^{(i)} = \mathbf{c}^{(1)} + \cdots + \mathbf{c}^{(k)}$ *with* $\mathbf{c}^{(1)}, \ldots, \mathbf{c}^{(k)} \in \mathrm{Ker}_{\mathbb{Z}}(A)$ *is a conformal decomposition. Then,* $\mathbf{b}' = (\mathbf{b}^{(1)}, \ldots, \mathbf{b}^{(i-1)}, \mathbf{c}^{(1)}, \ldots, \mathbf{c}^{(k)}, \mathbf{b}^{(i+1)}, \ldots, \mathbf{b}^{(r)})$ *belongs to the Graver basis of* $\Lambda^{(r+k-1)}(A)$.

Proof Suppose that $\mathbf{b}'$ does not belong to the Graver basis. Then, there exists a conformal decomposition $\mathbf{b}' = \mathbf{b}'_1 + \mathbf{b}'_2$ such that $\mathbf{0} \neq \mathbf{b}'_1, \mathbf{b}'_2 \in \mathrm{Ker}_{\mathbb{Z}}(\Lambda^{(r+k-1)}(A))$. Let

$$\mathbf{b}'_1 = (\mathbf{b}_1^{(1)}, \ldots, \mathbf{b}_1^{(i-1)}, \mathbf{c}_1^{(1)}, \ldots, \mathbf{c}_1^{(k)}, \mathbf{b}_1^{(i+1)}, \ldots, \mathbf{b}_1^{(r)}),$$

$$\mathbf{b}'_2 = (\mathbf{b}_2^{(1)}, \ldots, \mathbf{b}_2^{(i-1)}, \mathbf{c}_2^{(1)}, \ldots, \mathbf{c}_2^{(k)}, \mathbf{b}_2^{(i+1)}, \ldots, \mathbf{b}_2^{(r)}).$$

Then,

$$\mathbf{b}_1 = (\mathbf{b}_1^{(1)}, \ldots, \mathbf{b}_1^{(i-1)}, \mathbf{c}_1^{(1)} + \cdots + \mathbf{c}_1^{(k)}, \mathbf{b}_1^{(i+1)}, \ldots, \mathbf{b}_1^{(r)}),$$

$$\mathbf{b}_2 = (\mathbf{b}_2^{(1)}, \ldots, \mathbf{b}_2^{(i-1)}, \mathbf{c}_2^{(1)} + \cdots + \mathbf{c}_2^{(k)}, \mathbf{b}_2^{(i+1)}, \ldots, \mathbf{b}_2^{(r)}).$$

are nonzero vectors belonging to $\mathrm{Ker}_{\mathbb{Z}}(\Lambda^{(r)}(A))$ such that $\mathbf{b} = \mathbf{b}_1 + \mathbf{b}_2$ is a conformal decomposition. This contradicts that $\mathbf{b}$ belongs to the Graver basis. $\square$

Corollary 9.19 *Suppose that a nonzero vector* $\mathbf{b} = (\mathbf{b}^{(1)}, \ldots, \mathbf{b}^{(r)})$ *belongs to the Graver basis of* $\Lambda^{(r)}(A)$. *Then, there exists* $\mathbf{b}' = (\mathbf{b}'_1, \ldots, \mathbf{b}'_s)$ *belonging to the Graver basis of* $\Lambda^{(s)}(A)$ *for some* $s \geq r$ *such that:*

(i) *Each* $\mathbf{b}'_i$ *belongs to the Graver basis of* A;
(ii) $\mathbf{b}$ *is obtained by a conformal sum of the columns of* $\mathbf{b}'$.

In particular, we have $\mathrm{type}(\mathbf{b}) \leq \mathrm{type}(\mathbf{b}')$.

Proof (Theorem 9.17) Since the Graver basis of the Lawrence lifting $\Lambda(A)$ of A consists of integer vectors of type 2, we have $g(A) \geq 2$.

Let $B = (\mathbf{b}_1, \ldots, \mathbf{b}_k)$, where $\{\mathbf{b}_1, \ldots, \mathbf{b}_k\}$ be the Graver basis of A. By Corollary 9.19, we only need to consider the vector $\mathbf{u} = (\mathbf{u}^{(1)}, \ldots, \mathbf{u}^{(r)})$ belonging to the Graver basis of $\Lambda^{(r)}(A)$ such that each $\mathbf{u}^{(i)}$ belongs to $\{\pm\mathbf{b}_1, \ldots, \pm\mathbf{b}_k\}$. Let $\psi_{\mathbf{u}} = (\psi_{\mathbf{u}}^{(1)}, \ldots, \psi_{\mathbf{u}}^{(k)}) \in \mathbb{Z}^k$, where

$$\psi_{\mathbf{u}}^{(i)} = |\{j \in [r] : \mathbf{u}^{(j)} = \mathbf{b}_i\}| - |\{j \in [r] : \mathbf{u}^{(j)} = -\mathbf{b}_i\}|$$

for $1 \leq i \leq k$. Since $\sum_{j=1}^{r} \mathbf{u}^{(j)} = \mathbf{0}$, it follows that $\psi_{\mathbf{u}}$ belongs to $\mathrm{Ker}_{\mathbb{Z}}(B)$.

Case 1. $(\mathrm{Ker}_{\mathbb{Z}}(B) = \{\mathbf{0}\})$

In this case, $\psi_{\mathbf{u}} = \mathbf{0}$ for any $\mathbf{u}$, and hence there exists $1 \leq i < j \leq r$ such that $\mathbf{u}^{(i)} = -\mathbf{u}^{(j)} \neq \mathbf{0}$. Then, it follows that $\mathbf{u}^{(\ell)} = \mathbf{0}$ for all $\ell \neq i, j$, and hence $\mathrm{type}(\mathbf{u}) = 2$. Thus, we have $g(A) = 2$.

Case 2. $(\mathrm{Ker}_{\mathbb{Z}}(B) \neq \{\mathbf{0}\})$

If $\psi_{\mathbf{u}} = \mathbf{0}$, then type($\mathbf{u}$) $= 2$. Assume that $\psi_{\mathbf{u}} \neq \mathbf{0}$. Then, type($\mathbf{u}$) ≥ 3 and the 1-norm $|\psi_{\mathbf{u}}|$ of the vector $\psi_{\mathbf{u}}$ equals the type of $\mathbf{u}$. We will show that $\mathbf{u}$ belongs to the Graver basis if and only if the vector $\psi_{\mathbf{u}}$ belongs to the Graver basis of B.

Suppose that $\psi_{\mathbf{u}}$ does not belong to the Graver basis of B. Then, there exists a conformal decomposition $\psi_{\mathbf{u}} = \mathbf{c} + \mathbf{d}$ with $\mathbf{0} \neq \mathbf{c}, \mathbf{d} \in \mathrm{Ker}_{\mathbb{Z}}(B)$, which yields a conformal decomposition $\mathbf{u} = \mathbf{u}_1 + \mathbf{u}_2$ (where type($\mathbf{u}_1$) $= |\mathbf{c}|$ and type($\mathbf{u}_2$) $= |\mathbf{d}|$). Then, $\mathbf{u}$ does not belong to the Graver basis.

Suppose that $\mathbf{u}$ does not belong to the Graver basis. Then, there exists a conformal decomposition $\mathbf{u} = \mathbf{u}_1 + \mathbf{u}_2$ with $\mathbf{0} \neq \mathbf{u}_1, \mathbf{u}_2 \in \mathrm{Ker}_{\mathbb{Z}}(\Lambda^{(r)})$. Since each $\mathbf{u}^{(i)}$ belongs to the Graver basis, there are no nontrivial decomposition of $\mathbf{u}^{(i)}$. Thus, the conformal decomposition $\mathbf{u} = \mathbf{u}_1 + \mathbf{u}_2$ comes from some conformal decomposition $\psi_{\mathbf{u}} = \mathbf{c} + \mathbf{d}$ with $\mathbf{0} \neq \mathbf{c}, \mathbf{d} \in \mathrm{Ker}_{\mathbb{Z}}(B)$, as desired. $\qquad \square$

Let Δ_m be a simplicial complex whose facets are the $(m-1)$-subsets of $[m]$. A model of an m-way contingency table given by Δ_m (see Definition 9.5) is called *no m-way interaction model*. In the case of this model, for the $r_1 \times r_2 \times \cdots \times r_m$ contingency table ($r_1 \geq r_2 \geq \cdots \geq r_m \geq 2$)

$$T = \left(t_{i_1 i_2 \cdots i_m}\right)_{i_k = 1, 2, \ldots, r_k}, \quad 0 \leq t_{i_1 i_2 \cdots i_m} \in \mathbb{Z},$$

the model matrix is the configuration $A_{r_1 r_2 \cdots r_m} := A_{r_1 r_2 \cdots r_m}(\Delta_m)$.

For example,

$$A_{22} = \begin{pmatrix} 1 & 1 & 0 & 0 \\ 0 & 0 & 1 & 1 \\ 1 & 0 & 1 & 0 \\ 0 & 1 & 0 & 1 \end{pmatrix}, \quad A_{222} = \begin{pmatrix} 1 & 1 & & & & & & \\ & & 1 & 1 & & & & \\ & & & & 1 & 1 & & \\ & & & & & & 1 & 1 \\ \hline 1 & & & & 1 & & & \\ & 1 & & & & 1 & & \\ & & 1 & & & & 1 & \\ & & & 1 & & & & 1 \\ \hline 1 & & 1 & & & & & \\ & 1 & & 1 & & & & \\ & & & & 1 & & 1 & \\ & & & & & 1 & & 1 \end{pmatrix}.$$

In general, $A_{r_1 r_2 \cdots r_m}$ has $r_1 \cdots r_m (\sum_{k=1}^{m} 1/r_k)$ rows and $r_1 \cdots r_m$ columns.

Example 9.20 The configuration A_{333} is a 27×27 matrix. The toric ideal of A_{333} is the kernel of homomorphism

$$\pi : K[\{x_{ijk}\}_{1 \leq i, j, k \leq 3}] \rightarrow K[\{t_{jk}^{(1)}, t_{ik}^{(2)}, t_{ij}^{(3)}\}_{1 \leq i, j, k \leq 3}]$$

defined by $\pi(x_{ijk}) = t_{jk}^{(1)} t_{ik}^{(2)} t_{ij}^{(3)}$ for each $1 \le i, j, k \le 3$. By using computer, one can check that $I_{A_{333}}$ is generated by 27 binomials of degree 4 and 54 binomials of degree 6. On the other hand, there are 795 circuits of $I_{A_{333}}$. For example,

$$f_M = x_{112}x_{121}x_{133}x_{222}x_{231}x_{311}x_{323}x_{332}^2 - x_{111}x_{123}x_{132}x_{221}x_{232}x_{312}x_{322}x_{331}x_{333},$$

where

$$M = \begin{array}{|ccc|ccc|ccc|}
m_{111} & m_{112} & m_{113} & m_{211} & m_{212} & m_{213} & m_{311} & m_{312} & m_{313} \\
m_{121} & m_{122} & m_{123} & m_{221} & m_{222} & m_{223} & m_{321} & m_{322} & m_{323} \\
m_{131} & m_{132} & m_{133} & m_{231} & m_{232} & m_{233} & m_{331} & m_{332} & m_{333}
\end{array}$$

$$= \begin{array}{|ccc|ccc|ccc|}
-1 & 1 & 0 & 0 & 0 & 0 & 1 & -1 & 0 \\
1 & 0 & -1 & -1 & 1 & 0 & 0 & -1 & 1 \\
0 & -1 & 1 & 1 & -1 & 0 & -1 & 2 & -1
\end{array}$$

is a circuit of $I_{A_{333}}$. By Theorem 4.35, A_{333} is not unimodular.

Note that, by suitable row exchanges of $A_{r_1 r_2 \cdots r_m}$ one obtains $\Lambda^{(r_m)}(A_{r_1 r_2 \cdots r_{m-1}})$. Hence, we can apply Propositions 9.14, 9.15, Theorems 9.16, and 9.17 to study $A_{r_1 r_2 \cdots r_m}$ by using the properties of $A_{r_1 r_2 \cdots r_{m-1}}$. For example, we have the following immediately.

Example 9.21 We can compute the Graver complexity of $I_{A_{r_{33}}}$ ($r \ge 3$) as follows. Since $A_{r_{33}}$ is the r-th Lawrence lifting of A_{33}, we first compute the Graver basis of $I_{A_{33}}$. Then, $K[A_{33}]$ is the edge ring of a 3×3 complete bipartite graph $K_{3,3}$. It is known that the Graver basis of $I_{A_{33}}$ consists of the binomials arising from cycles of $K_{3,3}$. Thus, we have to compute the Graver basis $\{c_1, \ldots, c_\ell\}$ of the matrix

$$B = \begin{pmatrix}
1 & 1 & 1 & 1 & 0 & 0 & 0 & 0 & 0 & 0 & 1 & 0 & 1 & 1 & 1 \\
-1 & 0 & -1 & 0 & 1 & 1 & 0 & 0 & 0 & 1 & 0 & 1 & -1 & 0 & -1 \\
0 & -1 & 0 & -1 & -1 & -1 & 0 & 0 & 0 & -1 & -1 & -1 & 0 & -1 & 0 \\
-1 & -1 & 0 & 0 & 0 & 0 & 1 & 1 & 0 & -1 & 0 & 1 & -1 & -1 & 0 \\
1 & 0 & 0 & 0 & -1 & 0 & -1 & 0 & 1 & 0 & -1 & -1 & 0 & 1 & 1 \\
0 & 1 & 0 & 0 & 1 & 0 & 0 & -1 & -1 & 1 & 1 & 0 & 1 & 0 & -1 \\
0 & 0 & -1 & -1 & 0 & 0 & -1 & -1 & 0 & 1 & -1 & -1 & 0 & 0 & -1 \\
0 & 0 & 1 & 0 & 0 & -1 & 1 & 0 & -1 & -1 & 1 & 0 & 1 & -1 & 0 \\
0 & 0 & 0 & 1 & 0 & 1 & 0 & 1 & 1 & 0 & 0 & 1 & -1 & 1 & 1
\end{pmatrix}.$$

Here, the first 9 columns of B correspond to cycles of length 4, and the last 6 columns of B correspond to cycles of length 6. Then, $\max\{|c_1|, \ldots, |c_\ell|\} = 9$, and hence the Graver complexity of A_{33} is 9. Thus, the Graver basis of $I_{A_{r_{33}}}$ is computed by that of $I_{A_{933}}$.

By Theorem 4.42, we can check whether $A_{r_1 r_2 \cdots r_m}$ is unimodular.

Proposition 9.22 *The configuration* $A_{r_1 r_2 \cdots r_m}$ *is unimodular if and only if either* $m = 2$ *or* $r_3 = 2$.

Proof Since $K[A_{r_1 r_2}]$ is considered as the edge ring of a complete bipartite graph, Theorem 5.24 implies that $A_{r_1 r_2}$ is unimodular, and since $A_{r_1 r_2 2 \cdots 2}$ is obtained by taking the Lawrence lifting several times from $A_{r_1 r_2}$, Theorem 4.42 implies that $A_{r_1 r_2 2 \cdots 2}$ is unimodular as well.

As stated in Example 9.20, A_{333} is not unimodular. Hence, by Theorem 4.42, $K[A_{3332 \cdots 2}]$ is not unimodular. If $m \geq 3$ and $r_3 \geq 3$, then $K[A_{3332 \cdots 2}]$ is a combinatorial pure subring of $K[A_{r_1 r_2 \cdots r_m}]$, and hence $K[A_{r_1 r_2 \cdots r_m}]$ is not unimodular.   □

Next, we study $K[A_{r_1 r_2 \cdots r_m}]$ that is not very ample. We use the notion of fundamental binomials. A binomial f belonging to the toric ideal I_A of A is called *fundamental* if there exists a combinatorial pure subring $K[B]$ of $K[A]$ such that the toric ideal I_B of B is generated by f.

Lemma 9.23 *If* I_A *possesses a fundamental binomial* g *such that none of the monomials appearing in* g *is squarefree, then* $K[A]$ *is not very ample.*

Proof Since g is fundamental, there exists a combinatorial pure subring $K[B]$ of $K[A]$ such that $I_B = (g)$. It is enough to show that $K[B]$ is not very ample.

Let $g = x_1^2 u - x_2^2 v$. Since g is fundamental, g is irreducible, and hence u $(\neq 1)$ is not divided by x_2 and v $(\neq 1)$ is not divided by x_1. Let $\pi : S \to K[A]$ be defined as in (3.1). Since $\pi(x_1^2 u) = \pi(x_2^2 v)$, we have $\sqrt{\pi(uv)} = \pi(x_1 u)/\pi(x_2)$. Let x_k be a variable with $k \neq 1, 2$. Then, the monomial $\pi(x_k^m)\sqrt{\pi(uv)}$ belongs to the quotient field of $K[A]$ and is integral over $K[A]$ for all positive integer m. Suppose that there exists a monomial w such that $\pi(w) = \pi(x_k^m)\sqrt{\pi(uv)}$. It then follows that the binomial $g' = x_1 u x_k^m - x_2 w$ belongs to I_B. Since $I_B = (g)$ and $x_1 u x_k^m$ is divided by neither $x_1^2 u$ nor $x_2^2 v$, we have $g' = 0$. Hence, x_2 must divide u, which is a contradiction. Thus, $\pi(x_k^m)\pi(uv)$ corresponds to a hole in the sense of (4.3) for all m and $K[B]$ is not very ample.   □

Proposition 9.24 *If one of the following conditions holds, then* $K[A_{r_1 r_2 \cdots r_m}]$ *is not very ample (and hence not normal):*

(i) $m \geq 4$ *and* $r_3 \geq 3$;
(ii) $m = 3$ *and* $r_3 \geq 4$;
(iii) $m = 3$, $r_3 = 3$, $r_1 \geq 6$, *and* $r_2 \geq 4$.

Proof If condition (i) holds, then $K[A_{3332 \cdots 2}]$ is a combinatorial pure subring of $K[A_{r_1 r_2 \cdots r_m}]$. Since A_{333} is not unimodular, $K[A_{3332 \cdots 2}]$ is not very ample.

If condition (ii) holds, then $K[A_{444}]$ is a combinatorial pure subring of $K[A_{r_1 r_2 \cdots r_m}]$. If condition (iii) holds, then $K[A_{643}]$ is a combinatorial pure subring of $K[A_{r_1 r_2 \cdots r_m}]$. Thus, it is enough to show that $K[A_{444}]$ and $K[A_{643}]$ are not very ample. One can show that:

$$x_{111} x_{221} x_{331} x_{641} x_{212} x_{522} x_{432} x_{642} x_{413} x_{323} x_{633}^2 x_{143} x_{543}$$

$$-x_{211}x_{321}x_{631}x_{141}x_{412}x_{222}x_{632}x_{542}x_{113}x_{523}x_{333}x_{433}x_{643}^2 \tag{9.9}$$

is a fundamental binomial of the toric ideal $I_{A_{643}}$, and

$$x_{111}^2 x_{133}x_{144}x_{223}x_{224}x_{232}x_{242}x_{313}x_{322}x_{341}x_{414}x_{422}x_{431}$$

$$-x_{113}x_{114}x_{131}x_{141}x_{222}^2 x_{233}x_{244}x_{311}x_{323}x_{342}x_{411}x_{424}x_{432} \tag{9.10}$$

is a fundamental binomial of the toric ideal $I_{A_{444}}$ (see Problem 9.9). Thus, by Lemma 9.23, $K[A_{444}]$ and $K[A_{643}]$ are not very ample. □

A configuration A is said to be *compressed* if the initial ideal of I_A with respect to any reverse lexicographic order is squarefree.

Proposition 9.25 *The configuration $A_{r_1 r_2 \cdots r_m}$ is compressed if and only if one of the following holds:*

(i) $m = 2$;
(ii) $m \geq 3$ and $r_3 = 2$;
(iii) $m = 3$ and $r_2 = r_3 = 3$.

Proof If $A_{r_1 r_2 \cdots r_m}$ satisfies one of the conditions (i) and (ii), then $A_{r_1 r_2 \cdots r_m}$ is unimodular (Proposition 9.22), and hence it is compressed (Theorem 4.29). In Example 9.21, we checked that the Graver basis of $I_{A_{r33}}$ is computed by that of $I_{A_{933}}$. One can check that A_{933} is compressed by using a software, e.g., Normaliz, polymake. Thus, A_{r33} is compressed for all $r \geq 3$.

Suppose that $A_{r_1 r_2 \cdots r_m}$ is compressed and that $A_{r_1 r_2 \cdots r_m}$ satisfies none of the conditions (i), (ii), and (iii). Then, $K[A_{r_1 r_2 \cdots r_m}]$ is normal. By Proposition 9.24, we have $m = 3$ and $(r_1, r_2, r_3) \in \{(5, 5, 3), (5, 4, 3), (4, 4, 3)\}$. By Proposition 9.14, we may assume that $(r_1, r_2, r_3) = (4, 4, 3)$. However, one can check that A_{443} is not compressed by using a software, e.g., Normaliz, polymake. □

It was shown by software 4ti2 and Normaliz that $K[A_{r_1 r_2 r_3}]$ is normal if $(r_1, r_2, r_3) \in \{(5, 5, 3), (5, 4, 3), (4, 4, 3)\}$. Summing up, we obtain the following classification (Table 9.1) for the configurations $A_{r_1 r_2 \cdots r_m}$. It is not known whether A_{553}, A_{543}, and A_{443} have a squarefree initial ideal or not.

Table 9.1 Classification

$m = 2$	Unimodular
$r_1 \times r_2 \times 2 \times \cdots \times 2$	
$r_1 \times 3 \times 3$	Compressed, not unimodular
$5 \times 5 \times 3, 5 \times 4 \times 3, 4 \times 4 \times 3$	Normal, not compressed
Otherwise, i.e.,	
$m \geq 4$ and $r_3 \geq 3$	Not normal, not very ample
$m = 3$ and $r_3 \geq 4$	
$m = 3, r_3 = 3, r_1 \geq 6$ and $r_2 \geq 4$	

Problems

9.6 Prove Proposition 9.7.

9.7 Let C_ℓ be a cycle of length $\ell \geq 4$. Show that $K[A_{2222}(C_4)]$ is a combinatorial pure subring of $K[A_{2\cdots2}(C_\ell)]$

9.8 Show that $A_{322} = \Lambda^{(3)}(A_{22})$.

9.9 Show that binomials in (9.9) and (9.10) are fundamental.

9.5 Segre–Veronese Configurations

We start with a brief explanation of the Hardy–Weinberg model.

The contingency table T_0 in Table 9.2 shows the genotypes in the ABO blood types of 100 patients suffering from a particular disease. Then, the total number of each "allele" (A genes, B genes, and O genes) is one of the entries of the vector

$$\begin{pmatrix} 2 & 1 & 1 & 0 & 0 & 0 \\ 0 & 1 & 0 & 2 & 1 & 0 \\ 0 & 0 & 1 & 0 & 1 & 2 \end{pmatrix} \begin{pmatrix} 23 \\ 10 \\ 15 \\ 6 \\ 17 \\ 29 \end{pmatrix} = \begin{pmatrix} 71 \\ 39 \\ 90 \end{pmatrix}.$$

In this case, the null hypothesis is that a population being sampled is in the *Hardy–Weinberg equilibrium*, i.e., allele and genotype frequencies in a population will remain constant from generation to generation in the absence of other evolutionary influences. Then, the model matrix is a configuration

$$A = \begin{pmatrix} 2 & 1 & 1 & 0 & 0 & 0 \\ 0 & 1 & 0 & 2 & 1 & 0 \\ 0 & 0 & 1 & 0 & 1 & 2 \end{pmatrix}.$$

The toric ring $K[A]$ of A is known as the second Veronese subring of $K[t_1, t_2, t_3]$. The purpose to the present section is to introduce a more general notion which is called a Segre–Veronese configuration.

Fix integers $d, M \geq 1$ and sets of integers $\mathbf{a} = \{a_1, \ldots, a_M\}$, $\mathbf{b} = \{b_1, \ldots, b_M\}$, $\mathbf{r} = \{r_1, \ldots, r_M\}$, and $\mathbf{s} = \{s_1, \ldots, s_M\}$ such that:

Table 9.2 Genotypes in ABO blood type	$T_0 =$	Genotypes	AA	AB	AO	BB	BO	OO
		Total	23	10	15	6	17	29

(i) $0 \le b_i \le a_i$ for all $1 \le i \le M$;

(ii) $1 \le s_i \le r_i \le d$ for all $1 \le i \le M$.

Fix $\tau \ge 2$. Let $A_{\tau,\mathbf{a},\mathbf{b},\mathbf{r},\mathbf{s}}$ denote the configuration matrix whose columns are all nonnegative integer vectors $(f_1, f_2, \ldots, f_d) \in \mathbb{Z}_{\ge 0}^d$ such that:

(i) $\sum_{j=1}^d f_j = \tau$.

(ii) $b_i \le \sum_{j=s_i}^{r_i} f_j \le a_i$ for all $1 \le i \le M$.

Then, the toric ring $K[A_{\tau,\mathbf{a},\mathbf{b},\mathbf{r},\mathbf{s}}]$ is called an *algebra of Segre–Veronese type*.

Example 9.26 Several popular classes of semigroup rings are algebras of Segre–Veronese type.

(a) If $M = 2$, $\tau = 2$, $a_1 = a_2 = b_1 = b_2 = 1$, $s_1 = 1$, $s_2 = r_1 + 1$, and $r_2 = d$, then the toric ring $K[A_{\tau,\mathbf{a},\mathbf{b},\mathbf{r},\mathbf{s}}]$ is the Segre product of polynomial rings $K[q_1, \ldots, q_{r_1}]$ and $K[q_{r_1+1}, \ldots, q_d]$.

(b) If $M = d$, $s_i = r_i = i$, $a_i = \tau$, and $b_i = 0$ for all $1 \le i \le M$, then the toric ring $K[A_{\tau,\mathbf{a},\mathbf{b},\mathbf{r},\mathbf{s}}]$ is the classical τth Veronese subring of the polynomial ring $K[q_1, \ldots, q_d]$.

(c) If $M = d$, $s_i = r_i = i$, $a_i = 1$, and $b_i = 0$ for all $1 \le i \le M$, then the toric ring $K[A_{\tau,\mathbf{a},\mathbf{b},\mathbf{r},\mathbf{s}}]$ is the τth squarefree Veronese subring of the polynomial ring $K[q_1, \ldots, q_d]$.

(d) Algebras of Veronese type (i.e., $M = d$, $s_i = r_i = i$, and $b_i = 0$ for all $1 \le i \le M$).

The edge ring of a graph G discussed in Chapter 5 is an algebra of Segre–Veronese type if G is a complete multipartite graph.

Example 9.27 Let $q_1, \ldots, q_n$ denote a sequence of positive integers with $q_1 + \cdots + q_n = d$. Let $V_1, \ldots, V_n$ be a partition of $[d]$, say,

$$V_i = \left\{ 1 + \sum_{j=1}^{i-1} q_j, 2 + \sum_{j=1}^{i-1} q_j, \ldots, q_i + \sum_{j=1}^{i-1} q_j \right\}$$

for each $1 \le i \le n$. The *complete multipartite graph* of type $\mathbf{q} = (q_1, \ldots, q_n)$ is the finite graph $G_{\mathbf{q}}$ on the vertex set $[d]$ with the edge set

$$E(G_{\mathbf{q}}) = \{\{k, \ell\} : k \in V_i, \ell \in V_j, 1 \le i < j \le n\}.$$

Then, $K[G_{\mathbf{q}}]$ is an algebra of Segre–Veronese type with $\tau = 2$, $M = n$, $\mathbf{a} = (1, \ldots, 1)$, $\mathbf{b} = (0, \ldots, 0)$, $\mathbf{r} = (r_1, \ldots, r_n)$, and $\mathbf{s} = (s_1, \ldots, s_n)$, where $r_i = \max V_i$ and $s_i = \min V_i$ for each $1 \le i \le n$.

In order to construct a quadratic Gröbner basis, we need a notion of marked polynomials and reduction relations. A nonzero polynomial $f \in K[\mathbf{x}]$ is said to be *marked* if an *initial monomial* $\mathrm{in}(f)$ of f is specified. One can choose any of the monomials appearing in f as an initial monomial. Given a set $\mathscr{F}$ of marked polynomials, we define the *reduction relation modulo* $\mathscr{F}$ in the same way as it is done for monomial orders. The set $\mathscr{F}$ is said to be marked *coherently* if there exists a monomial order $<$ on $K[\mathbf{x}]$ such that $\mathrm{in}(f) = \mathrm{in}_<(f)$ for all $f \in \mathscr{F}$.

Example 9.28 In Example 1.18, it is shown that the set $\mathscr{F} = \{f_1, \ldots, f_5\}$ of marked binomials

$$f_1 = x_1 x_8 - x_2 x_6 \quad \text{with } \mathrm{in}(f_1) = x_1 x_8,$$
$$f_2 = x_2 x_9 - x_3 x_7 \quad \text{with } \mathrm{in}(f_2) = x_2 x_9,$$
$$f_3 = x_3 x_{10} - x_4 x_8 \quad \text{with } \mathrm{in}(f_3) = x_3 x_{10},$$
$$f_4 = x_4 x_6 - x_5 x_9 \quad \text{with } \mathrm{in}(f_4) = x_4 x_6,$$
$$f_5 = x_5 x_7 - x_1 x_{10} \quad \text{with } \mathrm{in}(f_5) = x_5 x_7$$

is marked *incoherently*. In this case, Theorem 9.29 below implies that there exists an infinite sequence of reductions modulo $\mathscr{F}$. In fact,

$$x_1 x_2 \cdots x_{10} \xrightarrow{f_1} x_2^2 x_3 x_4 x_5 x_6^2 x_7 x_9 x_{10} \xrightarrow{f_2} x_2 x_3^2 x_4 x_5 x_6^2 x_7^2 x_{10} \xrightarrow{f_3} x_2 x_3 x_4^2 x_5 x_6^2 x_7^2 x_8$$

$$\xrightarrow{f_4} x_2 x_3 x_4 x_5^2 x_6 x_7^2 x_8 x_9 \xrightarrow{f_5} x_1 x_2 \cdots x_{10} \xrightarrow{f_1} \cdots$$

yields an infinite sequence of reductions modulo $\mathscr{F}$.

Although the following theorem holds for any finite set $\mathscr{F} \subset K[\mathbf{x}]$ of marked polynomials, for the sake of simplicity, we confine ourselves to the case that $\mathscr{F}$ consists of binomials.

Theorem 9.29 *A finite set $\mathscr{F} \subset K[\mathbf{x}]$ of marked binomials is marked coherently if and only if the reduction relation modulo $\mathscr{F}$ is Noetherian, i.e., every sequence of reductions modulo $\mathscr{F}$ terminates.*

Proof It is trivial that if $\mathscr{F} \subset K[\mathbf{x}]$ is marked coherently, then the reduction relation modulo $\mathscr{F}$ is Noetherian.

Suppose that the set $\mathscr{F} = \{f_1, \ldots, f_\ell\} \subset K[\mathbf{x}]$ of binomials is marked incoherently. Let $f_i = \mathbf{x}^{\alpha_i} - \mathbf{x}^{\beta_i}$ and let γ_i be a nonzero integer vector $\beta_i - \alpha_i \in \mathbb{Z}^n$ for $1 \le i \le \ell$. Since $\mathscr{F}$ is marked incoherently, there exists no weight vector $\mathbf{w} = (w_1, \ldots, w_n) \in \mathbb{Q}^n$ such that $w_j > 0$ for each $1 \le j \le n$ and that $\mathrm{in}_{\mathbf{w}}(f_i) = \mathbf{x}^{\alpha_i}$ for all $1 \le i \le \ell$. This is equivalent to say that there exists no $\mathbf{w} \in \mathbb{Q}^n$ such that

$$\mathbf{w} \cdot (\mathbf{e}_1, \cdots, \mathbf{e}_n, -\gamma_1, \ldots, -\gamma_\ell) = (w_1, \ldots, w_n, \mathbf{w} \cdot (\alpha_1 - \beta_1), \ldots, \mathbf{w} \cdot (\alpha_\ell - \beta_\ell))$$

is a positive vector. By Linear Programming Duality [188, Section 7.3], there exists a nonzero, nonnegative integer vector

$$\mathbf{y} = (\widetilde{y}_1, \ldots, \widetilde{y}_n, y_1, \ldots, y_\ell)^t \in \mathbb{Z}_{\ge 0}^{n+\ell}$$

such that

$$(\mathbf{e}_1, \cdots, \mathbf{e}_n, -\gamma_1, \ldots, -\gamma_\ell) \cdot \mathbf{y} = \mathbf{0}.$$

Then, we have

$$\begin{pmatrix} \widetilde{y}_1 \\ \vdots \\ \widetilde{y}_n \end{pmatrix} = y_1\gamma_1 + \cdots + y_\ell\gamma_\ell,$$

and hence, $y_1\gamma_1 + \cdots + y_\ell\gamma_\ell$ is a nonnegative integer vector. Let $N = y_1 + \cdots + y_\ell$. Since $\mathbf{y}$ is nonzero, we have $N > 0$. Then, the vector $(y_1, \ldots, y_\ell)$ has a representation $(y_1, \ldots, y_\ell) = \mathbf{e}_{i_1} + \cdots + \mathbf{e}_{i_N}$ for some $1 \le i_1 \le \cdots \le i_N \le \ell$. Let $\mathbf{x}^\delta = \prod_{i=1}^\ell (\mathrm{in}(f_i))^{y_i}$. For each $1 \le r \le N$, since

$$\delta + \gamma_{i_1} + \cdots + \gamma_{i_{r-1}} - \alpha_r = \sum_{k=1}^{r-1} \beta_{i_k} + \sum_{k=r+1}^N \alpha_{i_k}$$

is a nonnegative integer vector, we have

$$\mathbf{x}^{\delta+\gamma_{i_1}+\gamma_{i_2}+\cdots+\gamma_{i_{r-1}}} = \mathbf{x}^{\delta+\gamma_{i_1}+\gamma_{i_2}+\cdots+\gamma_{i_{r-1}}-\alpha_r} f_{i_r} + \mathbf{x}^{\delta+\gamma_{i_1}+\gamma_{i_2}+\cdots+\gamma_{i_r}}.$$

Thus,

$$\mathbf{x}^\delta \xrightarrow{f_{i_1}} \mathbf{x}^{\delta+\gamma_{i_1}} \xrightarrow{f_{i_2}} \mathbf{x}^{\delta+\gamma_{i_1}+\gamma_{i_2}} \xrightarrow{f_{i_3}} \cdots \xrightarrow{f_{i_N}} \mathbf{x}^{\delta+\gamma_{i_1}+\gamma_{i_2}+\cdots+\gamma_{i_N}} = \mathbf{x}^{\delta+y_1\gamma_1+\cdots+y_\ell\gamma_\ell}$$

is a reduction sequence modulo $\mathscr{F}$. Moreover, since $y_1\gamma_1 + \cdots + y_\ell\gamma_\ell$ is a nonnegative integer vector, we have an infinite reduction sequence

$$\mathbf{x}^\delta \xrightarrow{\mathscr{F}} \mathbf{x}^{\delta+y_1\gamma_1+\cdots+y_\ell\gamma_\ell} \xrightarrow{\mathscr{F}} \mathbf{x}^{\delta+2(y_1\gamma_1+\cdots+y_\ell\gamma_\ell)} \xrightarrow{\mathscr{F}} \mathbf{x}^{\delta+3(y_1\gamma_1+\cdots+y_\ell\gamma_\ell)} \xrightarrow{\mathscr{F}} \cdots,$$

as desired. $\qquad \square$

Given a configuration $A_{\tau,\mathbf{a},\mathbf{b},\mathbf{r},\mathbf{s}} = (\mathbf{a}_1, \ldots, \mathbf{a}_n) \in \mathbb{Z}_{\ge 0}^{d \times n}$ of Segre–Veronese type, let $K[\mathbf{x}]$ denote a polynomial ring in the set of variables

$$\left\{ x_{j_1 \cdots j_\tau} : 1 \le j_1 \le \cdots \le j_\tau \le d, \sum_{k=1}^\tau \mathbf{e}_{j_k} \in \{\mathbf{a}_1 \ldots, \mathbf{a}_n\} \right\}.$$

The toric ideal $I_{A_{\tau,\mathbf{a},\mathbf{b},\mathbf{r},\mathbf{s}}}$ is the kernel of the surjective homomorphism $\pi : K[\mathbf{x}] \longrightarrow K[A_{\tau,\mathbf{a},\mathbf{b},\mathbf{r},\mathbf{s}}]$ defined by $\pi(x_{j_1 \cdots j_\tau}) = \prod_{k=1}^\tau t_{j_k}$. A monomial

$$x_{\alpha_1\alpha_2\cdots\alpha_\tau} x_{\beta_1\beta_2\cdots\beta_\tau} \cdots x_{\gamma_1\gamma_2\cdots\gamma_\tau}$$

is called *sorted* if

$$\alpha_1 \le \beta_1 \le \cdots \le \gamma_1 \le \alpha_2 \le \beta_2 \le \cdots \le \gamma_2 \le \alpha_\tau \le \beta_\tau \le \cdots \le \gamma_\tau.$$

Let sort($\cdot$) denote the operator which takes any string over the alphabet $\{1, 2, \ldots, d\}$ and sorts it into weakly increasing order.

Lemma 9.30 *Work with the same notation as above. A homogeneous binomial*

$$x_{\alpha_1\alpha_2\cdots\alpha_\tau}x_{\beta_1\beta_2\cdots\beta_\tau}\cdots x_{\gamma_1\gamma_2\cdots\gamma_\tau} - x_{\alpha_1'\alpha_2'\cdots\alpha_{\tau'}'}x_{\beta_1'\beta_2'\cdots\beta_{\tau'}'}\cdots x_{\gamma_1'\gamma_2'\cdots\gamma_{\tau'}'} \in K[\mathbf{x}]$$

belongs to $I_{A_{\tau,a,b,r,s}}$ *if and only if*

$$\mathrm{sort}(\alpha_1\cdots\alpha_\tau\beta_1\cdots\beta_\tau\cdots\gamma_1\cdots\gamma_\tau) = \mathrm{sort}(\alpha_1'\cdots\alpha_{\tau'}'\beta_1'\cdots\beta_{\tau'}'\cdots\gamma_1'\cdots\gamma_{\tau'}').$$

The proof is left as a problem to the reader (Problem 9.12). The following is a key lemma in order to construct a quadratic Gröbner basis of the toric ideal $I_{A_{\tau,a,b,r,s}}$.

Lemma 9.31 *Let* $x_{\alpha_1\alpha_2\cdots\alpha_\tau}x_{\beta_1\beta_2\cdots\beta_\tau}$ *be a quadratic unsorted monomial in* $K[\mathbf{x}]$ *such that* $\mathrm{sort}(\alpha_1\beta_1\cdots\alpha_\tau\beta_\tau) = \gamma_1\gamma_2\cdots\gamma_{2\tau}$. *Then,*

$$x_{\alpha_1\alpha_2\cdots\alpha_\tau}x_{\beta_1\beta_2\cdots\beta_\tau} - x_{\gamma_1\gamma_3\cdots\gamma_{2\tau-1}}x_{\gamma_2\gamma_4\cdots\gamma_{2\tau}}$$

is a binomial in $I_{A_{\tau,a,b,r,s}}$.

Proof First, we show that $x_{\gamma_1\gamma_3\cdots\gamma_{2\tau-1}}$ and $x_{\gamma_2\gamma_4\cdots\gamma_{2\tau}}$ are variables in $K[\mathbf{x}]$. For each $i = 1, 2, \ldots, n$, let $\rho_i = |\{j : s_i \le \gamma_{2j-1} \le r_i\}|$ and $\sigma_i = |\{j : s_i \le \gamma_{2j} \le r_i\}|$. Since $\gamma_1 \le \cdots \le \gamma_{2\tau}$, we have $|\rho_i - \sigma_i| \le 1$ for any i. If $\rho_i \le \sigma_i$, then $\sigma_i - \rho_i \in \{0, 1\}$. Since $2c_i \le \rho_i + \sigma_i \le 2b_i$, we have $\rho_i \le b_i + 1/2$ and $c_i - 1/2 \le \sigma_i$. Thus, $c_i \le \rho_i \le \sigma_i \le b_i$. If $\rho_i > \sigma_i$, then $\rho_i - \sigma_i = 1$. Since $2c_i \le \rho_i + \sigma_i \le 2b_i$, we have $\sigma_i \le b_i + 1/2$ and $c_i - 1/2 \le \rho_i$. Thus, $c_i \le \sigma_i < \rho_i \le b_i$. Hence, $x_{\gamma_1\gamma_3\cdots\gamma_{2\tau-1}}$ and $x_{\gamma_2\gamma_4\cdots\gamma_{2\tau}}$ are variables in $K[\mathbf{x}]$. By Lemma 9.30, $x_{\alpha_1\alpha_2\cdots\alpha_\tau}x_{\beta_1\beta_2\cdots\beta_\tau} - x_{\gamma_1\gamma_3\cdots\gamma_{2\tau-1}}x_{\gamma_2\gamma_4\cdots\gamma_{2\tau}}$ is a binomial in $I_{A_{\tau,a,b,r,s}}$. $\square$

A quadratic Gröbner basis of toric ideal $I_{A_{\tau,a,b,r,s}}$ is given as follows.

Theorem 9.32 *Let* $A_{\tau,a,b,r,s} = (\mathbf{a}_1, \ldots, \mathbf{a}_n) \in \mathbb{Z}_{\ge 0}^{d\times n}$ *be a configuration of Segre–Veronese type. Let* $\mathscr{G}$ *be the subset*

$$\left\{ x_{\alpha_1\cdots\alpha_\tau}x_{\beta_1\cdots\beta_\tau} - x_{\gamma_1\gamma_3\cdots\gamma_{2\tau-1}}x_{\gamma_2\gamma_4\cdots\gamma_{2\tau}} : \begin{array}{l} x_{\alpha_1\cdots\alpha_\tau}x_{\beta_1\cdots\beta_\tau} \text{ is not sorted} \\ \mathrm{sort}(\alpha_1\beta_1\cdots\alpha_\tau\beta_\tau) = \gamma_1\gamma_2\cdots\gamma_{2\tau} \end{array} \right\}$$

of $I_{A_{\tau,a,b,r,s}}$. *Then, there exists a monomial order on* $K[\mathbf{x}]$ *such that* $\mathscr{G}$ *is the reduced Gröbner basis of the toric ideal* $I_{A_{\tau,a,b,r,s}}$. *The initial ideal is generated by squarefree quadratic (unsorted) monomials.*

Proof Let the marking be such that the initial monomial of each binomial in $\mathcal{G}$ is the unsorted monomial. Since the reduction of a monomial by $\mathcal{G}$ with respect to this marking corresponds to a sort of the indices of monomials, the reduction always terminates. Hence, by Theorem 9.29, this marking is coherent, and hence given by a monomial order.

Suppose that $\mathcal{G}$ is not a Gröbner basis of the toric ideal $I_{A_{\tau,a,b,r,s}}$. By Theorem 3.11, there exists a binomial $0 \neq f \in I_{A_{\tau,a,b,r,s}}$ such that both monomials in f are sorted. This means that $f = 0$, which is a contradiction. Hence, $\mathcal{G}$ is a Gröbner basis of the toric ideal $I_{A_{\tau,a,b,r,s}}$. It is easy to see that the Gröbner basis $\mathcal{G}$ is reduced.

$\square$

Problems

9.10 Verify that the configuration

$$A = \begin{pmatrix} 1 & 1 & 1 & 0 & 0 & 0 & 2 & 1 & 1 & 0 & 0 & 2 & 1 & 1 & 0 & 0 \\ 0 & 0 & 0 & 1 & 1 & 1 & 0 & 1 & 0 & 2 & 1 & 0 & 1 & 0 & 2 & 1 \\ 0 & 0 & 0 & 0 & 0 & 0 & 0 & 1 & 0 & 1 & 0 & 0 & 1 & 0 & 1 \\ 2 & 1 & 0 & 2 & 1 & 0 & 1 & 1 & 1 & 1 & 0 & 0 & 0 & 0 & 0 \\ 0 & 1 & 2 & 0 & 1 & 2 & 0 & 0 & 0 & 0 & 1 & 1 & 1 & 1 & 1 \end{pmatrix}$$

is of Segre–Veronese type.

9.11 List up the binomials in the reduced Gröbner basis of the toric ideal of

$$A = \begin{pmatrix} 2 & 1 & 1 & 0 & 0 & 0 \\ 0 & 1 & 0 & 2 & 1 & 0 \\ 0 & 0 & 1 & 0 & 1 & 2 \end{pmatrix}$$

appearing in Theorem 9.32 by hand.

9.12 Prove Lemma 9.30.

Notes

A Markov chain Monte Carlo method via Markov bases was invented by Diaconis–Sturmfels [53]. A sequential importance sampling method for multiway contingency tables are introduced by Chen–Dinwoodie–Sullivant [34]. A relationship between a sequential importance sampling method and the normality of toric rings appears in, e.g., Hemmecke–Takemura–Yoshida[89].

Dobra [55] proved that the toric ideal of a decomposable graphical model is generated by quadratic binomials. Geiger–Meek–Sturmfels [79] extended this result to the existence of a quadratic Gröbner basis (Theorem 9.12). Hoşten–Sullivant [119] showed that the toric ideal of the hierarchical model arising from a quasi-forest has a quadratic Gröbner basis (Theorem 9.10).

The notions of the r-th Lawrence lifting and the Markov/Graver complexity were given by Santos–Sturmfels [185]. The notion of compressed polytopes was introduced by Stanley [196]. Sullivant [206] gave a complete characterization for compressed polytopes, and as an application, classified compressed $A_{r_1 r_2 \cdots r_m}$ (Theorem 9.25). The reduced Gröbner basis of $I_{A_{r_{33}}}$ is given in Bofi–Rossi [18]. Vlach [214] essentially showed that $K[A_{643}]$ is not normal. Ohsugi–Hibi showed that the toric rings of the configurations in the "otherwise" part in Table 9.1 are not normal [164] and not very ample [167]. The software normaliz and 4ti2 [26] verified that $K[A_{553}]$ is normal. Thus, the classification for the configuration $A_{r_1 r_2 \cdots r_m}$ in Table 9.1 can be obtained by the results in the papers [26, 164, 167, 206].

Algebras of Veronese type were studied in De Negri–Hibi [52] and Sturmfels [202]. Algebras of Segre–Veronese type were introduced in [161] and generalized in [4]. The notion of nested configurations is given in [3] and further developed in [166]. Toric fiber products [207] are a generalization of Segre products and proved to be an important concept for the study of the toric ideals arising from contingency tables. Especially, toric fiber products turn out to be very useful in the study of the toric ideals arising from cut polytopes of graphs, which are related with graph models of contingency tables [203]. Shibuta generalized both nested configurations and toric fiber products in his paper [189] on Gröbner bases of contraction ideals.

References

1. Adams, W.W., Loustaunau, P.: An Introduction to Gröbner Bases. Graduate Studies in Mathematics, vol. 3. American Mathematical Society, Providence (1994)
2. Anick, D.: A counterexample to a conjecture of Serre. Ann. Math. **115**, 1–33 (1982)
3. Aoki, S., Hibi, T., Ohsugi, H., Takemura, A.: Gröbner bases of nested configurations. J. Algebra **320**, 2583–2593 (2008)
4. Aoki, S., Hibi, T., Ohsugi, H., Takemura, A.: Markov basis and Gröbner basis of Segre–Veronese configuration for testing independence in group-wise selections. Ann. Inst. Stat. Math. **62**, 299–321 (2010)
5. Aramova, A., Herzog, J., Hibi, T.: Finite lattices and lexicographic Gröbner bases. Eur. J. Comb. **21**, 431–439 (2000)
6. Avramov, L.: Infinite free resolutions. In: Elias, J., Giral, J.M., Miró-Roig, R.M., Zarzuela, S. (eds.) Six Lectures on Commutative Algebra. Progress in Mathematics, vol. 166, pp. 1–118. Birkhäuser, Basel (1998)
7. Backelin, J., Fröberg, R.: Koszul algebras, veronese subrings and rings with linear resolutions. Rev. Roum. Math. Pures Appl. **30**, 85–97 (1985)
8. Badiane, M., Burke, I., Sköberg, E.: The universal Gröbner basis of a binomial edge ideal. Electron. J. Comb. **24**, #P4.11 (2017)
9. Banerjee, A., Núñez-Betancout, L.: Graph connectivity and binomial edge ideals. Proc. Am. Math. Soc. **145**, 487–499 (2017)
10. Bass, H.: On the ubiquity of Gorenstein rings. Math. Z. **82**, 8–28 (1963)
11. Bayer, D., Morrison, I.: Gröbner bases and geometric invariant theory I. J. Symb. Comput. **6**, 209–217 (1988)
12. Bayer, D., Popescu, S., Sturmfels, B.: Syzygies of unimodular Lawrence ideals. J. Reine Angew. Math. **534**, 169–186 (2001)
13. Becker, T., Weispfenning V.: Gröbner Bases. Springer, New York (1993)
14. Beerenwinkel, N., Eriksson, N., Sturmfels, B.: Conjunctive Bayesian networks. Bernoulli **13**, 893–909 (2007)
15. Bigdeli, M., Herzog, J., Hibi, T., Qureshi A. A., Shikama, A.: Isotonian Algebras. Nagoya Math. J. **230**, 83–101 (2018)
16. Billera, L.J., Lee, C.W.: A proof of the sufficiency of McMullen's conditions for f-vectors of simplicial convex polytopes. J. Combin. Theory Ser. A **31**, 237–255 (1981)
17. Birkhoff, G.: Lattice Theory. American Mathematical Society, Providence (1940)
18. Boffi, G., Rossi, F.: Lexicographic Gröbner bases for transportation problems of format $r \times 3 \times 3$. J. Symb. Comput. **41**, 336–356 (2006)

© Springer International Publishing AG, part of Springer Nature 2018
J. Herzog et al., *Binomial Ideals*, Graduate Texts in Mathematics 279,
https://doi.org/10.1007/978-3-319-95349-6

19. Brøndsted, A.: An Introduction to Convex Polytopes. Graduate Texts in Mathematics, vol. 90. Springer, Berlin (1983)
20. Brown, J., Lakshmibai, V.: Singular loci of Bruhat–Hibi toric varieties. J. Algebra **319**, 4759–4779 (2008)
21. Brown, J., Lakshmibai, V.: Singular loci of Grassmann–Hibi toric varieties. Mich. Math. J. **59**(2), 243–267 (2010)
22. Brown, J., Lakshmibai, V.: Arithmetically Gorenstein Schubert varieties in a minuscule G/P. Pure Appl. Math. Q. **8**, 559–587 (2012)
23. Bruns, W., Conca, A.: F-rationality of determinantal rings and their Rees rings. Mich. Math. J. **45**, 291–299 (1998)
24. Bruns, W., Conca, A.: Gröbner bases, initial ideals and initial algebras. In: Avramov, L.L., et al. (eds.) Homological Methods in Commutative Algebra, IPM Proceedings, Tehran (2004)
25. Bruns, W., Gubeladze, J.: Polytopes, Rings, and K-Theory. Springer Monographs in Mathematics. Springer, Dordrecht (2009)
26. Bruns, W., Hemmecke, R., Ichim, B., Köppe, M., Söger, C.: Challenging computations of Hilbert bases of cones associated with algebraic statistics. Exp. Math. **20**, (2011)
27. Bruns, W., Herzog, J.: Cohen–Macaulay Rings. Cambridge Studies in Advanced Mathematics, vol. 39. Cambridge University Press, Cambridge (1993)
28. Bruns, W., Herzog, J.: Semigroup rings and simplicial complexes. J. Pure Appl. Algebra **122**, 185–208 (1997)
29. Bruns, W., Vetter, U.: Determinantal Rings. Lecture Notes in Mathematics, vol. 1327. Springer, Berlin (1988)
30. Buchberger, B.: An algorithm for finding the basis elements of the residue class ring of a zero dimensional polynomial ideal. Ph.D. Dissertation, University of Innsbruck (1965)
31. Charalambous, H., Thoma, A., Vladoiu, M.: Markov bases and generalized Lawrence liftings. Ann. Comb. **19**, 661–669 (2015)
32. Charalambous, H., Thoma, A., Vladoiu, M.: Binomial fibers and indispensable binomials. J. Symbolic Comput. **74**, 578–591 (2016)
33. Chaudhry, F., Dokuyucu, A., Irfan, R.: On the binomial edge ideals of block graphs. An. Ştiinţ. Univ. Ovidius Constanţa **24**, 149–158 (2016)
34. Chen, Y., Dinwoodie, I.H., Sullivant, S.: Sequential importance sampling for multiway tables. Ann. Stat. **34**, 523–545 (2006)
35. Chiba, T., Matsuda, K.: Diagonal F-thresholds and F-pure thresholds of Hibi rings. Comm. Algebra **43**, 2830–2851 (2015)
36. Conca, A.: Ladder determinantal rings. J. Pure Appl. Algebra **98**, 119–134 (1995)
37. Conca, A, De Negri, E., Gorla, E.: Cartwright-Sturmfels ideals associated to graphs and linear spaces. arXiv:1705.00575
38. Conca, A., Rossi, M.E., Valla, G.: Gröbner flags and Gorenstein algebras. Compos. Math. **129**, 95–121 (2001)
39. Conca, A., Trung, N.V., Valla, G.: Koszul property for points in projective space. Math. Scand. **89**, 201–216 (2001)
40. Conca, A., Welker, V.: Lovasz-Saks-Schrijver ideals and coordinate sections of determinantal varieties. arXiv:1801.07916
41. Conti, P., Traverso, C.: Buchberger algorithm and integer progamming. In: Mattson, H., Mora, T., Rao, T. (eds.) Applied Algebra, Algebraic Algorithms and Error Correcting Codes. Lecture Notes in Computer Science, vol. 539, pp. 130–139. Springer, Berlin (1991)
42. Corso, A., Nagel, U.: Monomial and toric ideals associated to Ferrers graphs. Trans. Am. Math. Soc. **361**, 1371–1395 (2009)
43. Cox, D.A., Erskine, A.: On closed graphs I. Ars Combinatorica **120**, 259–274 (2015)
44. Cox, D.A., Little, J., O'Shea, D.: Ideals, Varieties, and Algorithms. Springer, New York (1992)
45. Crupi, M., Rinaldo, G.: Binomial edge ideals with quadratic Gröbner bases. Electron. J. Comb. **18**, ♯ P211 (2011)

46. Crupi, M., Rinaldo, G.: Closed graphs are proper interval graphs. An. Ştiinţ. Univ. Ovidius Constanţa **22**, 37–44 (2014)

47. D'Alì, A.: Toric ideals associated with gap-free graphs. J. Pure Appl. Algebra **219**, 3862–3872 (2015)

48. Danilov, V.I.: The geometry of toric varieties. Russ. Math. Surv. **33**(2), 97–154 (1978)

49. Decker, W., Greuel, G.-M., Pfister, G., Schönemann, H. Singular 4-1-0 – A computer algebra system for polynomial computations. http://www.singular.uni-kl.de (2016)

50. Dehy, R., Yu, R.W.T.: Degeneration of Schubert varieties of SL_n/B to toric varieties. Ann. Inst. Fourier (Grenoble) **51**, 1525–1538 (2001)

51. De Loera, A., Sturmfels, B., Thomas, R.: Gröbner bases and triangulations of the second hypersimplex. Combinatorica **15**, 409–424 (1995)

52. De Negri, E., Hibi, T.: Gorenstein algebras of Veronese type. J. Algebra **193**, 629–639 (1997)

53. Diaconis, P., Sturmfels, B.: Algebraic algorithms for sampling from conditional distributions. Ann. Statist. **26**, 363–397 (1998)

54. Dirac, G.A.: On rigid circuit graphs. Abh. Math. Sem. Univ. Hamburg **38**, 71–76 (1961)

55. Dobra, A.: Markov bases for decomposable graphical models. Bernoulli **9**, 1093–1108 (2003)

56. Dokuyucu, A.: Extremal Betti numbers of some classes of binomial edge ideals. Math. Rep. **17**, 359–367 (2015)

57. Eisenbud, D.: Commutative Algebra with a View Toward Algebraic Geometry. Graduate Texts in Mathematics. Springer, New York (1995)

58. Eisenbud, D., Sturmfels, B.: Binomial ideals. Duke Math. J. **84**, 1–45 (1996)

59. Ene, V.: Syzygies of Hibi rings. Acta Math. Vietnam. **40**, 403–446 (2015)

60. Ene, V., Herzog, J.: Gröbner Bases in Commutative Algebra. Graduate Studies in Mathematics. American Mathematical Society, Providence (2012)

61. Ene, V., Herzog, J., Hibi, T.: Cohen–Macaulay binomial edge ideals. Nagoya Math. J. **204**, 57–68 (2011)

62. Ene, V., Herzog, J., Hibi, T.: Koszul binomial edge ideals. In: Ibadula, D., Veys, W. (eds.) Bridging Algebra, Geometry, and Topology. Springer Proceedings in Mathematics and Statistics, vol. 96, pp. 125–136. Springer, Switzerland (2014)

63. Ene, V., Herzog, J., Hibi, T.: Linear flags and Koszul filtrations. Kyoto J. Math. **55**, 517–530 (2015)

64. Ene, V., Herzog, J., Hibi, T.: Linearly related polyominoes. J. Algebraic Comb. **41**, 949–968 (2015)

65. Ene, V., Herzog, J., Hibi, T., Mohammadi, F.: Determinantal facet ideals. Mich. Math. J. **62**, 39–57 (2013)

66. Ene, V., Herzog, J., Hibi, T., Madani, S.S.: Pseudo-Gorenstein and level Hibi rings. J. Algebra **431**, 138–161 (2015)

67. Ene, V., Herzog, J., Hibi, T., Qureshi, A.A.: The binomial edge ideal of a pair of graphs. Nagoya Math. J. **213**, 105–125 (2014)

68. Ene, V., Herzog, J., Mohammadi, F.: Monomial ideals and toric rings of Hibi type arising from a finite poset. Eur. J. Comb. **32**, 404–421 (2011)

69. Ene, V., Hibi, T.: The join-meet ideal of a finite lattice. J. Commut. Algebra **5**, 209–230 (2013)

70. Ene, V., Zarojanu, A.: On the regularity of binomial edge ideals. Math. Nachr. **288**, 19–24 (2015)

71. Engström, A., Norén, P.: Ideals of graph homomorphisms. Ann. Comb. **17**, 71–103 (2013)

72. Féray, V., Reiner, V.: P-partitions revisited. J. Commut. Algebra **4**, 101–152 (2012)

73. Fischer, K.G., Shapiro, J.: Mixed matrices and binomial ideals. J. Pure Appl. Algebra **113**, 39–54 (1996)

74. Fishburn, P.C.: Interval graphs and interval order. Discret. Math. **55**, 135–149 (1985)

75. Fröberg, R.: Determination of a class of Poincaré series. Math. Scand. **37**, 29–39 (1975)

76. Fröberg, R., Koszul algebras. In: Dobbs, D.E., Fontana, M., Kabbaij, S-E. (eds.) Advances in Commutative Ring Theory (Fez 1997). Lecture Notes in Pure and Applied Mathematics, pp. 337–350. M. Dekker, New York (1999)

77. Fulkerson, D.R., Hoffman, A.J., McAndrew, M.H.: Some properties of graphs with multiple edges. Can. J. Math. **17**, 166–177 (1965)
78. Gardi, F.: The Roberts characterization of proper and unit interval graphs. Discret. Math. **307**, 2906–2908 (2007)
79. Geiger, D., Meek, C., Sturmfels, B.: On the toric algebra of graphical models. Ann. Stat. **34**, 1463–1492 (2006)
80. Gelfand, I.M., Kapranov, M.M., Zelevinsky, A.V.: Discriminants, Resultants and Multidimensional Determinants. Modern Birkhäuser Classics, Reprint of the 1994 edn. Birkhäuser, Boston (2008)
81. Gitler, I., Reyes, E., Villarreal, R.H.: Ring graphs and complete intersection toric ideals. Discret. Math. **310** 430–441 (2010)
82. Golumbic, M.C.: Algorithmic Graph Theory and Perfect Graphs. Academic, New York (1980)
83. Gonciulea, N., Lakshmibai, V.: Schubert varieties, toric varieties, and ladder determinantal varieties. Ann. Inst. Fourier (Grenoble) **47**, 1013–1064 (1997)
84. Goto, S., Watanabe, K.-I: On graded rings, I. J. Math. Soc. Jpn **30**, 179–213 (1978)
85. Grünbaum, B.: Convex Polytopes. Graduate Texts in Mathematics, vol. 221. Springer, New York (2003)
86. Gulliksen, T.H.: A proof of the existence of minimal algebra resolutions. Acta Math. **120**, 53–58 (1968)
87. Hajós, G.: Über eine Art von Graphen. Internationale Mathematische Nachrichten **11**, 65–65 (1957)
88. Heggernes, P., Meister, D., Papadopoulos, C.: A new representation of proper interval graphs with an application to clique-width. DIMAP workshop on algorithmic graph theory 2009. Electron. Notes Discret. Math. **32**, 27–34 (2009)
89. Hemmecke, R., Takemura, A., Yoshida, R.: Computing holes in semi-groups and its applications to transportation problems. Contrib. Discret. Math. **4**, 81–91 (2009)
90. Herzog, J.: Generators and relations of abelian semigroups and semigroup rings. Manuscripta Math. **3**, 175–193 (1970)
91. Herzog, J.: Algebra retracts and Poincare series. Manuscripta Math. **21**, 307–314 (1977)
92. Herzog, J., Hibi, T.: Distributive lattices, bipartite graphs and Alexander duality. J. Algebraic Comb. **22**, 289–302 (2005)
93. Herzog, J., Hibi, T.: The depth of powers of an ideal. J. Algebra **291**, 534–550 (2005)
94. Herzog, J., Hibi, T.: Monomial Ideals. Graduate Texts in Mathematics. Springer, New York (2010)
95. Herzog, J., Hibi, T.: Finite lattices and Gröbner bases. Math. Nachr. **285**, 1969–1973 (2012)
96. Herzog, J., Hibi, T.: Ideals generated by adjacent 2-minors. J. Commut. Algebra **4**, 525–549 (2012)
97. Herzog, J., Hibi, T., Hreinsdóttir, F., Kahle, T., Rauh, J.: Binomial edge ideals and conditional independence statements. Adv. Appl. Math. **45**, 317–333 (2010)
98. Herzog, J., Hibi, T., Restuccia, G.: Strongly Koszul algebras. Math. Scand. **86**, 161–178 (2000)
99. Herzog, J., Hibi, T., Stamate, D.I.: The trace of the canonical module. arXiv:1612.02723
100. Herzog, J., Kiani, D., Madani, S.S.: The linear strand of determinantal facet ideals. Mich. Math. J. **66**, 10–123 (2017)
101. Herzog, J., Macchia, A., Madani, S.S., Welker, V.: On the ideal of orthogonal representations of a graph in $\mathbb{R}^2$. Adv. Appl. Math. **71**, 146–173 (2015)
102. Herzog, J., Madani, S.S.; The coordinate ring of a simple polyomino. Ill. J. Math. **58**, 981–995 (2014)
103. Herzog, J., Qureshi, A.A., Shikama, A.: Gröbner bases of balanced polyominoes. Math. Nachr. **288**, 775–783 (2015)

104. Hibi, T.: Distributive lattices, affine semigroup rings and algebras with straightening laws. In: Nagata, M., Matsumura, H. (eds.) Commutative Algebra and Combinatorics. Advanced Studies in Pure Mathematics, vol. 11, pp. 93–109. North–Holland, Amsterdam (1987)

105. Hibi, T.: Algebraic Combinatorics on Convex Polytopes. Carslaw Publications, Glebe (1992)

106. Hibi, T. (ed.): Gröbner Bases: Statistics and Software Systems. Springer, Berlin (2013)

107. Hibi, T., Li, N.: Chain polytopes and algebras with straightening laws. Acta Math. Vietnam. **40**, 447–452 (2015)

108. Hibi, T., Li, N.: Unimodular equivalence of order and chain polytopes. Math. Scand. **118**, 5–12 (2016)

109. Hibi, T., Li, N., Zhang, Y.X.: Separating hyperplanes of edge polytopes. J. Comb. Theory Ser. A **120**, 218–231 (2013)

110. Hibi, T., Matsuda, K., Ohsugi, H.: Strongly Koszul edge rings. Acta Math. Vietnam. **41**, 69–76 (2016)

111. Hibi, T., Mori, A., Ohsugi, H., Shikama, A.: The number of edges of the edge polytope of a finite simple graph. ARS Mat. Contemp. **10**, 323–332 (2016)

112. Hibi, T., Nishiyama, K., Ohsugi, H., Shikama, A.: Many toric ideals generated by quadratic binomials possess no quadratic Gröbner bases. J. Algebra **408**, 138–146 (2014)

113. Hibi, T., Qureshi, A.A.: Nonsimple polyominoes and prime ideals. Ill. J. Math. **59**, 391–398 (2015)

114. Hironaka, H.: Resolution of singularities of an algebraic variety over a field of characteristic zero. Ann. Math. **79**, 109–203, 205–326 (1964)

115. Hochster, M.: Rings of invariants of tori, Cohen–Macaulay rings generated by monomials, and polytopes. Ann. Math. **96**, 318–337 (1972)

116. Hochster, M.: Cohen–Macaulay rings, combinatorics, and simplicial complexes. In: McDonald, B.R., Morris, R.A. (eds.) Ring Theory II. Lecture Notes in Pure and Applied Mathematics, vol. 26. M. Dekker, New York (1977)

117. Hochster, M., Eagon, J.A.: Cohen–Macaulay rings, invariant theory and the generic perfection of determinantal loci. Am. J. Math. **93**, 1020–1058 (1971)

118. Hoşten, S., Shapiro, J.: Primary decomposition of lattice basis ideals. J. Symb. Comput. **29**, 625–639 (2000)

119. Hoşten, S., Sullivant, S.: Gröbner bases and polyhedral geometry of reducible and cyclic models. J. Comb. Theory Ser. A **100**, 277–301 (2002)

120. Hoşten, S., Sullivant, S.: Ideals of adjacent minors. J. Algebra **277**, 615–642 (2004)

121. Howe, R.: Weyl Chambers and standard monomial theory for poset lattice cones. Q. J. Pure Appl. Math. **1**, 227–239 (2005)

122. Howe, R., Lee, S.T.: Why should the Littlewood–Richardson rule be true? Bull. Am. Math. Soc. (N.S.) **49**, 187–236 (2012)

123. Kahle, T., Miller, E.: Decompositions of commutative monoid congruences and binomial ideals. Algebra Number Theory **8**, 1297–1364 (2014)

124. Kahle, T., Miller, E., O'Neill, C.: Irreducible decomposition of binomial ideals. Compos. Math. **152**, 1319–1332 (2016)

125. Kahle, T., Sarmiento, C. Windisch, T.: Parity binomial edge ideals. J. Algebr. Comb. **44**, 99–117 (2016)

126. Kaplansky, I.: Commutative Rings. Allyn and Bacon, Boston (1970)

127. Katzman, M.: Bipartite graphs whose edge algebras are complete intersections. J. Algebra **220**, 519–530 (1999)

128. Kiani, D., Madani, S.S.: Binomial edge ideals with pure resolutions. Collect. Math. **65**, 331–340 (2014)

129. Kiani, D., Madani, S.S.: The Castelnuovo-Mumford regularity of binomial edge ideals. J. Comb. Theory Ser. A. **139**, 80–86 (2016)

130. Kim, S.: Standard monomial theory for flag algebras of GL(n) and Sp($2n$). J. Algebra **320**, 534–568 (2008)

131. Kim, S.: The nullcone in the multi-vector representation of the symplectic group and related combinatorics. J. Comb. Theory Ser. A **117**, 1231–1247 (2010)

132. Kim, S.: Distributive lattices, affine semigroups, and branching rules of the classical groups. J. Comb. Theory Ser. A **119**, 1132–1157 (2012)
133. Kim, S., Lee, S.T.: Pieri algebras for the orthogonal and symplectic groups. Isr. J. Math. **195**, 215–245 (2013)
134. Kim, S., Yacobi, O.: A basis for the symplectic group branching algebra. J. Algebraic Comb. **35**, 269–290 (2012)
135. Kunz, E.: Introduction to Comuutative Algebra and Algebraic Geometry. Birkhäuser, New York (2013)
136. Lakshmibai, V., Mukherjee, H.: Singular loci of Hibi toric varieties. J. Ramanujan Math. Soc. **26**, 1–29 (2011)
137. Lakshmibai, V., Shukla, P.: Standard monomial bases and geometric consequences for certain rings of invariants. Proc. Indian Acad. Sci. Math. Sci. **116**, 9–36 (2006)
138. Laubenbacher, R.C., Swanson, I.: Permanental ideals. J. Symb. Comput. **30**, 195–205 (2000)
139. Lee, C.W.: Regular triangulations of convex polytopes. In: Applied Geometry and Discrete Mathematics, The Victor Klee Festschrift. DIMACS Discrete Mathematics and Theoretical Computer Science, vol. 4, pp. 443–456. American Mathematical Society, Providence (1991)
140. Löfwall, C.: On the subalgebra generated by one-dimensional elements in the Yoneda Ext-algebras. In: Roos, J.E. (ed.) Algebra, Algebraic Topology and their Interactions. Lecture Notes in Mathematics, vol. 1183, pp-291–338. Springer, New York (1986)
141. Looges, P.J., Olariu, S.: Optimal greedy algorithms for indifference graphs. Comput. Math. Appl. **25**, 15–25 (1993)
142. Lovász, L.: On the Shannon capacity of a graph. IEEE Trans. Inform. Theory **25**, 1–7 (1979)
143. Lovász, L., Saks, M., Schrijver, A.: Orthogonal representations and connectivity of graphs. Linear Algebra Appl. **114/115**, 439–454 (1989)
144. Matsuda, M., Murai, S.: Regularity bounds for binomial edge ideals. J. Commut. Algebra **5**, 141–149 (2013)
145. Matsumura, H.: Commutative algebra. Mathematics Lecture Note Series, vol. 56, 2nd edn. Benjamin/Cummings Publishing Co., Reading, MA (1980)
146. Miller, E., Sturmfels, B.: Combinatorial Commutative Algebra. Graduate Texts in Mathematics, vol. 227. Springer, New York (2005)
147. Milne, J.S.: Algebraic Geometry, Version 6.00, (2014), Available at www.jmilne.org/math/
148. Mohammadi, F., Sharifan, L.: Hilbert function of binomial edge ideals. Commun. Algebra **42**, 688–703 (2014)
149. Mora, T., Robbiano, L.: The Gröbner fan of an ideal. J. Symb. Comput. **6**, 183–208 (1988)
150. Munkres, J.R.: Elements of Algebraictopology. Springer, New York (1982)
151. Nagata, M.: Local Rings. Interscience, New York (1962)
152. Oda, T.: Convex Bodies and Algebraic Geometry. Springer, New York (1988)
153. Ogawa, M., Hara, H., Takemura A.: Graver basis for an undirected graph and its application to testing the beta model of random graphs. Ann. Inst. Stat. Math. **65**, 191–212 (2013)
154. Ohsugi, H.: Toric ideals and an infinite family of normal (0,1)-polytopes without unimodular regular triangulations. Discret. Comput. Geom. **27**, 551–565 (2002)
155. Ohsugi, H.: A geometric definition of combinatorial pure subrings and Gröbner bases of toric ideals of positive roots. Comment. Math. Univ. St. Pauli. **56**, 27–44 (2007)
156. Ohsugi, H., Herzog, J., Hibi, T.: Combinatorial pure subrings. Osaka J. Math. **37**, 745–757 (2000)
157. Ohsugi, H., Hibi, T.: Normal polytopes arising from finite graphs. J. Algebra **207**, 409–426 (1998)
158. Ohsugi, H., Hibi, T.: A normal (0, 1)-polytope none of whose regular triangulations is unimodular. Discret. Comput. Geom. **21**, 201–204 (1999)
159. Ohsugi, H., Hibi, T.: Toric ideals generated by quadratic binomials. J. Algebra **218**, 509–527 (1999)
160. Ohsugi, H., Hibi, T.: Koszul bipartite graphs. Adv. Appl. Math. **22**, 25–28 (1999)

161. Ohsugi, H., Hibi, T.: Compressed polytopes, initial ideals and complete multipartite graphs. Ill. J. Math. **44**, 391–406 (2000)

162. Ohsugi, H., Hibi, T.: Indispensable binomials of finite graphs. J. Algebra Appl. **4**, 421–434 (2005)

163. Ohsugi, H., Hibi, T.: Special simplices and Gorenstein toric rings. J. Comb. Theory Ser. A **113**, 718–725 (2006)

164. Ohsugi, H., Hibi, T.: Toric ideals arising from contingency tables. In: Commutative Algebra and Combinatorics. Ramanujan Mathematical Society Lecture Notes Series, vol. 4, pp. 91–115. Ramanujan Mathematical Society, Mysore (2007)

165. Ohsugi, H., Hibi, T.: Simple polytopes arising from finite graphs. In: Dehmer, M., Drmota, M., Emmert-Streib, F. (eds.) Proceedings of the 2008 International Conference on Information Theory and Statistical Learning (ITSL 2008), pp. 73–79. CSREA Press, Las Vegas (2008)

166. Ohsugi, H., Hibi, T.: Toric rings and ideals of nested configurations. J. Commut. Algebra **2**, 187–208 (2010)

167. Ohsugi, H., Hibi, T.: Non-very ample configurations arising from contingency tables. Ann. Inst. Stat. Math. **62**, 639–644 (2010)

168. Ohsugi, H., Hibi, T.: Toric ideals and their circuits. J. Commut. Algebra **5**, 309–322 (2013)

169. Ohtani, M.: Graphs and Ideals generated by some 2-minors. Commun. Algebra **39**, 905–917 (2011)

170. O'Neill, C.: Mesoprimary decomposition of binomial submodules. J. Algebra **480**, 59–78 (2017)

171. Priddy, S.B.: Koszul resolutions. Trans. AMS **152**, 39–60 (1970)

172. Qureshi, A.A.: Ideals generated by 2-minors, collections of cells and stack polyominoes. J. Algebra **357**, 279–303 (2012)

173. Qureshi, A.A., Shibuta, T., Shikama, A.: Simple polyominoes are prime. J. Commut. Algebra **9**, 413–422 (2017)

174. Rauf, A., Rinaldo, G.: Construction of Cohen-Macaulay binomial edge ideals. Commun. Algebra **42**, 238–252 (2014)

175. Rauh, J.: Generalized binomial edge ideals. Adv. Appl. Math. **50**, 409–414 (2013)

176. Reiner, V., Welker, V.: On the Charney–Davis and Neggers–Stanley conjectures. J. Combin. Theory Ser. A **109**, 247–280 (2005)

177. Reisner, G.A.: Cohen–Macaulay quotients of polynomial rings. Adv. Math. **21**, 30–49 (1976)

178. Reyes, E., Tatakis, C., Thoma, A.: Minimal generators of toric ideals of graphs. Adv. Appl. Math. **48**, 64–78 (2012)

179. Rigal, L., Zadunaisky, P.: Quantum analogues of Richardson varieties in the Grassmannian and their toric degeneration. J. Algebra **372**, 293–317 (2012)

180. Rigal, L., Zadunaisky, P.: Twisted semigroup algebras. Algebr. Represent. Theory **18**, 1155–1186 (2015)

181. Roberts, F.S.: Indifference graphs. In: Harary, F. (ed.) Proof Techniques in Graph Theory, pp. 139–146. Academic, New York (1969)

182. Roberts, F.S.: Graph Theory and Its Applications to Problems of Society. SIAM Press, Philadelphia (1978)

183. Saeedi Madani, S., Kiani, D.: Binomial edge ideals of graphs. Electron. J. Combin. **19**, ♯ P44 (2012)

184. Saeedi Madani, S., Kiani, D.: On the binomial edge ideal of a pair of graphs. Electron. J. Combin. **20**, ♯ P48 (2013)

185. Santos, F., Sturmfels, B.: Higher Lawrence configurations. J. Combin. Theory Ser. A **103**, 151–164 (2003)

186. Schenzel, P., Zafar, S.: Algebraic properties of the binomial edge ideal of a complete bipartite graph. An. St. Univ. Ovidius Constanţa **22**, 217–238 (2014)

187. Schoeller, C.: Homologie des anneaux locaux noethérien. C.R. Acad. Sci: Paris Sér. A **265**, 768–771 (1967)

188. Schrijver, A.: Theory of linear and integer programming. In: Wiley-Interscience Series in Discrete Mathematics. A Wiley-Interscience Publication/Wiley, Chichester (1986)
189. Shibuta, T.: Gröbner bases of contraction ideals. J. Algebr. Combin. **36**, 1–19 (2012)
190. Simis, A., Vasconcelos, W.V., Villarreal, R.H.: On the ideal theory of graphs. J. Algebra **167**, 389–416 (1994)
191. Simis, A., Vasconcelos, W.V., Villarreal, R.H.: The integral closure of subrings associated to graphs. J. Algebra **199**, 281–289 (1998)
192. Soprunova, E., Sottile, F.: Lower bounds for real solutions to sparse polynomial systems. Adv. Math. **204**, 116–151 (2006)
193. Sottile, F., Sturmfels, B.: A sagbi basis for the quantum Grassmannian. J. Pure Appl. Algebra **158**, 347–366 (2001)
194. Stanley, R.P.: The upper bound conjecture and Cohen–Macaulay rings. Stud. Appl. Math. **54**, 135–142 (1975)
195. Stanley, R.P.: The number of faces of a simplicial convex polytope. Adv. Math. **35**, 236–238 (1980)
196. Stanley, R.P.: Decomposition of rational convex polytopes. Ann. Discret. Math. **6**, 333–342 (1980)
197. Stanley, R.P.: Two poset polytopes. Discret. Comput. Geom. **1**, 9–23 (1986)
198. Stanley, R.P.: Ordered Structures and Partitions. Memoirs of the American Mathematical Society, vol. 119. American Mathematical Society, Providence (1972)
199. Stanley, R.P.: Combinatorics and Commutative Algebra. Progress in Mathematics, vol. 41, 2nd edn. Birkhäuser, Boston (1996)
200. Stanley, R.P.: Enumerative Combinatorics, vol. 1. Cambridge Studies in Advanced Mathematics, vol. 49. Cambridge University Press, Cambridge (1997)
201. Sturmfels, B.: Gröbner bases of toric varieties. Tohoku Math. J. **43**, 249–261 (1991)
202. Sturmfels, B.: Gröbner Bases and Convex Polytopes. American Mathematical Society, Providence (1996)
203. Sturmfels, B., Sullivant, S.: Toric geometry of cuts and splits. Special volume in honor of Melvin Hochster. Mich. Math. J. **57**, 689–709 (2008)
204. Sturmfels, B., Thomas, R.R.: Variation of cost functions in integer programming. Math. Program. **77**, 357–387 (1997)
205. Sturmfels, B., Weismantel, R., Ziegler, G.M.: Gröbner bases of lattices corner polyhedra and integer programming. Beiträge zur Geometrie und Algebra **36**, 281–298 (1995)
206. Sullivant, S.: Compressed polytopes and statistical disclosure limitation. Tohoku Math. J. **58**, 433–445 (2006)
207. Sullivant, S.: Toric fiber products. J. Algebra **316**, 560–577 (2007)
208. Tatakis, C., Thoma, A.: On the Universal Gröbner bases of toric ideals of graphs. J. Combin. Theory Ser. A **118**, 1540–1548 (2011)
209. Tatakis, C., Thoma, A.: On complete intersection toric ideals of graphs. J. Algebr. Combin. **38**, 351–370 (2013)
210. Tate, J.: Homology of noetherian rings and local rings. Ill. J. Math. **1**, 14–25 (1957)
211. Thomas, H.: Order-preserving maps from a poset to a chain, the order polytope, and the Todd class of the associated toric variety. Eur. J. Combin. **24**, 809–814 (2003)
212. Tran, T., Ziegler, G.M.: Extremal edge polytopes. Electron. J. Combin. **21**, P2.57 (2014)
213. Villarreal, R.: Rees algebras of edge ideals. Commun. Algebra **23**, 3513–3524 (1995)
214. Vlach, M.: Conditions for the existence of solutions of the three-dimensional planar transportation problem. Discret. Appl. Math. **13**, 61–78 (1986)
215. Wang, Y.: Sign Hibi cones and the anti-row iterated Pieri algebras for the general linear groups. J. Algebra **410**, 355–392 (2014)
216. Weibel, C.A.: An Introduction to Homological Algebra. Cambridge Studies in Advanced Mathematics vol. 38. Cambridge University Press, Cambridge (1994)

217. Weispfenning, V.: Constructing universal Gröbner bases. In: Huguet, L., Poli, A. (eds.) Applied Algebra, Algebraic Algorithms and Error-Correcting Codes. Lecture Notes in Computer Science, vol. 356, pp. 408–417. Springer, Berlin (1987)

218. Woodroofe, R: Matching, coverings, and Castelnuovo-Mumford regularity. J. Commut. Algebra **6**, 287–304 (2014)

219. Zahid, Z., Zafar, S.; On the Betti numbers of some classes of binomial edge ideals. Electron. J. Combin. **20**, ♯ P37 (2013)

220. Zariski, O., Samuel, P.: Commutative Algebra. vol 1 & vol. 2. Springer Science and Business Media, New York (1960)

221. Ziegler, G.M.: Lectures on Polytopes. Springer, New York (1995)

Index

© Springer International Publishing AG, part of Springer Nature 2018
J. Herzog et al., *Binomial Ideals*, Graduate Texts in Mathematics 279,
https://doi.org/10.1007/978-3-319-95349-6